CINCINNATI STATE
WITHDRAWN

D0849402

Integrated Circuits: materials, devices, and fabrication

Integrated Circuits: materials, devices, and fabrication

William C. Till
Electrical Engineering Department
General Motors Institute

James T. Luxon
Science and Mathematics Department
General Motors Institute

Prentice-Hall, Inc., Englewood Cliffs, NJ 07632

Library of Congress Cataloging in Publication Data

TILL, WILLIAM C.
Integrated circuits.

Includes bibliographical references and index.
1. Microelectronics. 2. Integrated circuits. I. Luxon, James T. II. Title.
TK7874.T55 621.381'7 81-5929
ISBN 0-13-469031-1 AACR2

Editorial/production supervision
and interior design by Karen Skrable
Manufacturing buyer: Joyce Levatino
Cover design by Frederick Charles, Ltd.

Printed in the United States of America

10 9 8 7 6 5 4 3 2 1

Prentice-Hall International, Inc., *London*
Prentice-Hall of Australia Pty. Limited, *Sydney*
Prentice-Hall of Canada, Ltd., *Toronto*
Prentice-Hall of India Private Limited, *New Delhi*
Prentice-Hall of Japan, Inc., *Tokyo*
Prentice-Hall of Southeast Asia Pte. Ltd., *Singapore*
Whitehall Books Limited, *Wellington, New Zealand*

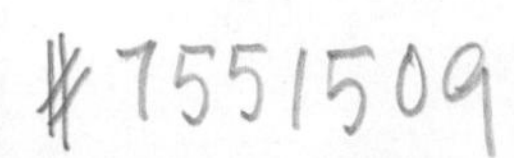

To Our Wives

PAULINE and SALLY

contents

preface

There appear to be two major schools of thought within undergraduate electronics curricula regarding the relationships between the general subject of "electronics" and more specific subjects that might be entitled "solid-state materials," "solid-state devices," and "integrated-circuit design and fabrication."

One school of thought is that solid-state materials and devices should be taught as a first course in an electronics sequence. This course is intended to provide a sound theoretical basis for later electronics courses. It can be followed by courses in digital and nondigital electronic circuits, looking toward the design of electronic systems that use integrated circuits as building blocks. Engineering students who are interested in pursuing the more specialized area of integrated-circuit design and fabrication are then in a position to take senior/M.S.-level courses in this subject area.

A second school of thought is that electronic circuits should be taught without a first course in solid-state materials and devices. The intent here is to emphasize the education of integrated-circuit users rather than integrated-circuit makers. With this approach, devices are introduced descriptively, as needed, in electronic circuit courses. Device characteristics and equations are presented and device models are used to predict circuit operation. Device fabrication is also discussed, but briefly, as needed. A basic electronic circuits course can be followed by courses in digital and nondigital electronic circuits, again leading toward the design of electronic systems that use integrated circuits as building blocks.

This text has been written with the second school of thought in mind. It is primarily intended for junior- and senior-level students who have had one or more electronic circuits courses and want to learn more about solid-state materials, solid-state devices, and integrated-circuit fabrication.

We lay no claim to originality in any of the fundamental aspects of the subject matter. If there is any originality, it is in the emphasis, format, and level of presentation of the subject matter. Such excellent books as *The Theory and Practice of Microelectronics* by Ghandhi, *Physics and Technology of Semiconductor Devices* by Grove, and *Integrated Circuit Engineering* by Glaser and Subak-Sharpe have been relied on heavily.

We have attempted to produce a readable textbook, not a recipe book for practicing designers or fabricators of microelectronic circuits. It is hoped that the book can be read by junior/senior-level engineering students with a minimum of interpretation by the professor. We feel, as well, that this text provides a good starting point for the graduate engineer just getting into the field of microelectronics.

The chapters of this book are organized in the following fashion: Chapter 1 is an overview of microelectronics, in which we present a brief introduction to all aspects of the subject covered later so that the reader has an appreciation of why certain seemingly isolated topics, such as phase diagrams or diffusion, are covered prior to any actual discussion on microelectronic circuits or their fabrication. In Chapter 1, the properties of silicon are tabulated and the various fabrication processes are briefly discussed. A short glossary of terms is presented to establish a starting point for learning the rather extensive "jargon" of microelectronics. Chapters 2 through 5 deal with the materials science concepts fundamental to silicon and silicon epitaxial films. In Chapters 6 through 9 we deal with the physics and properties of diodes, bipolar transistors, the metal–insulator system, and field-effect devices. In Chapters 10 through 13, lithography, fabrication, bonding, testing, and packaging of silicon integrated circuits and hybrid thick-film and thin-film circuits are covered. In Chapter 14 the basic digital logic families are presented in relation to the earlier chapters on devices and integrated-circuit fabrication.

The entire field of microelectronics could not be covered, even cursorily, in a text of this magnitude. Consequently, the coverage in this book has been limited to the fabrication and properties of silicon structures, including bipolar devices and circuits, metal–oxide–semiconductor devices and circuits, and junction field-effect devices. Somewhat more material has been presented on hybrid-circuit technology than is common for texts of this type, since many colleges have hybrid-circuit laboratories and related courses. No mention is made of silicon-controlled rectifiers, diacs, triacs, photodiodes, or light-emitting diodes.

The text may be used for a one- or two-semester course. In a two-semester course, several different approaches are possible. If a heavy emphasis on materials properties and device physics is desired, Chapters 1 through 9 could be covered. If a strong emphasis on fabrication and circuits is desired, Chapters 1, 10, 11, 13, and 14 could be covered, with occasional reference to Chapters 2 through 5. Chapter 12, which deals with hybrid circuits, can be covered or left out of any scheme. At GMI, our emphasis is on materials properties, device physics, and fabrication, so we cover Chapters 1 through 11, and parts of Chapter 14. Some sections of various chapters must be omitted because of the limited time.

The authors are greatly indebted to many people. We want to express our appreciation to our families for their patience and encouragement. We want to thank Professor Harley Anderson, the retired chairman of the Electrical Engineering Department at GMI and Dr. Jack Olin, the present chairman of this department, for their support and technical assistance. We want to acknowledge our indebtedness to our colleagues at Delco Electronics and General Motors Research Laboratories, whose continual assistance over the years has been invaluable. Particularly, we would like to express our thanks to Dr. Frank Stein, Chief Engineer, Research and Development at Delco Electronics, and the Delco volunteers that he enlisted to proofread, revise, and correct the original manuscript. Special thanks are also due to John Hile, Senior Research Engineer at GM Research Laboratories, for proofreading and correcting the original manuscript. We would like to thank many of our students, who provided constructive criticism and assistance with drawing figures. Finally, our thanks go to Mrs. Roberta Green, who typed the manuscript and its many revisions with great skill and accuracy.

Integrated Circuits: materials, devices, and fabrication

integrated-circuit technology

1

This chapter provides an overview of most of the subject matter presented in the remainder of the text. The purpose of providing this overview is to give the reader, in a rather rapid fashion, a basic understanding of the subject of microelectronics and an appreciation for why such topics as crystal properties, solid-state diffusion, and binary phase diagrams are covered prior to discussions of device physics, fabrication, and various types of integrated circuits. It is expected that this will enable the reader to perceive the subject in a cohesive manner at the outset rather than being required to wait until near the end of the book to grasp the overall picture.

The term *microelectronics,* as used in this book, refers to the modern technologies involved in the design, fabrication, and use of monolithic integrated circuits (ICs), thick-film hybrid circuits, and thin-film hybrid circuits. The prefix "micro" alludes to the tremendous size reductions that have been accomplished through the integration of complete circuits on a single chip of silicon and the combination of these chips with thick- and thin-film passive circuitry (hence the term hybrid thick- or thin-film circuit).

Microelectronic devices are also incorporated into printed circuits either as dual-in-line packaged (DIP) monolithic ICs or as thick- or thin-film hybrids. Figure 1-1 presents the various microelectronic classifications and shows how they are related.

Functional devices include Gunn diodes, IMPATT diodes, charge injection devices (CIDs), and charge-coupled devices (CCDs), which are basically discrete or single devices that perform a complex function.

Perhaps the most remarkable aspect of the modern technology is the capability of packing a tremendous number of functions into a physically small

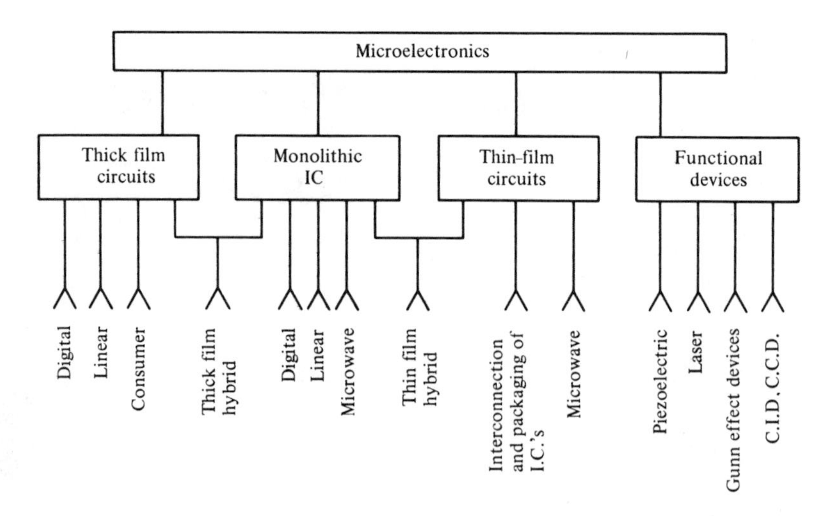

Fig. 1-1 The branches of microelectronics.

area. As a result of this, a nomenclature has developed in the digital area to describe the degree of packing density. Small-scale integration (SSI) refers to ICs with fewer than 10 to 12 devices per chip. Medium-scale integration (MSI) refers to designs with about 12 to 100 devices per chip. Large-scale integration (LSI) refers to chips with more than 100 devices. Very-large-scale integration (VLSI) refers to digital chips with 10,000 or more devices. LSI chips with several thousand devices or memory elements, plus associated input and output circuitry for such items as digital watches, pocket calculators, and microprocessors, are common. Further significant increases in packing density will be achieved through the use of electron-beam and X-ray lithographic techniques.

There is ample evidence of the impact of LSI on calculators and computers. Large-scale integration is bringing about great changes in industrial process control, automotive electronics, and other fields in which data acquisition, computation, or controls are necessary.

To provide an overview of microelectronics, we will briefly discuss silicon and list some of its properties as they pertain to integrated circuits and their fabrication. We will also discuss the fabrication of integrated circuits and hybrid circuits to a sufficient extent that the reader will have a basic understanding of these devices and the concepts involved in their fabrication.

There is an extensive vocabulary associated with microelectronics because of the great variety of devices and fabrication techniques. Table 1-1 is a glossary of a few of the key terms and abbreviations that are used. Most of the terms we use will be defined within the context of their discussion. However, owing to limitations of space, it is not possible to define each term the first time it is used. The glossary is intended to bridge this gap by providing a short definition

TABLE 1-1 Glossary

Acceptor: An impurity atom which may replace an atom of a semiconductor but which has one less valence electron than required to complete the usual number of covalent bonds (four in Si) per atom. Hence, it readily *accepts* an additional electron. In the process a "hole" (effectively a positive free charge) is produced.
Bipolar Junction Transistor (BJT): A device that has an emitter and a collector between which charge flows through a thin base region. The primary current is a diffusion current from the heavily doped emitter, and this current is controlled by voltage (current) applied to the base. Use of the term "bipolar" is based on the fact that both positive and negative charge carriers (two "polarities") carry current in such devices.
Complementary Symmetry Metal–Oxide–Semiconductor (C-MOS): A type of MOS circuit in which both N-MOS and P-MOS devices are used to reduce power consumption.
Diffusion Current: A motion of charge resulting from a concentration gradient. The diffusion current is in the opposite direction of increasing concentration.
Donor: An impurity atom that may replace an atom of the semiconductor but which has one more valence electron than required to complete the usual number of covalent bonds (four in Si) per atom. This extra electron is easily dissociated from the donor; thus the impurity donates a free electron.
Drift Current: A current that results from a movement of charge under the influence of an electric field.
Electron volt (eV): The amount of energy acquired by one electronic charge (1.6×10^{-19} C) when accelerated through a potential difference of 1 V. Thus, 1 eV = 1.6×10^{-19} J.
Epitaxial: Literally, epitaxial means "arranged upon." In microelectronics, epitaxial layers are grown on substrate wafers to produce an abrupt junction between *n*- and *p*-type semiconductor materials.
Field-Effect Transistor (FET): A transistor in which drift currents flow between a source and drain through a channel, under the influence of an electric field. The conductance of the channel is varied by application of a voltage to a gate.
Flip Chip: An integrated circuit on which metal (solder) bumps have been placed on the pads of the chip to make contact to pads on a thick-film or other type of circuit. Since the chip must be inverted to make contact to the bonding pads, it is called a flip chip.
Hole: A descriptive term given to the effective positive free charge that is produced in a semiconductor when an "acceptor" picks up an electron, thereby creating a positively charged incomplete covalent bond elsewhere in the crystal.
Integrated Circuit (IC): IC actually refers to monolithic integrated circuits, or complete circuits built in a single chip of Si.
Metal–Oxide–Semiconductor (MOS): MOS devices are FETs in which an insulator (oxide) separates the channel and the gate metallization.
Microelectonics: The entire field of modern electronics, which incorporates circuits and devices of greatly reduced size. (See Fig. 1-1 for a more complete breakdown.)
n-Channel Metal–Oxide–Semiconductor (N-MOS): An MOS device in which the channel is *n*-type.
n-Type: A semiconductor material in which the predominant charge carriers are electrons.
p-Channel Metal–Oxide–Semiconductor (P-MOS): An MOS device in which the channel is *p*-type.
p-n Junction: Formulation of a contiguous *p*-type and *n*-type region in a semiconductor, leading to a junction between them. This is the basis for the semiconductor diode, and such junctions are found in all types of ICs, as well as in many other solid-state devices.
p-Type: Semiconductor materials in which the predominant charge carriers are holes (positive charges).
Silicon-on-Sapphire (SOS): A fabrication technique in which Si is deposited on sapphire to make MOS devices. The sapphire is used to isolate the devices; because it is a good insulator with low dielectric constant, parasitic capacitances are minimized and speeds improved.
Thick-Film Circuit: These circuits are printed by an adaptation of silk-screen printing. Thicknesses

TABLE 1-1 (Cont.)

are of the order of 25 μm. Resistors, conductors, and dielectrics can be deposited on a ceramic substrate in this manner. When ICs and other components are attached to the circuit, it is called a hybrid thick-film circuit.

Thin-Film Circuit: These circuits are deposited by one of several vacuum techniques or by an electrolytic process. Thicknesses are usually hundreds of angstroms. Resistors, conductors, and dielectrics may be deposited. When ICs or other components are attached, these circuits are called hybrid thin-film circuits.

of the most commonly used terms, particularly those used in this and other early chapters.

Table 1-2 lists some of the important units and universal constants that will be used later in the book.

TABLE 1-2 Units and Universal Constants

Units
1 angstrom (Å) = 10^{-8} cm
1 cm = 1/2.54 in. = 0.3937 in.
0.001 in. (mil) = 25.4 μm
1 eV = 1.6×10^{-19} J
Universal Constants
Permittivity of free space $\varepsilon_0 = 8.85 \times 10^{-12}$ F/m
Boltzmann's constant $k = 1.38 \times 10^{-23}$ J/K $= 8.63 \times 10^{-5}$ eV/K
Planck's constant $h = 6.63 \times 10^{-34}$ J
Speed of light in free space $C = 3.0 \times 10^{8}$ m/s
Electronic charge (q) $= 1.6 \times 10^{-19}$ C
Electron mass (m) $= 9.11 \times 10^{-31}$ kg

1-1 SILICON

Monolithic IC technology, including bipolar junction transistor (BJT) circuits and field-effect transistor (FET) devices, is at present based entirely on silicon (Si). For that reason, a few introductory remarks and relevant data will be presented in this overview.

The name *silicon* comes from the latin "silex" or "silicis," meaning flint. Silicon, second only to oxygen in abundance, constitutes 25.7% of the earth's crust by weight. Crystalline silicon has the diamond structure. Table 1-3 lists the values of some of the important physical quantities relating to the electronic properties of silicon.

Germanium (Ge) is the second most commonly used semiconductor material, and is used primarily for diodes and transistors. It is not used in integrated-circuit fabrication for two principal reasons. First, the energy required to produce an electron–hole pair (the charge carriers) in Ge is 0.77 eV, compared with

TABLE 1-3 Properties of Silicon

Physical Quantity	Units	Conditions	Silicon
Atomic number			14
Atomic weight			28.06
Melting point	°C		1420
Atoms/cm^3	cm^{-3}		5×10^{22}
Density	g–cm^{-3}	300 K	2.329
Energy gap (indirect)	eV	300 K	1.11
Mobility, electrons (μ_n)	cm^2/V-s	300 K	1350
Mobility, holes (μ_p)	cm^2/V-s	300 K	480
Diffusion coefficient, electrons (D_n)	cm^2/s	300 K	35
Diffusion coefficient, holes (D_p)	cm^2/s	300 K	12.5
Relative dielectric constant (ε_R)			12
Dopants, acceptors		0.1 eV	B, Al
Dopants, donors		0.1 eV	P, As, Sb
Field for avalanche (approx.)	V/cm	300 K	4×10^5
Intrinsic resistivity (approx.)	Ω-cm	300 K	2.5×10^5
Intrinsic carrier density (N_i)	cm^{-3}	300 K	1.5×10^{10}
Increase of N_i per °C		300 K	9%
Lattice constant (cube edge)	Å	300 K	5.430
Electron mass (relative to free electron)			1.1
Hole mass (relative to free electron)			0.59
Refractive Index			3.5
Dielectric constant (relative permittivity ε_R)		300 K	12.0
Dielectric breakdown strength	V/cm		2×10^5

1.1 eV in silicon. This results in much greater temperature sensitivity on the part of germanium. Second, it is relatively simple to grow silicon dioxide (SiO_2) on silicon to serve as a passivating layer. Germanium oxide (GeO_2) has tetragonal and hexagonal crystalline forms. The tetragonal is difficult to form and the hexagonal is water-soluble; hence GeO_2 is useless as a passivating surface. Consequently, Ge devices must be hermetically sealed, whereas it is not normally necessary to hermetically seal Si devices. In addition, silicon nitride (Si_3N_4) can be deposited on Si, and this provides an extremely high quality passivating layer. Although many high-voltage rectifiers and power transistors are fabricated from germanium, silicon, because of its lower sensitivity to temperature and the ability of its thermally grown oxide layers to function as a mask during IC processing steps, is used almost solely for integrated circuits.

1-2 INTEGRATED-CIRCUIT FABRICATION

More details will be presented later on all the processes that are discussed in the remainder of this chapter. A brief description of the processes is presented here simply to provide the uninitiated reader with a clearer understanding of

TABLE 1-4 Silicon Epitaxial Slice Specifications

Type configuration: n^+ on p
Substrate description
1. Resistivity range: *5–10* Ω-cm
2. Conductivity type: p
3. Required dopant: *Boron*

Layer description
1. Resistivity and tolerance: *1.0* Ω-cm ± 20%
2. Conductivity type: n^+
3. Required dopant: *arsenic*
4. Thickness and tolerance: *8–10μm* ± 20%

General description
1. Overall thickness: *0.5* ± *0.05* mm (±0.03 mm across wafer)
2. Orientation: *1-1-1* ± *2*°
3. Slice diameter: *100* ± *2* mm
4. Flat: *standard*
5. Surface finish on epitaxial layer side: *mirror polish*

why certain topics are presented in subsequent chapters. We will limit this discussion to BJT fabrication at this point, as that should be sufficient to provide the reader with the desired overview.

Single-crystal silicon is generally grown by the Czochralski method. This consists of pulling a roughly cylindrical boule of single-crystal silicon from a melt by inserting a properly oriented "seed" crystal into the melt and withdrawing it under carefully controlled conditions. The boules may be several meters in length and up to 12.5 cm in diameter (see Fig. 2-20 for a photograph of single-crystal Si boules). The melt is initially doped to make the crystal either p-type or n-type. The boules are sawed into wafers 0.25 to 0.4 mm thick and one side is mechanically and chemically etched to a mirror finish. An epitaxial layer may be grown on top of the original crystal such as to provide an n-on-p configuration. An n^+ buried layer[1] may be added to the substrate prior to the epitaxial growth to reduce collector series resistance. Some typical silicon epitaxial slice specifications are listed in Table 1-4.

Figure 1-2 is a flowchart of the basic steps involved in fabricating a Si BJT integrated circuit, assuming that the proper Si wafers are available. We will comment briefly here on each of these steps.

The integrated-circuit design procedure is unique. Traditional attitudes toward discrete-component circuit designing, where passive devices are maximized, are discarded since transistors require less area (real estate) on the IC chip than do passive devices. Inductors simply do not exist, capacitors are large in area and low in value (picofarads/cm²), and large-value resistors require too much space.

[1] The + sign indicates very heavy doping, approaching the solid solubility limit.

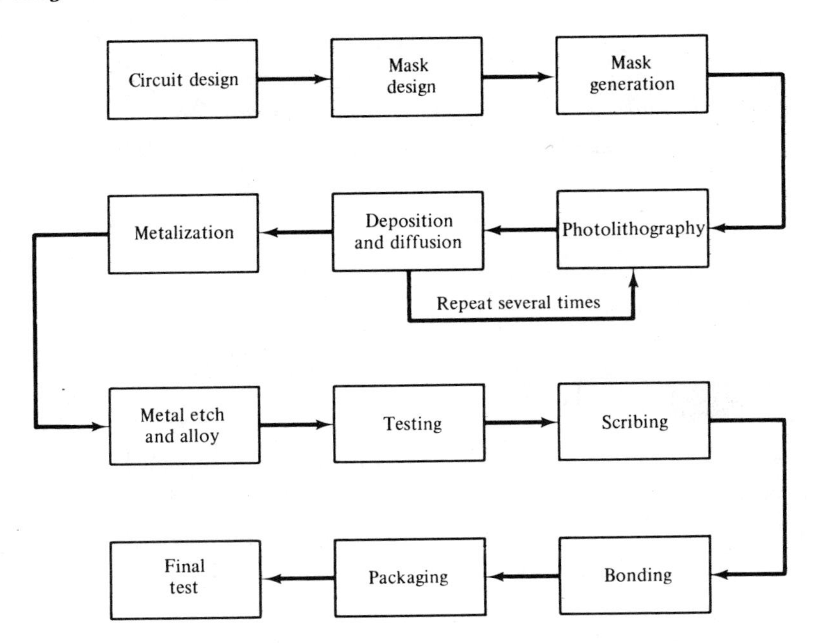

Fig. 1-2 Flowchart of major steps in the BJT integrated circuit fabrication process.

After the circuit has been designed, a set of masks must be fabricated. These masks are used to process the silicon devices. Usually five or six separate masks are required (sometimes 10 or more in MOS LSI circuits). Very precise design criteria must be followed in generating a mask. It may be done by computer, and subsequently a computer-controlled X-Y plotter controls the exposure of a tenfold-size mask (reticle). A further tenfold reduction is made at the same time that a step-and-repeat process is introduced to generate a mask consisting of hundreds to thousands of identical patterns arranged very precisely in rows and columns. Test patterns are introduced periodically to provide a means for checking the process steps. An alternative approach is to use a scanning electron microscope beam to "write" the mask one times size under computer control.

The masks are then used to expose photoresist which has been spun on the top surface of a SiO_2-covered wafer. Positive photoresist hardens where not exposed to light; negative photoresist hardens where exposed to light. After exposure, a development and rinse process leaves a SiO_2-covered wafer with a patterned photoresist on top. The photoresist acts as a resist to a hydrofluoric acid etchant which is used to etch windows in the SiO_2, exposing bare silicon. This process is depicted in Fig. 1–3.

The photoresist is stripped and dopants, such as boron or phosphorus, can now be deposited on the Si in the desired areas and diffused into the silicon

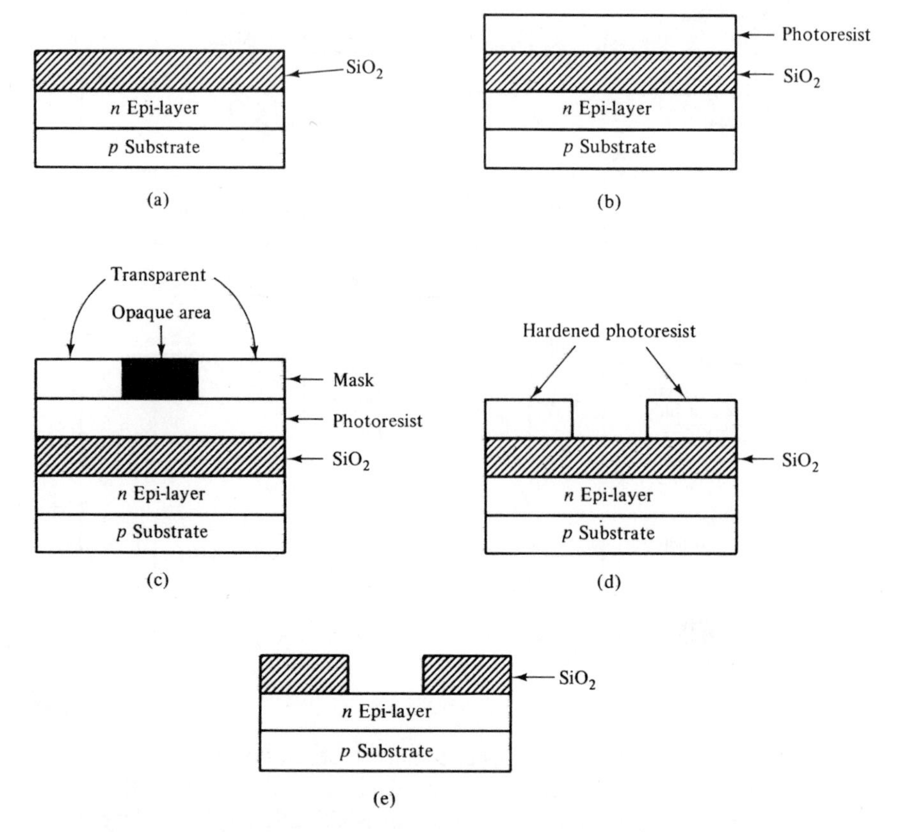

Fig. 1-3 Photoprocessing steps: (a) wafer; (b) photoresist applied; (c) exposure through mask; (d) developed photoresist; (e) SiO_2 etched and photoresist removed.

under carefully controlled atmospheric and temperature conditions. A new layer of SiO_2 is grown on the Si to protect it and to act as a mask, to prevent the diffusion of unwanted dopants during further processing. Extensive cleaning procedures are involved at the various stages. The lithographic process is repeated as often as necessary to build transistors, diodes, and resistors in the epitaxial layer. Windows are then etched in the SiO_2 in order to evaporate metal contacts (usually aluminum) where desired. Aluminum is deposited in vacuum on the entire wafer by means of a vapor or electron-beam process. By a masking technique similar to that discussed above, metal is etched away to leave the desired interconnecting metalization and connections to the outside world. Figure 1-4 summarizes the various mask steps required to complete a circuit. Figure 1-5 is a picture of a completed wafer.

Every circuit on the entire wafer is tested and defective ones are marked with ink. The wafer is then scribed by a diamond stylus or a laser and "broken"

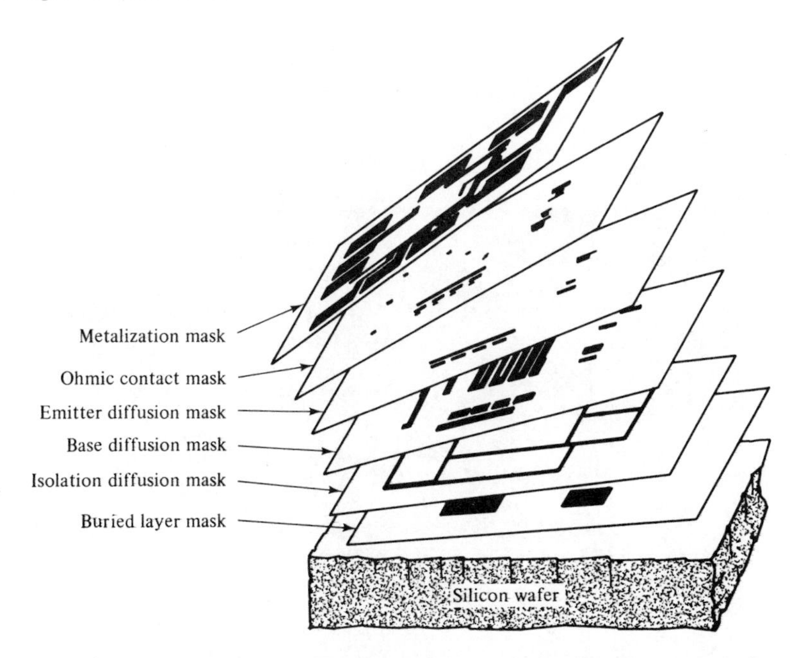

Fig. 1-4 Series of photo masks required to fabricate a bipolar integrated circuit. (From Integrated Circuit Engineering, *Basic Technology,* Copyright 1966, Integrated Circuit Engineering, with permission of I. C. E.)

(cleaved) into hundreds to thousands of individual chips. Wafers may alternatively be sawed into chips by means of a diamond saw. (An individual chip is often called a die; the plural of die is dice.) Figure 1-6 is a picture of an integrated circuit chip with wire bonds attached to connect to the outside world. This is

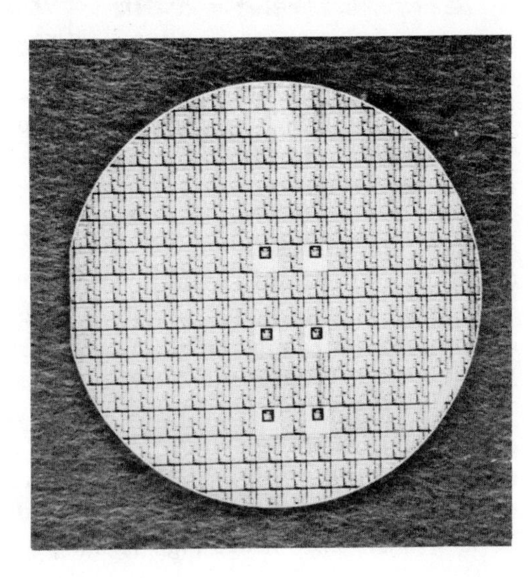

Fig. 1-5 Completed Si IC wafer. This is a 6.35-cm wafer with chips 4.2 × 4.2 mm. Note the test patterns located on the wafer.

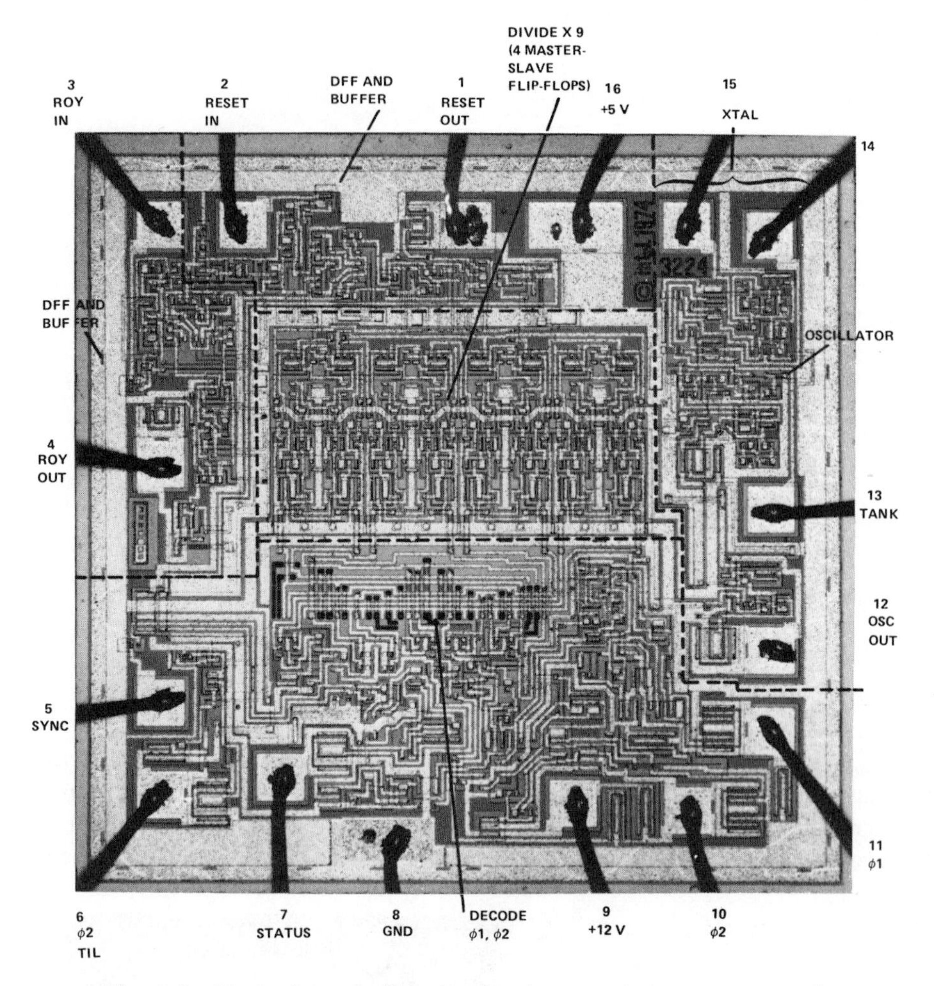

Fig. 1-6 Bipolar integrated-circuit chip (2 × 2 m). (From Integrated Circuit Engineering, *Basic Technology,* Copyright 1966, Integrated Circuit Engineering, with permission of I.C.E.)

a fairly complex bipolar circuit; however, most metal–oxide–semiconductor (MOS) circuits contain several times as many transistors.

Monolithic integrated circuits have substantial advantages over their discrete-component counterparts: (1) they represent at least a 1000-fold reduction in size, (2) ICs are cheaper in large volume, and (3) they are much more reliable. The desire to make circuits smaller is motivated not only by a desire to design circuits that require less space, but by the fact that the smaller the chips, the fewer will be affected by crystal defects, hence the greater the yield (percentage of good chips per wafer). On the other hand, in the case of memory or imaging devices, small size—hence high packing density—is very desirable because it results in increased speed as well as in greater memory storage and computational capability.

1-3 THICK-FILM CIRCUITS

The manufacture of thick-film circuits is an adaptation and refinement of the ancient art of silk-screen printing. Passive circuits are literally printed in this process. Conductors, resistors, and, less commonly, capacitors are produced by this technique. Discrete components such as ceramic chip capacitors, diodes, transistors, or integrated circuits may be attached to the passive thick-film circuit to complete its functional capability.

A brief outline of the design and fabrication of thick-film circuits will be presented here. Greater detail is provided in a subsequent chapter.

Thick-film circuits may have an advantage where moderately high power dissipation is required, say 2.5 to 4.0 W/cm^2. They are also practical where a large number of resistors but relatively few active devices are required. Actually, one 1×1 mm IC chip on a 2.5×2.5 cm hybrid thick-film circuit may contain more components than the rest of the circuit. Nevertheless, large-value resistors requiring fairly tight tolerances are often best manufactured by the thick-film process.

After a circuit is designed, it must be laid out 5 to 20 times actual size in a thick-film format according to layout rules that consider such factors as line width, power dissipation, geometrical restrictions, and many other considerations. The layouts, one for each conductor, resistor, dielectric glass, or solder printing, are then translated into Rubylith.[2] The Rubylith master is photographically reduced in one or two steps to produce a transparency which serves as a mask. The mask is used to expose an emulsion on a screen, generally stainless steel. In this way, openings are defined in the emulsion of the screen. Inks are then printed through these openings onto a ceramic substrate, usually alumina. Obviously, computer-aided mask design can be used to supplement or replace some of these steps.

Conductor, resistor, dielectric, glass, and solder inks are available for printing. The conductor and resistor inks are compositions of glass frit, metal oxides, and metal particles, with an organic binder and other organic materials to control rheology.

The dielectric inks are glass, barium titanate ($BaTiO_3$), and organics. The glass inks are used for low-dielectric crossovers, solder dams, and protective coatings. The solder inks consist of fine particles of solder dispersed in an appropriate flux and organic materials.

Conductor, resistor, dielectric, and glass inks are dried after printing at about 125°C for 15 min and fired in a conveyor furnace. The firing process

[2] Rubylith is a two-layer strippable plastic—one layer clear, the other opaque to the exposing light. Patterns are cut into one layer, which is then stripped from the other.

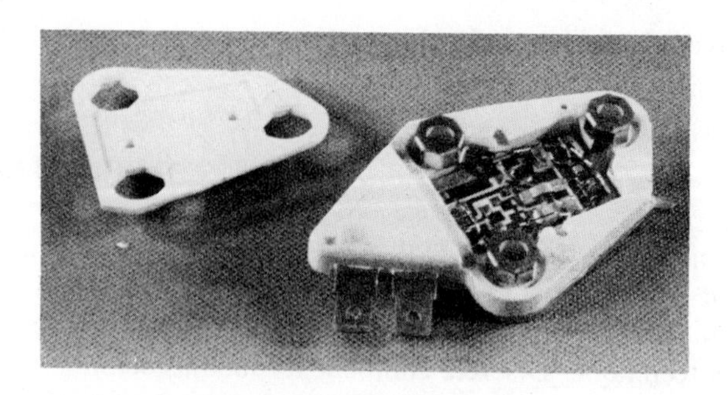

Fig. 1-7 Hybrid thick-film circuit (maximum dimension 7 cm). (Courtesy of Delco Electronics Division of General Motors Corporation.)

requires about 45 min and the peak temperature reaches about 850°C for 8 to 10 min for certain conductor compositions. Resistor and conductor firing is performed separately, but some conductor and dielectric firing can be performed simultaneously.

Resistors are printed 15% or more below the desired resistance value and are air-abraded or, more commonly, laser-trimmed to obtain the final value. Tolerances of ±1 to ±5% are normal, with ±0.1% possible. Printed capacitors may also be trimmed.

Figure 1-7 is a hybrid thick-film circuit containing an IC flip-chip and ceramic capacitors. A Darlington power transistor is located on a heat sink adjacent to the hybrid thick-film circuit.

1-4 THIN-FILM CIRCUITS

Thin-film circuits are fabricated by various vacuum deposition techniques. Although thin-film transistors and capacitors are a reality, thin-film technology is used primarily for the production of resistor networks where small size (generally smaller than for thick film), close tolerances, and precise definition are required. Thin-film is a costlier process than thick-film, and the power-handling capability and resistance values are more limited. However, in certain applications, especially in the microwave area where very tight tolerances on the geometry and edge definition of the conductors and resistors are required, thin film has a definite advantage.

Although many different materials are used, Nichrome (a nickel–chromium alloy) and gold, for resistors and conductors, respectively, are the most common. The most prevalent technique is to deposit about 150 nm of Nichrome on a glass, alumina, or beryllia substrate and about 500 nm of gold on top of that. Conductor and resistor patterns are then selectively etched to form Nichrome resistors and gold on Nichrome conductors. The design and production of masks for thin-film circuits is quite similar to that for thick-film circuits. The rules governing such design criteria as line width, line separation, and power density are, of course, different. Appropriate photoresists exist that can be used as etchant masks.

Evaporation, sputtering, or electron-beam film deposition may be used in the vacuum process. Many other conductor, resistor, and dielectric materials are used. Additional details of materials and processes will be presented later.

Once the thin-film circuit is complete, resistors may be trimmed, or trimming may be performed functionally after discrete devices (ICs, capacitors) have been added to the circuit. Flip chips, beam-lead IC chips, and wire-bonded transistors and ICs may be used.

Tantalum, which must be sputtered because of its high melting temperature, can be passivated and its resistance adjusted by anodic oxidation of its surface. Nitride deposition achieves the same effect. Tantalum may be used to form complete thin-film circuits, including resistors, conductors, and capacitor dielectrics.

energy bands and crystal properties

2

A review of the electronic structure of the atom, including spectroscopic notation, is presented in this chapter. The concept of energy bands in crystals is introduced in such a way as to establish their physical plausibility. Crystal properties, including crystallographic concepts and defects, are discussed. Wherever possible the concepts discussed are related to the properties of semiconductors, particularly silicon (Si), as this book deals primarily with the fabrication and physical understanding of the operation of silicon devices.

It is felt that the review, and perhaps introduction of new material for some students, is essential for a proper understanding of the development of fabrication topics and device-operation concepts introduced in subsequent chapters.

2-1 ELECTRONIC STRUCTURE OF THE ATOM

Our present concept of the hydrogen atom consists of a fairly massive nucleus (a proton) orbited[1] by a nebulous and very light negative cloud of charge (the electron). If it is assumed that electrostatics gives the correct attractive force, and therefore the correct potential energy, and that the kinetic energy is correctly given by the usual classical definition, $\frac{1}{2}(mv^2)$, then application of Schrödinger's wave equation[2] leads to some very remarkable results. With some

[1] The term "orbit" should not be taken literally; theory predicts only probabilities of the electron's locations, not actual paths of motion.

[2] Schrödinger's wave equation is generally accepted as a valid equation for nonrelativistic atomic-level systems.

qualifications, these results can then be generalized to more complicated atoms. Only the results pertinent to later discussions will be presented here.

It turns out that an atom has certain allowable states, meaning that as long as the atom remains in that state, no radiation of energy occurs. These states are best described in terms of allowed energy states. However, it has been traditional, for historical reasons, to think of the electrons as occupying certain allowed orbits or shells surrounding the nucleus. Although essentially meaningless, it is still useful as a mnemonic device to think in terms of orbits or shells. Energy is either absorbed or given off when electrons undergo transitions between energy states.

The allowed states are characterized in terms of quantum numbers, which have direct physical significance with respect to the dynamical properties of an atom. There are four such quantum numbers, three of which result from the boundary conditions imposed in the solution of Schrödinger's equation. These three are often given the symbols n, l, and m_l. The fourth, m_s, is treated separately because it results from the purely quantum relativistic nature of the electron. The names and allowed values of these quantum numbers are: the principal quantum number $n = 1, 2, 3, 4, \ldots$; the orbital angular momentum number $l = 0, 1, 2, \ldots, n - 1$; the orbital magnetic quantum number $m_l = -l \ldots 0 \ldots l$; and the intrinsic magnetic quantum number (often called spin, although no such classical analogy exists), $m_s = \pm\frac{1}{2}$.

The principal quantum number n primarily determines the energy of an electron, but l, m_l, and m_s contribute to this as well in atoms with more than one electron. The quantum number n determines the shell or orbit, and in spectroscopic notation, which has its conveniences, the numbers $n = 1, 2, 3, 4, \ldots$ correspond to the $K, L, M, N, \ldots$ shells. The quantum number l determines the orbital angular momentum, which is given by $\sqrt{l(l+1)}\, h/2\pi$, where $h = 6.62 \times 10^{-34}$ J-s is Planck's constant, a universal constant. Note that $l = 0$ represents zero angular momentum, which points up the invalidity of the concept of orbits as such. The l values determine what are called subshells within the shells: for example, the K shell has an $l = 0$ subshell; the L shell has $l = 0$ and $l = 1$ subshells. There is a useful spectroscopic notation for l. In this notation the values of $l = 0, 1, 2, 3, 4, \ldots$ are given the symbols *s, p, d, f, g*. The first three are of historical origin; the remaining ones are alphabetical. Spectroscopic notation has been developed so that it is possible to talk and write intelligibly about quantum-number values without having to repeatedly state which quantum number you are referring to—it is implicit in the notation.

Quantum numbers m_l and m_s relate to the possible ways in which the magnetic moments associated with the angular moments can align in the presence of a magnetic field, either an externally applied field or an internally produced field.

It turns out that not more than one electron can occupy a given quantum state in an atom, where by "quantum state" we mean any set of four quantum numbers, such as $n = 3$, $l = 2$, $m_l = 1$, and $m_s = \frac{1}{2}$. That is, no two electrons can have the same set of quantum numbers in the same atom. Known as the Pauli exclusion principle, this principle applies also to energy levels in molecules and crystals. What this principle tells us is that if we know what quantum states are allowed and how energy levels relate to these, we can predict which of these states are occupied except for some excitation of the outermost electrons at room temperature. We can predict precisely which states are occupied at $T = 0$ K, since the lowest possible states will be occupied.

Given the quantum numbers, the spectroscopic notation, and the Pauli exclusion principle, we can write an expression for the electronic configuration of an atom in a very compact fashion. As an example, sodium (Na) has an atomic number of 11 and therefore has 11 electrons when neutral. Its electronic configuration is written $1s^2 2s^2 2p^6 3s^1$. In this notation the large numbers stand for values of n, s and p represent values of l, and the superscripts indicate how many electrons occupy the available states or subshells. Notice that for $n = 2$, $l = 0, 1$ and for $l = 0$, $m_l = 0$ and $m_s = +\frac{1}{2}$ and $-\frac{1}{2}$, whereas for $l = 1$, $m_l = -1, 0, 1$ and $m_s = \pm\frac{1}{2}$ for each of the m_l values. Thus, there are two s levels and six p levels. If this is pursued to higher n and l values, you can show that there are 10 d levels, 14 f levels, and so on.

Energy levels in atoms relate directly to the type of bonding that occurs in crystals—metallic, covalent, or ionic—and to the existence of bands of allowed energy states in crystals. In the next section this relationship will be elaborated upon.

2-2 ENERGY-BAND STRUCTURE OF CRYSTALS

To illustrate the origin of energy bands in crystals, sodium (Na) will be considered. The electronic configuration of atomic Na is $1s^2 2s^2 2p^6 3s^1$. Imagine N Na atoms sufficiently far away from each other that they do not interact. As the atoms are assembled to form a crystal, the outer orbitals will begin to overlap from atom to atom, which means that the electron clouds overlap. This will cause perturbations or distortions of the orbitals. We start with N $3s$ electrons occupying half of the available $3s$ states (there are two $3s$ states per atom). When the crystal is formed, the $3s$ states spread to form a band of $2N$ distinct but closely spaced states. hence, there are $2N$ available states with N electrons to occupy them. This is represented pictorially in Fig. 2-1. Notice that in addition to the fact that the $3s$ band is only half-filled, the $3p$ band overlaps the $3s$ band. For electrical conduction to take place in a crystal, electrons must have available to them a reservoir of unoccupied energy states. This is clearly the case for Na, a metallic[3] crystal, as it is for all the alkali metals—Li, K, Rb, Cs, Fr—which have an s^1 outer-shell configuration. The alkaline earths (group II of the periodic table)—Be, Mg, Ca, Sr, Ba, and Ra—are also

[3] The word "metal" implies "conductor," although not all conductors are metals.

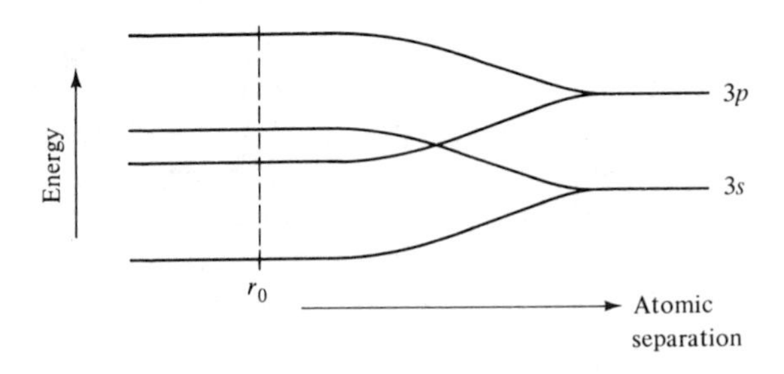

Fig. 2-1 Energy levels of Na as a function of atomic separation. r_0 is the equilibrium atomic separation.

metals. Although they have a closed outer subshell (e.g., Mg, which is $1s^2 2s^2 2p^6 3s^2$), they are metals because of the overlap between the s and p energy bands. This approach to explaining energy bands in crystals is referred to as the valence-band model.

The transition metals, except for Cu, Ag, and Au, are characterized by an unfilled d subshell. Titanium (Ti) has the configuration $1s^2 2s^2 2p^6 3s^2 3p^6 4s^2 3d^2$. The $4s^2$ is written ahead of the $3d^2$ to indicate lower energy. The d and s levels overlap for the transition metals and they too are all conductors, the s electrons being the primary current carriers. Copper, Ag, and Au have filled d subshells and are good conductors because of the unfilled s states.

The valence (conduction) electrons in a metal are literally shared by a large number of atoms, perhaps 1000 or more. It is this mutual sharing of electrons that provides the cohesiveness (bonding forces) of the crystal and, in fact, it is referred to as metallic bonding.

If the energy-band picture for a material—elemental or compound—appears as in Fig. 2-2, it is an insulator or semiconductor, depending on the size of the gap between the lower and upper bands. In this example, there are N electrons to occupy N states; hence, at low temperatures all the states in

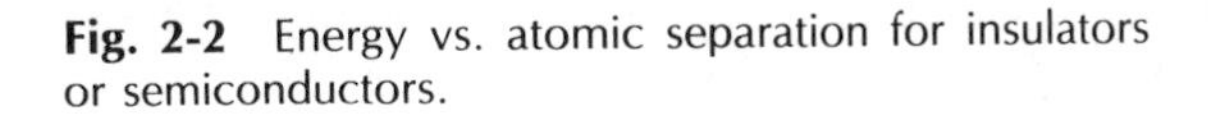

Fig. 2-2 Energy vs. atomic separation for insulators or semiconductors.

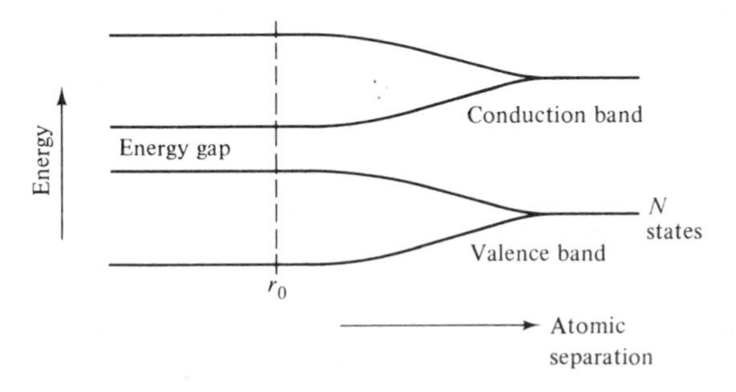

the lower band, the valence band, will be occupied. At higher temperatures, electrons will be able to acquire enough thermal energy to jump the energy gap to the unoccupied states in the upper (conduction) band if the gap is not too wide.[4] If at room temperature there are a substantial number of electrons in the conduction band, the material is a semiconductor. If there are no (or very few) electrons in the conduction band, the material is an insulator. Silicon has an energy gap of 1.1 eV and is a semiconductor. Diamond has an energy gap of 6.0 eV and is an electrical insulator at room temperature.

The origin of the energy gap in Si is interesting and important to an understanding of the properties of Si devices and other semiconductors. Atomic Si has the electronic configuration $1s^2 2s^2 2p^6 3s^2 3p^2$. Diamond (carbon) and Ge are similar, with outer configurations $2s^2 2p^2$ and $4s^2 4p^2$, respectively. All three of these elements form crystals of the diamond structure. When crystalline Si, Ge, or diamond is formed, the *s* and *p* levels combine to form what is known as hybrid (sp^3) levels.[5] This results in four electrons per atom and eight states, but what is unique about this hybridization is that two (sp^3) bands are formed separated by an energy gap, each containing half the states. Hence, at low temperature the lower band is filled and the upper one empty. If the gap between bands is not too great, the result is a semiconductor; otherwise, it is an insulator. Figure 2-3 illustrates the formation of energy bands in this type of material.

From a crystal-bonding point of view, the sp^3 hybridization results in covalent bonding, which means the sharing of two electrons by different atoms to achieve a minimum energy configuration. Each carbon, Si, or Ge atom thus has four sp^3 electrons to share with four surrounding atoms. This is the strongest form of crystalline bonding. To get an electron from the valence band to the conduction band, the electron pair bond must be broken. In the process, a positive charge is left at the site of the errant electron. Other electrons can occupy the vacant bond, and in the process the positive charge moves to another location in the crystal. The positive charge is referred to as a hole. Although this physical picture of the movement of a hole in a crystal is too simplistic, it does at least suggest how both positive and negative charge carriers can occur in semiconductors. Actually, a hole is a quasi-particle which represents the accumulative effect of all the remaining electrons. Physically, it behaves just like a real positively charged particle.

Pure semiconductors (i.e., undoped) are called intrinsic semiconductors. Charge carriers in intrinsic semiconductors must originate in pairs (electron–hole pairs); thus, there is an equal number of electrons and holes, although the holes generally do not move around as easily as electrons do. Semiconductors with controlled amounts of impurities (dopants) are called extrinsic semiconduc-

[4] The average thermal energy available to an electron at room temperature is of the order of 0.03 eV.

[5] It is called sp^3 because the tendency to create half-filled subshells results in one *s* electron moving to a *p* level. This is what gives Si its +4 valence.

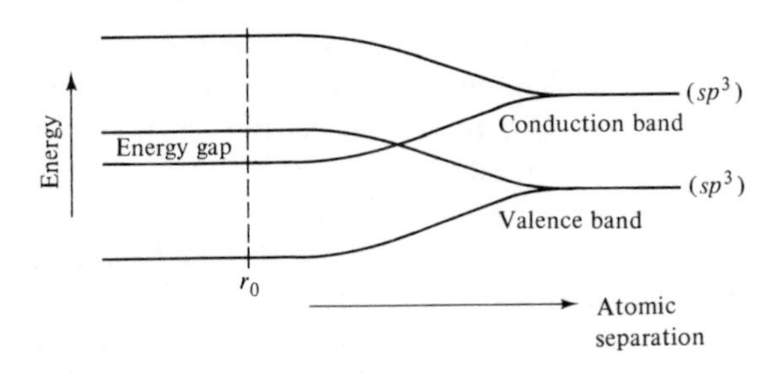

Fig. 2-3 Energy bands for *sp* hybridization.

tors. The purpose of the doping is to tailor the electrical properties of the material.

Figure 2-4 is a simplified two-dimensional representation of the atoms in a Si crystal. Figure 2-4(a) is intrinsic Si; (b) is extrinsic Si doped with something from group V of the periodic table, such as phosphorus (P), which has five valence electrons, one more than is needed to complete the four electron pair bonds. The fifth electron is easily dissociated from the parent atom. This leaves the P atom positively charged, but there is no mobile positive charge created, just a fixed ionic charge. The energy state corresponding to the fifth electron lies just below the conduction band (a few hundredths of an electron volt). Such energy states are called donor states and the impurities are called donors. A semiconductor doped in this fashion (i.e., so that the predominant charge carriers are electrons) is referred to as *n*-type, the *n* standing for negative charge carrier. Figure 2-4(c) represents Si with an impurity atom from group III of the periodic table. An element such as boron (B) has only three valence electrons to share, and consequently one bond is incomplete. An electron from somewhere else in the crystal can occupy this unshared site and make the B atom a negatively charged ion. This negative charge is fixed but the incomplete bond (hole) can move, and only a few hundredths of an electron volt is required to accomplish the dissociation of the hole from the B atom. The energy states therefore lie

Fig. 2-4 Intrinsic and extrinsic silicon: (a) intrinsic Si; (b) extrinsic Si, *n*-type dopant (○, Si atoms; ●, dopant atom); (c) extrinsic Si, *p*-type dopant.

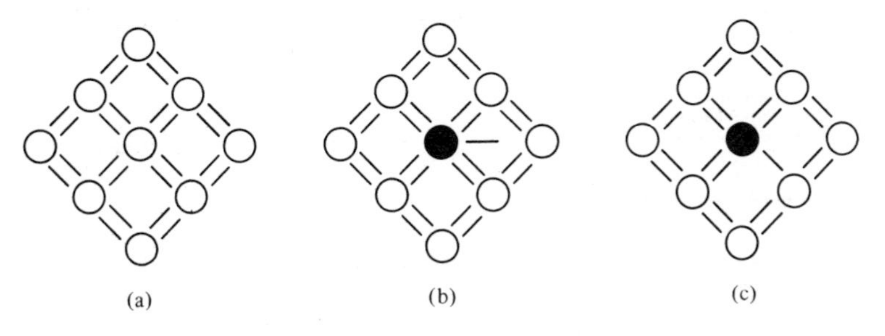

just above the valence band and are called acceptor states; the impurity atoms are called acceptors. A semiconductor doped in this fashion is called *p*-type for positive charge carriers. Figure 2-5 depicts the location of the donor and acceptor states in *n*- and *p*-type silicon.

At room temperature, nearly all the donor states are empty in *n*-type material (i.e., all the electrons are in the conduction band). Similarly, in *p*-type material, all the acceptor sites are occupied, and thus nearly all the holes are in the valence band. Generally, in a doped semiconductor there are many more electrons (holes) from the dopant than from breaking covalent bonds. Hence, the extrinsic semiconductor usually has either electrons or holes as the predominant charge carrier.

In addition to electron–hole pairs being created by thermal energy in a semiconductor, pairs can be created by absorption of energy from photons if the photons have energy exceeding that of the band gap. Consequently, semiconductors make excellent photodetectors as well as being useful for direct conversion of solar energy into electrical energy. When electrons and holes recombine, all or part of the energy may be given off as photons. If a large momentum change is required when the electron makes the transition from the conduction band to the valence band, photons of energy comparable to the band-gap energy cannot be emitted. Photons have very little momentum. The situation is analogous to a shotput and a BB. The shotput can impart but a small fraction of its energy to a BB in a head-on collision because of the extreme difference in mass and the requirement of conservation of momentum. The photon has an extremely small effective mass compared with an electron. In semiconductors of this type (i.e., requiring momentum change upon electron–hole recombination), most of the energy is absorbed by the crystal lattice as thermal vibration. Semiconductors of this type are referred to as indirect-band-gap semiconductors. Si and Ge are of this type.

In semiconductors, such as GaAs and $GaAs_{1-x}P_x$ (up to $x = 0.44$), little or no momentum change is required for the transition of the electron from the conduction band to the valence band. In this case, the probability for photon emission with energy nearly equal to the band gap is very high. Such semiconductors make excellent light-emitting diodes and diode lasers. These materials are

Fig. 2-5 Energy bands and impurity states for extrinsic silicon.

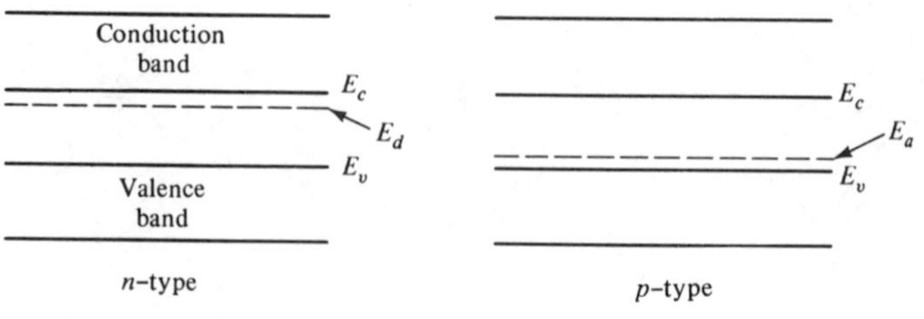

referred to as direct-band-gap semiconductors. Some indirect-band-gap semiconductors, such as GaP, can be doped so that light is emitted during transitions between states introduced into the band gap by the doping. In this case, the photon energy is less than the band-gap energy.

It is difficult to conceptualize the energy-band structure of *p-n* junctions without knowledge of a parameter known as the Fermi level E_F. In conductors the Fermi level is the highest energy state occupied at absolute zero of temperature. As the temperature is raised, electrons lying within a few hundredths of an electron volt of the Fermi level can acquire energy and move up to unoccupied energy states. The Fermi level then represents that level at which the probability of occupancy by an electron is $\frac{1}{2}$. It is not necessary that an allowed energy state actually exist at the Fermi level. In fact, in intrinsic semiconductors, the Fermi level lies, to a sufficient approximation, in the middle of the energy gap. For extrinsic semiconductors, it lies nearer the conduction or valence band for *n* and *p* types, respectively. It may lie in the bands if the doping is very heavy; these are called degenerate semiconductors. The Fermi level tends to approach the intrinsic level with increasing temperature for all semiconductors.

Figure 2-6 is a schematic representation for a metal of the energy levels and the Fermi probability distribution *f(E)*, which gives the probability of a state being occupied by an electron. It is impossible graphically to represent the energy levels accurately because there are on the order of 10^{22} levels per cubic centimeter of crystal with a spacing between levels on the order of 10^{-22} eV. Also, the spacing between levels decreases with increasing energy. It is clear that if more electrons could be added to the metal, the Fermi level would have to shift to a higher energy. This indeed happens in semiconductors through the introduction of impurities which add electrons to the material.

Notice in Fig. 2-6 that the Fermi level for $T > 0$ K is at that energy

Fig. 2-6 Energy levels and the Fermi probability distribution for a metal.

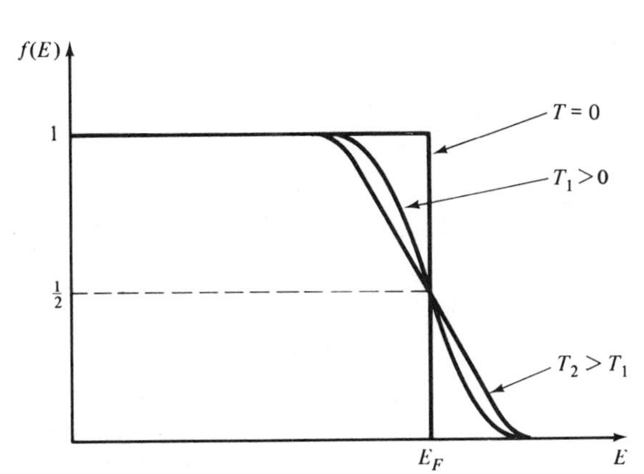

level that has a probability of 0.5 of being occupied. Figure 2-7 depicts the energy bands of a semiconductor and the corresponding Fermi probability function for $T > 0$ K in an intrinsic semiconductor.

As can be seen in Fig. 2-7, the Fermi level lies in the middle of the gap and $f(E)$ is slightly less than unity just below the valence band edge and has finite values near the bottom of the conduction band. To evaluate the number of electrons per unit volume in the conduction band (also the number of holes per unit volume in the valence band), one needs a quantitative expression for $f(E)$:

$$f(E) = \frac{1}{\exp\left[\frac{(E - E_F)}{kT} + 1\right]} \tag{2-1}$$

where E = energy in electron volts from some arbitrary reference level
E_F = Fermi level measured from the same reference level
T = absolute temperature, K
k = Boltzmann's constant = 8.63×10^{-5} eV/K

Also needed for the calculation of the number of electrons or holes per unit volume is a quantitative expression for the number of energy states per unit energy per unit volume, called the energy density of states $N(E)$. This expression can be determined for a simple model of a crystal from quantum mechanics and will not be presented here. However, it is worthwhile to point out that the density of states tends to increase as the one-half power of E moving away from either band edge.

For energy levels above the Fermi level, when $(E - E_F)$ is several times kT, the exponential term dominates, and

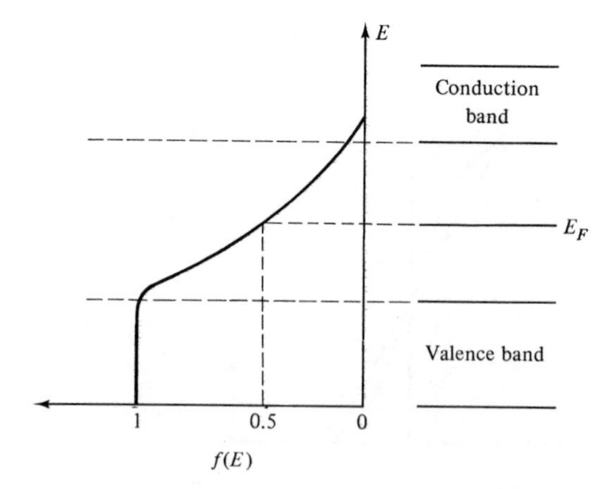

Fig. 2-7 Energy bands and Fermi probability distribution for an intrinsic semiconductor.

$$f(E) = \frac{1}{1+\exp[(E-E_F)/kT]} \approx \frac{1}{\exp[(E-E_F)/kT]} = \exp\left[-\frac{(E-E_F)}{kT}\right] \quad (2\text{-}2)$$

This is referred to as the Boltzmann approximation.

For energies below the Fermi level ($E - E_F < 0$), Eq. (2-1) is approximated by

$$f(E) = \frac{1}{1+\exp[(E-E_F)/kT]} \approx 1 - \exp\left(\frac{E-E_F}{kT}\right) \quad (2\text{-}3)$$

Figure 2-8 shows the Fermi function at a specific temperature with the two approximations as dashed lines.

With the generation of an electron–hole pair, a conduction state is occupied by an electron. The absence of an electron in the valence state, as pointed out previously, is referred to as a hole. The hole plays just as significant a part in supporting current as the electron does. In an intrinsic crystal, there must be a valence state occupied by a hole for every conduction state occupied by an electron ($n_i = p_i$).

The probability of finding a hole at a particular energy level in the valence band is the same as the probability of that state not being occupied by an electron. Thus, the probability of finding a hole at a particular energy level is

$$P(E) = 1 - f(E) \quad (2\text{-}4)$$

For $E << E_F$,

Fig. 2-8 Fermi function at $T > 0$ K and the two approximations.

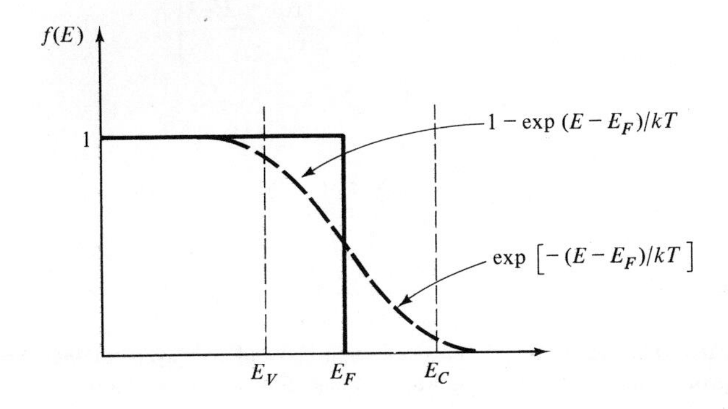

$$P(E) = 1 - \frac{1}{1 + \exp{(E - E_F)/kT}} = \frac{\exp{(E - E_F)/kT}}{1 + \exp{(E - E_F)/kT}} \quad \text{or}$$

$$P(E) \approx \exp\left(\frac{E - E_F}{kT}\right) \tag{2-5}$$

An expression for the number of electrons in the conduction band of an intrinsic semiconductor can be developed by taking the product of the energy density of states in the conduction band, $N(E)$, and the probability of occupancy $f(E)$ and integrating over the conduction band:

$$n_i = \int_{E_C}^{E_{top}} N(E)f(E)\,dE \tag{2-6}$$

where E_{top} is the energy level at the top of the conduction band, which can be set equal to infinity since $f(E)$ drops off to zero rapidly for $E > E_C$; and E_C is the energy level at the bottom of the conduction band. This integration yields the expression

$$n_i = k_1 T^{3/2} \exp\left[-\left(\frac{E_C - E_F}{kT}\right)\right] \tag{2-7}$$

where

$$k_1 = 2\left(\frac{2\pi m_e k}{h^2}\right)^{\frac{3}{2}} \tag{2-8}$$

In this expression, n_i provides a count of the electrons per unit volume in the conduction band of a semiconductor as a function of temperature and the energy difference between the edge of the conduction band and the Fermi level, m_e is the effective mass[6] of an electron in the semiconductor, and h is Planck's constant.

Performing similar operations for holes in the valence band yields

$$p_i = k_2 T^{3/2} \exp\left[-\frac{(E_F - E_V)}{kT}\right] \tag{2-9}$$

where

$$k_2 = 2\left(\frac{2\pi m_h k}{h^2}\right)^{\frac{3}{2}} \tag{2-10}$$

[6] The effective mass of an electron is a quantum mechanical parameter that takes into account the interaction of the electron with the crystal lattice. For Si, $m_e = 1.1 m_0$, where m_0 is the free-electron rest mass. The effective mass of a hole is a convenient way of taking into account the collective effect of all the electrons in the valence band. For Si, $m_h = 0.55 m_0$.

Here p_i provides a count of holes per unit volume in the valence band, m_h represents the effective mass of holes in the semiconductor, and E_V denotes the edge of the valence band.

Since n_i must equal p_i, taking the ratio yields

$$\frac{n_i}{p_i} = 1 = \left(\frac{m_e}{m_h}\right)^{\frac{3}{2}} \frac{\exp\,[(E_F - E_V)/kT]}{\exp\,[(E_C - E_F)/kT]} \tag{2-11}$$

Since m_e is not equal to m_h, $(E_F - E_V)$ differs from $(E_C - E_F)$ by just the right amount to compensate for the ratio of effective masses. In other words, the Fermi level is shifted very slightly from the center of the forbidden band to account for the differences in m_e and m_h.

One of the most important features of the Fermi level is the fact that for a semiconductor at equilibrium (no thermal gradients or electric fields applied to it) the Fermi level is constant throughout the crystal, even if the semiconductor is a device with one or more *p-n* junctions.

If the Fermi level differed between two regions of a crystal at equilibrium, this would imply that different energy levels would have the same probability of occupancy. If this were true, higher energy electrons moving from a region of higher Fermi level would constitute a continuous energy flow, which is impossible.

Another way of looking at the invariance of the Fermi level is by realizing that a higher Fermi level implies a higher concentration of electrons. Electrons, like any other species, will diffuse away from a region of high concentration, just like a drop of dye in a glass of water. If the electrons were neutral, they would distribute themselves uniformly throughout the crystal. Since electrons and holes are not neutral, equalization of the Fermi level in *p-n* junctions occurs in a somewhat different manner. Consider separate *p* and *n* regions as shown in Fig. 2-9(a). The Fermi level is indicated appropriately for *p* and *n* materials. When *p* and *n* regions are juxtaposed in a *p-n* junction, the only way the Fermi level can remain constant is by deformation of the energy-band edges in the junction region. Notice that at some distance from the junction on either side the Fermi level has the same location relative to the band edges as in a

Fig. 2-9 (a) Fermi levels in *p*- and *n*-type semiconductors. (b) *p-n* Junction at equilibrium.

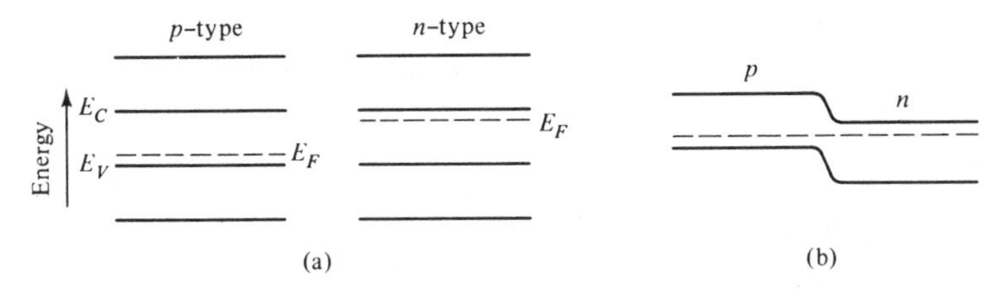

single *p*- or *n*-type semiconductor. Keeping in mind this important feature of the Fermi level makes it relatively simple to sketch the energy-band diagrams for *p-n* junctions in diodes and transistors. The effect of forward- or reverse-biasing *p-n* junctions will be discussed in Chapter 6.

2-3 CRYSTAL STRUCTURE—SILICON

The variety of elemental and compound semiconductors is great, but silicon, exclusively, is used for the manufacture of monolithic integrated circuits. There is some work being done on gallium–arsenide (GaAs) MOS (metal–oxide–semiconductor) integrated circuits. Germanium is used for special-purpose transistors, particularly power transistors. Compound semiconductors, such as GaAs, gallium–phosphide (GaP), and lead–telluride (PbTe), are used for light emitters and photodetectors. In this text the primary concern is with the properties of Si, although reference to Ge and other semiconductors will be made on occasion.

Because of the extensive application of Si and Ge, a great deal is known about their crystal properties. Silicon has an energy gap of 1.1 eV compared with 0.8 eV for Ge. Si microcircuits can be operated at higher temperatures than Ge because higher temperatures are required to generate electron–hole pairs. The upper temperature range for Si is 125 to 175°C. Another advantage of Si over Ge has long been the ease with which a passivating surface layer[7] can be grown on Si, specifically SiO_2, and its ability to prevent dopants such as B and P from diffusing to the Si surface. Silicon nitride (Si_3N_4) is used to overcome some of the drawbacks of SiO_2 passivation, particularly in the manufacture of MOS devices. Both passivation techniques will be discussed later.

The remainder of the chapter is devoted to crystal structure, defects, and related electrical properties. This discussion is initiated with a brief description of the crystal structure and dimensions of Si so that the reader may relate the later, more basic, discussion primarily to Si, although it certainly is not necessarily so restricted.

Silicon and Ge have the diamond crystal structure. The atomic number of Si is 14, which might be expected to lead to the electronic structure $1s^2 2s^2 2p^6 3s^2 3p^2$. However, as has been pointed out previously, a hybridization of the *s* and *p* levels (sp^3) leads to a valence of 4 rather than 2. This hybridized state is, of course, a lower-energy state for the crystal lattice and is characteristic of covalent bonding (i.e., the sharing of electrons between a pair of atoms).

The diamond structure belongs to the cubic class of crystals. The diamond lattice (lattice refers to the points at which the atoms are located) consists of two interpenetrating face-centered cubic (fcc) sublattices. Figure 2-10 is an illustration of a fcc lattice. In the diamond structure a second fcc lattice interpene-

[7] A passivating surface layer is one used to protect the semiconductor from contamination from the atmosphere.

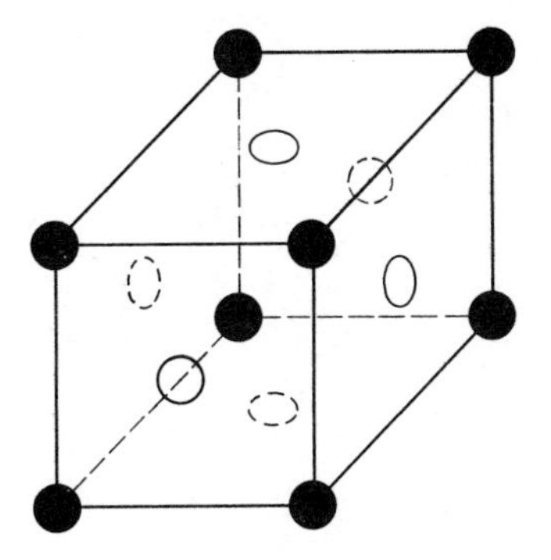

Fig. 2-10 Face-centered cubic lattice. Face-centered points are drawn as open circles for ease of identification only.

trates the first, with a corner point of the second lattice one-fourth of the distance along with a major diagonal of the first lattice. Figure 2-11 shows a ball-and-stick model of the diamond structure. The unit cell is outlined in thread. The light and dark balls are on different fcc lattices. Each atom is tetrahedrally coordinated. The simplest way to visualize this is to imagine an atom at the center of a cube with its four nearest neighbors at corners, as shown by the model in Fig. 2-12.

In Si the distance between nearest neighbors is 1.18 Å, the cube edge $a = 5.428$ Å, there are eight atoms per unit cell, and the packing density (volume of atoms divided by volume of unit cell) is 34%. The most commonly used dopants in Si are phosphorus (P), arsenic (As), antimony (Sb), boron (B), aluminum (Al), and gold (Au).

Fig. 2-11 Model of diamond structure.

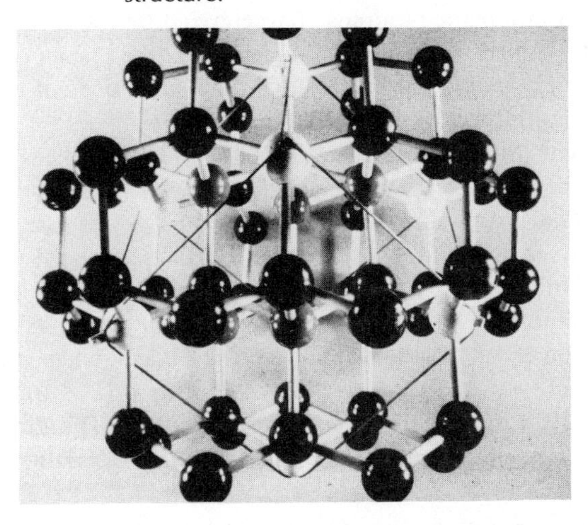

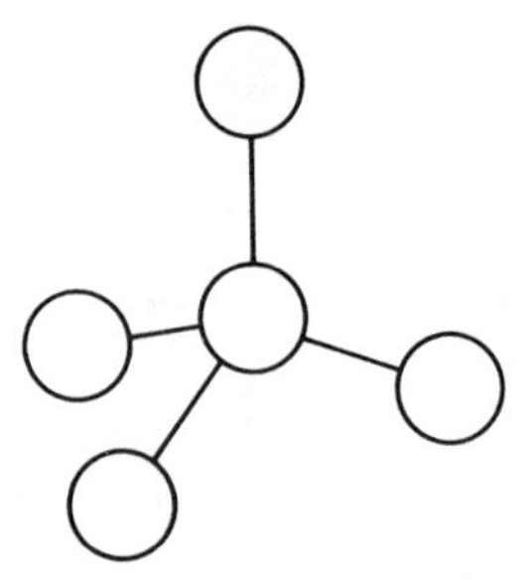

Fig. 2-12 Tetrahedral coordination in the diamond structure.

2-3-1 Miller Indices

In crystals we wish to measure distances in units of the length of the unit cell edges, a, b, c. It may be that $a \neq b \neq c$, but we are assuming that the axes are mutually orthogonal. Any plane in space may be described by the equation

$$\frac{x}{sa} + \frac{y}{tb} + \frac{z}{uc} = 1 \tag{2-12}$$

This is the intercept form where sa, tb, and uc are intercepts of the x, y, and z axes, respectively. The reciprocals of the multipliers of the unit cell edges are written as $h = 1/s$, $k = 1/t$, and $l = 1/u$; thus,

$$h\frac{x}{a} + k\frac{y}{b} + l\frac{z}{c} = 1 \tag{2-13}$$

is the equation of any plane of atoms characterized by the Miller indices *(hkl)*. Generally, Miller indices are given as integers, thus the plane $(1\ \frac{1}{2}\ 0)$ is equivalent (parallel) to (2 1 0). Figure 2-13 illustrates a number of crystallographic planes and their appropriate Miller notation.

A convenient physical interpretation of the Miller indices is the fact that they represent the number of times planes of a particular type, say (2 1 0), intercept the crystal axes within one unit cell. Thus, in this case, (2 1 0) planes intercept the a axis twice, the b axis once, and are parallel to the c axis. To calculate Miller indices it is usually easiest to find the intercepts, invert them, and then determine the set of smallest integers. For example, intercepts at 1 on the a axis, 2 on the b axis, and infinity on the c axis when inverted become 1, $\frac{1}{2}$, and 0, respectively, and the lowest set of integers is (2 1 0). Similar planes, such as (100), (010), and $(0\bar{1}0)$,[8] are denoted by {001}. The directions normal to lattice planes are important, and the indices characterizing them are just

[8] The notation $\bar{1}$ signifies a negative index (i.e., the plane intercepts the negative axis).

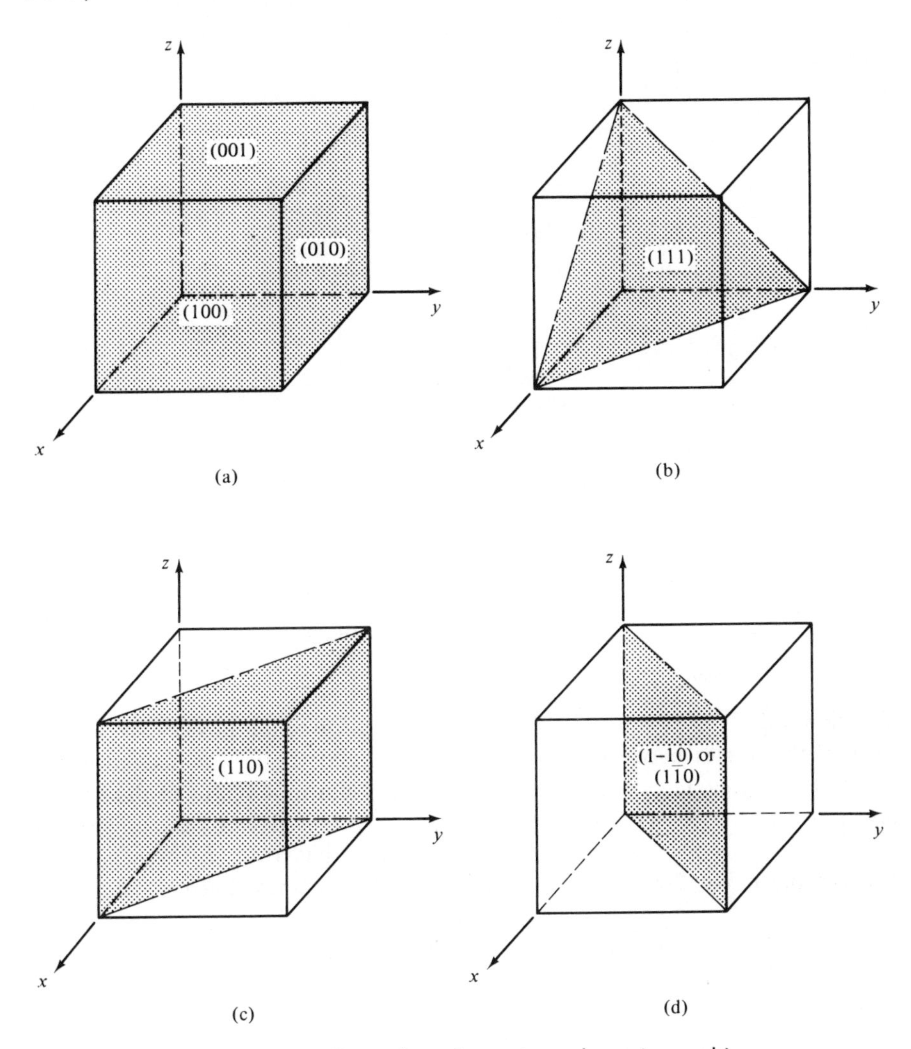

Fig. 2-13 Miller indices for various planes in a cubic crystal.

the components of a vector normal to the planes. Thus, the plane (111) will have a direction [111]. This could be written $[\frac{1}{2}\ \frac{1}{2}\ \frac{1}{2}]$ or [222], but the convention is to use the smallest integral representation possible. For planes such as (100) and (010) the directions are simply [100] and [010], and since they are similar may be denoted <001>.

The direction indices for planes other than {001} planes are still the same as the Miller indices for cubic crystals, but it is not so obvious. To determine the direction indices for the general case it is necessary to realize that the

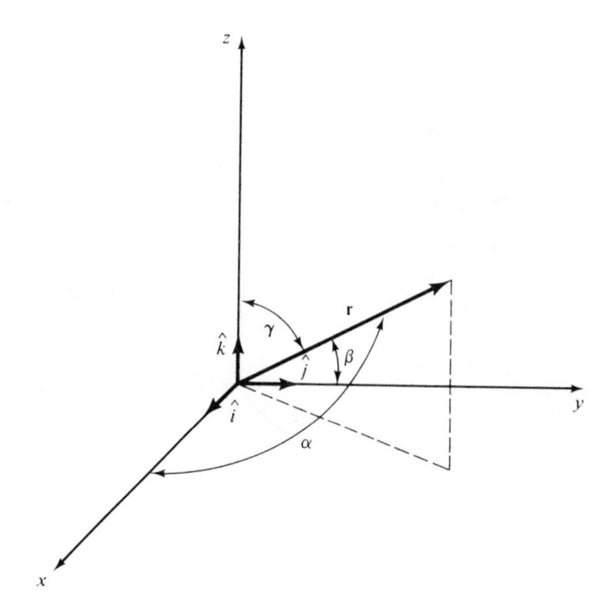

Fig. 2-14 Direction cosines.

direction indices are essentially direction cosines. Direction cosines are the cosines of the angles between the coordinate axes and the vector normal to the plane in question. They are also the components of a unit vector in the normal direction (see Fig. 2-14). The vector **r** has the direction given by the unit vector $\mathbf{i}\cos\alpha + \mathbf{j}\cos\beta + \mathbf{k}\cos\gamma$.[9] Let us assume that $|\mathbf{r}| = d$, the separation between the *(hkl)* planes measured normal to these planes. Then

$$\mathbf{r} = d(\mathbf{i}\cos\alpha + \mathbf{j}\cos\beta + \mathbf{k}\cos\gamma) \tag{2-14}$$

Now

$$\cos\alpha = \frac{d}{a/h}$$

$$\cos\beta = \frac{d}{b/k}$$

$$\cos\gamma = \frac{d}{c/l}$$

where a/h is the intercept of the x axis, b/k the intercept of the y axis, and c/l the intercept of the z axis. For a cubic crystal such as Si, $a = b = c$. It is

[9] **i**, **j**, and **k** are Cartesian unit vectors.

not necessary that $a = b = c$; however, these symbols do represent the length of the unit cell edge for the respective directions. Hence,

$$\mathbf{r} = d\left(\mathbf{i}\frac{dh}{a} + \mathbf{j}\frac{dk}{b} + \mathbf{k}\frac{dl}{c}\right) \tag{2-15}$$

which gives

$$|r| = d = d^2\sqrt{\left(\frac{h}{a}\right)^2 + \left(\frac{k}{b}\right)^2 + \left(\frac{l}{c}\right)^2}$$

or

$$d = \frac{1}{\sqrt{(h/a)^2 + (k/b)^2 + (l/c)^2}}$$

which for cubic crystals becomes

$$d = \frac{a}{\sqrt{h^2 + k^2 + l^2}} \tag{2-16}$$

As can be seen from Eq. (2-15), the direction cosines for a cubic crystal are hd/a, kd/a, ld/a, or substituting for d, they become

$$\frac{h}{\sqrt{h^2 + k^2 + l^2}}, \qquad \text{etc.}$$

Since only the direction is of interest, the denominator common to all three direction cosines can be ignored giving the direction numbers $[hkl]$. Thus, if the edge length of the unit cell is known, the spacing of any set of planes can easily be calculated and the direction of the normal to the planes can be specified. It is notable that the planes of higher Miller indices are the most closely spaced.

The angle between any two sets of planes is the angle between the normals to those planes. Hence, if $\mathbf{r}_1$ and $\mathbf{r}_2$ are the normal vectors for two planes, $\mathbf{r}_1 \cdot \mathbf{r}_2 = d_1 d_2 \cos\theta$ where θ is the angle between the planes. For a cubic crystal

$$\cos\theta = \frac{h_1 h_2 + k_1 k_2 + l_1 l_2}{[(h_1^2 + k_1^2 + l_1^2)(h_2^2 + k_2^2 + l_2^2)]^{1/2}} \tag{2-17}$$

2-4 CRYSTAL DEFECTS

There are various types of defects in the regularity of single crystals which influence electrical and optical properties. These defects can be generally classified as point defects, line or dislocation defects, and gross or large-scale defects.

Point defects include vacancies, and interstitial and substitutional impurities. Dislocations are one-dimensional defects in the crystal lattice. Gross defects include slip and twinning, which occur along crystallographic planes.

2-4-1 Point Defects

Vacancies are classified as Schottky defects if the missing atom is far removed from the vacancy site. An atom may be dislodged from its crystal site by thermal energy and migrate to the crystal surface. The energy required to accomplish this in Si is a fairly large 2.3 eV. If the atom dislodged remains in the vicinity of the vacancy as an interstitial impurity, it is called a Frenkel defect and the activation energy is much lower.

The diamond lattice has five interstitial voids per unit cell, all large enough for Si atoms; consequently, interstitial atoms are quite common. Impurities are deliberately introduced into Si to control its properties. These impurities, if they come from group III or V of the periodic table, are usually substitutional in nature at moderate doping levels. On the other hand, Au is about 10% interstitial and the rest substitutional, and Ni is about 99.9% interstitial. Unintentional impurities, such as Zn, Cu, Co, Fe, and Mn, are usually located at interstitial sites.

The equilibrium concentration of Schottky defects in a crystal can be determined from

$$n_s = N \exp\left(\frac{-E_s}{kT}\right) \tag{2-18}$$

where N is the number of atoms per unit volume, n_s the number of Schottky defects per unit volume, E_s the activation energy, k Boltzmann's constant, and T the absolute temperature. Equation (2-18) is an application of the Maxwell–Boltzmann distribution function,

$$\frac{n}{N} = \exp\left(\frac{-E}{kT}\right)$$

which gives the fraction of particles (whatever they might be) with energy E. The quantity n/N is also the probability that a given particle has energy E.

Frenkel defects are a little more difficult to analyze. The probability that an atom occupies an interstitial site will be the product of the probability that the atom is freed from its lattice site times the probability that it is trapped by an interstitial site. Assume that an energy E_1 is required to break the bond; then the probability for a bond being broken is

$$\frac{n_f + \Delta n}{N} = \exp\left(\frac{-E_1}{kT}\right) \tag{2-19}$$

where $(n_f + \Delta n)$ is the number of atoms with broken bonds and n_f is the number of atoms that make it to interstitial sites. Let E_2 be the energy required for the atom, once freed, to reach an interstitial site and let N_i be the number of interstitial sites. Thus,

$$\frac{n_f + \Delta n}{N} \frac{n_f}{N_i} = \exp\left(\frac{-E_1}{kT}\right) \exp\left(\frac{-E_2}{kT}\right)$$

Assuming that $\Delta n << n_f$, then

$$n_f = \sqrt{NN_i} \exp\left(\frac{-E_f}{2kT}\right) \tag{2-20}$$

where $E_f = E_1 + E_2$ is the Frenkel defect activation energy and n_f the number of Frenkel defects per unit volume. Concentrations of point defects can exceed the equilibrium values as the result of quenching or radiation damage.

Isolated Si vacancies are uncommon. Complexes of vacancies are formed at low temperatures.

2-4-2 Dislocations

An edge dislocation in a crystal lattice appears as a crystallographic plane terminated along a line. This is illustrated in Fig. 2-15.

Screw dislocations are caused by shear stresses that tend to cause rotation. Figure 2-16 schematically depicts this process. The energy of formation of edge dislocations in Si is 10 to 19 eV per atom length along the dislocation line. The energy of formation of screw dislocations is about 15 to 30 eV per atom length.

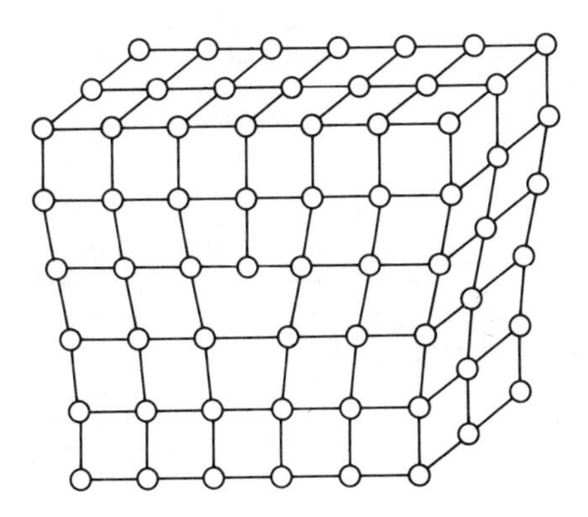

Fig. 2-15 Edge dislocation.

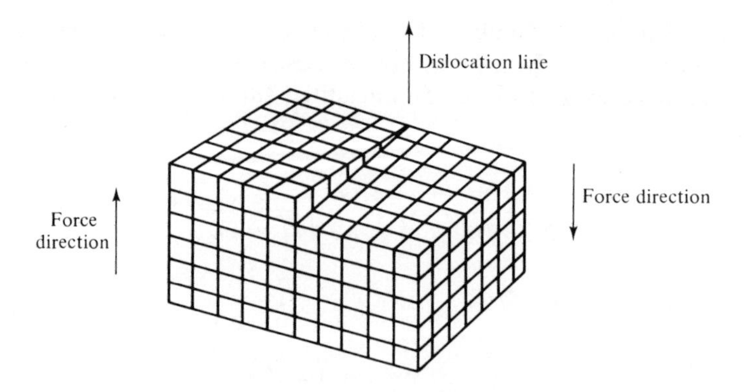

Fig. 2-16 Screw dislocation.

2-4-3 Movement of Dislocations

The movement of an edge dislocation is depicted by the series of diagrams in Fig. 2-17. As indicated in Fig. 2-17(e), the dislocation will eventually reach the surface if it is not pinned by other dislocations or imperfections. This process of movement is called slip. Screw dislocations move by a similar process. Another process of movement is called climb, which occurs at 90° to the slip plane. Climb involves atoms entering or leaving the dislocation lines. Most dislocation movement is a combination of slip and climb. The energy of movement for dislocations is about 0.15 eV per atomic spacing per atom length in Si.

Fig. 2-17 Dislocation movement.

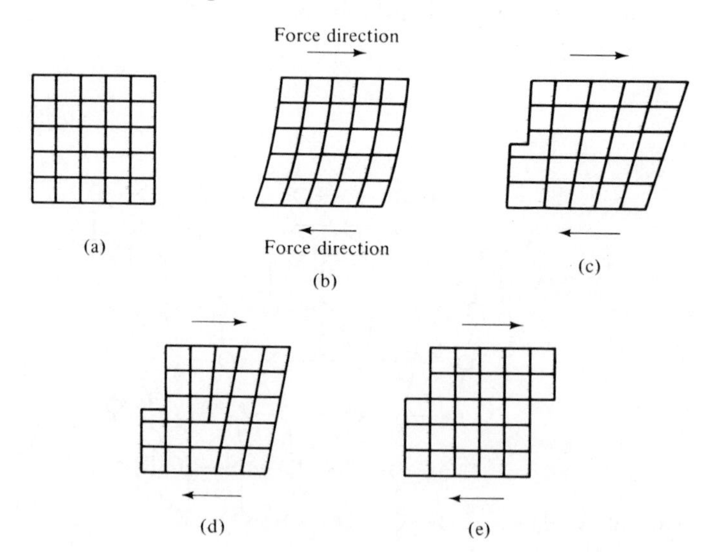

Dislocations are relatively difficult to form in Si and generally are not too serious a problem. The low energy of movement of dislocations makes it quite easy to move the dislocations out of the crystal. This is discussed in more detail in Section 2-6.

It is possible for dislocations to generate new dislocations. This happens in distorted or highly stressed crystals. The process by which this occurs is depicted in Fig. 2-18. A dislocation line terminated inside the crystal (the ends

Fig. 2-18 Frank–Read source.

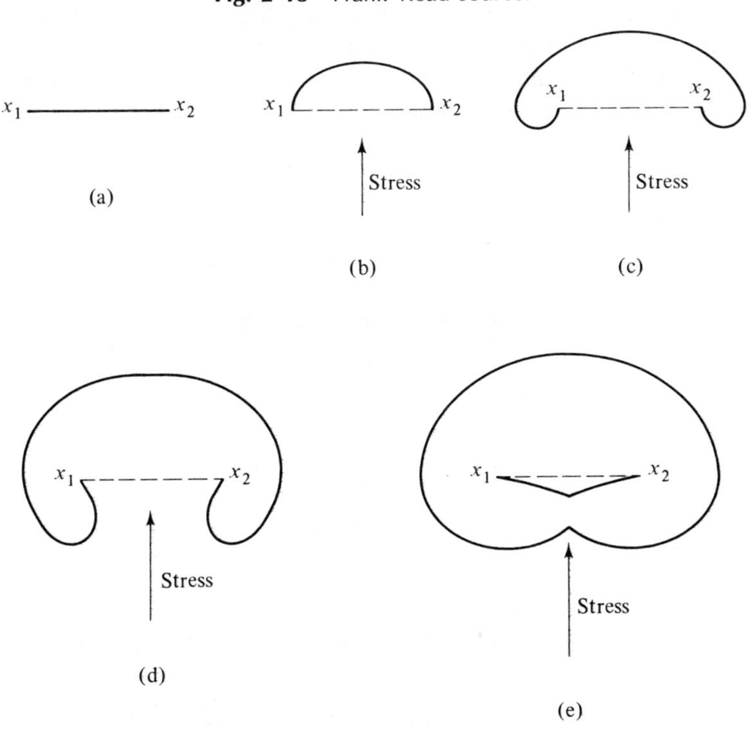

are usually pinned by imperfections, which prevent easy movement of the dislocation) can act as what is called a Frank–Read source, named in honor of its discoverers. As a stress F is applied, the dislocation line bows out of its slip plane. If the stress is sufficiently large, the bowing out progresses as shown in Fig. 2-18(a) to (e). The original source (dislocation line), pinned between x_1 and x_2, is still there and can produce more dislocations until the stress is removed or otherwise relieved.

2-4-4 Twinning

Twinning is a mismatch defect that occurs in crystals. This defect involves the presence of two different orientations of the crystal lattice occurring on opposite sides of a separating plane. The twinning plane is called the composition plane; atoms along this plane are common to both sections of the crystal. A schematic illustration of twinning is given in Fig. 2-19. Twinning occurs in Si whenever it is confined during growth due to expansion upon freezing; hence, defect-free Si cannot be grown in a crucible.

Another important defect that occurs in the production of Si wafers is called a spike. This is a protuberance above the surface of the wafer that occurs during epitaxial growth (a process discussed in Chapter 3). In epitaxial growth, a new layer of crystal is grown on the original wafer to produce, for example, a *p*-type substrate with an *n*-type layer on top. Scratches or inclusions at the interface of the original wafer surface can cause these spikes to occur on the

Fig. 2-19 Twinning.

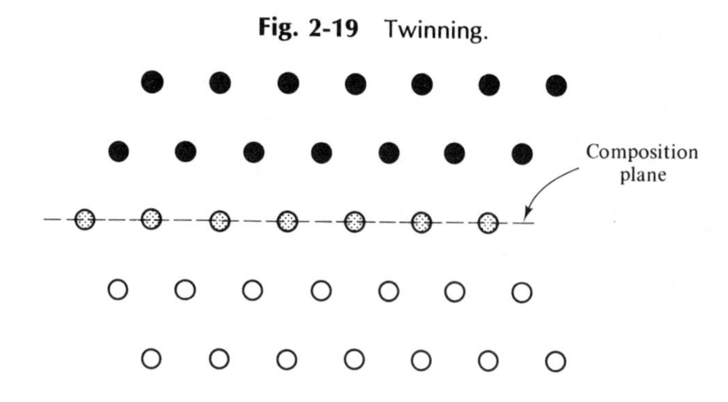

epitaxial surface. They may protrude several micrometers above the surface and thereby cause damage to masks used in contact mask alignment.

2-5 INFLUENCE OF DEFECTS ON ELECTRONIC PROPERTIES

A vacancy in the Si lattice means that four unpaired bonds are left dangling. The vacancy thus behaves like an acceptor, trapping free electrons. It is not likely that more than one or two electrons are trapped, because of the large electrostatic repulsion that occurs between these unneutralized charges. Interstitial Si atoms behave like donors because of the excess electrons carried by such an atom, from the point of view of the crystal lattice. It is not likely that interstitial Si atoms lose more than one or two electrons, because of the large electrostatic attractive force of the positive Si ion for nearby electrons.

The energy levels for the donor and acceptor levels referred to here are in the energy gap as they usually are for dopants.

Dislocations act weakly as acceptors in Si due to a few dangling bonds. However, the major effect of dislocations results from their tendency to enhance diffusion in their vicinity and segregate metallic impurities near them. Silicon with a high concentration of dislocations exhibits excessive leakage and premature breakdown in *p-n* junctions. Concentrations of impurities around a dislocation tend to reduce charge mobility. Of course, any type of imperfection in the crystal lattice tends to reduce mobility.

It is expected, and has been verified by experiment, that carrier lifetime is decreased by the presence of a large number of point defects due to their trapping characteristics. It should be mentioned that modern Si-growing technology provides large quantities of extremely low defect concentration material.

2-6 CRYSTAL GROWTH

Silicon expands about 10% when solidified from the melt. Because of this, silicon cannot be confined in a boat for crystal growth. Even if the boat survives the expansion, the stresses exerted on the Si will cause an excess of dislocations. Most Si is grown by the Czochralski method, although the zone process is also used.

The Czochralski process is illustrated in Fig. 2-20. In this process, a pure, dislocation-free seed crystal oriented in the <111> or <100> direction is brought into contact with the surface of a melt of doped Si. By proper control of the temperature at the crystal–melt interface, rotation rate, and pulling rate, Si crystals over 1 m in length and 12.5 cm in diameter may be grown (pulled) from the melt.

The <111> direction is chosen for crystal growth because of rapid growth. Growth actually occurs by a laying down of atoms in the <001> directions which lie in the <111> plane. Crystals grown in the <001> direction exhibit about one-third of the undesirable energy states at the Si–SiO_2 interface when thermally oxidized, which could make them perferable for some MOS devices; however, with the advent of ion implantation and polysilicon gate technology, this is not necessarily a significant factor. Because of the necessity for carefully controlling crystal growth, the interface between crystal and melt must be closely monitored and held at a desirable shape. Pull and rotation rate can be set to achieve this. Dislocations can be influenced to grow out of the crystal by rapid rotation, but this also tends to corrode the crucible, so a compromise must be made. A rapid initial rotation rate is generally used, followed by a slower rotation rate for the bulk of the crystal growth. It is possible to grow dislocation-free Si under proper conditions.

Impurities are introduced into the melt by adding a known amount of a highly doped Si pellet. One difficulty encountered in growing doped crystals

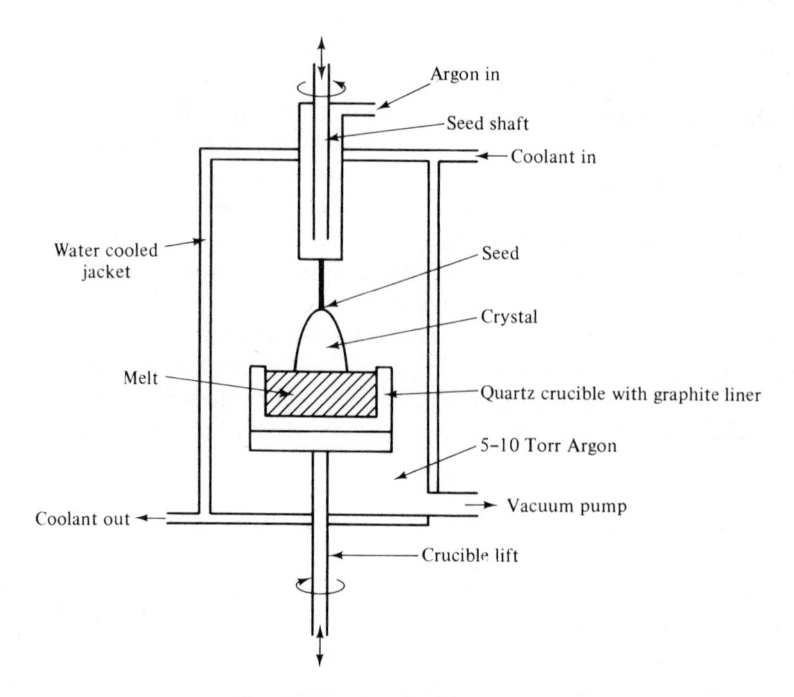

Fig. 2-20 Czochralski process.

is that the concentration of dopant in the solid and the liquid are different. The ratio of the dopant concentration by weight in the solid, C_s, to the concentration in the liquid, C_l, is called the segregation coefficient k. The approximate segregation coefficients for common dopants are listed in Table 2–1. For all dopants of interest, the melt becomes solute-rich. Under normal conditions the melt becomes solute-rich near the crystal–melt interface. Care must be taken to program the pull rate so that the concentration of dopant in the crystal does not fluctuate too widely along the direction of growth. The best dopants for use in a melt are those with high segregation coefficients, all other things being equal.

Concentrations of dislocations may run as high as 10^4 per square centimeter in Czochralski-grown crystals. Variation of resistivity in the radial direction can be a problem. Radial resistivity may vary as much as 30%, whereas 5 to 10% is tolerable in microcircuit fabrication. The presence of oxygen in Si is

TABLE 2-1 Segregation Coefficients for Si Dopants

Dopant	P	As	Sb	B	Al
$K = \frac{C_S}{C_L}$	0.35	0.30	0.023	0.80	1.8×10^{-3}

also a problem. The oxygen comes from the corrosion of quartz crucibles. Concentrations of 10^{16} to 10^{18} atoms/cm^3 are obtained, depending on the rotation rate. Most of the oxygen ends up as SiO_2, which tends to segregate along dislocations and cause defective circuits built in that part of the crystal. Pure oxygen tends to pin the ends of dislocations, giving rise to Frank–Read sources. Some of the oxygen ends up as SiO_4, which is donor-like. Heat treatment alters the SiO_4/SiO_2 ratio, causing variable electrical properties with varying operating temperatures. Oxygen concentrations are generally held low enough so that these problems are not serious. However, in dislocation-free Si, vacancies tend to pile up and attract oxygen and heavy metals. This has deleterious local effects on such things as leakage current and mobility. An annealing step at 700°C is performed to precipitate the oxygen so that it does not affect later process steps. This lowers the bulk resistivity somewhat.

Single crystals can be grown that are over 12 cm in diameter and 2 m long. However, the larger the diameter, the thicker the wafers must be for mechanical strength and to prevent warpage during high-temperature processing.

2-7 FLOAT ZONE PROCESS

The zone process is illustrated in Fig. 2-21. A rod of predoped cast polycrystalline Si is attached at one end to a seed crystal with the proper orientation. A small molten zone is initiated by radio-frequency (RF) or electron-beam heating at the seed crystal–polysilicon interface and is slowly moved to the other end of the rod. Single-crystal Si freezes out at the back side (seed side) of the molten zone. Several passes may be made to refine the crystal. There is a variation of dopant concentration along the rod because of the segregation coefficient. The problem is most serious with low-segregation-coefficient dopants and is enhanced with repeated passes of the molten zone. This can be compensated for by reversing the direction of travel of the zone.

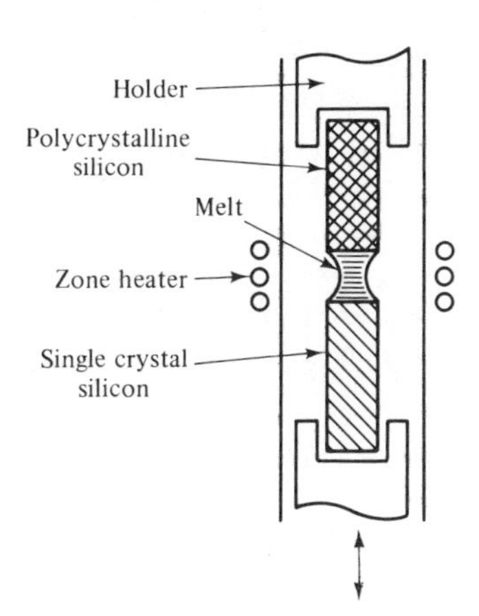

Fig. 2-21 Float zone process.

Zone leveling is a technique whereby a charge of dopant is added to the molten zone. The dopant is then distributed at the same time the crystal is refined.

Dislocation concentrations are higher in the zone process, owing to stresses occurring around the molten zone. Surface dislocation concentrations of 10^3 to $10^5/cm^2$ are common. The composition of high resistivity materials (up to 2000 Ω-cm) is more controllable owing to the low contaminant levels that can be achieved with this method because of the absence of a boat to hold molten Si. Oxygen content can be held to a negligible level for this reason.

The radial variation of resistivity is about the same as in the Czochralski method.

As larger- and larger-diameter crystals have been grown, it has been found that wafer thickness had to be increased to prevent warpage during high-temperature diffusion processing. The early 2.5-cm wafers were about 0.250 mm thick. The 7.5-cm and 10.0-cm wafers have to be about 0.500 to 0.550 mm thick to prevent warpage. Computer control of furnace heat-up and cool-down has, however, greatly reduced the problem of wafer warpage during processing.

It turns out that epitaxial[10] layers grown on {111} surfaces contain high defect concentrations related to the slow growth rate in a <111> direction. Consequently, wafers obtained from <111>-oriented crystals are cut with a slight misorientation to reduce the incidence of defects. It has been shown that the optimum misorientation direction, to minimize pattern shifting effects due to nonuniform diffusion of impurities, is toward the [110] direction parallel to a $(1\bar{1}0)$ plane $3.0 \pm 0.5°$.[11] There is usually a reference flat ground on the boule parallel to the $(1\bar{1}0)$ plane to within 1°. Wafers with <100> orientation should be grown as close to <100> orientation as possible because this is a fast-growth direction and epitaxial layers can be grown relatively defect free.

The mechanical properties of crystals associated with their crystallographic planes lead to optimum scribing directions for minimal chip loss. When using mechanical scribing, even the direction of tool travel can be important. In Si, the {111} planes, because of large bond distances between them, are the easy-break planes and are the most easily etched. Etch patterns may be used to check crystal orientation. A standard etch solution such as Sirtl etch will expose a triangular pattern on a <111>-oriented crystal and a square etch pattern on <100>-grown crystals. This is illustrated in Fig. 2-22, which also shows the relation of the etch pattern to the flat. In the case of the <111> crystal, one scribe direction is perpendicular to the $(1\bar{1}0)$ flat and there is no symmetry difference in either direction. For the scribe parallel to the $(1\bar{1}0)$ flat, tool motion in the direction shown in Fig. 2-22(a) is preferred. Tool travel in the opposite

[10] Epitaxial literally means "arranged upon" and as used here refers to a doped layer with identical crystal structure grown on a differently doped substrate.

[11] C. M. Drum and C. A. Clark, *Journal of the Electrochemical Society,* Vol. 115, No. 6, June 1968.

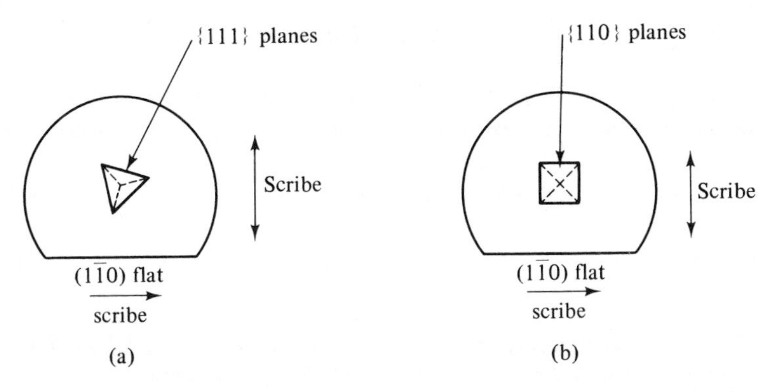

Fig. 2-22 Silicon etch patterns for <111> and <100> grown crystals with reference flats and appropriate scribe directions.

direction will cause irregular cleavage along the intersecting {111} planes. In the case of the [100]-grown crystal, tool travel can be in either direction as long as it is perpendicular or parallel to the (1$\bar{1}$0) reference flat. In all cases the scribing occurs parallel to oblique {111} planes and breakage occurs along those planes.[12]

In the remaining chapters most of the concepts presented in this chapter will be applied directly or expanded upon in discussions of integrated circuit processing and/or device fundamentals.

BIBLIOGRAPHY

[1] STREETMAN, B. G., *Solid State Electronic Devices.* Englewood Cliffs, N.J.: Prentice-Hall, Inc., 1972

[2] VAN VLACK, L. H., *Elements of Materials Science,* 2nd ed. Reading, Mass.: Addison-Wesley Publishing Company, Inc., 1964.

[3] KITTEL, C., *Introduction to Solid State Physics,* 5th ed. New York: John Wiley & Sons, Inc., 1976.

[4] POHL, H. A., *Quantum Mechanics for Science and Engineering.* Englewood Cliffs, N.J.: Prentice-Hall, Inc., 1967.

[5] WEHR, M. R., AND J. A. RICHARDS, *Physics of the Atom.* Reading, Mass.: Addison-Wesley Publishing Company, Inc., 1960.

[6] LEVINE, S. N., *Quantum Physics of Electronics.* New York: Collier-Macmillan Ltd., 1965.

[7] ADLER, R. B., A. C. SMITH, AND R. L. LONGINI, *Introduction to Semiconductor Physics,* Vol. 1. New York: John Wiley & Sons, Inc., 1964.

[12] D. O. Townly, "Optimum Crystallographic Orientation for Silicon Device Fabrication," *Solid State Technology,* January 1973.

PROBLEMS

2-1 Write down the electronic structure of magnesium (Mg). On the basis of the valence-bond model, would you predict that it should be a metal or an insulator? Why is it a metal?

2-2 The energy density of states $N(E)$ is given by $N(E) = CE^{\frac{1}{2}}$, where

$$C = \frac{1}{2\pi^2}\left(\frac{2m_0}{\hbar^2}\right)^{\frac{3}{2}}$$

m_0 is the electron mass, and $\hbar$ is Planck's constant divided by 2π for metals. Determine the value of the Fermi level E_F at absolute zero of temperature in terms of C.

2-3 Calculate the number of electrons per unit volume in the conduction band of intrinsic Si at $T = 300$ K.

2-4 Determine the probability of an energy state 0.863 eV above the Fermi level being occupied in intrinsic Si at $T = 1000$ K.

2-5 It is observed that a plane in a cubic crystal intersects the axes at $3a$, $2a$, $2a$, where a is the unit cell size. Determine the Miller indices and the direction numbers for this plane.

2-6 (a) Calculate the minimum spacing between the {120} planes in Si. The edge length of the Si unit cell is 5.43 Å.

(b) Determine the angle between the (110) and $(1\bar{1}2)$ planes in Si. (*Note:* $\bar{1} = -1$ represents an intersection on a negative axis.)

2-7 Determine the Miller indices and direction numbers for the planes shown.

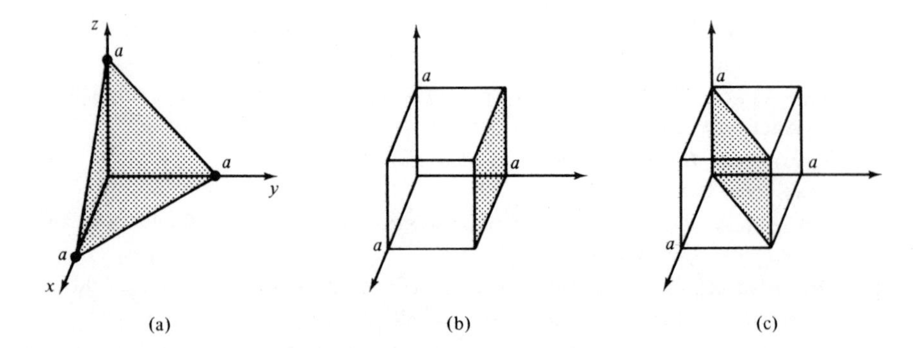

2-8 (a) Calculate the distance between the (100) planes, the (110) planes, and the (111) planes in Si ($a = 5.43$ Å).

(b) Calculate the angle between the (111) and (110) planes.

2-9 Determine the Miller indices and direction numbers for the planes shown.

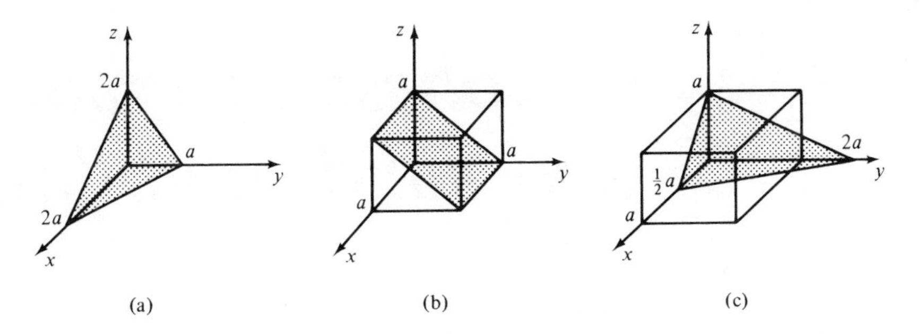

2-10 Sketch the (001), (110), (111), and $(1\bar{1}1)$ planes for a cubic crystal.

2-11 Explain what a Frank–Read source is and sketch a case where a combination of screw and edge dislocation might give rise to such a source.

2-12 Are any of the {111} planes normal to any of the {011} planes? If so, give a specific example.

2-13 List the various defects that occur in Si, and discuss their effects on the electronic properties of Si devices within the context of what you presently know about Si-device properties.

2-14 Consider intrinsic silicon at 300K. If the effective mass of an electron in the conduction band were two times greater than the effective mass of a hole in the valence band, find the energy shift (ΔE_F) of the Fermi level from the center of the forbidden band.

phase diagrams and solid solubility

3

Phase[1] diagrams are an essential feature of any discussion of material properties where interactions between various materials are involved. This is especially important in microelectronics because a wide variety of different materials are utilized for such things as bonding pads and passivation layers. Since Si comes into intimate contact with a variety of metals in the processing of Si integrated circuits, particular emphasis is placed on phase diagrams involving Si as one component.

3-1 TYPES OF PHASE DIAGRAMS

Unary phase diagrams involve only one material and its phases plotted as a function of pressure, temperature, and volume. Instead of attempting to depict a three-dimensional plot in two dimensions, a projection of the diagram onto the pressure vs. temperature plane is usually shown. It could just as well be volume vs. temperature. Figure 3-1 depicts a representative case. In certain regions of the diagram, the material exists only as a solid, a liquid, or vapor. Along the lines, the material can have two phases in contact with each other: either a solid–liquid, a solid–vapor, or a liquid–vapor combination. At one point, referred to as the triple point, all three phases can exist in intimate contact with each other at the same time. The condition occurs for precisely one temperature only and therefore serves as an excellent reference temperature. The triple point of water is commonly used as a reference temperature, particularly in

[1] A phase may be defined as a structurally homogeneous part of a material system. This may be amorphous, crystalline, liquid, or gas.

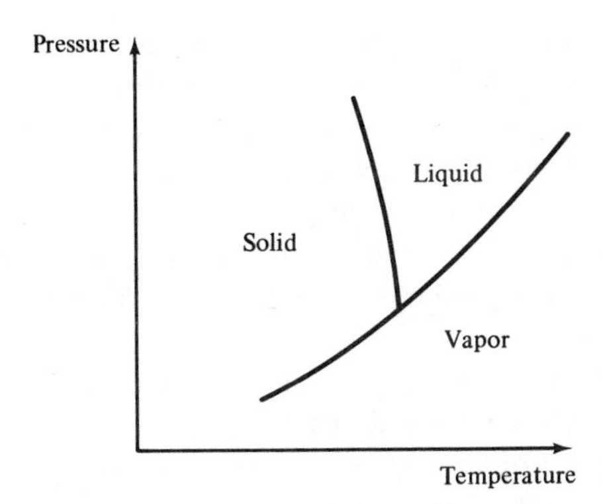

Fig. 3-1 Unary phase diagram.

precise measurements employing thermocouples. The reference junction is placed in contact with an ice–water–vapor bath.

Systems with two components are represented by binary phase diagrams. In these diagrams the ordinate is temperature and the abscissa is percent component.[2] Pressure is generally taken to be 1 atm, and volume changes are considered negligible if only the liquid and solid phases are depicted. Figure

Fig. 3-2 Binary phase diagram.

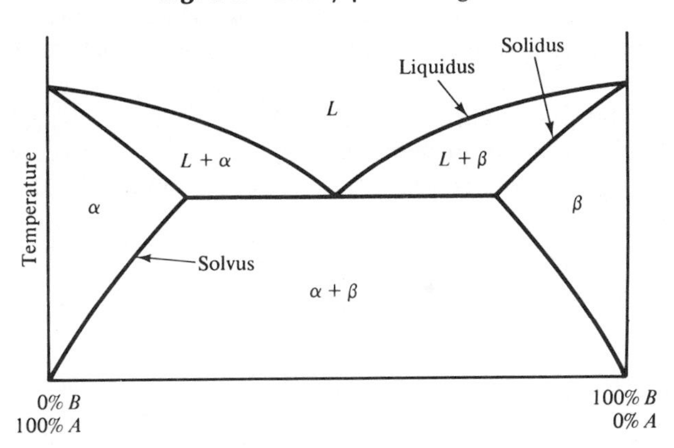

3-2 represents a typical binary diagram for components A and B by either mass or atomic percent.

The Greek letter α represents a phase of A with B dissolved in it, β represents a phase of B with A the solute, and $\alpha + \beta$ implies a mixture of the two phases. L stands for liquid, and $L + \beta$ and $L + \alpha$ stand for liquid plus β phase and liquid plus α phase, respectively. Various lines separating

[2] This is usually either a percentage of the total mass (also called weight percent) or a percentage of the total number of atoms (atomic percent).

phases, where different phases may coexist, are called: solidus, where the α or β phases are in contact with the $L + \alpha$ or $L + \beta$ mixtures, respectively; solvus, where α and $\alpha + \beta$ or β and $\alpha + \beta$ coexist, respectively; and liquidus, where the liquid phase coexists with either the liquid and β or liquid and α phases.

The point at which two liquidus curves come together often represents the lowest possible melting point of any combination of A and B and is then called the eutectic point; the mixture at that point is called the eutectic composition.

It is instructive to consider how a composition solidifies from the liquid state (the melt) and how a phase diagram enables us to predict what the equilibrium composition of each phase present at any given temperature will be. Referring to Fig. 3-3, suppose that the original composition is C_M at T_1. From T_1 down to T_2 only the liquid phase is present. At T_2 two phases, L and β, may exist together. The composition of L present is C_M, the composition of β is $C_{\beta 1}$. If the temperature is reduced further, to T_3, the composition of L slides down the liquidus curve and the composition of β slides down the solidus curve to the points of intersection with the horizontal line from T_3 (isotherm). The composition of L present is now C_L, the composition of β is $C_{\beta 2}$. It should be noted that not only is the β forming at T_2 of $C_{\beta 2}$ composition, but all the old β formed at higher temperatures must also now be $C_{\beta 2}$ composition. This will have occurred by solid-state diffusion of component A into the old β to bring it to $C_{\beta 2}$ by the time T_2 is reached. Upon further cooling, the eutectic temperature is reached where the liquid and the α and β phases exist together. Below the eutectic temperature, only the α and β phases exist. There will be a mixture of α and β of composition C_E with aggregates of the β phase with composition $C_{\beta 3}$ initially. Given sufficient time at a temperature below the eutectic temperature, the solid will consist of two phases, α and β, with the concentra-

Fig. 3-3 Binary phase diagram to illustrate the freezing process.

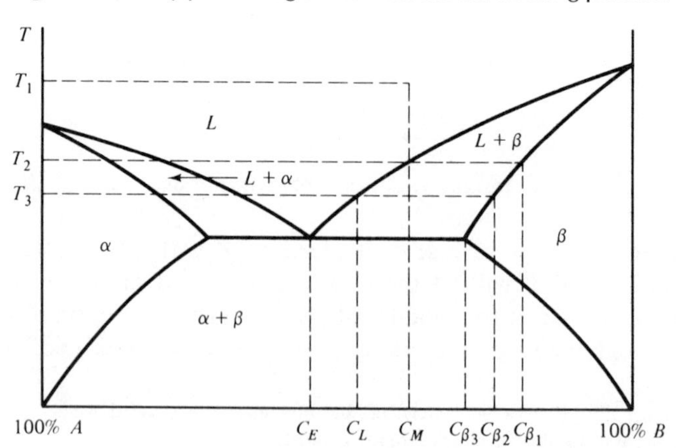

tion of each given by the intercept of the isotherm with the appropriate solvus lines.

It has been shown how the compositions of the various phases present can be determined. The problem of determining how much of each phase is present will now be discussed. The simple binary phase diagram is repeated in Fig. 3-4 to avoid confusion. Assume that the composition of the melt at temperature T_1 is C_M (C_M is in percent B), and at T_2 the L phase will have composition C_L; the β phase will have composition C_S. Let M_L be the mass of liquid and M_S the mass of solid. The following equation can be written based on conservation of mass:

$$(M_L + M_S)C_M = M_L C_L + M_S C_S$$

This expresses the fact that the total mass at T_1 times the percentage of B is the total mass of B present and is equal to the sum of the mass of B present in the liquid and solid phases at T_2. Solving this equation leads to

$$\frac{M_S}{M_L} = \frac{C_M - C_L}{C_S - C_M} = \frac{l}{s} \tag{3–1}$$

This expression is known as the lever rule. This rule, combined with knowledge of the original composition of the melt and total mass, are all that is necessary to determine the mass of the two phases and the amount of B present in either phase of *any* two-phase region of the phase diagram. This also makes it possible to calculate the amounts of A present in each phase.

Fig. 3-4 Lever rule.

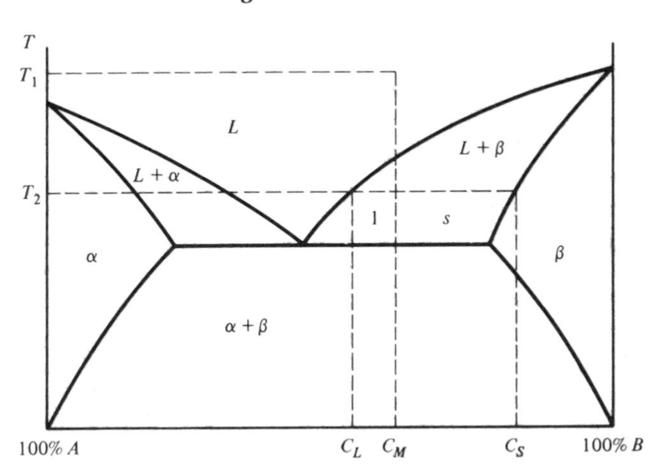

Another example of freezing from the melt is illustrated in Fig. 3-5. Lowering the temperature from T_1 to T_2 results in a mixture of L and β with compositions C_M and C_β, respectively. As the temperature is lowered further, the L composition slides down the liquidus and the β composition down the solidus as before. When T_3 is reached, the composition of β becomes C_M, and there will be no liquid below T_3, as can be seen from the lever rule. Below T_4 the α and β phases coexist as aggregates of α phase and β phase. For example, at T_5 the β aggregates will have a composition given by the intersection of the isotherm T_5 and the β solvus. Similarly, the α composition is given by the intersection of the isotherm and the α solvus.

The regions of the binary phase diagrams seen so far that are labeled either α or β are regions of solid solubility: that is, in a β region, A is dissolved in B, for example. The maximum amount of A that can be dissolved in B is given by the solvus or solidus line at any given temperature, whichever is appropriate. Rapid alloying can take place between A and B at the eutectic temperature and above. If the resulting alloy is quenched, the A atoms can be "locked into" the B lattice. However, since the solid solubility is much lower at room temperature (this is essentially due to increased misfit factor at this temperature), there will be extreme stresses developed in the alloy, seriously affecting its properties. In some cases the effect might be desirable, such as in strain hardening in the martensite reaction in steel. In microelectronics, the results would be disastrous. Consequently, depositions of dopants on Si, prior to diffusion, are carried out at higher temperatures to prevent surface damage due to excessive alloying. Also, if the solid solubility limit of the dopant in the substrate is exceeded at any temperature, a second phase, with associated strain, will appear.

Fig. 3-5 Binary phase diagram to illustrate the freezing process.

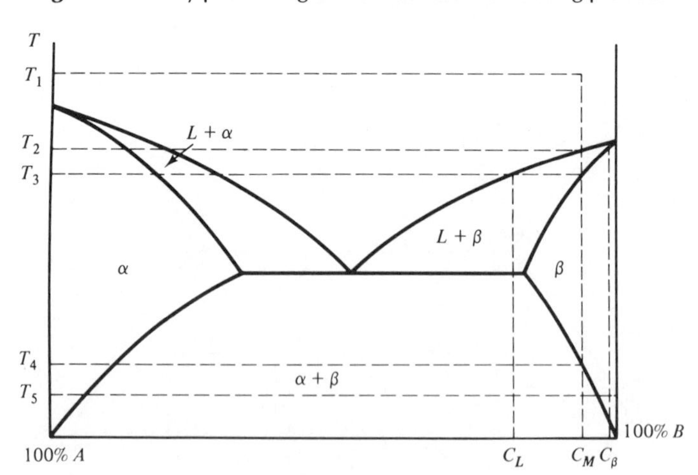

3-2 IMPORTANT SYSTEMS IN MICROELECTRONICS

Some materials are completely soluble in one another. One very significant example of this is the Si–Ge system shown in Fig. 3-6. This type of diagram, which exhibits no eutectic point, is called an isomorphous diagram. To be isomorphous, elements must subscribe to the Hume–Rothery rules by having atomic radii within 15% of each other, and have the same valence, the same crystal structure, and no significant difference in electronegativity.[3] Other examples of isomorphous systems are Cu–Ni, Au–Pt, and Ag–Pd.

The Pb–Sn system is an excellent example of a simple binary system which

Fig. 3-6 Germanium–silicon system. (From *Constitution of Binary Alloys* by M. Hansen and K. Anderko. Copyright 1958, McGraw-Hill Book Company. Used with permission of McGraw-Hill Book Company.)

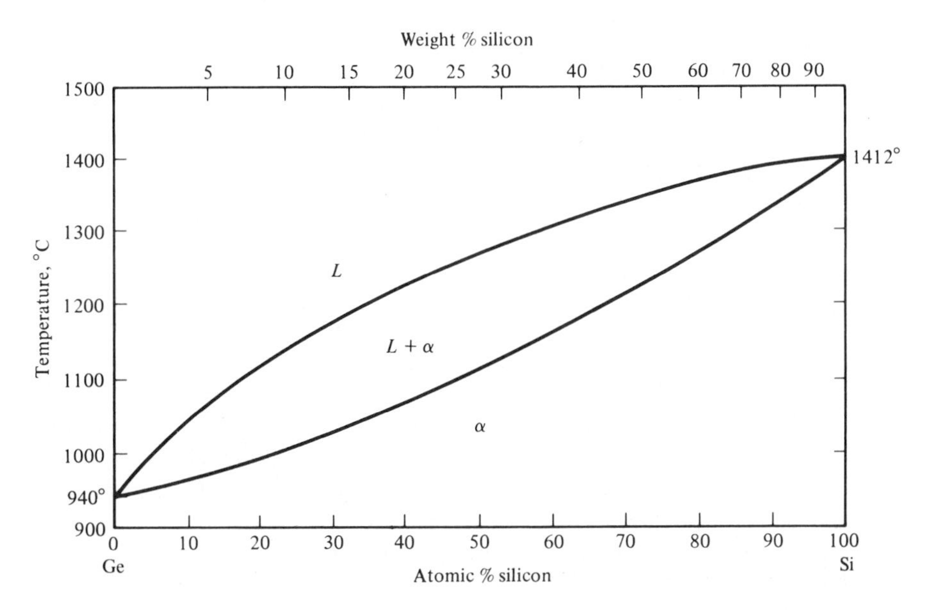

exhibits substantial, although limited, solid solubility. Figure 3-7 is the phase diagram for this system. The point at which the solidus and solvus meet is called the terminal solubility. The terminal solubility for both Sn in Pb and Pb in Sn is substantial. This system is important in microelectronics because of the extensive use of Pb–Sn solder. It can be seen from the Pb–Sn phase diagram how variation of composition can control the melting point of Pb–Sn solder. When successive soldering operations are required in microcircuit fabrica-

[3] Electronegativity of an atom is a measure of its ability to attract or pick up an extra electron. As one would expect, elements such as F have very high electronegativity; and Li, very low.

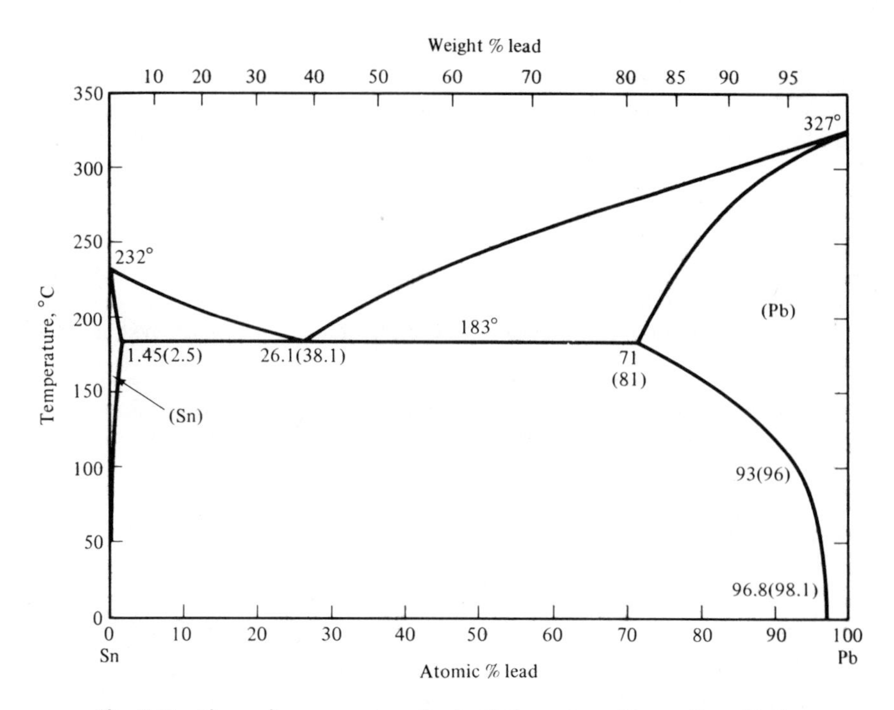

Fig. 3-7 Phase diagram system for lead–tin system. (From *Constitution of Binary Alloys* by M. Hansen and K. Anderko. Copyright 1958, McGraw-Hill Book Company. Used with permission of McGraw-Hill Book Company.)

tion, Pb–Sn solders of successively lower melting point (reflow temperature) can be used to avoid reflowing the solder of previously soldered parts.

The Al–Si system is very important in microelectronics. The metalization on monolithic Si-integrated circuits is done with Al, and some wire bonding is done with Al wire. The phase diagram for the Al–Si system is given in Fig. 3-8. The terminal solid solubility of Al in Si is less than 0.1% and is too small to be depicted in Fig. 3-8.

The Au–Si system is very important in Si microcircuit fabrication because of an extremely low eutectic temperature compared with the melting points of either pure Au or pure Si (see Fig. 3-9). The solubilities of Au in Si and Si in Au are too small to be indicated on a regular phase diagram. The low eutectic point makes practical the bonding of Si integrated-circuit chips to Au headers, preforms, or printed pads using the Au-Si eutectic reaction as the chief bonding (or soldering) mechanism. Gold, with a few percent germanium, has also been used to solder Si chips.

Elements that combine to form compounds have more complex phase diagrams, which may be broken down into two or more simpler diagrams representing binary combinations of compounds or a compound and an element. For example, the compound $AuAl_2$ (Fig. 3-10) is formed at a composition of

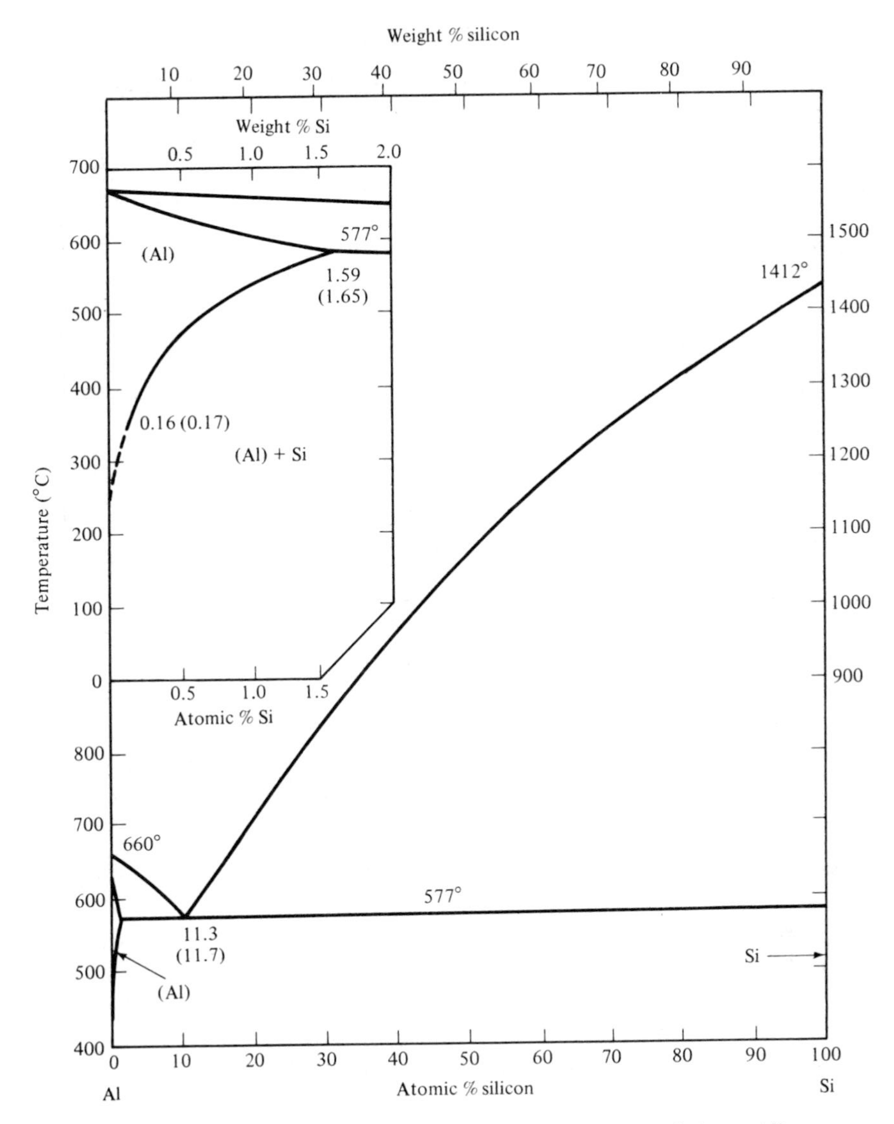

Fig. 3-8 Aluminum–silicon system. (From *Constitution of Binary Alloys* by M. Hansen and K. Anderko. Copyright 1958, McGraw-Hill Book Company. Used with permission of McGraw-Hill Book Company.)

33 atomic % Au below 1060°C. To the left of this line there is $AuAl_2$ plus an Al phase. Compounds such as $AuAl_2$ are called intermetallic compounds and occur when there is a proper stochiometric ratio of the two elements. Intermetallics characteristically have high melting points, complex crystal structures, and are hard and brittle.

The Au–Al phase diagram may be separated into two or more diagrams;

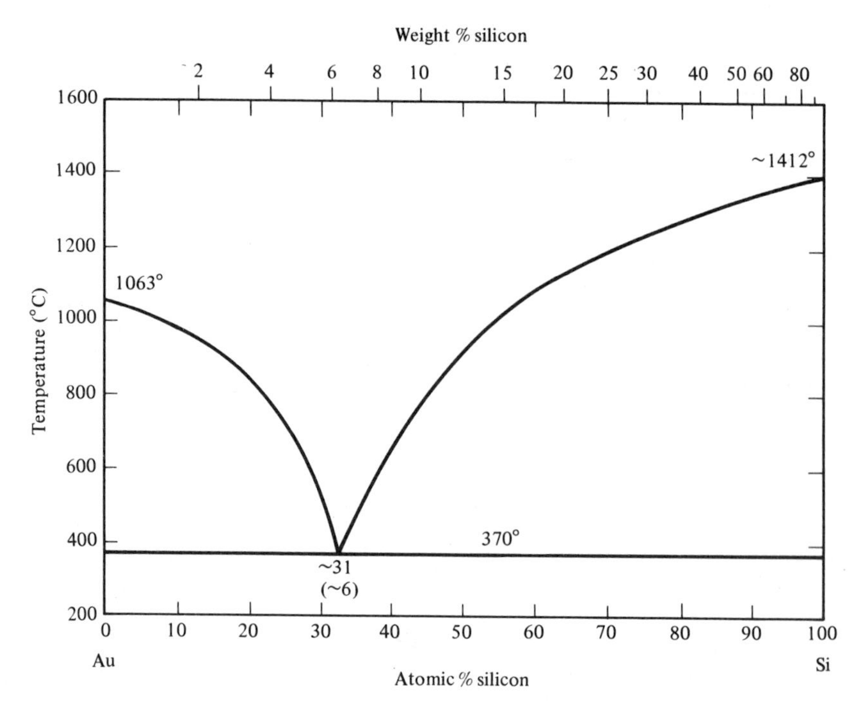

Fig. 3-9 Gold–silicon system. (From *Constitution of Binary Alloys* by M. Hansen and K. Anderko. Copyright 1958, McGraw-Hill Book Company. Used with permission of McGraw-Hill Book Company.)

for example, it can be divided into an Al–$AuAl_2$ diagram and an $AuAl_2$–Au diagram.

The Au–Al system of Fig. 3-10 is extremely important in microelectronics work, because most gold wire bonding is done to Al on Si metalizations. There are several important intermetallic compounds shown here: $AuAl_2$ (actually a ternary compound with Si), AuAl, Au_2Al, Au_5Al_2, and Au_4Al. All of these may be present in an Au–Al wire bond. Problems associated with these phases. variously referred to as purple and tan plague, will be discussed in Chapter 13.

3-3 SOLID SOLUBILITY

The terminal solubility of most Si dopants is extremely small and is not, in fact, the maximum solubility. Figure 3–11 contains a representative solidus curve for a dopant in Si. Notice that the solubility increases up to some temperature and then decreases to zero as Si approaches its melting point. This is called retrograde solubility. An even more expanded version of this diagram in the vicinity of the melting point of Si is given in Fig. 3-12. If the composition

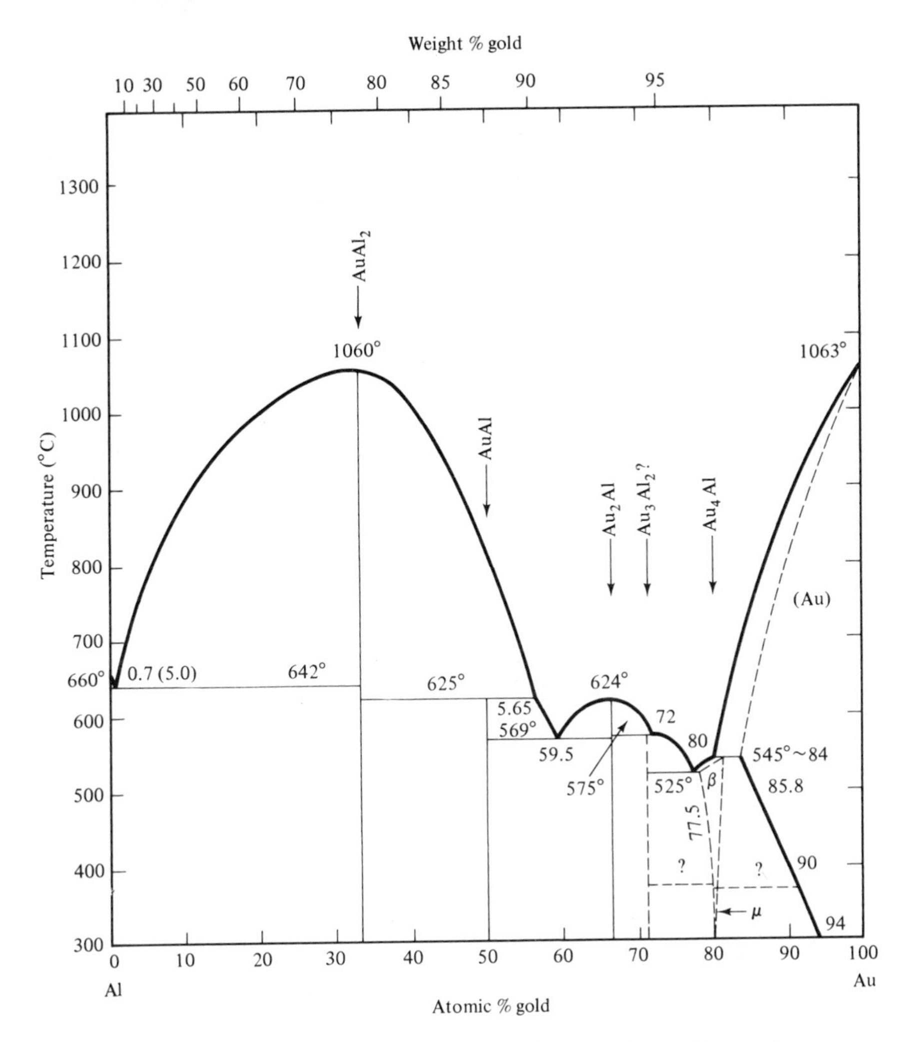

Fig. 3-10 Gold–aluminum system. (From *Constitution of Binary Alloys* by M. Hansen and K. Anderko. Copyright 1958, McGraw-Hill Book Company. Used with permission of McGraw-Hill Book Company.)

of the melt is C_M in percentage mass of the solute, the Si will begin to freeze out with composition kC_M, where k is the segregation coefficient ($k = C_s/C_L$). When the concentration of the solid reaches C_M, upon freezing, C_M/k is the composition of the liquid since the ratio of the concentration of solute in the solid to that of the liquid must equal k. The slope of the solidus line is, therefore,

$$S_s = \frac{T_1 - T_2}{kC_M - C_M}$$

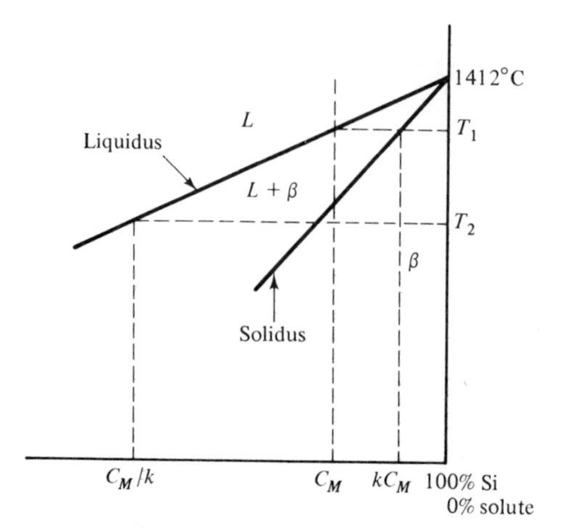

Fig. 3-11 Retrograde solubility in Si.

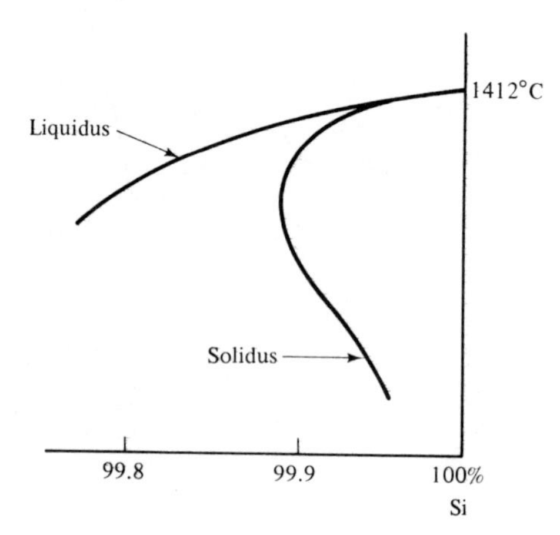

Fig. 3-12 Doping diagram for Si.

and the slope of the liquidus line is

$$S_L = \frac{T_1 - T_2}{C_M - C_M/k}$$

The ratio of slope of the solidus line to the slope of the liquidus line turns out to be the segregation coefficient:

$$\frac{S_L}{S_s} = \frac{kC_M - C_M}{C_M - C_M/k} = k \tag{3-2}$$

BIBLIOGRAPHY

[1] VAN VLACK, L. H., *Elements of Materials Science,* 3rd ed. Reading, Mass.: Addison-Wesley Publishing Company, Inc., 1974.

[2] GHANDHI, S. K., *The Theory and Practice of Microelectronics.* New York: John Wiley & Sons, Inc., 1968.

[3] GLASER, A. B., AND G. E. SUBAK-SHARPE, *Integrated Circuit Engineering.* Reading, Mass.: Addison-Wesley Publishing Company, Inc., 1977.

PROBLEMS

3-1 A Pb–Sn solder is 70% by weight Pb at 300°C. Determine the concentration of the liquid and the solid at 200°C. For a total mass of 5 g, determine the mass of the liquid and of the solid at 200°C.

3-2 A Pb–Sn solution that is 80% by weight Pb is cooled from 300°C to 150°C.
(a) Describe the cooling process in detail.
(b) Determine the mass of the liquid and the solid at 200°C if the total mass is 25 g.

3-3 Sketch a eutectic phase diagram. Label all the phases and the various boundary lines. Describe in detail how freezing from the melt occurs at three distinct concentrations, including the eutectic point but not zero or 100% concentration.

3-4 What is the maximum solubility of Si in Al, and at what temperature does this occur?

3-5 To bond bare-backed Si chips to Au headers, what is the minimum bonding temperature? What is the concentration of Si in Au at this optimum point?

diffusion and ion implantation

4

The selective control of impurity atoms in single-crystal silicon makes possible the mass production of extremely compact electronic circuits.

Consider the lattice structure of single-crystal silicon. Ideally, we would like to be able to replace specific rows and columns of silicon atoms with *p*- and *n*-type acceptor and donor atoms. Although this would be the ultimate in impurity control, it is not possible. The two impurity-control techniques that are available to us are diffusion and ion implantation.

Diffusion involves making impurity atoms available at the surface of single-crystal silicon and then heating the crystal so that the impurity atoms diffuse in, replacing silicon atoms that have been shaken loose from their crystal sites.

Ion implantation involves directing a beam of high-velocity impurity ions at the surface of single-crystal silicon. The ions penetrate the surface, slowing and coming to rest through interaction with the lattice structure.

The parameters that we wish to control with these process techniques are the concentration and physical distribution of impurity atoms. Although diffusion and ion implantation control these parameters differently, our degree of control is very limited in either case.

The properties of crystalline silicon that permit the infusion of impurities, and the nature of the infusion processes, have necessarily become the concern of many practical-minded people. How do impurities move through a crystal lattice? What laws govern the processes of diffusion and ion implantation? What dopants are most useful and how soluble are they in the silicon lattice?

In this chapter we discuss the kinds of diffusion that occur in silicon, the laws governing the diffusion process, the behavior of the diffusion coefficient, and the solubilities of impurities in silicon. The diffusion equation (Fick's second law) will be solved for two special cases, and some of the ways in which diffused layers are evaluated will be discussed. The ion implantation process will be described. The impurity concentration and the impurity profile will be related to the implantation system parameters: the beam current and the acceleration voltage.

4-1 KINDS OF DIFFUSION

To diffuse impurity atoms into a crystal lattice, we have to heat it. Thinking of temperature as thermal vibration, we can get a crystal lattice shaking so violently that some of the atoms shake loose and others move in and substitute for them. If impurity atoms of roughly the same size and valence are in the vicinity, they may well replace the missing crystal atoms in *lattice* sites. Impurity atoms substitute for missing *host* atoms. This process is called *substitutional diffusion.* Impurity atoms jump from one lattice site to the next, moving randomly in all directions. If we allow this random process to begin with a concentration of impurity atoms in one region, then fewer of these atoms will be available to jump back to regions of higher concentration than are available to jump forward into regions of lower concentration. Thus, a net motion will take place into regions of lower impurity concentration. But in a crystal it is necessary that the next site be vacant. Thus, the speed with which substitutional diffusion takes place depends on the creation of vacancies within the lattice.

It is possible for an impurity atom to occupy a position in a crystal lattice without substituting for a host atom. Locations exist in the lattice (between atoms) which are most favorable for occupancy. These locations are called *interstitial sites.* At high temperatures, impurities can jump between interstitial sites, moving about in the lattice. This motion is called *interstitial diffusion.* Interstitial diffusion occurs at a much higher rate than substitutional diffusion because most interstitial sites are ordinarily vacant, whereas substitutional diffusion must wait for vacancies to occur.

Combinations of these two types of diffusion can occur in a crystal lattice. A certain fraction of a given impurity may diffuse interstitially while the remainder diffuses substitutionally. In order to model the interstitial diffusion process so that the model agrees with experimental evidence, it is convenient to specify that a complete interstitial diffusion process can start at lattice sites or interstitial sites and can end at either, but the jumping from site to site must occur between interstitial sites.

All the donors and acceptors that are normally used in silicon diffuse substitutionally. A number of interstitial diffusers in silicon are gold, iron, copper, and lithium. Whereas iron, copper, and lithium are not usually desirable, gold has, on occasion, been intentionally diffused into some high-speed integrated circuits. The gold atoms diffuse interstitially, but most of them terminate in substitutional sites. They do not fit in the lattice very well and they act somewhat like discontinuities. Since hole–electron pairs recombine more readily in the vicinity of discontinuities, the gold atoms cause an increase in the recombination rate and a corresponding decrease in the *average lifetime* of mobile charge carriers. The switching speed of a transistor is inversely related to carrier lifetime. The gold atoms decrease carrier lifetime and therefore increase switching speed. The use of gold to increase device switching speeds has, for the most part, been superseded by other methods. Modifications in device geometries, fabrication techniques, and circuit design are presently used to improve switching speeds.

4-2 FICK'S FIRST AND SECOND LAWS

Because the silicon crystal lattice is cubic, the diffusion of impurity atoms into it is an *isotropic* process. Isotropic diffusion processes can be described by *Fick's first law* of diffusion, which is

$$\mathbf{J} = -D \nabla N \tag{4-1}$$

where $\mathbf{J}$ is the flux density of diffusing atoms in atoms per unit area and unit time, D the diffusion coefficient in area per unit time, N the concentration of diffusing atoms per unit volume, and ∇ the gradient operator. In words, the equation says that *the flux density of diffusing atoms is proportional to the concentration gradient.* D is the constant of proportionality, and the minus sign indicates that the diffusion proceeds in the direction of decreasing concentration.

Our interest is restricted to the processing of silicon integrated circuits, or in the diffusion of impurity atoms to alter the electronic properties of the silicon lattice structure. Although we will concern ourselves with lateral diffusion later, for now we restrict the analysis to one dimension: downward diffusion into plane parallel structures. For one-dimensional analysis, Eq. (4-1) takes the form

$$J = -D \frac{\partial N}{\partial x} \tag{4-2}$$

Figure 4-1(a) represents a circular wafer of silicon with a diffusion of impurity atoms taking place from left to right. At room temperature, the depth of diffusion is negligible. However, at temperatures around 1000°C, significant diffusion depths are achieved in a matter of hours. Controlling the speed of diffusion is therefore a matter of controlling the temperature and time. Figure 4-1(b) represents the impurity distribution in atoms per cm^3 as a function of distance into the wafer after a specific diffusion time. A distance dx between plane 1 and plane 2 is as shown.

Fick's second law will be more useful to us than his first law in calculating

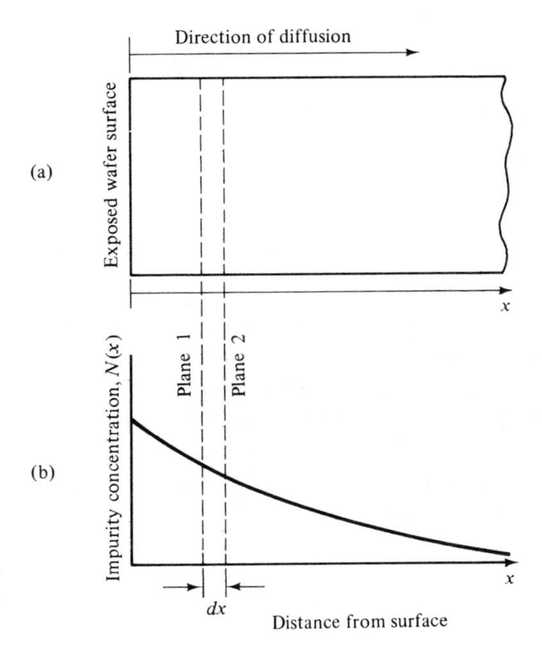

Fig. 4-1 Diffusion into a silicon wafer: (a) a wafer with diffusion occurring from left to right; (b) the impurity concentration as a function of distance.

diffusion parameters. It can be derived from the first law by a continuity argument and consideration of Fig. 4-1. The number of particles per unit area between plane 1 and plane 2 in this figure will be $N\,dx$. Letting N be a function of time as well as distance, the rate of accumulation of particles in this region will be

$$\text{rate of particle accumulation} = \frac{\partial N}{\partial t}\,dx \tag{4-3}$$

This can also be written as the difference between the flow across the two planes. If the entering flux into plane 1 is J, and the departing flux at plane 2 is $J + (\partial J/\partial x)\,dx$, the net flux into the region between planes is

$$J - \left(J + \frac{\partial J}{\partial x}\,dx\right) = -\frac{\partial J}{\partial x}\,dx \tag{4-4}$$

We can now substitute $-D\,(\partial N/\partial x)$ from Eq. (4-2) into Eq. (4-4), so that

$$-\frac{\partial J}{\partial x}\,dx = -\frac{\partial}{\partial x}\left(-D\frac{\partial N}{\partial x}\right)dx = D\frac{\partial^2 N}{\partial x^2}\,dx \tag{4-5}$$

This, in turn, must be equal to the rate of accumulation of particles in the region as given by Eq. (4-3), or

$$D\frac{\partial^2 N}{\partial x^2}dx = \frac{\partial N}{\partial t}dx$$

Then

$$\frac{\partial N}{\partial t} = D\frac{\partial^2 N}{\partial x^2} \tag{4-6}$$

This is *Fick's second law.* It says that *the time rate of change of impurity concentration is proportional to the second derivative of concentration with respect to distance.* It is a linear, second-order, partial differential equation in which D can, with certain restrictions, be treated as a constant. When appropriate boundary conditions are applied and the equation is solved, the result is an expression for impurity concentration as a function of the distance x into the wafer and of the diffusion time.

4-3 DIFFUSION COEFFICIENT

If crystal temperature is increased, diffusing impurity atoms have more thermal energy; thus, the speed with which they move through the lattice increases. The equation relating diffusion coefficient and temperature is

$$D = D_0 \exp\left(-\frac{\Delta H}{RT}\right) \tag{4-7}$$

where D_0 and R are constants, T is temperature in degrees Kelvin, and ΔH is the activation energy for the diffusion of an impurity into a crystal. For a specific impurity and a specific type of crystal, ΔH is also constant. Changing Eq. (4-7) to $\ln D$, we obtain

$$\ln D = \left(\frac{1}{T}\right)\left(-\frac{\Delta H}{R}\right) + \ln D_0 \tag{4-8}$$

If we plot $\ln D$ vs. $1/T$, we obtain a straight line with a negative slope. Fuller and Ditzenberger [1] experimentally determined the diffusion coefficients of substitutional diffusers in silicon as a function of reciprocal temperature. Figure 4-2 indicates some of their results.

Since temperature appears in the exponent in Eq. (4-7), the diffusion coefficient (and therefore the speed of diffusion) is very temperature-sensitive. A temperature variation of just a few degrees can destroy the base of a transistor. For this reason, in the semiconductor industry, furnace temperatures are controlled to within $\pm\frac{1}{4}$°C up to 1250°C.

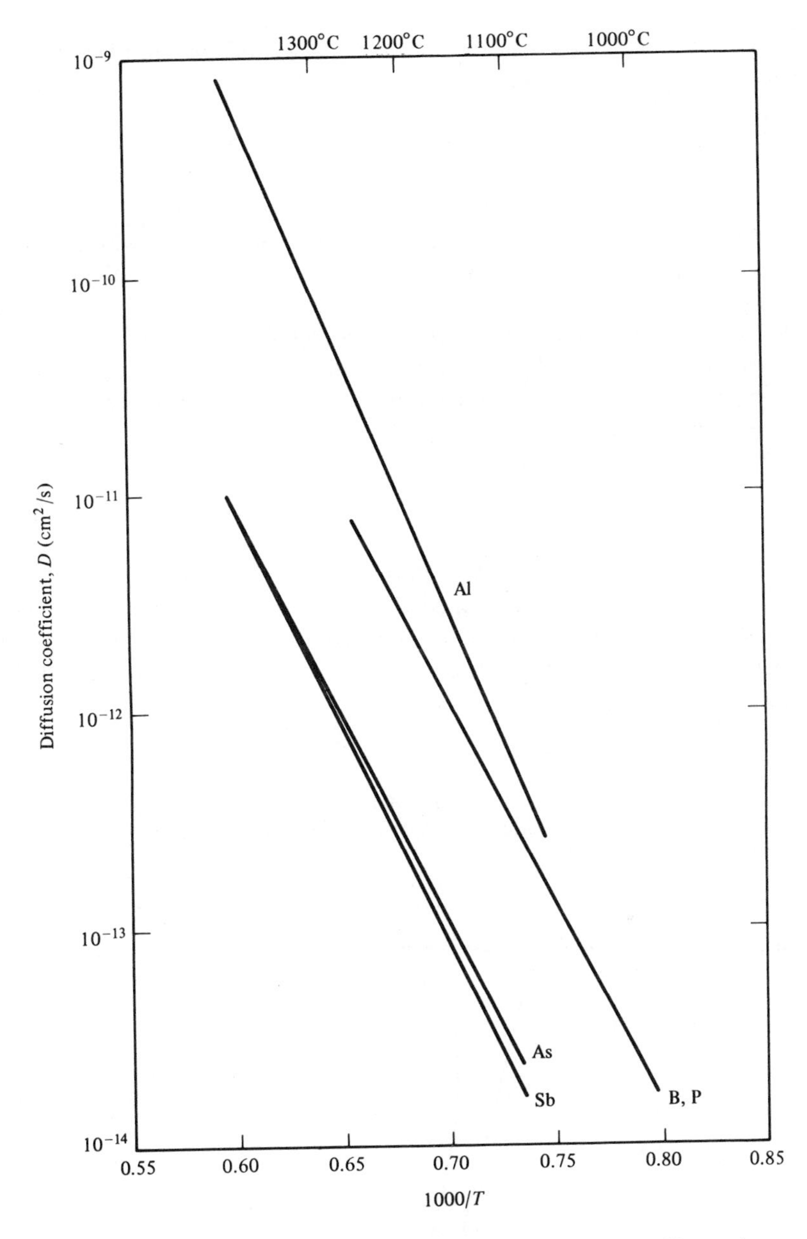

Fig. 4-2 Diffusion coefficients of substitutional diffusers in silicon plotted against reciprocal of absolute temperature. (Adapted from C. S. Fuller and J. A. Ditzenberger. Copyright 1956, *Journal of Applied Physics*, Vol. 27, by permission [1].)

At normal doping levels, the diffusion coefficient is independent of concentration. At very high doping levels, it is found to increase with increasing concentration.

The possible *n*-type dopants for use in IC processing are phosphorus, antimony, and arsenic. All three are useful. Figure 4-2 shows that at any fixed temperature, the diffusion coefficient of phosphorus is about 10 times larger than that of antimony or arsenic. This is one of the reasons why phosphorus is used for emitter diffusions. The larger diffusion coefficient allows a much faster diffusion. Either antimony or arsenic, on the other hand, may be used for the *n*-type impurity in the epitaxial layers (epi-layers) of IC wafers because they are slow diffusers. Thus, during isolation, base, and emitter diffusions, the *n*-type epi-layer will diffuse slightly into the *p*-type substrate because it is doped with slow diffusers.

4-4 SOLID SOLUBILITY

The property *solid solubility* refers to the *maximum concentration of an element that can be dissolved in a solid at any specified temperature.* Figure 4-3 shows solid-solubility curves of a number of common diffusants in silicon as a function of temperature [2]. These curves set an upper limit on the diffusant surface concentration.

Fig. 4-3 Solid solubilities of impurity elements in silicon. (Adapted from F. A. Trumbore. Copyright 1960, American Telephone and Telegraph Co., by permission [2].)

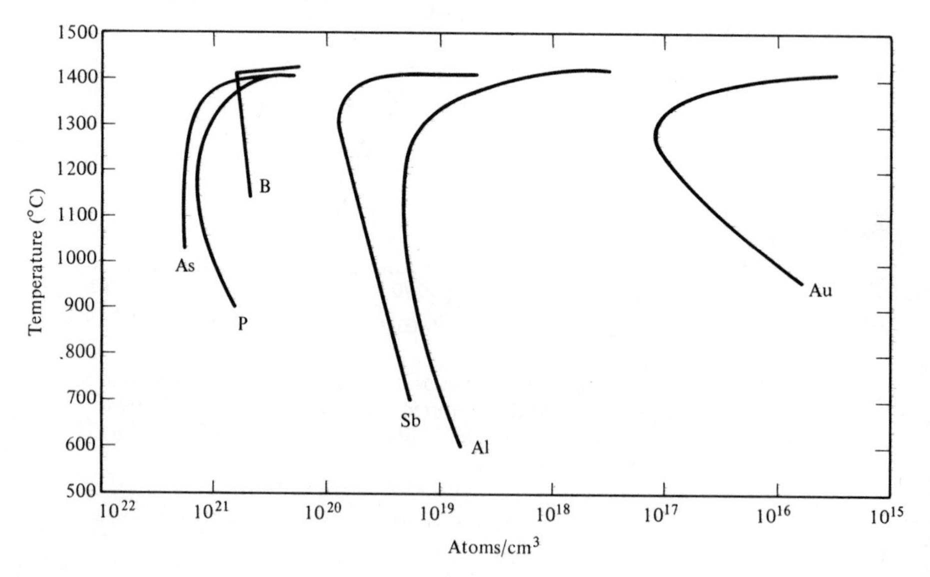

4-5 SOLUTIONS TO FICK'S SECOND LAW

Most IC processing is done by first depositing the desired impurity on the wafer at a specific temperature and time duration *(predeposition),* and then allowing the dopant to diffuse into the silicon at a second temperature and time duration (diffusion). During the predeposition, the impurity layer can be controlled so that, within the limits of solid solubility, a heavy or light deposition can be obtained.

In order to have manageable solutions to Fick's second law, we have to make simplifying assumptions. The assumptions are then written as boundary conditions and used to solve the equation. The solution will not be exact because the assumptions are not strictly correct, but it will be a good approximation.

4-5-1 Complementary Error Function Distribution

In the case of a heavy predeposition, we make the following simplifying assumptions:

1. The diffusion will not significantly alter the amount of dopant on the surface, even though it is diffusing into the bulk of the wafer. Then the surface concentration (N_0) is independent of diffusion time.
2. At the beginning of the diffusion, the impurity concentration exists in an infinitesimally thin layer of silicon at the surface. Then $N(x) = 0$ at $t = 0$, where $N(x)$ for $x > 0 =$ the impurity concentration in the silicon.

With these two boundary conditions applied, the solution to Fick's second law is

$$N(x, t) = N_0 \left[1 - \frac{2}{\sqrt{\pi}} \int_0^{\frac{x}{2\sqrt{Dt}}} \exp(-\lambda^2)\, d\lambda \right] \tag{4-9}$$

where $N(x,t)$ is the impurity concentration at any given point in the silicon for a fixed diffusion time and λ is an integration variable. The integral term in Eq. (4-9) is commonly referred to as the *error function.* This equation is also written as[1]

$$N(x, t) = N_0 \left(1 - \operatorname{erf} \frac{x}{2\sqrt{Dt}} \right) \tag{4-10}$$

[1] A table of erf *(z)* vs. *(z)* can be found in S. K. Ghandi, *The Theory and Practice of Microelectronics,* John Wiley & Sons, Inc., New York, 1968, pp. 451–452.

or

$$N(x, t) = N_0 \operatorname{erfc} \frac{x}{2\sqrt{Dt}} \tag{4-11}$$

where erfc is the abbreviation for *error function complement,* x is in centimeters, and t is in seconds. The impurity distribution obtained from an error function type of diffusion is shown in Fig. 4-4.

If the diffusing impurity is of an opposite type from the background impurity concentration, a *p-n* junction is formed where the two concentrations are equal.

In IC processing, the isolation diffusion and emitter diffusion are approximated by the complementary error function model.

4-5-2 Gaussian Distribution

In the case of a light predeposition of impurities, we make the assumptions that the impurity atoms are dissolved into the silicon lattice before the diffusion begins, and that the impurity atoms lie in an extremely thin layer. The Gaussian impurity concentration resulting from this type of diffusion is given by

$$N(x, t) = \frac{Q}{\sqrt{\pi Dt}} \exp\left[-\left(\frac{x}{2\sqrt{Dt}}\right)^2\right] \tag{4-12}$$

where Q is the amount of impurity placed on the surface prior to diffusion (in atoms per cm^2) and the other terms are as previously shown.

Gaussian diffusion profiles for specific diffusion times are as shown in Fig. 4-5. In this case, the surface concentration decreases with time and the

Fig. 4-4 Impurity profiles for erfc diffusion.

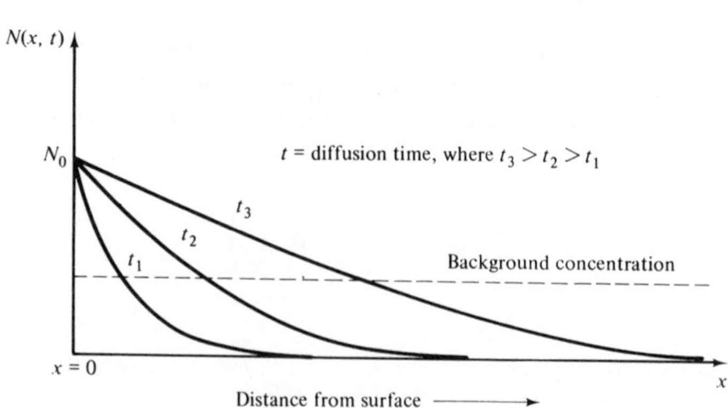

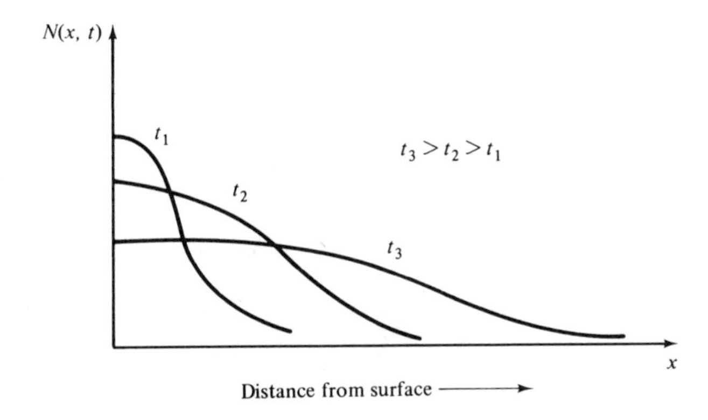

Fig. 4-5 Impurity profiles for Gaussian diffusion.

area under the curve remains constant. A Gaussian diffusion provides a lower surface concentration and a higher diffusion depth than the erfc diffusion. The base diffusion of a transistor is normally approximated by a Gaussian distribution.

We are discussing a two-step Gaussian diffusion process. A predeposition is performed at temperature T_1, during which impurities are deposited on the silicon surface. This is followed by a diffusion at temperature T_2, during which the impurities substitutionally diffuse into the silicon wafer. In Eq. (4-12), Q is given as

$$Q = 2N_{01}\left(\frac{D_1 t_1}{\pi}\right)^{\frac{1}{2}} \tag{4-13}$$

where N_{01} is the impurity concentration on the surface at the predeposition temperature in atoms per cm^2, D_1 is the diffusion coefficient at predeposition temperature in cm^2/s, and t_1 is the predeposition time in seconds. Letting D_1 and t_1 represent predeposition parameters and D_2 and t_2 represent diffusion parameters, we can rewrite Eq. (4-12) as

$$N(x, t_1, t_2) = \frac{2N_{01}}{\pi}\left(\frac{D_1 t_1}{D_2 t_2}\right)^{\frac{1}{2}} \exp\left[-\left(\frac{x}{2\sqrt{D_2 t_2}}\right)^2\right] \tag{4-14}$$

This equation is valid as long as $\sqrt{D_1 t_1} << \sqrt{D_2 t_2}$.

Given that an erfc or a Gaussian diffusion is to be performed, it is important to be able to predict the diffusion depth as a function of the diffusion temperature and diffusion time. This prediction can be accomplished by means of a calculator, a computer, or a graphical look-up process. For the purpose of working examples and homework problems, it is convenient to use a graphical technique. For this purpose, a normalized logarithmic plot of the complementary error function and Gaussian distributions is shown in Fig. 4-6.

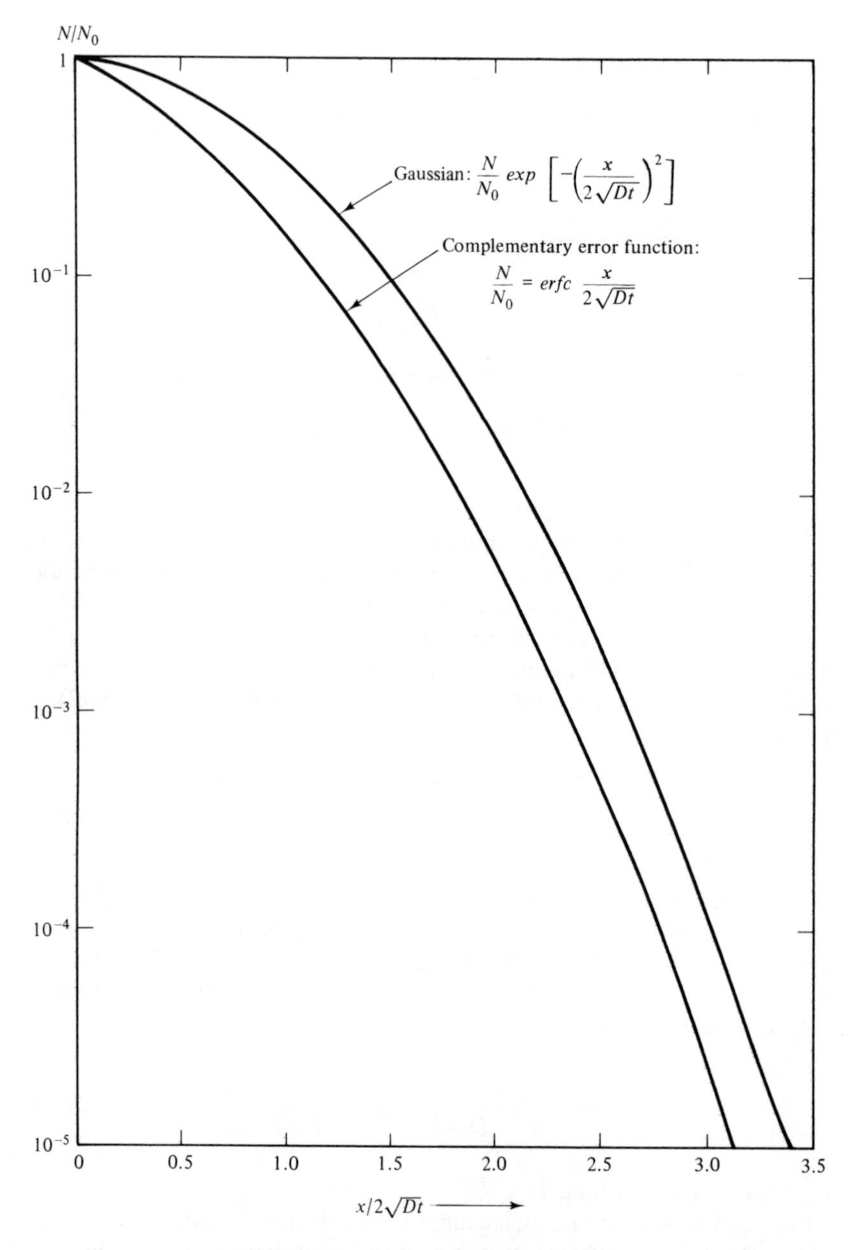

Fig. 4-6 erfc and Gaussian profiles.

Example I. Boron–nitride wafers are used in conjunction with silicon wafers in a deposition step. $Q = 2.25 \times 10^{13}$ boron atoms/cm^2 are deposited on the surface of a silicon slice. The slice is subsequently placed in a diffusion furnace tube at 1145°C for 2 h. The *n*-type epi-layer that we are diffusing into has an impurity concentration $N_D = 1 \times 10^{16}$ atoms/cm^3. Assuming that the diffusion is Gaussian, find the depth of the *p-n* junction in micrometers.

Solution

(a) 1145°C = 1418 K; $1000/T = 1000/1418 = 0.705$. From Fig. 4-2, the diffusion coefficient for boron $= D = 9.2 \times 10^{-13}$ cm^2/s.

(b) $t = 2 \text{ h} = 7.2 \times 10^3$ s.

(c) $N(x, t) = (Q/\sqrt{\pi Dt}) \exp[-(x/2\sqrt{Dt})^2]$.

The junction will occur where the background concentration is equal to the boron concentration. Then

$$1 \times 10^{16} = \frac{2.25 \times 10^{13}}{\sqrt{(3.14)(9.2 \times 10^{-13})(7.2 \times 10^3)}} \exp\left[-\left(\frac{x}{2\sqrt{Dt}}\right)^2\right]$$

or

$$\exp\left[-\left(\frac{x}{2\sqrt{Dt}}\right)^2\right] = 6.4 \times 10^{-2}$$

Then, by calculation or from Fig. 4-6, $(x/2\sqrt{Dt}) = 1.66$, or

$$x = 3.26\sqrt{Dt} = 2.7 \times 10^{-4} \text{ cm} = 2.7 \ \mu\text{m}$$

Example II. An emitter–base junction is formed by diffusing *n*-type phosphorus atoms into *p*-type silicon for 1 h at 1000°C. The phosphorus concentration at the silicon surface is maintained at the limit of solid solubility. Assuming that the base has a uniform *p*-type background concentration of 10^{17} atoms/cm^3, locate the emitter–base junction.

Solution

$$N(x, t) = N_0 \operatorname{erfc} \frac{x}{2\sqrt{Dt}}$$

From Fig. 4-3, the solid solubility of phosphorus at 1000°C is approximately 10^{21} atoms/cm^3. Then

$$\frac{10^{17}}{10^{21}} = 10^{-4} = \text{erfc}\,\frac{x}{2\sqrt{Dt}}$$

From Fig. 4-6, $x/2\sqrt{Dt} = 2.8$. From Fig. 4-2, $D = 3 \times 10^{-14}$ cm²/s at 1000°C. Thus, $x = 5.6\sqrt{Dt} = 5.6\sqrt{(3\times10^{-14})(3.6\times10^{3})}$, or $x = 5.82 \times 10^{-5}$ cm = 0.582 μm.

4-6 DIFFUSED-LAYER EVALUATION

A diffused layer can be characterized by four parameters:

N_0 = surface impurity concentration
N_B = background concentration
X_j = junction depth
R_s = sheet resistance of the layer

A knowledge of the relationship between these parameters is necessary in the establishment of device-processing techniques and in the evaluation of diffusion techniques. The experimental data that can be most easily obtained during processing involve the measurement of sheet resistance and junction depth. From X_j and R_s, N_0 can be determined. From N_0 and N_B, the diffusion profile is established.

Data are not usually taken from the epitaxial wafers being processed. "Dummy" wafers are purchased with the same impurity and impurity concentration as the epitaxial layer. These wafers are much less expensive. They can be processed before, or together with the epitaxial wafers and then used for sheet resistance and junction depth measurement. Dummy wafers that have had base depositions and diffusions are later used to evaluate emitter diffusions.

4-6-1 Sheet Resistance

Where the resistivity or conductivity of a specimen is constant, Eq. (4-15) is used to calculate resistance:

$$R = \frac{\rho l}{A} = \frac{l}{\sigma A} \tag{4-15}$$

where ρ = resistivity in Ω-cm
σ = conductivity in $(\Omega\text{-cm})^{-1}$

A diffusion, however, leaves a very nonuniform impurity distribution. We have to resort to average resistivity ($\bar{\rho}$) and sheet resistance at the surface *(R_s)*. To understand R_s, consider a rectangular sheet of silicon after an impurity diffusion:

$$R = \frac{\rho(x)l}{xw} = \frac{\rho(x)}{x}\left(\frac{l}{w}\right) = R_s\left(\frac{l}{w}\right) \tag{4-16}$$

where $\rho(x)$ = resistivity as a function of distance x into the wafer
l = length
w = width
R_s = sheet resistance = the ratio of bulk resistivity and thickness for a thin uniform layer

R_s is measured in *ohms per square* (i.e., when $l = w$, $R = R_s$). The ratio of bulk resistivity and thickness will be the same regardless of the size of the square. R_s acts like a material property.

Sheet resistance is normally measured by using a *four-point probe*, as shown in Fig. 4-7. Basically, four spring-loaded and equally spaced probe needles are used to make electrical contact with the silicon wafer. A constant current that is small enough to avoid resistive heating (typically 1 mA) is passed through the outer two probes. The resulting potential difference between the inner probes is measured by a potentiometer. The relationship between wafer resistivity, voltage, and current has been developed by analogy with electrostatic charge. The two outer probes are modeled as positive and negative charge and the semiconductor surface is treated as a plane of symmetry. The potential is determined at the position of the inner probes. The resulting equation is transferred back to the ohmic flow case, and the wafer resistivity is given by

$$\rho = 2\pi s \frac{V}{I} \tag{4-17}$$

where s is probe spacing in centimeters, and V and I are volts and amperes measured at the four-point probe as shown in Fig. 4-7.

How do we get from here to sheet resistance? Well, first we will be interested in the surface sheet resistance of a thin diffused layer (n on p, or p on n). To handle this change in the electrostatic model, image charges are introduced to provide symmetry about the bottom of the layer. Requiring symmetry about both top and bottom surfaces has led to an infinite array of image charges and a more complex solution [4,5]. The layer thickness (x) and the probe spacing (s) are usually such that $x << s$. It is also assumed that the substrate-layer

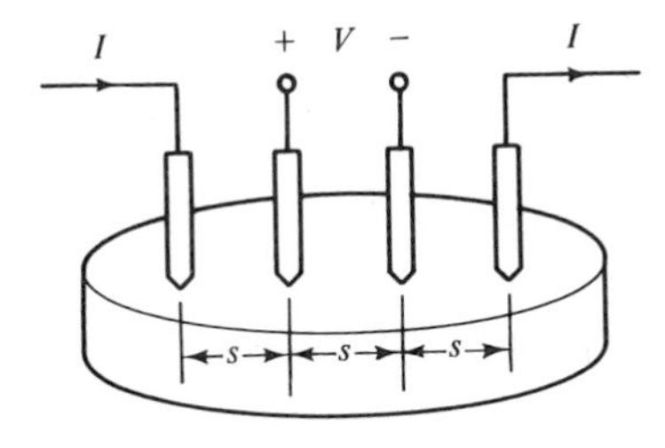

Fig. 4-7 Four-point probe on silicon sample.

diode is slightly reverse-biased due to lateral voltage drops in the layer. Under these circumstances, the average layer resistivity is given as

$$\bar{\rho} = 4.5 \frac{V}{I} x \tag{4-18}$$

where $\bar{\rho}$ is in Ω-cm, V is in volts, I is in amperes, and x is in centimeters. Then sheet resistance is given by

$$R_s = \frac{\bar{\rho}}{x} = 4.5 \frac{V}{I} \ \Omega/\square \tag{4-19}$$

Since a 1-mA constant-current source is typically used for I, R_s is numerically equal to 4.5 times the voltage reading in millivolts. The base doping is usually arranged so that R_s is about 200 Ω/□ after the base diffusion. Then the value of a resistor made during the base diffusion would be $200l/w$.

4-6-2 Junction Depth

If a *p*-type impurity is diffused into a uniformly doped *n*-type silicon sample (or vice versa), a *p-n* junction will be located at a depth x_j where the two impurity concentrations are equal. A fine polishing abrasive (Al_2O_3) is used to expose the junction. Since *p* and *n* regions can be made to take on different colors or shades through a chemical staining process, the *p-n* junction can be made visible. Then the junction depth x_j can be measured. Two of the most frequently used techniques for abrading samples are angle lapping and grooving.

Angle Lapping. Using the angle-lapping technique, a small sample is removed from a dummy wafer by scribing. The sample is attached with hot wax to a beveling post and beveled by moving the post and sample over a flat glass plate covered with an abrasive slurry. The sample is shown in Fig. 4-8. To accurately measure the stained junction depth, the sample is usually abraded at a small angle θ (1 to 5°). This results in an increase in the length exposed

Fig. 4-8 Angle lapping fixture: (a) before abrasion; (b) after abrasion; (c) detail.

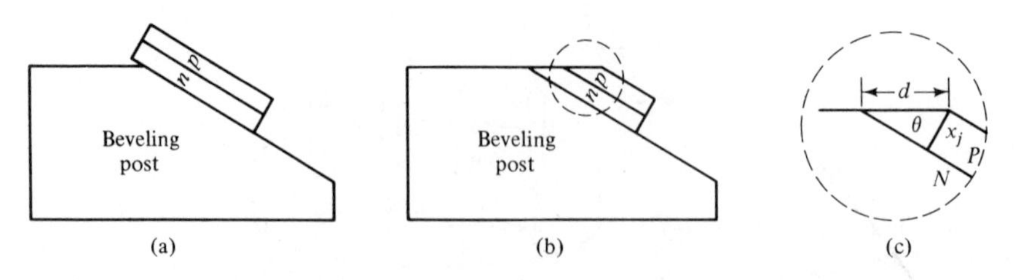

for staining $(d > x_j)$. For x, θ, and d as shown in Fig. 4-8(c), $x_j = d \sin \theta$. The beveling post must have the desired angle machined onto one surface.

Wafer Grooving [6]. A grooving method can be used for abrading dummy wafers. It consists of polishing with a large cylinder or sphere which cuts an arc through the cross section of the diffused layers. The cylinder or sphere (bar or ball mandrill) is made of stainless steel, phenolic, or bonded glass. To allow direct measurements, the diameter of the mandrill is maintained at 1.0000 ± 0.0005 in. [7]. A vacuum chuck holds the wafer in place at the end of a lever arm. The weight of the lever arm holds the wafer against the mandrill, which rotates. The abrasive washes by, keeping both surfaces coated. This method is intended for shallow junctions. With a deep junction, it may take hours to obtain a groove of sufficient depth. When the groove is completed, the junction depth is determined from Fig. 4-9 as follows:

$$x_j = [(R^2 - b^2)^{\frac{1}{2}} - (R^2 - a^2)^{\frac{1}{2}}] \tag{4-20}$$

Factoring out R, we have

$$x_j = R\left[\left(1 - \frac{b^2}{R^2}\right)^{\frac{1}{2}} - \left(1 - \frac{a^2}{R^2}\right)^{\frac{1}{2}}\right] \tag{4-21}$$

For a, $b << R$, x_j can be approximated by the first two terms of a series expansion:

$$x_j \approx R\left(1 - \frac{1}{2}\frac{b^2}{R^2} - 1 + \frac{1}{2}\frac{a^2}{R^2}\right) = \frac{1}{2}\frac{a^2 - b^2}{R} \tag{4-22}$$

In practice, it is much easier to measure x and y, where

$$\begin{aligned} x &= a - b \\ y &= a + b \\ xy &= a^2 - b^2 \end{aligned} \tag{4-23}$$

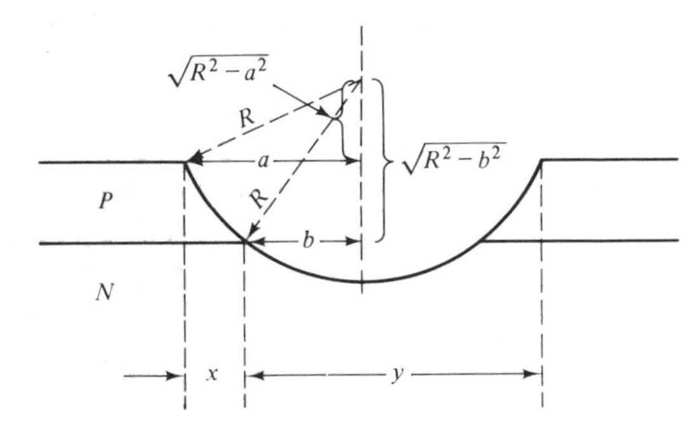

Fig. 4-9 Wafer cross section after grooving with a bar mandrill.

Then

$$x_j \approx \frac{1}{2}\frac{xy}{R} = \frac{xy}{D} \tag{4-24}$$

This method is supposed to provide better edge definition, better staining, and therefore more accurate junction measurements.

Staining. The *p-n* junction is made visible by means of a chemical staining procedure. Since none of the staining solutions and procedures in use have been completely satisfactory, a large number of them exist. One of the most common stains consists of about 0.1% concentrated HNO_3 in concentrated hydrofluoric acid. The action of the stain is to darken the *p* regions and leave the *n* regions unchanged. When the proper stain is obtained, the sample is rinsed with water and dried. The stain is permanent.

4-6-3 Surface Concentration

When junction depth x_j and sheet resistance R_s have been determined experimentally, surface impurity concentration N_0 can be calculated. With N_0 known, the diffusion profile can be obtained from

$$N(x, t) = N_0 \exp\left[-\left(\frac{x}{2\sqrt{Dt}}\right)^2\right] \tag{4-25}$$

or

$$N(x, t) = N_0 \operatorname{erfc}\left(\frac{x}{2\sqrt{Dt}}\right) \tag{4-26}$$

Irvin [8] performed the computations to find surface concentration from R_s and x_j for various levels of background impurity concentration. He plotted average conductivity ($\overline{\sigma} = 1/R_s x_j$) vs. surface concentration for *p* and *n* materials, Gaussian and erfc distributions, and many different background concentrations. Some of the results that are of immediate interest here involve *p*-type Gaussian diffusions with *n*-type background concentrations on the order of $N_B = 10^{16}$ atoms/cm^3 (transistor bases) and *n*-type erfc diffusions into *p*-type background concentrations on the order of $N_B = 10^{18}$ atoms/cm^3 (transistor emitters). These results are shown in Figs. 4-10 and 4-11.

From the measurement of x_j and R_s, the base and emitter diffusion profiles can be determined; and from these and a knowledge of component geometries, the electrical parameters of passive and active IC components can be approximated.

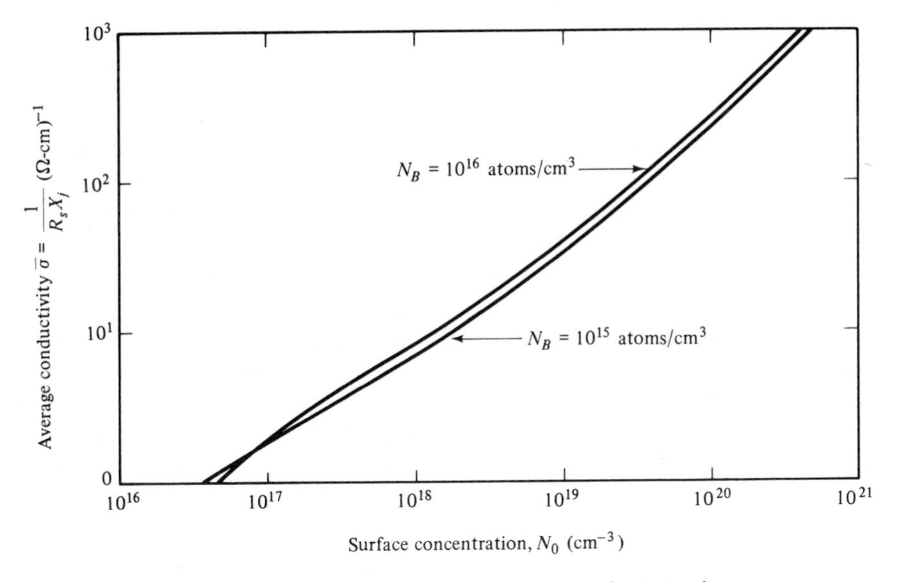

Fig. 4-10 Surface concentration vs. average conductivity for *p*-type Gaussian diffusions in silicon. (Adapted from John C. Irvin. Copyright 1962, American Telephone and Telegraph Co., by permission [8].)

Fig. 4-11 Surface concentration vs. average conductivity for *n*-type erfc diffusions in silicon. (Adapted from John C. Irvin. Copyright 1962, American Telephone and Telegraph Co., by permission [8].)

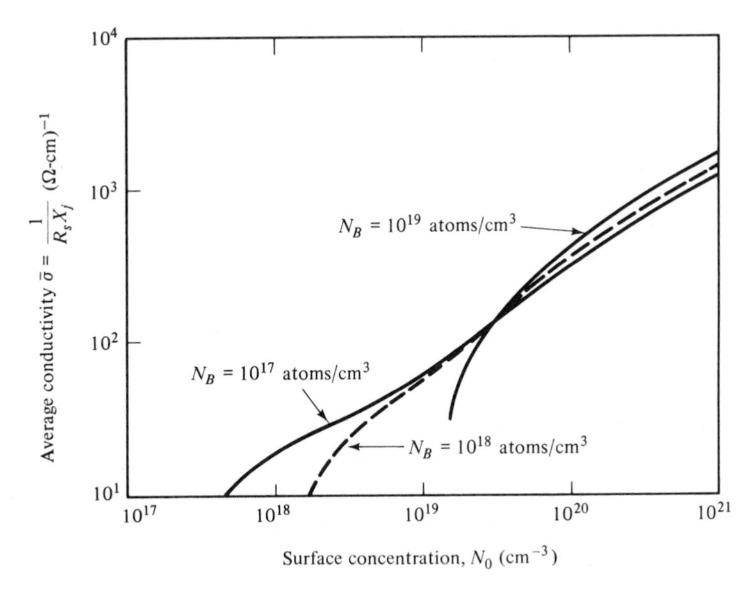

4-7 ION IMPLANTATION

Ion implantation is a widely used alternative to diffusion in situations where the desired impurity concentration is less than 1.0 μm from the silicon surface. This process consists of ionizing impurity atoms and then accelerating them to high kinetic energies in an electric field. The high-energy impurity ions impinge on the host material (silicon wafers), penetrating for tenths of micrometers beneath the surface before coming to a stop.

Through the Gaussian and erfc impurity profiles, *diffusion* provides some control over impurity concentration in silicon between the surface and depths on the order of 25 μm. *Ion implantation* provides more precise control, with a different set of boundary conditions, but only to depths on the order of 0.5 μm. Thus, it is more accurate to think of diffusion and ion implantation as complementary technologies than as competing technologies. To the extent that direct comparisons between the two are valid, however, ion implantation has the following advantages:

1. The impurity profile can be accurately and reproducibly controlled. To a large extent, control of the impurity concentration is determined by the implant system rather than by the physical properties of the substrate. Impurity layers can be placed completely below the silicon surface. A shallow layer can be implanted before a deeper lying one of a different type. Also, the dopant concentration is not necessarily limited by considerations of solid solubility. Implantation thus allows a wide distribution of impurity elements to be used in silicon for purposes such as controlling carrier lifetime and improving radiation resistance.
2. Implanted impurities are directed downward into the substrate with very little lateral motion. The absence of lateral impurity motion leads to smaller device geometries and therefore to higher packing densities.
3. High-temperature processing tends to warp and physically distort silicon wafers. Since integrated-circuit fabrication requires accurate pattern alignment over the entire surface of a wafer, warpage and distortion reduce the number of good chips obtained from a wafer, or lower the yield. Ion implantation is inherently a *low-temperature* process.

In comparison with diffusion, ion implantation has the following disadvantages:

1. Only shallow impurity profiles can be obtained.
2. As an impurity ion moves through the silicon lattice and comes to rest, it causes crystal damage along its path. The resulting lattice disorder must be annealed. Temperatures between 450 and 1000°C are used to heal the crystal damage and activate the implanted ions.

3. Ion implantation equipment is large, expensive, and complex. (Fortunately, microprocessor-based controllers have reduced the importance of system complexity.)

4-7-1 Ion Implantation Equipment

The basic components of an ion implantation system are shown in Fig. 4-12. The system consists of an ion source, a magnetic mass filter which separates unwanted ions from the beam, a charged path to accelerate the ions, electrostatic lenses and plates to focus the beam and provide x and y beam deflection, and a target. The target is usually a metal disk that can hold up to 50 or 60 silicon wafers in concentric rings. The interior of the entire system is a vacuum chamber.

The most commonly used ions are boron, phosphorus, and arsenic. Since these elements are not gases at room temperature, molecular compound gases are used as ion sources. BF_3 or BCl_3 is used for boron ions, PH_3 or PF_3 is used for phosphorus ions, and AsH_3 or AsF_3 is used for arsenic ions. The ions are typically formed through bombardment by electrons from an arc discharge or a cold cathode source. The collisions create more than one kind of ion. The mass filter or ion separator in Fig. 4-12 uses a magnetic field to deflect the ion beam. Since ions with different masses are deflected to different degrees, the exit path from the magnetic field selects only the desired ions.

Many variations from the configuration of Fig. 4-12 are commercially available, depending on the desired application. In systems using high beam currents, the beam position may be fixed while scanning motion is provided

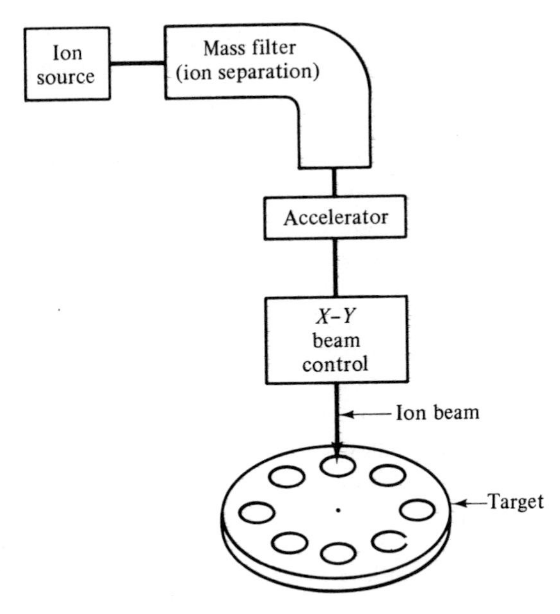

Fig. 4-12 Basic components of an ion implantation system.

by the target, which rotates with a simultaneous back-and-forth motion. In other systems, the beam scans in one direction while the target rotates.

Acceleration potentials range between 10 and 300 keV for most implant systems. The ion beam is typically focused to a diameter of about 1 cm at the target. The beam current, which can be accurately measured, ranges between about 10 μA and 2 mA. The number of implanted ions per unit area is called the *ion dose* Φ. Typical ion doses range between 10^{11} and 10^{16} atoms/cm^2. The ion dose is related to the beam current I, the beam area A, and the dose time t by the equation

$$\Phi = \frac{tI}{qA} \tag{4-27}$$

where q represents the charge per ion (1.6×10^{-19} C). If the beam current were set at 100 μA for 1.6 ms, and the beam area was 1 cm^2, the resultant dose would be

$$\Phi = \frac{(1.6 \times 10^{-3}\ \text{s})(1 \times 10^{-4}\ \text{A})}{(1.6 \times 10^{-19}\ \text{C})(1\ \text{cm}^2)} = 1 \times 10^{12}\ \text{ions/cm}^2.$$

If the ion dose and the beam area are known, and the total target area to be scanned is known, an appropriate scan pattern and scan rate can be determined.

4-7-2 Distribution and Range of Implanted Ions

When positive ions moving at high velocities penetrate a solid target material, they lose energy due to collisions with atomic nuclei and due to interaction with electrons in the target material. In ion–nuclei collisions, energy is transmitted to the target atoms; the ions lose velocity and change direction. Interaction between ions and electrons, on the other hand, causes the ions to lose energy without corresponding changes in direction.

Single-crystal targets such as silicon present many channels that ions can follow. Along these channels, the open lattice structure offers few opportunities for ion–nuclei collisions. Ions will thus travel two to three times farther when the incident beam is aligned with a low-index crystallographic direction. *Channeling* is generally avoided in semiconductor processing because the resulting impurity distribution is sensitive to many factors that are difficult to control. Because of the control problems, channeled distributions are not easily reproduced. Channeling, and channeled components of ion implants, are minimized by aligning the crystal and the incident ion beam to deliberately avoid channels in the lattice structure.

When channeling effects are minimized, collisions between ions and nuclei are almost random in nature. This randomness causes the final distribution of implanted ions to be approximately *Gaussian.* It is convenient to characterize this Gaussian distribution in terms of a mean, called the range, and a standard deviation called the *straggle.* The *range R* of an ion is the total distance it travels in the target material before coming to rest. Owing to collisions between ions and nuclei, this total distance is not in a straight line, however. We need to know the *projection of the total path length on the direction of incidence.* This distance is called the *projected range* R_p. For most implantations, the direction of incidence is nearly normal to the wafer surface. The projected range thus corresponds to the depth of the implanted ion. When a large number of ions is implanted, R_p corresponds to the depth at which the distribution is maximum.

The range R and the projected range R_p are approximately related by the equation

$$R_p \approx R\left(1+\frac{M_2}{3M_1}\right)^{-1} \tag{4-28}$$

where M_1 and M_2 denote the masses of the projectile and target atoms, respectively. From Eq. (4-28), the projected ranges for $^{11}B^+$, $^{31}P^+$, and $^{75}As^+$ in ^{28}Si are given by $0.54R$, $0.77R$, and $0.89R$, respectively.

When impurities are *diffused* into silicon, we can choose either a Gaussian or an error function distribution. In either case, the impurity concentration and impurity profile are closely interrelated.

Ion implantation permits more independent control of the impurity concentration and the impurity profile. The accelerating voltage controls the profile while the ion dose controls the concentration.

Figure 4-13 shows profiles of boron ions implanted in silicon with accelerating voltages of 50, 100, and 200 keV. The ion dose is held constant at 1×10^{12} atoms/cm^2. With increased ion energy, both the projected range (R_p) and the straggle (ΔR_p) are seen to increase. The straggle is a measure of the spread of the distribution. The profiles of Fig. 4-13 were calculated assuming Gaussian distributions. In practice, some channeling usually occurs, leaving exponential tails on the distributions which are larger than indicated. Projected range R_p depends on ion mass and ion energy. For a fixed acceleration potential, the relative width of the profile depends on the ion mass/silicon mass ratio. Thus, heavy ions create narrow impurity profiles.

The concentration of implanted ions can be calculated from the Gaussian equation

$$N(x)=\frac{\phi}{\sqrt{2\pi}\,\Delta R_p}\exp\left[-\frac{1}{2}\left(\frac{x-R_p}{\Delta R_p}\right)^2\right] \tag{4-29}$$

Here ϕ denotes the ion dose, R_p and ΔR_p denote the projected range and the straggle, x denotes the ion depth into the substrate, and $N(x)$ represents the

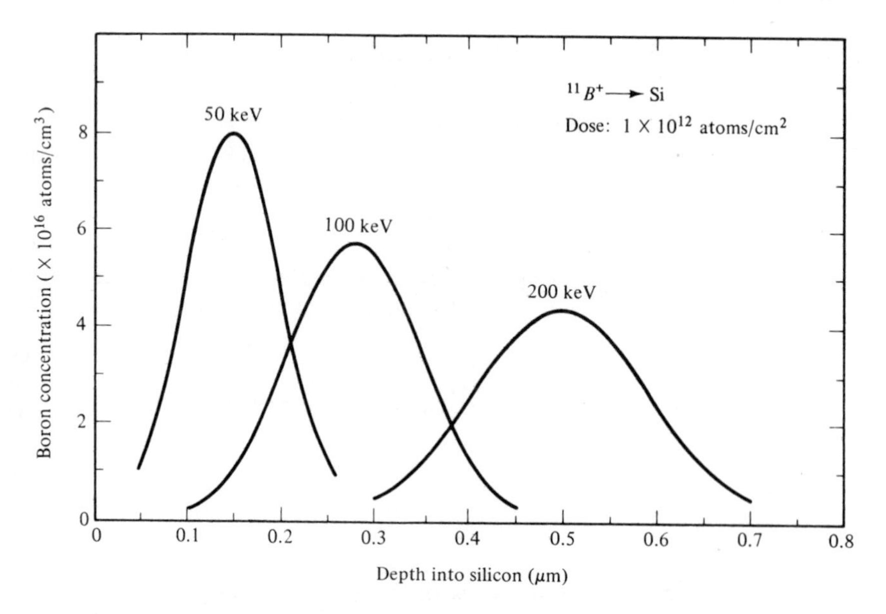

Fig. 4-13 Calculated depth distributions for boron implanted in silicon.

ion concentration in atoms/cm³ at that depth. The concentration is maximum at $x = R_p$. At this depth, Eq. (4-29) reduces to

$$N(x = R_p) = \frac{\phi}{\sqrt{2\pi}\,\Delta R_p} \tag{4-30}$$

Lee and Mayer [11] have compiled projected range and range straggling data for boron, phosphorus, and arsenic in silicon. Curves indicating R_p and ΔR_p versus ion energy are shown in Fig. 4-14.

Given the type of impurity and the implant energy, Fig. 4-14 can be used to find the projected range and straggle. Given the ion dose, the concentration as a function of depth $N(x)$ can be found from Eq. (4-29).

Example Problem. We wish to implant boron atoms into the channel of a metal–oxide–semiconductor (MOS) transistor. A maximum boron concentration of about 8×10^{16} atoms/cm³ is needed at a depth of about 1500 Å. Find the accelerating potential, the ion dose, and the straggle (ΔR_p).

Solution. From Fig. 4-14(a), $R_p = 0.15$ μm at approximately *50* keV. At 50 keV, the straggle $\Delta R_p =$ *0.05 μm*. From Eq. (4-30), $\phi = \sqrt{2\pi}\,\Delta R_p\,N_{max}$, or $\phi = (\sqrt{2\pi})(5 \times 10^{-6}\text{ cm})(8 \times 10^{16}\text{ atoms/cm}^3) = 1 \times 10^{12}$ atoms/cm².

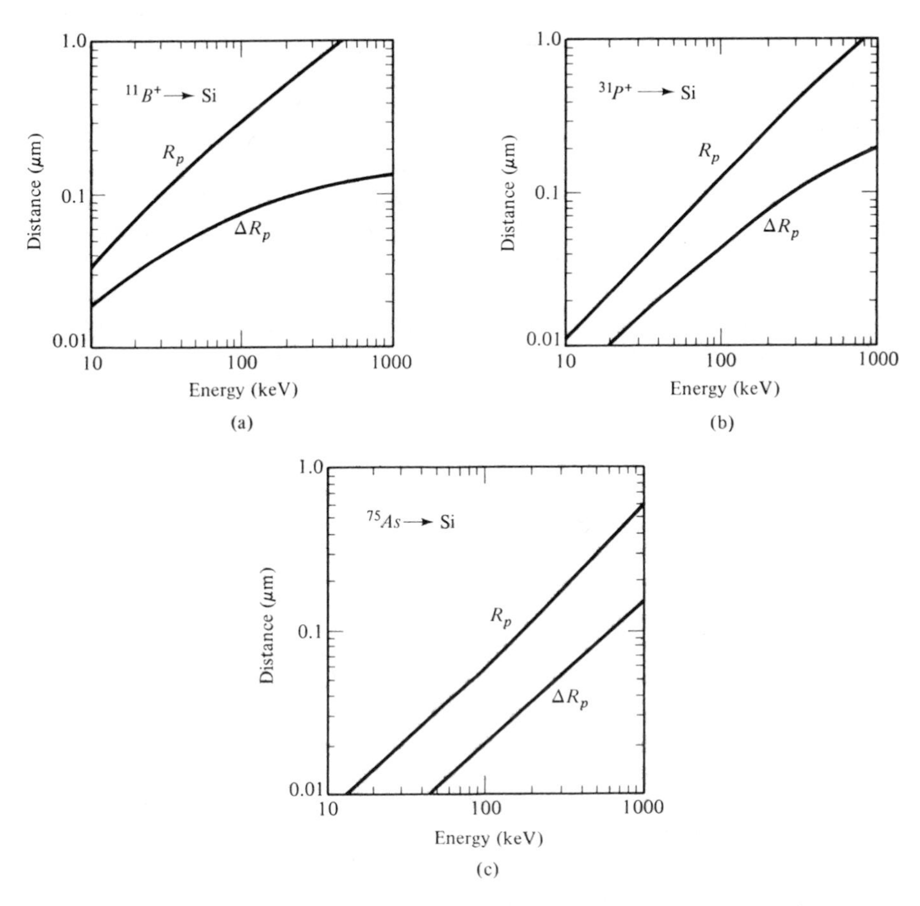

Fig. 4-14 Projected range and range straggling data for (a) boron, (b) phosphorus, and (c) arsenic in silicon. (Adapted from D. H. Lee and J. W. Mayer. Copyright 1974, IEEE, Inc., by permission [11].)

4-7-3 Lattice Damage and Annealing

As ions move through a crystal lattice, collisions occur with substrate atoms. The ions leave trails of displaced substrate atoms along their paths. Ion doses on the order of 10^{11} ions/cm^2 leave individual damage tracks. Lattice damage increases with increasing dosage. If the ion dose is large enough, the damage tracks overlap to the extent that the damaged region becomes amorphous.

Another somewhat separate problem is that implanted ions do not ordinarily come to rest where we would like. We want them in substitutional lattice sites, donating holes or electrons to the lattice structure, and acting as trapped ions.

Fortunately, thermal annealing at relatively low temperatures can solve both problems. Thermal vibration heals the lattice damage and activates the implanted ions. A heavily damaged amorphous layer will recrystallize, following the underlying crystal pattern, at about 550°C. Almost all of the implanted ions will become electrically active during this process. Under conditions where an amorphous layer was not formed, temperatures as high as 1000°C are needed to anneal the crystal damage and activate the implanted ions. Some broadening of the implant profile occurs due to diffusion in this temperature range, but the lattice damage is removed. Typical anneal times are on the order of 30 min.

4-7-4 Masking Techniques

Picture an ion beam scanning across a silicon wafer in a raster pattern. Some areas of the wafer are *masked* to prevent ion penetration, whereas in other areas the silicon surface is exposed. Ions can only be implanted in the exposed areas. The depth of ion penetration is controlled primarily by the acceleration potential. The concentration is controlled primarily by the ion dose. The areas of ion penetration are controlled by the mask.

Photoresist is commonly used as a mask against ion implantation. Silicon dioxide and silicon nitride are not good masks against implants unless the layer is very thick and/or the implant is very shallow. Ion implantation is often performed through a layer of silicon dioxide on the order of 700 Å thick. The oxide layer prevents surface contamination during the implantation process.

Polycrystalline silicon and metals such as aluminum, molybdenum, and gold can also be used to mask against ion implantation.

4-7-5 Applications

Ion implantation is commonly used in the fabrication of both bipolar and MOS (metal–oxide–semiconductor) integrated circuits.

Some of the standard *MOS* applications of ion implantation are as follows:

1. It provides a means of accurately adjusting the threshold voltages of MOS devices. Historically, this has been the most important application because it made possible the low-cost production of *n*-channel MOS circuits.
2. It is commonly used as an alternative to predeposition in the fabrication of MOS circuits.

Some *bipolar* applications of ion implantation are as follows:

1. It is used to form high-value resistors. Conventionally, resistors are formed with transistor bases and have sheet resistances of about 200 Ω/□. Large resistor values thus require large areas. With ion implantation, sheet resistances on the order of 100 kΩ/□ can be obtained.
2. It can be used as an alternative to predeposition.
3. Accurate impurity profiles can be obtained for special-purpose bipolar devices. Through accurate profile control, ion implantation is often used to "fine tune" device parameters.

BIBLIOGRAPHY

[1] FULLER, C. S. AND J. A. DITZENBERGER, "Diffusion of Donor and Acceptor Elements in Silicon," *Journal of Applied Physics,* Vol. 27, pp. 544–553, May 1956.

[2] TRUMBORE, F. A., "Solid Solubilities of Impurity Elements in Germanium and Silicon," *The Bell System Technical Journal,* Vol. 39, pp. 205–233, January 1960.

[3] GHANDHI, S. K., *The Theory and Practice of Microelectronics.* New York: John Wiley & Sons, Inc., 1968.

[4] VALDES, L. B., "Resistivity Measurements on Germanium for Transistors," *Proceedings of the IRE,* Vol. 42, pp. 420–427, February 1954.

[5] UHLIR, A., "The Potentials of Infinite Systems of Sources and Numerical Solutions of Problems in Semiconductor Engineering," *The Bell System Technical Journal,* Vol. 37, pp. 105–128, January 1955.

[6] BURGER, R., AND R. DONOVAN, *Fundamentals of Silicon Integrated Device Technology,* Vol. 1: *Oxidation, Diffusion and Epitaxy.* Englewood Cliffs, N.J.: Prentice-Hall, Inc., 1967.

[7] *Operating Instructions for Wafer Grooving Machines Models G-1070-VAC and G-1080-WM.* Instrumentation Division, Signatone Company, Sunnyvale, Calif.

[8] IRVIN, J. C., "Resistivity of Bulk Silicon and of Diffused Layers in Silicon," *The Bell System Technical Journal,* Vol. 41, pp. 387–410, March 1962.

[9] WARNER, R. M., J. N. FORDEMWALT, C. S. MEYER, D. K. LYNN, ET AL., *Integrated Circuits—Design Principles and Fabrication.* New York: McGraw-Hill Book Company, 1965.

[10] MAYER, J. W., L. ERIKSSON, AND J. A. DAVIES, *Ion Implantation in Semiconductors.* New York: Academic Press, Inc., 1970.

[11] LEE, D. H. AND J. W. MAYER, "Ion-Implanted Semiconductor Devices," *Proceedings of the IEEE,* pp. 1241–1255, September 1974.

PROBLEMS

4-1 A base region is diffused by depositing *p*-type impurities on the silicon surface ($Q = 5 \times 10^{15}$ atoms/cm^2) and placing the slices in a diffusion furnace for 1 h. The diffusion coefficient at the furnace temperature is $D = 3.0 \times 10^{-12}$ cm^2/sec.

The diffusion is done into an n-type epitaxial layer with a concentration of $N_D = 1 \times 10^{16}$ atoms/cm^3. Plot the impurity concentration profile and locate the junction.

4-2 A diffusion is made through an epitaxial layer that is 10 μm thick and has an n-type concentration of 1×10^{16} atoms/cm^3. The effective concentration at the surface of the diffused region is 5×10^{19} atoms/cm^3, and is constant throughout the diffusion. Calculate the time required for this diffusion at 1000°C ($D = 4 \times 10^{-14}$), 1100°C ($D = 4 \times 10^{-13}$), and 1200°C ($D = 3 \times 10^{-12}$). Indicate diffusion times in hours.

4-3 We need to do a Gaussian boron diffusion into an arsenic-doped epitaxial layer. Before the diffusion, the epi-layer has an arsenic concentration of 1×10^{16} atoms/cm^3. The boron diffusion constant at the diffusion temperature that we will use is $D = 1 \times 10^{-12}$ cm^2/s. Our goal is to achieve a sheet resistance $R_s = 200$ $\Omega/\square$ and a junction depth $x_j = 2 \times 10^{-4}$ cm. Find the time in hours needed for the diffusion.

4-4 A p-type isolation diffusion is made through a uniformly doped n-type epi-layer which is 10 μm thick. The epi-layer impurity concentration is 5×10^{15} atoms/cm^3. The boron impurity concentration at the surface of the wafer is 1×10^{19} atoms/cm^3. It is kept constant throughout the diffusion. Find the temperature to be used such that the diffusion takes place in 10 h.

4-5 Phosphorus ions are implanted into the channels of MOS devices with an acceleration potential of 100 keV. The beam current is 1.0 mA, the beam exposure time is 100 μs, and the beam area is 1 cm^2. Find the projected range, the straggle, and the maximum impurity concentration.

4-6 Arsenic ions are to be implanted in silicon substrates. A maximum concentration of 1×10^{17} atoms/cm^3 is needed at a depth of about 1000 Å. Find the accelerating potential, the ion dose, and the straggle.

4-7 In the fabrication of complementary MOS devices (CMOS), a p region or a p tub is formed in a lightly doped n-type substrate ($N_D = 1 \times 10^{15}$ atoms/cm^3). A boron implantation is used rather than a predeposition step. This is followed by a diffusion. The boron is implanted through a thin oxide layer (600 Å) as shown. The thicker field oxide prevents ions from getting through. For the boron implantation, the acceleration voltage is 50 keV and the ion dose $\phi = 1 \times 10^{13}$ ions/cm^2. The implantation is followed by a 6-h diffusion at 1200°C.

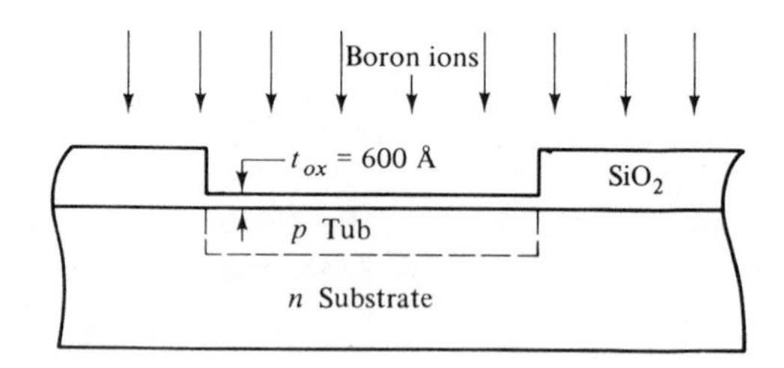

(a) Find the depth of the p tub by locating the p-n junction.
(b) Find the boron concentration at the silicon surface after the diffusion.

Assume that the silicon dioxide behaves like silicon as far as the ions are concerned. Also assume that the ion density remaining trapped in the thin SiO_2 layer is negligibly small.

growth and properties of epitaxial films

5

In this chapter we deal with the nature of epitaxial films and the manner in which they are grown. Epitaxial films are an essential part of the IC fabrication process. The introduction of the process of growing a uniformly doped epitaxial layer on a uniformly doped silicon substrate resulted in about a tenfold increase in the yield of IC circuit manufacturing over the old triple-diffusion technique. The term "epitaxial" literally means "arranged upon." In vapor-phase epitaxial growth, gases containing a silicon compound are passed over the substrate that is maintained at an elevated temperature and reactions take place in the gas stream and at the Si surface which result in the deposition of Si atoms. These atoms snap into their appropriate places in the crystal structure, resulting in a continuation of the crystal with whatever orientation the substrate has. By including appropriate doping gases in the gas stream, impurities such as P, B, and As can be uniformly distributed in the epi-layer. The process is carried out in an epitaxial reactor. This is either a horizontal or vertical furnace tube. Heating is by induction heating of a susceptor made of material such as graphite or by infrared heating of the susceptor and wafer surface. In the manufacture of monolithic bipolar ICs, the substrate is p-type. A prediffused n^+ (heavily doped) buried layer in the form of pockets in the wafer surface is generally included to minimize collector resistance. The n-type epi-layer is then grown on the p-type substrate with the n^+ buried layer. The epi-layer is typically from 1 to 25 μm thick and all the circuitry is fabricated in this layer. The substrate serves as a means of providing electrical isolation.

Silicon may also be grown on some insulating crystals with related crystal structures such as sapphire (Al_2O_3 single crystal) or spinel (Al_2O_2–MgO single crystal). In these cases, electrical isolation is achieved by means of the insulating substrate.

5-1 GROWTH OF EPITAXIAL FILMS

Epitaxial film growth is carried out in a furnace tube which, for this discussion, is assumed to be horizontal. Figure 5-1 is a schematic representation of such a system.

Two main deposition processes are used in epitaxial systems; one is the reduction of silicon tetrachloride ($SiCl_4$) and the second is the pyrolitic decomposition of silane (SiH_4). Other deposition techniques that may be useful at times are the reduction of $SiBr_3$ and $SiHCl_3$.

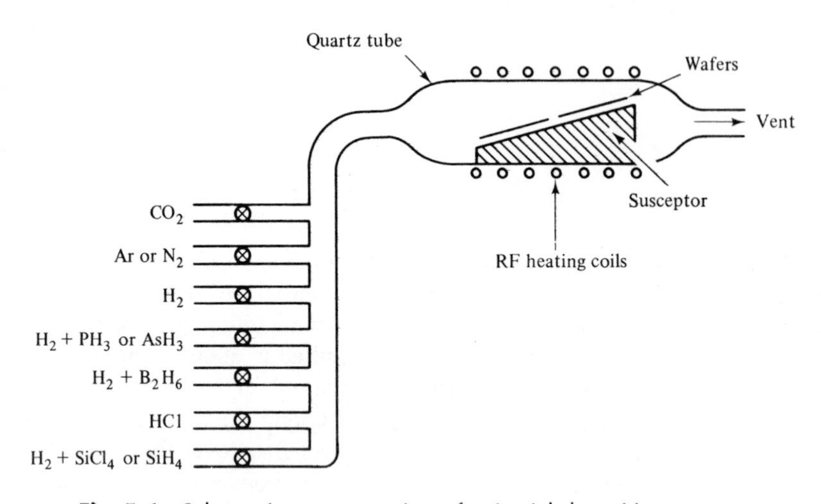

Fig. 5-1 Schematic representation of epitaxial deposition system.

Figure 5-2 is an illustration of processes going on during the reduction of $SiCl_4$. $SiCl_4$ is added to H_2 gas by bubbling H_2 through liquid $SiCl_4$, usually held at a controlled temperature. The $SiCl_4$ is reduced to $SiCl_2$ and HCl downstream in the gas above the wafers at 800 to 1100°C. $SiCl_2$ is absorbed on the Si wafers, where it is reduced to Si and $SiCl_4$. Since two molecules of $SiCl_2$ are involved, reactions are not likely to occur in the gas phase. The reaction equations are

$$SiCl_4 + H_2 \rightleftharpoons SiCl_2 + 2HCl \tag{5-1}$$

and

$$2SiCl_2 \longrightarrow Si + SiCl_4 \tag{5-2}$$

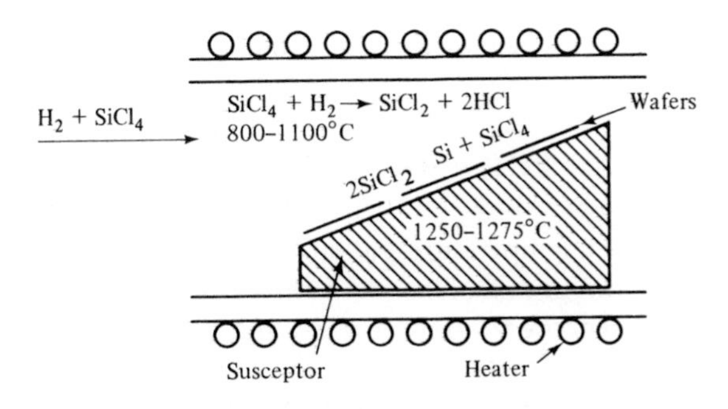

Fig. 5-2 Silicon tetrachloride epitaxial process.

with an overall reaction of

$$2SiCl_4 + 2H_2 \longrightarrow Si + SiCl_4 + 4HCl \tag{5-3}$$

Figure 5-3 is a plot of silicon deposition rate vs. mole fraction of $SiCl_4$ in H_2. Deposition rates in excess of 2 μm/min result in polycrystalline film formation due to insufficient time for proper nucleation and crystal growth to take place.

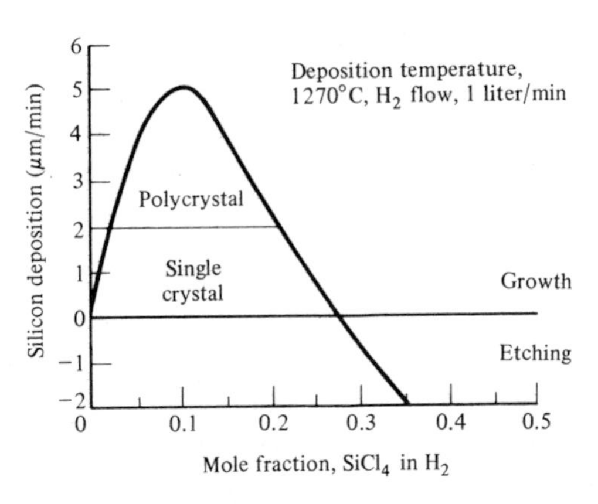

Fig. 5-3 Silicon deposition rate vs. mole fraction of $SiCl_4$ in H_2 for epitaxial growth of Si. (Adapted from H. C. Theuerer, "Epitaxial Silicon Films by the Hydrogen Reduction of SiCl," *J. Electrochem. Soc.*, Vol. 108, 1961, pp. 649–653. Adapted by permission of the publisher, The Electrochemical Society, Inc.)

A concentration of $SiCl_4$ in excess of 0.28 mole fraction results in net removal (etching) of the Si surface because of the reversible nature of the $SiCl_4$ reduction reaction. The thermally induced mobility of Si atoms on the surface helps to prevent premature nucleation and polycrystalline growth.

Dopants are added to the epitaxial film by way of gases such as diborane (B_2H_6), phosphine (PH_3), and arsine (AsH_3). Most *n*-type epitaxial layers are doped with As because of its substantially lower diffusion constant, which minimizes diffusion of donor atoms from the epi-layer during subsequent processing steps.

The other primary reaction used in epitaxial growth, pyrolysis of SiH_4, is given

$$SiH_4 \xrightarrow{1000°C} Si + 2H_2 \tag{5-4}$$

A balance between temperature and mole fraction SiH_4 must be maintained to achieve single-crystal growth. The temperature must be high enough to give the Si atoms enough energy to move around and find a "seat" in the growing crystal lattice. The mole fraction must be high enough to give a usable film growth rate, yet balanced with temperature to minimize gas-phase nucleation and "raining" Si particles onto the wafer surface.

Silane is pyrophoric (ignites spontaneously in air) and therefore must be handled with more care than most compressed gases. All gas lines must be leak-tight.

The epitaxial layer dopant concentration can be estimated from the dilution formula

$$N_D = CN_t\left(\frac{r_t}{r_t + r_h}\right)\left(\frac{r_d}{r_d + r_m}\right) \tag{5-5}$$

Here N_D is the dopant concentration in the film, N_t the dopant concentration in the tank, r_t the rate of flow from the doping tank, r_h the rate of flow of diluting hydrogen, r_d the rate of flow of diluted dopant in the main flow, r_m the main flow rate before introduction of the dopant, and C is a constant dependent on processing conditions which can be empirically determined. Since $r_m >> r_d$,

$$N_D \simeq C\frac{N_t}{r_m}\frac{r_t r_d}{r_t + r_h} \tag{5-6}$$

Prior to the growth of the epitaxial film, the wafers are etched (in the reactor) by a $HCl–H_2$ gas mixture and finally cleaned with a flow of dry H_2 at 1200°C to reduce any remaining oxides on the surface. Residual oxides and surface contaminants are a primary cause of defects in epitaxial films.

The flux (j) of gas molecules arriving at the Si surface (net number of atoms or molecules striking the surface per unit area per unit time) is related to the gas-phase mass-transfer coefficient (h) and the concentrations of the Si-bearing gas in the bulk gas phase (C_g) and at the Si surface (C_s):

$$j = h(C_g - C_s) \tag{5-7}$$

At equilibrium this flux equals the chemical reaction rate per unit area at the surface give by Eq. (5-8),

$$j = kC_s \tag{5-8}$$

Here k is the surface-reaction rate constant. Equating these and solving for C_s gives

$$C_s = \frac{C_g}{1 + k/h} \tag{5-9}$$

From Eq. (5-9) it is seen that when $h >> k$, $C_s \approx C_g$. The other extreme, where $k >> h$, leads to $C_s = 0$.

The number of atoms or molecules that arrive at a surface area A in time dt is $jA\,dt$. The volume of Si grown as a result of this flux on area A in time dt is $dV = jA\,dt/n$, where n is the number of Si atoms per cubic centimeter (5×10^{22} cm^{-3}). The rate of growth of the film is therefore given by

$$\frac{dZ}{dt} = \frac{1}{A}\frac{dV}{dt} = \frac{j}{n} \tag{5-10}$$

Substitution for C_s in Eq. (5-8) in terms of C_g, Eq. (5-9), leads to the following equation for the rate of film growth:

$$\frac{dZ}{dt} = \frac{C_g}{n}\frac{hk}{h + k} \tag{5-11}$$

For the case $h >> k$, we see that the growth rate is controlled by k. This is referred to as surface-reaction control. When $k >> h$, mass-transfer control occurs. The mass-transfer coefficient is essentially independent of temperature, whereas the surface-reaction coefficient is given by

$$k = Ae^{-E_a/kT} \tag{5-12}$$

where A is a constant and E_a is an activation energy. Grove [1] has shown, based on data of Shepherd and Theuerer, that for $SiCl_4$ with a mole fraction of 0.005, (h) lies between 5 and 10 cm/s and $k = 10^7 e^{-1.9\,eV/kT}$ cm/s. At 1250°C, $k = 5.6$ cm/s. At the temperatures at which Si epitaxial films are grown, the mass-transfer coefficient h, and the surface-reaction rate constant k are comparable in value. Stated in another way, there are two mechanisms that affect epitaxial growth; the rate at which Si atoms arrive at the surface, h, and the rate at which the atoms move after they are at the surface, k.

There are many assumptions involved in the model of film growth described here. For example, no account has been made for h variation with gas flow rate. This parameter is extremely sensitive to flow rate for vertical reactors but relatively insensitive for horizontal reactors. Also, no account has been taken of temperature gradients or the effect of the flux of reaction products. In spite of these omissions, the model does provide correct order of magnitude results.

5-2 DOPING PROFILES IN EPI-LAYERS

The ideal doping profiles for an n-type epi-layer on a p-type substrate and an n-type epi-layer on an n^+ buried layer are depicted in Fig. 5-4. There are two reasons why actual doping profiles do not appear this way. The first is due to an etch-back (or autodoping) effect during the film deposition; the

Fig. 5-4 Ideal dopant concentrations for n on n^+ and n-on-p epi-layers: (a) n-on-n^+ epi-layer (ideal); (b) n-on-p epi-layer (ideal).

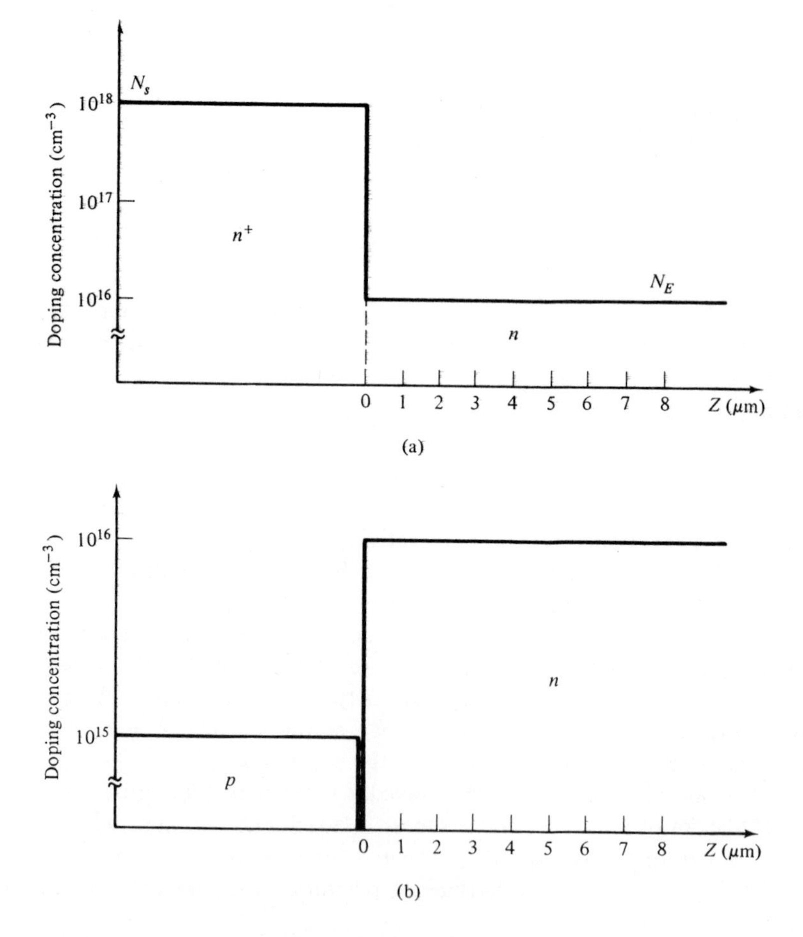

second is a diffusion effect that occurs during film deposition and subsequent high-temperature processing.

The cause of etch-back is as follows. The dopant concentration of the gas in the reactor does not remain constant, owing to the reversible nature of the reaction

$$SiCl_4 + 2H_2 \rightleftharpoons Si + 4HCl$$

As the reaction proceeds to the right, Si is being deposited. As the reaction proceeds to the left, Si and dopant are being etched from the surface and alter the dopant concentration of the gas mixture. The $SiCl_4$ reaction can go either direction, depending on localized temperature and concentration gradients. If a sufficiently thick epi-layer is grown, an equilibrium gas mixture concentration is reached; but for thin (~10-μm) layers, a continuous variation in dopant concentration can result. Most Si epi-layers are from 5 to 15 μm thick.

The resultant doping profile is the superposition of two cases [2]: (1) intrinsic gas and a doped substrate, and (2) doped gas and an intrinsic substrate. It has been experimentally established that case (1) is governed by

$$N_z = N_s e^{-\phi z} \tag{5-13}$$

and case (2) by

$$N_z = N_E(1 - e^{-\phi z}) \tag{5-14}$$

where N_z is the dopant concentration in the epi-layer, N_s the dopant concentration in the substrate, N_E the equilibrium concentration in a sufficiently thick epi-layer, z the distance measured from the substrate epi-layer interface normally outward from the substrate, and ϕ is an experimentally determined growth factor [5]. Growth factors vary with the system in use but generally vary inversely with temperature. Figure 5-5 contains a plot of ϕ as a function of T for As and B doped substrates. Superimposing these two cases leads to

$$N_z = (N_s - N_E)e^{-\phi z} + N_E \tag{5-15}$$

where N_s and N_E have similar signs for like doping and opposite signs for opposite dopants. Figure 5-6 is a plot of the superposition of the two cases to describe n on n^+ and n-on-p situations. For n on n^+ with a substrate concentration of 10^{18} cm^{-3} and an ideal epi-layer concentration of 10^{16} cm^{-3}, the etch-back effect extends well into the epi-layer, causing a larger-than-normal donor concentration. For an n-on-p case with a substrate doping of 10^{15} cm^{-3} and an ideal epi-layer doping of 10^{16} cm^{-1}, the junction is displaced slightly (about 0.2 μm) into the epi-layer. In these calculations, on which Fig. 5-6 is based, it was

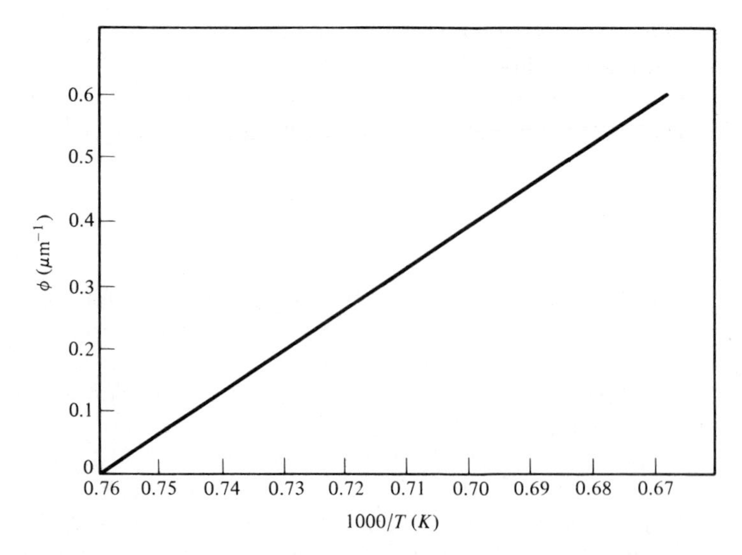

Fig. 5-5 Etch-back growth parameter as a function of temperature. (Adapted from D. Kahng et al., "Epitaxial Silicon Junctions," *J. Electrochem. Soc.*, Vol. 110, No. 5, 1963, pp. 394–400. Adapted by permission of the publisher, The Electrochemical Society, Inc. [5].)

assumed that $\phi = 0.5\ \mu\text{m}^{-1}$. Most epitaxial growth is carried out at over 1200°C, which corresponds to a somewhat higher value of ϕ.

In the case of the *n*-on-*p* epi-layer, the junction occurs in the epi-layer a distance z_j, known as junction lag, from the substrate surface:

$$z_j = \frac{1}{\phi} \ln \left(1 - \frac{N_s}{N_E}\right) \tag{5-16}$$

The higher the growth factor, the less the junction lag; hence, it is desirable to grow epi-layers at as high a temperature as possible to minimize z_j.

Since the deposition of an epi-layer takes place at elevated temperatures, and subsequent dopant deposition and diffusion processes are also carried out at high temperatures, it is to be expected that the doping profile in the vicinity of the substrate epi-layer interface will be affected. As in the etch-back phenomenon, the general case can be treated as the superposition of two special cases: (1) diffusion from a doped substrate into an undoped epi-layer and (2) diffusion from a doped epi-layer into an updoped substrate. It may be assumed that both cases are approximated by diffusion from a uniformly doped semiinfinite slab into an undoped semiinfinite slab. Equation (5-17) is the solution of the diffusion equation for case (1). This gives the impurity concentration N_i in an intrinsic epi-layer in terms of the initial impurity concentration of the substrate

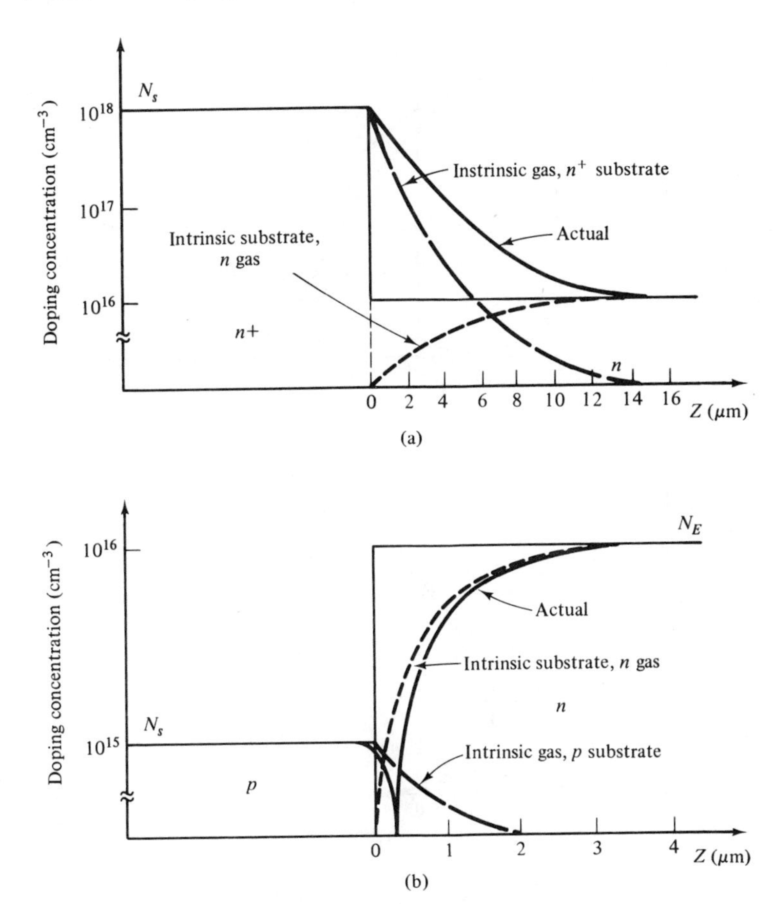

Fig. 5-6 Dopant distribution in epi-layers due to etch-back: (a) n-on-n^+ epi-layer; (b) n-on-p epi-layer.

N_s, the distance into the epi-layer z, and the diffusion constant for the substrate dopant D_s [3]:

$$N_i(z) = \frac{N_s}{2}\left(1 - \operatorname{erf}\frac{z}{2\sqrt{D_s t}}\right) \tag{5-17}$$

To determine the doping profile in the epi-layer for case (2) we subtract the previous result [Eq. (5-17)] from the ideal concentration in the epi-layer. This leads to

$$N_D(z) = \frac{N_E}{2}\left(1 + \operatorname{erf}\frac{z}{2\sqrt{D_E t}}\right) \tag{5-18}$$

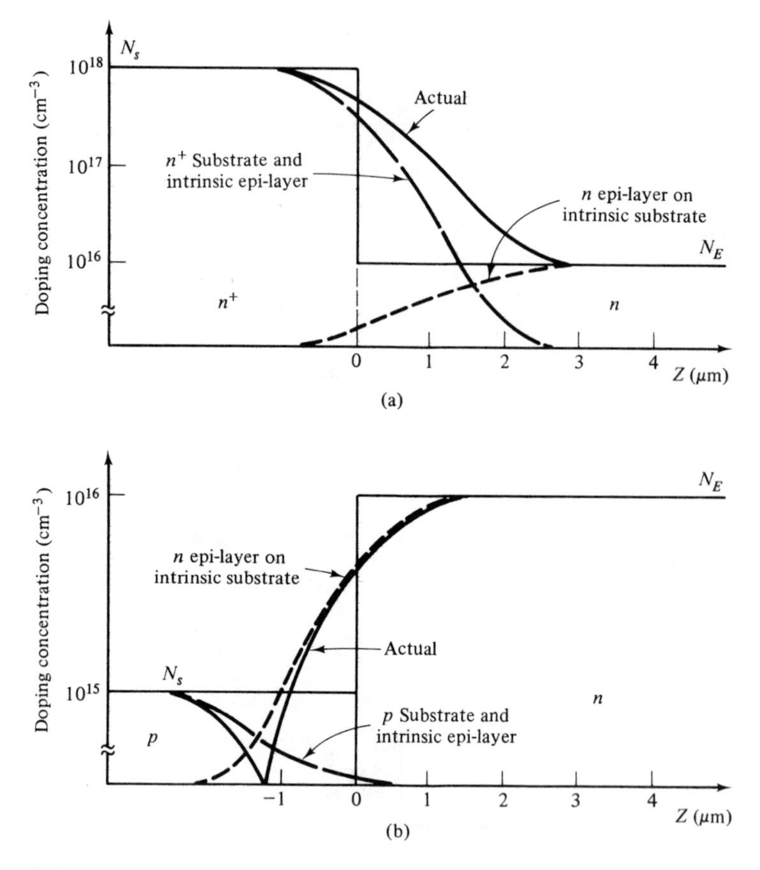

Fig. 5-7 Dopant distribution in epi-layers due to diffusion: (a) n-on-n^+ epi-layer; (b) n-on-p epi-layer.

Here N_D is the concentration in the doped epi-layer on an intrinsic substrate, and N_E is the concentration at the surface of the epi-layer assuming that it is sufficiently thick. If $z >> 2\sqrt{D_E t}$, then $N_D \approx N_E$.

Redistribution of the dopants during epitaxial growth is negligible compared with the redistribution that takes place during subsequent processing. To calculate the redistribution due to diffusion during processing, an effective diffusion constant time product must be calculated. This is done simply by equating $(Dt)_{\text{eff}}$ to the sum of the products of the appropriate diffusion constants and time for each high-temperature process step. Figure 5-7 contains plots of redistribution due to diffusion for n on n^+ with As the dopant and for n on p with As and B the dopants. For the n-on-n^+ case, a graded junction between substrate and epi-layer results. For the n-on-p case the junction tends to recede into the substrate, tending to compensate for junction lag due to the etch-back effect. To minimize diffusion effects when buried layers are employed, dopants with small diffusion constants, such as antimony or arsenic, are used for both the buried layer and the epi-layer.

5-3 DEFECTS IN EPITAXIAL LAYERS

Epitaxial layers grow by random nucleation to bond sites on the substrate. They do not grow by completing one whole atomic layer and then starting another. This would require renucleation for each monolayer. Instead, a terracing or stepped effect occurs at each original nucleation site. This is schematically depicted in Fig. 5-8. After a certain number of nucleation sites are established, their number does not change substantially. They simply grow upward by starting new layers at each site, and eventually grow together due to the lateral motion of atoms on the surface. An atom landing on the surface must have enough energy to move to a good bonding site, but once an atom reaches a step (see Fig. 5-8) it is tightly bound and the likelihood of escaping is greatly reduced. This is basically what happens in the vapor deposition of epitaxial films. Figure 5-8 is an oversimplification because it does not show the chemical reaction taking place at the surface during gas-phase epitaxial growth.

A major defect in epi-layers is the formation of stacking faults. These faults originate at the interface between the substrate and epi-layer. If there is an imperfection on the Si substrate surface, it may act both as a nucleation site and as a means of disrupting the orderly laying down of atomic layers in their proper stacking sequence. The stacking at most nucleation sites will be in the proper sequence. When the boundaries of the proper and improperly stacked nucleation regions meet, a discontinuity will result in the crystal structure. Even though subsequent stacking is in the proper sequence, the atomic layers above the improperly started planes will remain out of sequence with the areas originating from the properly nucleated sites. The diamond lattice is an interpenetrating face-centered cubic lattice with the stacking sequence *aa' bb' cc' aa' bb'*. For simplicity, we will consider the fcc lattice with the stacking *abcabcab,* which leads to the same results. The stacking in a fcc lattice is shown

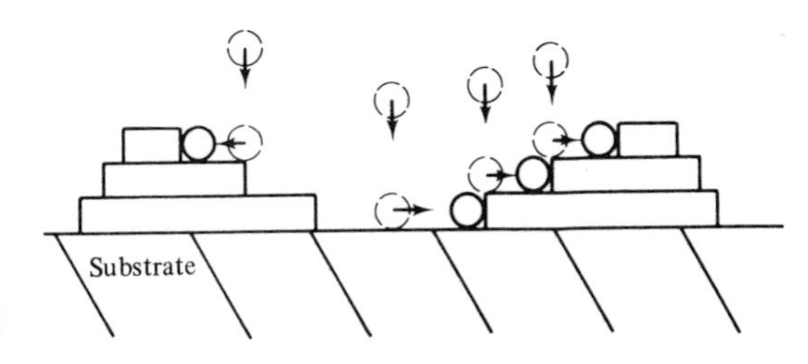

Fig. 5-8 Nucleation during epitaxial growth.

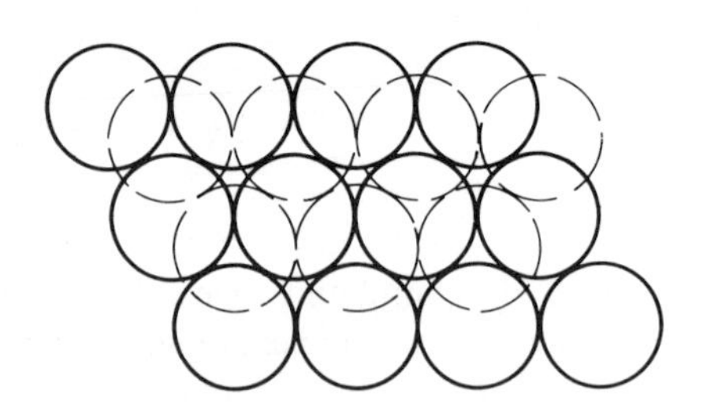

Fig. 5-9 Stacking in face centered cubic lattice top view. Only two planes of atoms are shown (dashed circles represent top plane; solid circles represent bottom plane). Fcc lattice results when a third plane is placed so that atoms are not directly over atoms of bottom plane. If the atoms were placed directly over atoms of bottom plane, the lattice would be hexagonal.

in Fig. 5-9. To avoid confusion, only the *ab* part of the sequence is shown. The third plane of atoms is positioned in the voids, not directly over the atoms of the bottom plane, thus giving the sequence *abc,* which is then repeated over and over. In a stacking fault, the stacking might start off *aca* or *aba.* In any event, the atoms in this sequence of planes will be positioned differently than those surrounding them.

When silicon is grown in the <111> direction, stacking faults are bound by {111} planes. When looking down at the surface, the {111} planes intersect at 60°, and so when an epitaxial stacking fault is formed on the surface of the wafer, it appears to be a triangle. In reality it is a tetrahedron with its apex truncated at the epi-layer–substrate interface. If the area of the truncated part of the tetrahedron at the epi-layer–substrate interface is negligible, the epi-layer thickness t can be simply related to the length of the edge of the tetrahedron. The stacking fault tetrahedron is illustrated in Fig. 5-10. The epi-layer thickness and edge length are related by Eq. (5-19) from geometrical considerations:

$$t = \sqrt{\tfrac{2}{3}}\, l \tag{5-19}$$

Equation (5-19) provides a means of determining epi-layer thickness as long as the maximum-size faults are used and it can be correctly assumed that the largest faults start at the substrate–epi-layer interface.

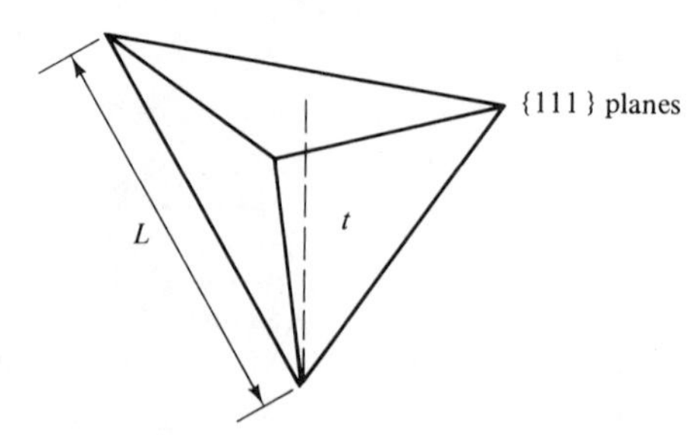

Fig. 5-10 Stacking fault tetrahedron.

In <100>-oriented Si crystals, the stacking faults are terminated by a {100} plane at the surface and bounded by {111} planes producing a four-sided pyramid with the base being a square. Figure 5-11 contains a scanning electron micrograph of an etched stacking fault for <100>-grown epitaxial Si. For <100>-grown Si, Eq. (5-20) relates epi-layer thickness to the size of the etch figure,

$$t = \frac{l}{\sqrt{2}} \tag{5-20}$$

where l is the length of one side of the square terminating the stacking fault at the surface.

Stacking faults by themselves are not a serious defect. However, they do cause anomalies in the diffusion characteristics of impurities. They also act as nucleation sites for metal precipitates and sources of microplasmas at their edges. Consequently, *p-n* junctions may be shorted out or exhibit extremely low voltage reverse-bias breakdown when there is a high concentration of stacking faults. Dislocations occur in epi-layers, but their density is generally not such as to cause serious trouble. However, stresses introduced during high-temperature processing, particularly where a wafer contacts a boat, although localized, can be excessive and produce large numbers of dislocations in the epi-layer.

Poor epitaxial-layer growth techniques can give rise to defects in the epi-layer. These may be exhibited as lines of stacking faults due to substrate scratches;

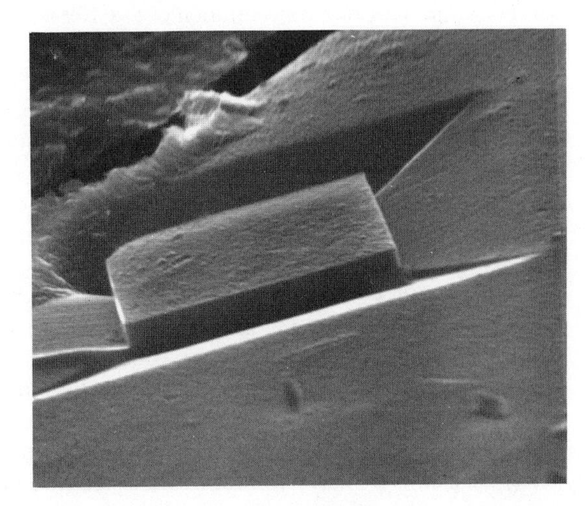

Fig. 5-11 Scanning electron micrograph of stacking fault in epitaxial layer on <100> grown silicon.

pits, voids, or spikes due to small particles of Si or SiO_2 present during the growth process; pyramids due to too high a growth rate or improperly oriented substrate; and haze due to leaks in the system or a poor cleaning procedure. Spikes can be particularly troublesome because of the damage they do to photomasks during subsequent wafer processing.

Since the fast growth direction for Si is actually the <100>, it turns out that epitaxial growth on a <111> crystal is enhanced if the wafer orientation is not exactly <111>. The misorientation of <111>-grown Si wafers is discussed in Chapter 2. For crystals whose growth direction is <100>, the wafer should be kept as close to <100> as possible to maximize epitaxial growth, although pattern shift ("washout") is a more important consideration in determining the extent of misalignment.

BIBLIOGRAPHY

[1] GROVE, A. S., *Physics and Technology of Semiconductor Devices.* New York: John Wiley & Sons, Inc., 1967.

[2] GHANDI, S. K., *The Theory and Practice of Microelectronics.* New York: John Wiley & Sons, Inc., 1968.

[3] BOLTAKS, B. I., *Diffusion in Semiconductors.* New York: Academic Press, Inc., 1963.

[4] HANSEN, M., AND K. ANDERKO, *Constitution of Binary Alloys.* New York: McGraw-Hill Book Company, 1958.

[5] KAHNG, D., ET AL., "Epitaxial Silicon Junctions," *Journal of the Electrochemical Society,* Vol. 110, No. 5, pp. 394–400, 1963.

PROBLEMS

5-1 For an n^+ substrate concentration of 5×10^{18} cm^{-3}, an ideal epi-layer concentration of 5×10^{15} cm^{-3} and using an etch-back constant of $\phi = 0.55\ \mu$m^{-1}, plot the concentration as a function of distance into the epi-layer.

5-2 Calculate the junction lag for an n-on-p epi-layer–substrate arrangement for $\phi = 0.5\ \mu$m^{-1} given ideal substrate and epi-layer concentrations of 5×10^{15} cm^{-3} and 10^{17} cm^{-3}, respectively.

5-3 Plot the electron concentration as a function of distance into the substrate and the epi-layer for an n-on-n^+ arrangement given concentrations of 5×10^{18} cm^{-3} and 2×10^{16} cm^{-3} for the substrate and epitaxial layers, respectively, and an effective diffusion constant time product of 7×10^{-8} cm^2.

5-4 Repeat Problem 5–3 for an n-on-p system with a substrate concentration of 8×10^{14} cm^{-3} and an ideal epi-layer concentration of 5×10^{16} cm^{-3}. Use the same effective diffusion constant time product. How far is the junction displaced?

5-5 For a growth rate of 1 μm/min of Si in an epitaxial growth process, calculate the concentration of Si in the gas phase assuming that h, the gas-phase mass-transfer coefficient, is 6 cm/s and k, the surface-reaction rate constant, is 5 cm/s.

the p-n *junction diode*

6

In Chapter 4 we discussed methods of selectively locating donor and acceptor impurities in single-crystal silicon (i.e., by diffusion and ion implantation). The growth and properties of epitaxial films were discussed in Chapter 5. Epitaxial layers provide uniform impurity distributions that improve both device behavior and device yields. Both of these chapters concentrated on how solid-state devices are made. In this chapter we begin a discussion of how electronic devices work. This discussion must begin with the *p-n* junction diode, which is both a fundamental building block for other devices and a solid-state device in its own right. A useful explanation of junction-diode behavior thus serves a twofold purpose: it explains how the device works, and does it in a way that leads naturally to later explanations of bipolar and field-effect transistor behavior.

The most important characteristic of a junction diode can be simply stated. It allows the passage of current in only one direction. How a diode does this is not simple, however. To explain this, and other diode characteristics, we must start with basic *p*-type and *n*-type materials and logically progress to the behavior of *p-n* junction devices with externally applied voltages.

This chapter begins, then, with a description of extrinsic silicon in equilibrium. The electrical balance that exists between holes, electrons, and trapped impurity ions is described. When equilibrium conditions are disturbed by an excess or deficit of charge carriers, or by an applied electric field, the disturbance results in hole and electron motion. (This subject is discussed later in this chapter.) The *p* and *n* regions of silicon are then joined and equilibrium conditions are restored. A new balance exists between the motions of holes and electrons across the metallurgical junction. Regions of trapped impurity ions are left on either side of the junction. When external voltages are applied across the device, equilibrium conditions are disturbed. The imbalance created between the relative motions of holes and electrons, and between drift and diffusion currents, leads to an explanation of why the device passes current in one direction and not the other.

6-1 *n* AND *p*-TYPE SILICON

Impurity concentrations in silicon range from light concentrations of about 1×10^{14} impurity atoms/cm^3 to heavy concentrations of about 1×10^{20} atoms/cm^3. A silicon crystal contains about 5×10^{22} silicon atoms/cm^3. A light impurity concentration of 1×10^{14} dopant atoms/cm^3 thus provides about two impurity atoms for every billion silicon atoms. A heavy impurity concentration of 1×10^{20} dopant atoms/cm^3 provides about two impurity atoms for every thousand silicon atoms. This is the range of impurity concentrations ordinarily used. Practically all of the impurity atoms are ionized over the normal range of operating temperatures.

Because of thermal vibration within the crystal lattice, covalent bonds break and re-form. Hole–electron pairs are continuously being generated and recombining. At room temperature, the concentration of electrons and holes in intrinsic silicon (n_i and p_i, respectively) is only about 1.5×10^{10} cm^{-3}. This is a small number when compared to the concentration of holes or electrons typically provided by impurity atoms. Thus, even though impurity concentrations are small compared to the density of silicon atoms, they can control the conductivity of the silicon crystal.

The concept of Fermi levels and the equations for the intrinsic concentrations of electrons and holes in semiconductors were developed in Chapter 2. Equations (2-7) and (2-9) provide the concentrations of electrons and holes in an intrinsic semiconductor as a function of temperature and the energy intervals between the band edges and the Fermi level. These equations are repeated here for convenience:

$$n_i = K_1 T^{3/2} \exp\{-[(E_C - E_F)/kT]\} \qquad (6\text{-}1)$$

and

$$p_i = K_2 T^{3/2} \exp\{-[(E_F - E_V)/kT]\} \qquad (6\text{-}2)$$

The donor impurity atoms which are used in silicon are the group V elements phosphorus, arsenic, and antimony. As discussed in Chapter 4, arsenic and antimony are normally used for buried layers and epitaxial layers because they are slow diffusers. Phosphorus, which diffuses more rapidly, is used for n^+ emitters. When valence V elements substitute for silicon atoms in the crystal lattice, each impurity atom has an extra valence electron (or donor electron) which does not fit into the covalent bond structure. The ionization energy re-

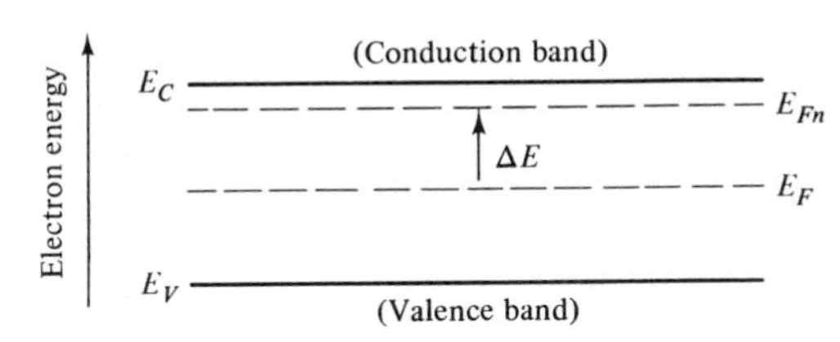

Fig. 6-1 Energy levels in *n*-type silicon.

quired to set a donor electron free is about 0.044 eV. At normal operating temperatures, sufficient thermal energy is present to ensure that practically all of the donor electrons are in the conduction band. When a donor electron moves to the conduction band, it leaves behind a positive ion that is permanently lodged in the silicon lattice and is not free to move. Charge neutrality is maintained because there is a donor electron around for every trapped positive ion.

Figure 6-1 shows an energy-level diagram for *n*-type silicon indicating the shift in energy (ΔE) from E_F to E_{Fn} due to the addition of donor impurity atoms. E_F represents the Fermi level in intrinsic silicon and E_{Fn} represents the Fermi level in *n*-type silicon. Referring to Fig. 6-1, we have

$$E_{Fn} = E_F + \Delta E \tag{6-3}$$

The relationship between ΔE and the concentration of electrons in *n*-type material (n_{no})[1] can be found by substituting Eq. (6-3) in Eq. (6-1). Thus,

$$n_{no} = K_1 T^{3/2} \exp\{-[(E_C - E_F - \Delta E)/kT]\} \tag{6-4}$$

or

$$n_{no} = n_i \exp\left(\frac{\Delta E}{kT}\right) \tag{6-5}$$

Thermally generated hole–electron pairs still exist in *n*-type silicon. The increased concentration of donor electrons causes an increase in hole–electron recombination, however, so that the hole concentration diminishes. The minority concentration of holes in *n*-type material (p_{no}) can be found by substitution E_{Fn} in Eq. (6-3) for E_F in Eq. (6-2). Then

$$p_{no} = n_i \exp\left(\frac{-\Delta E}{kT}\right) \tag{6-6}$$

The acceptor impurity atoms that are used in silicon are the group III elements boron and aluminum. Boron is used extensively as a dopant, and aluminum is used for metalization. Valence III elements cannot satisfy the cova-

[1] The second subscript in n_{no} is used to denote that this is an equilibrium concentration

lent bonding requirements of the silicon crystal—an electron is missing, or alternatively, a hole is present. At normal operating temperatures the acceptor atoms will each capture an electron, thus becoming negative ions, trapped in the crystal lattice. Holes are released to the valence band. Charge neutrality is maintained since a hole exists in the valence band for every trapped negative ion in the crystal lattice. Under the influence of an electric field, the holes can conduct current.

Figure 6-2 shows an energy-level diagram for *p*-type silicon, indicating the shift in energy ($\Delta E'$) from E_F to E_{Fp} due to the addition of acceptor atoms. E_F represents the Fermi level in intrinsic silicon and E_{Fp} represents the Fermi level in *p*-type silicon. Referring to Fig. 6-2, we obtain

$$E_{Fp} = E_F - \Delta E' \tag{6-7}$$

The relationship between $\Delta E'$ and the concentration of holes in *p*-type material (p_{po}) can be found by substituting E_{Fp} in Eq. (6-7) for E_F in Eq. (6-2). Then

$$p_{po} = n_i \exp\left(\frac{\Delta E'}{kT}\right) \tag{6-8}$$

The minority concentration of electrons in *p*-type silicon (n_{po}) can be found by substituting E_{Fp} in Eq. (6-7) for E_F in Eq. (6-1). Then

$$n_{po} = n_i \exp\left(-\frac{\Delta E'}{kt}\right) \tag{6-9}$$

It is useful to remember that *n* and *p* represent electrons and holes, respectively. As subscripts, they denote the type of silicon; that is, n_{no} and p_{no} represent equilibrium concentrations of electrons and holes in *n*-type silicon, while n_{po} and p_{po} represent equilibrium concentrations of electrons and holes in *p*-type silicon. If we take the product of hole and electron concentrations in *n*- or *p*-type silicon, we find that

$$n_{no}p_{no} = n_{po}p_{po} = n_i^2 \tag{6-10}$$

In general, when a doped semiconductor is in *equilibrium,* the product of hole and electron concentrations per unit volume is always equal to n_i^2. Thus,

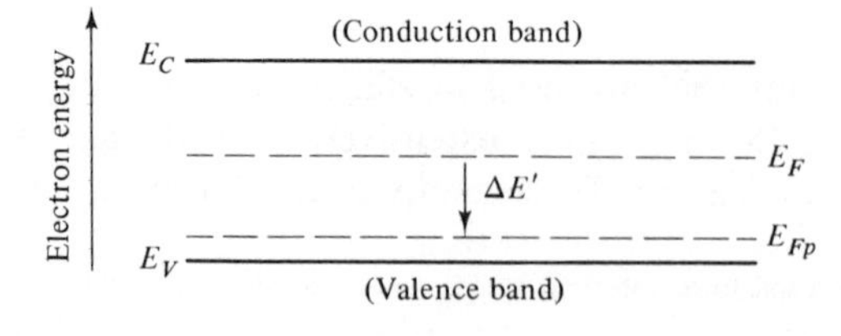

Fig. 6-2 Energy levels in *p*-type silicon.

$$pn = n_i^2 \tag{6-11}$$

where p and n represent the generalized hole and electron densities in a given semiconductor sample. This is known as the *law of mass action.*

Let N_D represent the concentration of *donor* atoms per unit volume in silicon. Let N_A represent the concentration of *acceptor* atoms per unit volume. Since practically all of the donor or acceptor atoms are ionized, N_D and N_A also represent the concentrations of trapped positive or negative ions, respectively. Impurity concentrations are usually controlled by counter-doping. Donor atoms may be added to a selected region of a p-type substrate until that region has the desired n-type properties. The net ion concentration of this region will then be given by N_D minus N_A.

If the impurity atoms are distributed within a volume element of a semiconductor, a condition of charge neutrality will exist. In other words, the net charge density ρ will be zero in that volume element. Adding all the positive and negative charges, we have that

$$\rho = q(p - n + N_D - N_A) = 0 \tag{6-12}$$

or

$$p - n = N_A - N_D \tag{6-13}$$

We can use the law of mass action to replace p, and solve the resulting quadratic equation. This leads to the expression

$$n_{n0} = \tfrac{1}{2}\left[N_D - N_A + \sqrt{(N_D - N_A)^2 + 4n_i^2}\right] \tag{6-14}$$

for electrons in an n-type semiconductor. By analogy, the concentration of holes in a p-type semiconductor is

$$p_{p0} = \tfrac{1}{2}\left[N_A - N_D + \sqrt{(N_A - N_D)^2 + 4n_i^2}\right] \tag{6-15}$$

Typically, $4n_i^2$ is negligibly small when compared to $N_A - N_D$. Also, most impurity concentrations in solid-state devices are such that N_D is orders of magnitude larger than N_A in n-type semiconductor regions, and $N_A >> N_D$ in p-type semiconductor regions. Thus,

$$n_{n0} \approx N_D \tag{6-16}$$

and

$$p_{p0} \approx N_A \tag{6-17}$$

Combining Eqs. (6-16) and (6-17) with the law of mass action leads to the minority carrier concentrations:

$$p_{n0} \approx \frac{n_i^2}{N_D} \tag{6-18}$$

and

$$n_{p0} \approx \frac{n_i^2}{N_A} \tag{6-19}$$

6-2 MOBILITY

Mobile charge carriers in a semiconductor are in a process of random thermal motion, interrupted by many collisions. In silicon at room temperature, electrons have thermal velocities of about 10^7 cm/s. If an electric field is applied to the semiconductor, an average velocity component, referred to as the *drift velocity,* v_d, will be superimposed on the random thermal motion. The drift velocity of the charge carriers is directly proportional to the electric field as

Fig. 6-3 Drift velocity vs. electric field for holes and electrons in silicon at room temperature. (Adapted from E. J. Ryder. Copyright 1953, American Institute of Physics, by permission [1].)

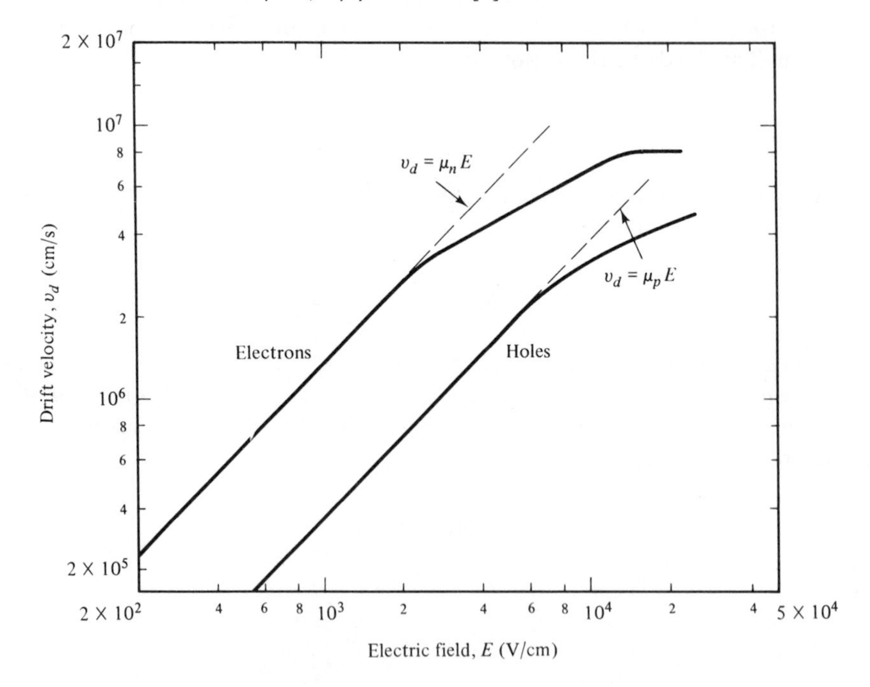

long as v_d is small compared to the thermal velocity of the charge carriers. The proportionality factor is defined as the carrier *mobility* μ. Thus,

$$V_d = \mu E \tag{6-20}$$

Mobility is given as drift velocity per unit electric field. If velocity is given in cm/s, and the field in V/cm, μ has dimensions of $cm^2/V\text{-}s$.

When drift velocities approach thermal velocities in magnitude, a fundamental change takes place in the interaction between the charge carriers and the lattice structure. As the electric field increases, mobility decreases, until the drift velocity reaches a maximum saturated value. In this region, charge carriers are referred to as being "hot." The relationship between electric field and drift velocity for holes and electrons in silicon is shown on the log-log graph of Fig. 6-3. If we had plotted v_d vs. E on a linear graph, then electron or hole mobility would appear as the slope of the v_d vs. E curve in the straight-line region.

Figure 6-4 shows a graph of electron and hole mobilities vs. total impurity concentration for silicon at room temperature. The data of Fig. 6-4 are valid for electric field magnitudes of 10^3 V/cm or less. As shown, electron mobility

Fig. 6-4 Mobility vs. impurity concentration for silicon at room temperature. (Adapted from E. M. Conwell. Copyright 1958, IEEE, Inc., by permission [2].)

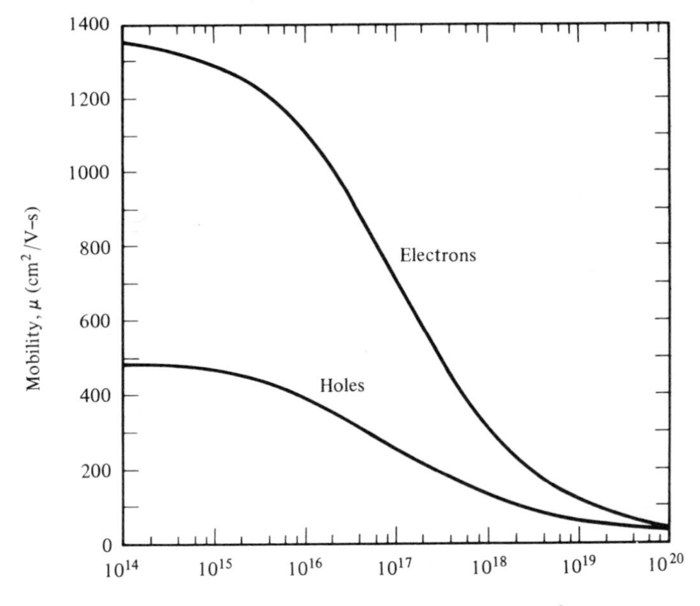

values are greater than hole mobility values by a factor of about 3. This factor creates a preference for electronic devices that employ electrons. Because of the higher electron mobilities, devices that use electrons rather than holes to carry information can be made smaller and can operate at higher frequencies. Figure 6-4 also indicates that electron and hole mobilities decrease as impurity concentrations are increased. Holes and electrons moving about in the crystal lattice are deflected by ionized impurity atoms, both because these atoms are in thermal motion and because they are charged. As the concentration of impurity atoms is increased, more collisions between charge carriers and the crystal lattice

Fig. 6-5 Mobility vs. temperature in silicon for various impurity concentrations: (a) electron mobility vs. temperature; (b) hole mobility vs. temperature. (Adapted from *Transistors: Principles, Design, and Applications* by Wolfgang W. Gartner. Copyright 1960, Litton Educational Publishing, Inc. Reprinted by permission of Van Nostrand Reinhold Company [3].)

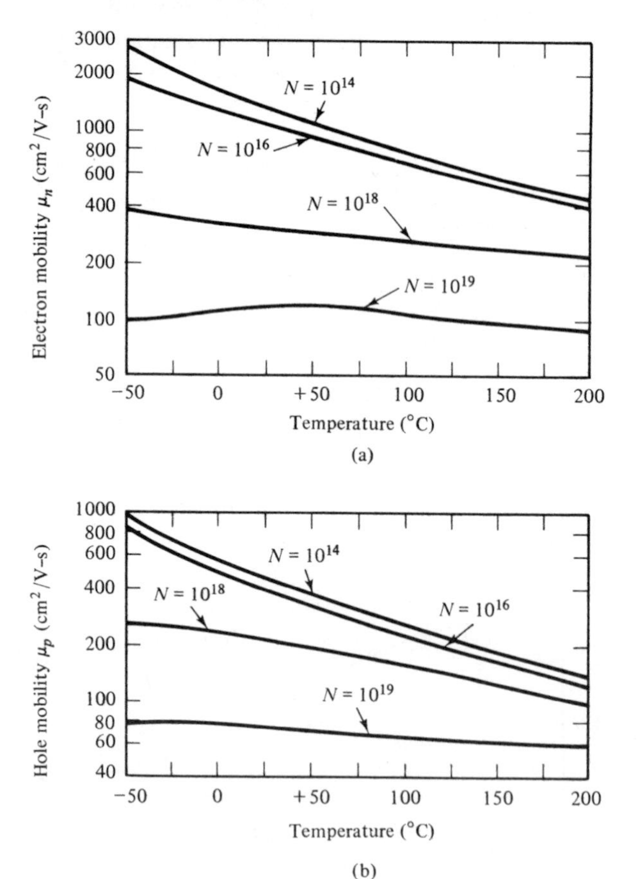

will occur, causing charge carrier mobilities to decrease. Mobility is thus a function of the total impurity concentration N_D plus N_A. At impurity concentrations on the order of 10^{14} atoms/cm^3 and at room temperature, μ_n is approximately 1350 cm^2/V-s, and μ_p is approximately 480 cm^2/V-s.

Two major *scattering mechanisms* limit hole and electron mobilities in silicon. *Lattice scattering* is caused by thermal vibration of the atoms in the crystal lattice. This vibration disrupts the periodicity of the lattice and thus impedes the motion of charge carriers. *Impurity scattering* is caused by the field interaction between charged impurity ions and charge carriers. Hole and electron paths in the crystal lattice are deflected by the trapped impurity ions.

Charge carrier mobilities are dependent on the effectiveness of these scattering mechanisms. Lattice and impurity scattering, in turn, are temperature-dependent. Figure 6-5 shows the resulting variations in electron and hole mobilities in silicon as a function of temperature and impurity concentration. At temperatures above 50°C, mobility values are seen to decrease with increasing temperature. This mobility decrease is primarily caused by lattice scattering. At low temperatures and heavy impurity concentrations (10^{19} cm^{-3}), mobility values are seen to decrease with *decreasing* temperature. This mobility decrease is caused by impurity scattering. This happens because charge carriers moving at lower speeds (corresponding to lower temperatures) will tend to be deflected more by the proximity of charged impurity atoms. Heavily doped silicon thus shows a broad peak in charge carrier mobility values in the temperature range 25 to 100°C [see Fig. 6-5(a)]. At temperatures below 25°C, impurity scattering dominates. Above 100°C, lattice scattering dominates.

The operation of most field-effect devices involves the motion of charge carriers near a silicon–silicon dioxide interface, where the crystal lattice becomes very irregular and makes a transition to amorphous SiO_2. In Chapter 9, we discuss *surface mobility* values (μ_{ns} and μ_{ps}) which are distinctly different from the bulk mobility values treated above.

6-3 CHARGE CARRIER LIFETIMES

Charge carriers are generated in pairs, exist separately for a while, and recombine in pairs. *Charge carrier lifetime can be defined as the average length of time that charge carriers exist between the time that they are generated and the time that they recombine.* Both generation and recombination take place in the vicinity of defects in the crystal lattice or irregularities at the crystal surface. The magnitudes of carrier lifetimes are determined by bulk defects and surface interface conditions which cause energy states near the center of the energy gap. These energy states, also called traps or recombination centers, act as stepping stones and allow transitions between the conduction and valence bands in silicon with energies on the order of 0.55 eV.

We can discuss carrier lifetimes in terms of equilibrium or nonequilibrium concentrations of charge carriers. If the carrier concentrations are in equilibrium, the *rates* of generation and recombination must be equal. These rates can be determined by dividing carrier concentration by carrier lifetime. In intrinsic silicon at room temperature, for example, carrier concentrations are denoted by n_i. The average lifetime of a hole–electron pair might be denoted by T_i. If we arbitrarily assume that T_i equals 2 μs, the rate of generation and recombination of hole–electron pairs is given by

$$\frac{n_i}{T_i} = \frac{1.5 \times 10^{10}\ \text{carriers/cm}^3}{2\ \mu\text{s}} = 0.75 \times 10^{10}\ \text{carriers/cm}^3/\mu\text{s}$$

In order to discuss nonequilibrium carrier concentrations (situations in which a surplus or deficit of charge carriers exist in a silicon sample), we must distinguish between majority and minority concentrations. If equilibrium carrier concentrations are momentarily disturbed in an n-type sample, for example, the mechanisms by which the system restores itself to equilibrium are quite different for excess electrons than for excess holes.

To begin with nonequilibrium *majority* concentrations, imagine a long bar of uniformly doped n-type silicon. The equilibrium concentrations of electrons and holes are given by n_{no} and p_{no}, respectively. The top of the bar is connected to a source of electrons and the bottom is connected to ground, as shown in Fig. 6-6.

Suddenly, at $t = 0$, we inject a number of electrons into the top of the bar. The local electron concentration is increased from n_{no} to n_n. This represents only a small fractional change in the total electron concentration. The increase in negative charge is given by $-q(n_n - n_{no})$. This charge increase causes an electric field to propagate through the sample at the speed of light (in silicon). The electric field causes a slight displacement of electrons throughout the sample. The displacement results in a number of electrons (equal to the number injected) being pushed out the bottom of the bar. The process is not instantaneous. The

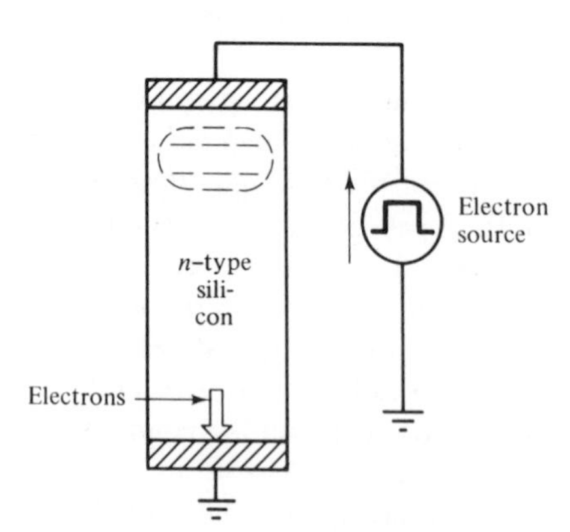

Fig. 6-6 Nonequilibrium majority carrier concentrations and dielectric relaxation time.

excess electron concentration at the top of the bar exponentially decays with a characteristic time constant T. The electron concentration at the top of the bar (n) at any time (t) is given by

$$(n - n_{no}) = (n_n - n_{no})e^{-t/T} \tag{6-21}$$

The time constant, T, which characterizes the exponential change in majority carrier concentration, is called the *dielectric relaxation time.* It is equal to the ratio of the permittivity and the conductivity of the silicon sample involved $(T = \epsilon/\sigma)$. For a 1-mho/cm silicon sample, T is about 10^{-12} s.

Whereas dielectric relaxation times are on the order of picoseconds, carrier lifetimes range between nanoseconds and microseconds. When we injected majority carriers into the top of the sample, an equal number were pushed out the bottom in a matter of picoseconds. The excess majority carriers were not there long enough for carrier lifetimes to be a factor in the process.

To discuss nonequilibrium *minority* carrier concentrations, we can inject electrons into a p-type bar of silicon, as shown in Fig. 6-7. If at $t = 0$, we suddenly inject a number of electrons into the top of the silicon sample, the local electron concentration will be increased from n_{po} to n_p. Because we are dealing with p-type silicon, this represents a large fractional change in the total electron concentration. The increase in negative charge is given by $-q(n_p - n_{po})$. An electric field again propagates through the sample, resulting in a slight electron displacement. A number of electrons (equal to the number injected) are pushed out the bottom of the bar. When the dielectric relaxation process is completed (in about 5 ps), there is an excess of electrons in the top region of the bar and an electron deficit in the bottom region. Although the diffusion process will tend to spatially redistribute the charge carriers, the essential processes in regaining equilibrium conditions are generation and recombination.

The rate of hole–electron pair *generation* involves breaking covalent bonds. It depends on the thermal energy of lattice vibration (barring generation by other, external means). The rate of *recombination,* on the other hand, depends

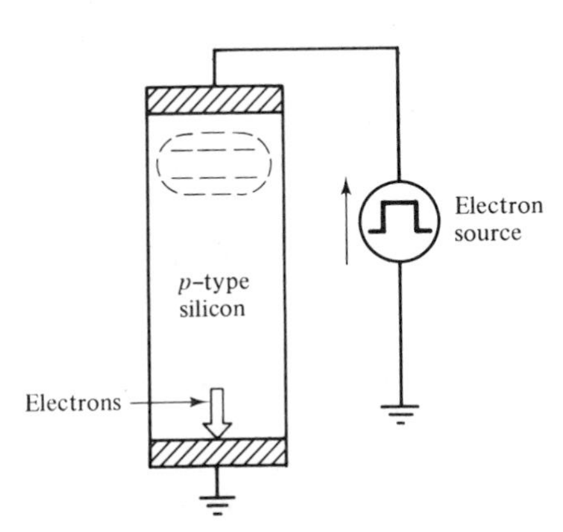

Fig. 6-7 Nonequilibrium minority carrier concentrations and minority carrier lifetime.

on the availability of charge carriers. If excess carriers exist, recombination exceeds generation. If a charge carrier deficit exists, generation exceeds recombination. The rate of change of carrier concentration due to generation and recombination is given by

$$\frac{dn}{dt} = -\frac{n_p - n_{po}}{T_n} \tag{6-22}$$

where $(n_p - n_{po})$ represents the excess (or deficit) electron concentration and T_n represents the average lifetime of electrons in p-type silicon. The minus sign indicates a concentration decrease when excess electrons are present, or a concentration increase when a deficit exists. Integration of Eq. (6-22) yields the expression

$$(n - n_{po}) = (n_p - n_{po})e^{-t/T_n} \tag{6-23}$$

where n represents the electron concentration at any time t. It can be seen from Eq. (6-23) that minority carrier lifetime can also be defined as the *characteristic time constant involved in the exponential change back to an equilibrium concentration.* T_n in Eq. (6-23) is thus the time required for an excess or deficit concentration to change to $1/e$ of its initial value.

Referring back to Fig. 6-7, recombination exceeds generation in the top region of the bar, whereas generation exceeds recombination in the bottom region. Equilibrium is restored in about five carrier lifetimes ($5T_n$).

The rate of change of the nonequilibrium hole concentration in n-type silicon is given by

$$\frac{dp}{dt} = -\frac{p_n - p_{no}}{T_p} \tag{6-24}$$

Integration of this expression yields the minority concentration as a function of time:

$$p - p_{no} = (p_n - p_{no})e^{-t/T_p} \tag{6-25}$$

T_p denotes the average hole lifetime in n-type silicon.

The dielectric relaxation mechanism occurs so quickly when compared with minority carrier lifetimes that it is often disregarded in explanations of solid-state-device operation. The deliberate creation of excess or deficit concentrations of minority carriers, however, is fundamental to device operation. Understanding device operation becomes a matter of understanding equilibrium, transient, and steady-state conditions in p- and n-type materials.

6-4 CONDUCTIVITY AND DRIFT CURRENT

When an electric field is applied across a uniformly doped semiconductor, a *drift current* results due to the net motion of electrons and holes. Consider a uniformly doped *n*-type silicon bar as shown in Fig. 6-8. We will ignore the net motion of minority holes and also the metal–semiconductor contact effects at the ends of the bar. Referring to Fig. 6-8(a), an electron at the edge of the conduction band is assumed to be at rest (to have a reference value of zero kinetic energy). An electron a distance ΔE_K above the conduction-band edge will then have kinetic energy ΔE_K. At normal temperatures, the great majority of the donor electrons will have total energies very near the conduction-band edge. An applied electric field causes the energy levels to bend, as shown in Fig. 6-8(b). Because of the field, conduction electrons accelerate (gain kinetic energy). Frequent collisions cause some or all of the kinetic energy of the conduction electrons to be transferred to the lattice, where it is converted to heat. Conduction electrons thus gain and lose kinetic energy as they drift from left to right. We are dealing with uniformly doped silicon in which a small net drift velocity is superimposed on the random thermal motion of charge carriers. Since the electron concentration is uniform throughout the device, the Fermi level remains parallel to the conduction band [Fig. 6-8(b)]. Electron energy is lost as collisions increase thermal vibration within the crystal lattice.

The relationship between the electric field and the drift current density in a semiconductor is given by

$$J = \sigma E \tag{6-26}$$

where σ denotes the conductivity of the semiconductor. The current density can also be expressed as

$$J = qnv_{dn} + qpv_{dp} \tag{6-27}$$

Fig. 6-8 Energy level diagrams for *n*-type silicon: (a) without an externally applied field; (b) with an externally applied field.

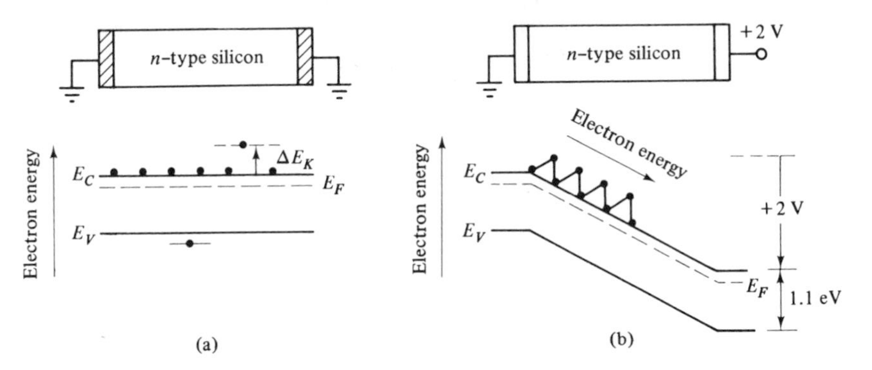

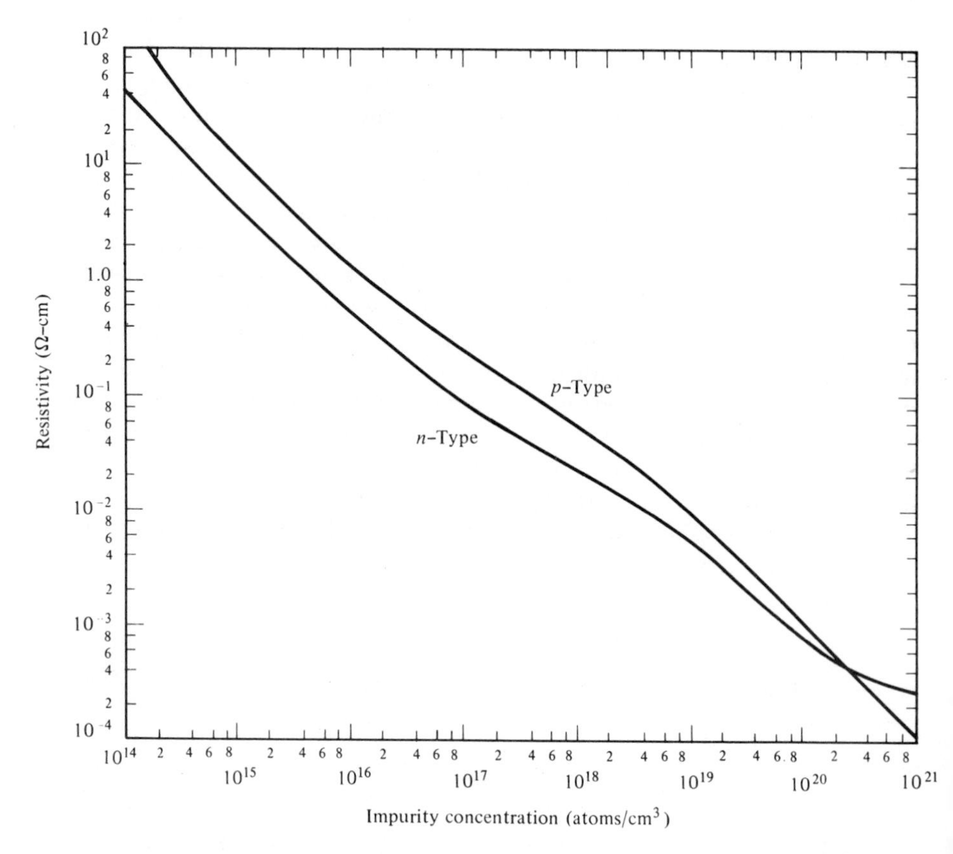

Fig. 6-9 Resistivity of silicon at room temperature vs. the donor or acceptor impurity concentration. (Reprinted with permission from *The Bell System Technical Journal.* Copyright 1962, The American Telephone and Telegraph Company [4].)

where v_{dn} and v_{dp} denote the drift velocities of electrons and holes as shown in Fig. 6-3. Regions exist in Fig. 6-3 for which v_d equals μE. Thus, Eq. (6-27) can be rewritten as

$$J = (qn\mu_n + qp\mu_p)E \tag{6-28}$$

Comparing Eqs. (6-26) and (6-28), conductivity can then be written

$$\sigma = qn\mu_n + qp\mu_p \tag{6-29}$$

Since resistivity is the reciprocal of conductivity,

$$\rho = \frac{1}{qn\mu_n + qp\mu_p} \tag{6-30}$$

The resistivity of silicon at room temperature as a function of the donor or acceptor impurity concentration is shown in Fig. 6-9. This graph is based on extensive resistivity measurements of silicon samples containing either donor or acceptor impurities, and was prepared by Irvin [4]. It is widely used in the semiconductor industry.

6-5 DIFFUSION CURRENT

Let us start with a uniformly doped p-type semiconductor in equilibrium $(pn = n_i^2)$. In a specific region of this semiconductor we disturb the equilibrium conditions by somehow creating a cloud of excess electrons. Through the mechanisms of drift, diffusion, generation, and recombination, the system will return to equilibrium. Let us concentrate on the diffusion process. Electrons will *diffuse* outward from the region of excess concentration because a concentration gradient exists and because the entire system is in thermal motion. The diffusion current density is proportional to the concentration gradient as described by Fick's first law of diffusion [see Eq. (4-1)]. For one-dimensional analysis, Fick's first law has the form

$$J_n = -qD_n \frac{\partial n}{\partial x} \tag{6-31}$$

where J_n denotes the diffusion current density of electrons, $\partial n/\partial x$ the concentration gradient in the x direction, and D_n represents the *diffusivity* of electrons. The minus sign indicates that the direction of the diffusion current is from a higher to lower concentration of excess electrons. Electron diffusivity D_n is related to electron mobility by the Einstein relation,

$$\frac{D_n}{\mu_n} = \frac{kT}{q} \tag{6-32}$$

As a mnemonic aid, this can be stated, "D is to μ as kT is to q."

If we begin again with a uniformly doped n-type semiconductor and introduce a cloud of excess holes, the hole diffusion current density will be given by

$$J_p = -qD_p \frac{\partial p}{\partial x} \tag{6-33}$$

and the diffusivity of holes will be given by

$$\frac{D_p}{\mu_p} = \frac{kT}{q} \tag{6-34}$$

The diffusion of charge carriers is an important process in the operation of semiconductor devices. It is a dominant process in the operation of diodes and bipolar transistors.

6-6 *p-n* JUNCTION IN EQUILIBRIUM

Consider two uniformly doped pieces of silicon, one n type and one p type. Each region is in equilibrium. For every trapped positive impurity ion in the n-type material, an electron exists in the conduction band. A similar condition exists in the p-type material. Suppose that these two pieces of silicon could be joined in such a way that a single crystal suddenly existed throughout both regions. The large concentration gradients in the vicinity of the junction (a sea of electrons on one side and a sea of holes on the other) would give rise to large hole and electron diffusion currents. In the vicinity of the junction, electrons would diffuse into the p-type material, *uncovering trapped positive ions.* Holes would diffuse into the n-type material, *uncovering trapped negative ions.* The trapped ions on either side of junction would cause an electric field to exist across the junction. The electric field (positive on the n side) would cause hole and electron drift currents. Electrons would drift to the n side of the junction and holes would drift to the p side. With no external voltages applied, the drift and diffusion components of electron current would become precisely equal in magnitude and opposite in direction, resulting in zero net current. The same condition would apply to hole currents.

Actual p-n junctions are formed by impurity diffusion or ion implantation, as discussed in Chapter 4, or by epitaxial growth, as discussed in Chapter 5. The resulting depletion region and the balance between drift and diffusion components of current are the same, however, as in the foregoing mental experiment.

The junction region in which the electric field exists is called the *depletion region* or *space-charge region.* It is depleted of charge carriers because they are swept out by the electric field. The drift and diffusion current densities and the depletion region are shown in Fig. 6-10. In this figure, x is intended to be zero at the metallurgical junction (i.e., where N_A equals N_D). The edges of the depletion region are then shown as $-x_p$ or $+x_n$ (the width of the depletion region will be discussed later in this chapter). The n and p regions distant from the depletion region are not altered by the existence of the junction. In the n-type material to the right of the junction, there is an electron in the conduction band for every trapped positive impurity ion.

Referring again to Fig. 6-10, if we suppose that a net current exists at a particular instant, the equations for drift and diffusion current densities may be written as follows:

$$J_p(\text{net}) = -qp\mu_p \frac{\partial v}{\partial x} - q\,D_p \frac{\partial p}{\partial x}$$
$$J_n(\text{net}) = +qn\mu_n \frac{\partial v}{\partial x} - q\,D_n \frac{\partial n}{\partial x} \tag{6-35}$$

The signs on the diffusion currents are both negative because diffusion occurs in the direction of decreasing concentration gradient. As shown in Fig. 6-10, the hole drift current has a negative sign because holes are moving down the potential gradient. Electron drift current is moving up the potential gradient, so the sign is positive.

The energy-level diagrams for the p and n regions are shown in Fig. 6-11. Figure 6-11(a) shows the p region alone, (b) shows the n region alone, and (c) shows the p-n junction in equilibrium.

The drift and diffusion currents discussed above may be said to cause the Fermi levels to line up as shown in Fig. 6-11(c). Saying that the system is in equilibrium, with no net current, is equivalent to saying that the Fermi level is constant throughout the system.

As shown in Fig. 6-11(c), the total change in the height of the potential hill across the p-n junction is seen to be $\Delta E + \Delta E'$. Since $E = qV$, the contact potential across the junction is given as

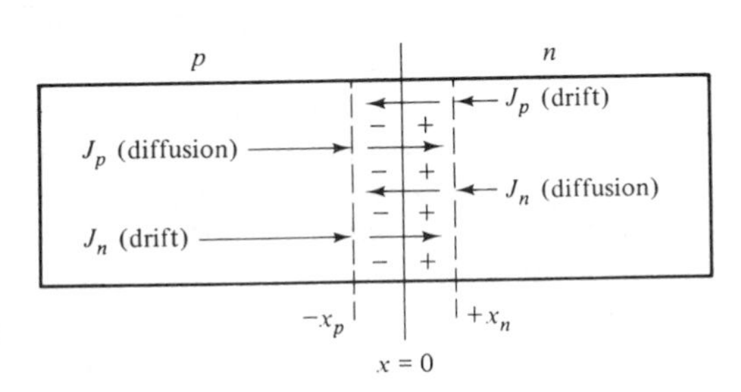

Fig. 6-10 Depletion region and junction currents.

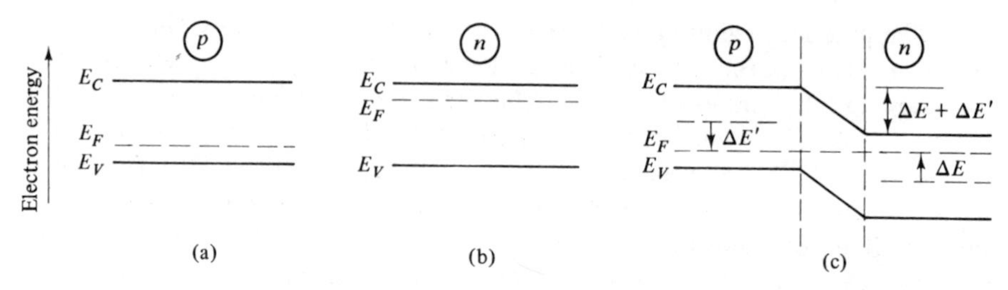

Fig. 6-11 Energy-level diagrams: (a) *p*-type Si; (b) *n*-type Si; (c) *p-n* junction.

$$V_B = \frac{\Delta E + \Delta E'}{q} \tag{6-36}$$

where ΔE and $\Delta E'$ may be obtained from the equations

$$p_{po} = n_i \exp\left(\frac{\Delta E'}{kT}\right) \tag{6-8}$$

and

$$n_{no} = n_i \exp\left(\frac{\Delta E}{kt}\right) \tag{6-5}$$

Solving Eqs. (6-5) and (6-8) for ΔE and $\Delta E'$, and substituting these values in Eq. (6-36), we find that the contact potential is

$$V_B = \frac{kT}{q} \ln \frac{n_{no} p_{po}}{n_i^2} \tag{6-37}$$

6-7 *p-n* JUNCTION WITH AN APPLIED VOLTAGE

With zero external voltage applied to a *p-n* junction diode, a contact potential V_B exists across the depletion region such that drift and diffusion currents are precisely balanced. When an external voltage is applied to the terminals of a junction diode, the equilibrium conditions are disturbed. An imbalance is created between the drift and diffusion components of current, causing charge carriers to move across the junction. If the external voltage is applied in a direction that lowers the junction potential, as shown in Fig. 6-12(a), the drift components of current are reduced and there is a net diffusion of holes and electrons across the junction. The diode is said to be *forward-biased.* If the external voltage is applied in a direction that raises the junction

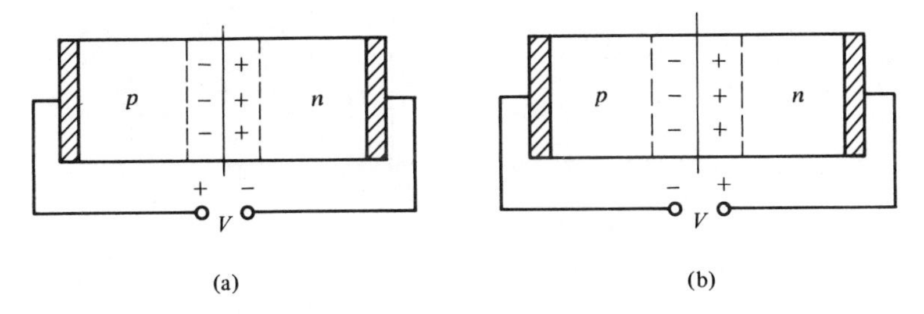

Fig. 6-12 *p-n* Junction diode with an externally applied voltage: (a) forward bias; (b) reverse bias.

potential [Fig. 6-12(b)], only small leakage currents exist, owing to the thermal generation of hole–electron pairs in the vicinity of the junction. The diode is said to be *reverse-biased.*

6-7-1 Forward Bias

In a forward-biased diode, holes diffuse across the depletion layer into the n region and electrons diffuse across the depletion layer into the p region. The directions of hole and electron motion can be thought of as an attempt to restore the junction potential to its equilibrium value. Of course, the externally connected battery that lowered the junction potential prevents this from happening. In the n and p regions in the vicinity of the depletion layer, there is an excess of injected minority carriers. As these minority carriers diffuse deeper into the "neutral" n and p regions, recombination takes place until, far from the junction, equilibrium is again restored. The minority carriers disappearing through recombination are replenished by more minority carriers diffusing across the junction, resulting in a constant forward current. Far from the junction, the hole and electron concentrations are denoted by p_{no} and n_{no} in the n region, and p_{po} and n_{po} in the p region. In the vicinity of the junction, the system is not in equilibrium. The Fermi level is *not* constant in this region, and the law of mass action does *not* apply. For now, we will refer to the nonequilibrium concentrations of holes and electrons in the n region as $p_n(x)$ and $n_n(x)$. In the vicinity of the junction, we will refer to the hole and electron concentrations in the p region as $p_p(x)$ and $n_p(x)$.

Energy-level diagrams for an unbiased diode and a forward-biased diode are shown in Fig. 6-13. With the application of a forward bias voltage V, the junction potential is reduced to $V_B - V$, as shown in Fig. 6-13(b). Since the energy gap must remain at 1.1 eV, and the Fermi levels in the neutral regions are determined by impurity concentrations, the Fermi level in Fig. 6-13(b) is not constant across the depletion layer. In the vicinity of the depletion layer, we assume that the system is in *quasi-equilibrium.* In other words, we assume that the pn product is a constant in this region, larger than n_i^2. The concept

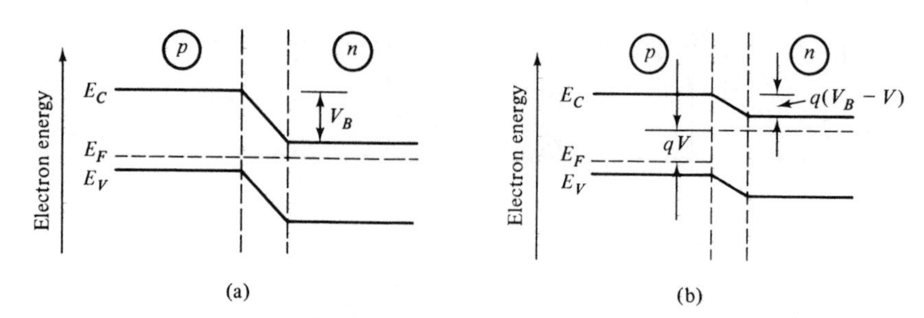

Fig. 6-13 Idealized *p-n* junction: (a) in equilibrium; (b) forward-biased with an applied voltage *V*.

of Fermi levels is replaced by that of *quasi-Fermi levels.* The quasi-Fermi level is then defined as *that quantity which, when substituted into the place of the Fermi level, gives the carrier concentration under nonequilibrium conditions* [5].

With forward bias, recombination of hole–electron pairs occurs in the depletion region as well as in the *p* and *n* regions. For the time being, however, we will *assume that the effects of recombination in the depletion region are negligibly small.* We will develop a conceptual diode model in which we consider only the effects of recombination in the *p* and *n* regions. After this model is developed, we will briefly discuss the effects of recombination in the depletion region.

With forward bias, holes are injected into an *n* region that contains a sea of electrons. Although this process has a considerable effect on the minority hole concentration in the *n* region, it has little effect on the electron concentration over a wide range of bias conditions. Thus, for conditions of *low-level injection,* we will *assume that majority carrier concentrations are relatively unaffected by minority carrier injection.* Then

$$p_p(x) \approx p_{po} \tag{6-38}$$

at the *p*-region depletion-layer interface, and

$$n_n(x) \approx n_{no} \tag{6-39}$$

at the *n*-region depletion-layer interface. This will necessarily limit the useful range of our model. After this model is developed, we will briefly discuss the effects of high-level injection.

With the foregoing assumptions in mind, carrier concentrations in the *p* and *n* regions of a junction diode are as shown in Fig. 6–14. Figure 6-14(a) shows the *p-n* junction diode in equilibrium, and (b) shows the junction diode under forward bias conditions. Since $p_n(x)$ is roughly two orders of magnitude smaller than n_{no} at the edge of the depletion layer in Fig. 6-14(b), it will have little effect on the majority electron concentration.

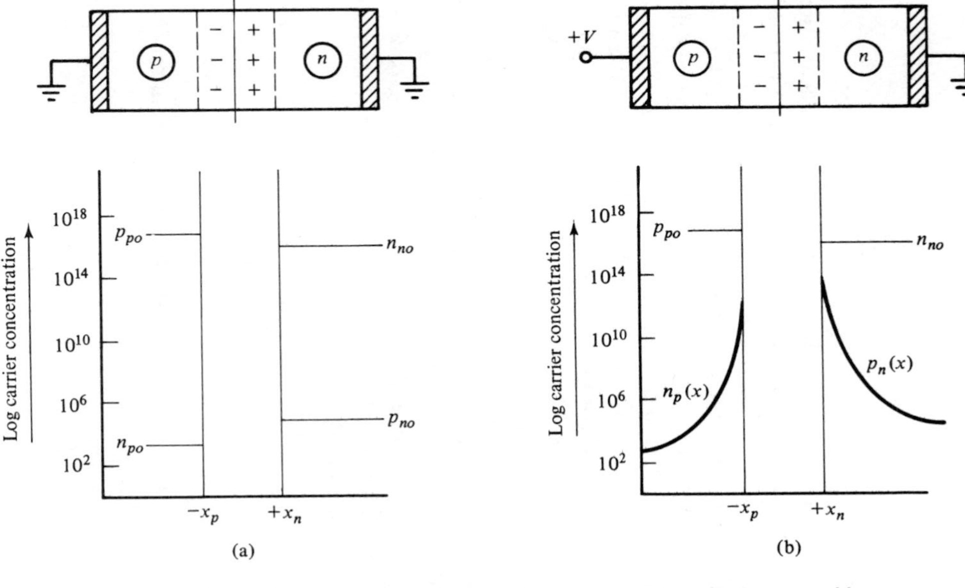

Fig. 6-14 (a) Carrier concentrations for a *p-n* junction diode in equilibrium. (b) Carrier concentrations for a *p-n* junction diode with forward bias.

It is useful to model *p-n* junction behavior in terms of the drift and diffusion conduction mechanisms. It would be physically impossible to create a situation in which one of these current mechanisms existed without the other across a *p-n* junction. But it is possible to *calculate* a drift or diffusion current density alone. Equilibrium electron diffusion current density might, for example, be approximated by

$$qD_n\frac{dn}{dx} \approx qD_n\frac{n_{no} - n_{po}}{x_p + x_n} \tag{6-40}$$

where $n_{no} - n_{po}$ represents the change in electron concentration across the depletion layer and $x_p + x_n$ represents the width of the depletion layer. This calculation, with $n_{no} = 10^{18}$ cm^{-3}, $n_{po} = 2.25 \times 10^6$ cm^{-3}, and $x_p + x_n = 10^{-4}$ cm, yields a current density of 56,000 A/cm^2. Physically, the situation is equivalent to suddenly removing the electric field from the depletion layer and then determining the electron current density diffusing across the junction (neglecting any resistive voltage drops). The calculation does serve a purpose, however. It shows that we cannot expect to forward-bias a *p-n* junction diode to the extent that the contact potential is canceled or reversed. All we can do is lower it. It also suggests that when a diode is forward-biased to the extent that milliamperes of current are present, only a slight imbalance has been created between the drift and diffusion components of current. We can make use of this fact to find the minority carrier concentrations at the edges of the depletion layer: that is, $p_n(x)$ at $x = x_n$ and $n_p(x)$ at $x = -x_p$ [see Fig. 6-14(b)]. Assuming that the drift and diffusion hole current densities are approximately equal with forward bias, we can state that

$$qp\mu_p\frac{dv}{dx} \approx -qD_p\frac{dp}{dx} \tag{6-41}$$

By rearranging terms and making use of the Einstein relation (Eq. 6-34), this can be written

$$\int_{V(-x_p)}^{V(+x_n)} dV = -\frac{kT}{q}\int_{p_{po}}^{p_n(x_n)} \frac{dp}{p} \tag{6-42}$$

Integrating both sides of this equation, we have that

$$V_B - V = -\frac{kT}{q}\ln\frac{p_n(x_n)}{p_{po}} \tag{6-43}$$

This can be written

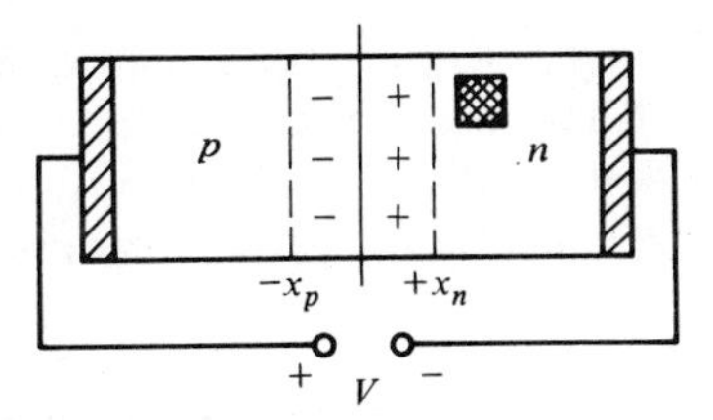

Fig. 6-15 Differential volume in the *n*-region of a *p-n* junction diode.

$$\frac{p_n(x_n)}{p_{po}} = \exp\left(\frac{-qV_B}{kT}\right)\exp\left(\frac{qV}{kT}\right) \tag{6-44}$$

where, from Eq. (6-37),

$$\exp\left(\frac{-qV_B}{kT}\right) = \frac{p_{no}}{p_{po}} \tag{6-45}$$

Substituting Eq. (6-45) in Eq. (6-44), we have that

$$p_n(x_n) = p_{no}\exp\left(\frac{qV}{kT}\right) \tag{6-46}$$

A similar treatment for electrons injected into the *p* region yields

$$n_p(-x_p) = n_{po}\exp\left(\frac{qV}{kT}\right) \tag{6-47}$$

Minority Carrier Diffusion. Consider a differential volume in the *n* region of a *p-n* junction, close to the depletion region. This is indicated in Fig. 6-15. The factors that might cause the minority hole concentration to change within this differential volume are as follows:

1. Generation and recombination
2. Diffusion due to a concentration gradient
3. Drift due to an applied electric field
4. A net motion of charge carriers due to an applied magnetic field
5. A net motion of charge carriers due to a temperature gradient

We will assume that no temperature gradients or magnetic fields are present. Under low-level injection conditions, the electric field in the *n* region is assumed to be negligibly small. Thus, hole generation, recombination, and diffusion remain to be considered.[2]

[2] It should be kept in mind that we are using a one-dimensional model with an abrupt junction and uniformly doped *p* and *n* regions.

Minority hole lifetime T_p has been defined as the *average lifetime of holes in an n-type material.* The rate of change of hole concentration in the differential volume due to generation and recombination is given by the excess or deficit hole concentration divided by the carrier lifetime:

$$\frac{\partial p_n(x)}{\partial t}\text{(due to generation and recombination)} = \frac{p_{no} - p_n(x)}{T_p} \qquad (6\text{-}48)$$

where, under forward bias, $p_n(x)$ is assumed to be greater than p_{no}. An expression for the change in hole concentration within this volume due to diffusion can be found by applying Fick's second law to the diffusion of minority carriers [see Eq. (4-6)]. Then

$$\frac{\partial p_n(x)}{\partial t}\text{(due to diffusion)} = D_p\frac{\partial^2 p_n(x)}{\partial x^2} \qquad (6\text{-}49)$$

Since the model contains a uniformly doped n region, D_p is a constant. Combining Eqs. (6-48) and (6-49), we have

$$\frac{\partial p_n(x)}{\partial t} = \frac{p_{no} - p_n(x)}{T_p} + D_p\frac{\partial^2 p_n(x)}{\partial x^2} \qquad (6\text{-}50)$$

Under steady-state conditions, with a dc bias voltage applied to the diode, the concentration of holes in the differential volume will not change with time. Thus,

$$\frac{\partial p_n(x)}{\partial t} = 0 \qquad (6\text{-}51)$$

and Eq. (6-50) can be written

$$\frac{d^2 p_n(x)}{dx^2} - \frac{p_n(x) - p_{no}}{D_p T_p} = 0 \qquad (6\text{-}52)$$

It can be shown that the *average diffusion length* of minority holes is denoted by L_p, where

$$L_p = \sqrt{D_p T_p} \qquad (6\text{-}53)$$

L_p represents the average distance a hole diffuses in an n region before disappearing through recombination. Substituting Eq. (6-53) in Eq. (6-52), we have

$$\frac{d^2 p_n(x)}{dx^2} - \frac{p_n(x) - p_{no}}{L_p^2} = 0 \qquad (6\text{-}54)$$

Equation (6-54) is linear second-order differential equation for which we can assume an exponential solution:

$$p_n(x) - p_{no} = A \exp\left(\frac{-x}{L_p}\right) + B \exp\left(\frac{+x}{L_p}\right) \tag{6-55}$$

where A and B are constants of integration that can be determined by applying boundary conditions.

Far from the junction, the equilibrium hole concentration will be restored. Thus,

$$\lim_{x\to\infty} p_n(x) = p_{no} \tag{6-56}$$

This boundary condition requires that $B = 0$. From Eq. (6-46), $p_n(x_n)$ is equal to $p_{no} \exp\,(qV/kT)$. This boundary condition requires that

$$A = p_{no}\left[\exp\left(\frac{qV}{kT}\right) - 1\right] \exp\left(\frac{x_n}{L_p}\right) \tag{6-57}$$

Thus,

$$p_n(x) - p_{no} = p_{no}\left[\exp\left(\frac{qV}{kT}\right) - 1\right] \exp\left[\frac{(x_n - x)}{L_p}\right] \tag{6-58}$$

for $(x_n - x) \leq 0$. A similar treatment of minority electrons in the p region yields the equation

$$n_p(x) - n_{po} = n_{po}\left[\exp\left(\frac{qV}{kT}\right) - 1\right] \exp\left[\frac{(x_p + x)}{L_n}\right] \tag{6-59}$$

where L_n is the average diffusion length of minority electrons. Equation (6-59) is valid for $x_p + x \leq 0$. Equations (6-58) and (6-59) describe the exponential decrease in minority carrier concentrations, as these carriers diffuse away from the depletion-layer boundaries as shown in Fig. 6-14(b).

Under conditions of low-level injection, the excess minority hole current density in the n region is caused by diffusion. Thus, this hole current density is given by

$$J_p(x) = -q D_p \frac{dp_n(x)}{dx} \tag{6-60}$$

or

$$J_p(x) = \frac{qD_p p_{no}}{L_p}\left[\exp\left(\frac{qV}{kT}\right) - 1\right]\exp\left[\frac{(x_n - x)}{L_p}\right] \tag{6-61}$$

Electron current density in the p region is given by

$$J_n(x) = -qD_n \frac{dn_p(x)}{dx} \tag{6-62}$$

or

$$J_n(x) = \frac{qD_n n_{po}}{L_n}\left[\exp\left(\frac{qV}{kT}\right) - 1\right]\exp\left[\frac{(x_p + x)}{L_n}\right] \tag{6-63}$$

The sign of $J_n(x)$ is positive in Eq. (6-63) because the electron charge is negative.

The variation in minority carrier current densities with distance is shown in Fig. 6-16(a) for an idealized p-n junction diode. It is assumed here that no recombination takes place within the depletion layer. Also, the depletion layer is relatively much thinner than shown in this figure.

As holes diffuse into the n region, they recombine with electrons. This gives rise to an electron recombination current. In effect, majority electrons in the n region move toward the junction. Within a few diffusion lengths of the junction, they recombine with the minority holes diffusing into the n region. The total hole and electron current densities must sum to a constant value at every point within the device. Thus, the electron current density in the n region is just the difference between the total (constant) current density and the hole diffusion current density. Hole and electron current densities are as shown in Fig. 6-16(b).

The total diode current density can be found by summing the minority current densities at the edges of the depletion layer, where the recombination process has not yet begun. Thus, from Eqs. (6-61) and (6-63),

$$J_p(x_n) = \frac{qD_p p_{no}}{L_p}\left[\exp\left(\frac{qV}{kT}\right) - 1\right] \tag{6-64}$$

and

$$J_n(-x_p) = \frac{qD_n n_{po}}{L_n}\left[\exp\left(\frac{qV}{kT}\right) - 1\right] \tag{6-65}$$

The total current density is then given by

$$J_{total} = \left(\frac{qD_p p_{no}}{L_p} + \frac{qD_n n_{po}}{L_n}\right)\left[\exp\left(\frac{qV}{kT}\right) - 1\right] \tag{6-66}$$

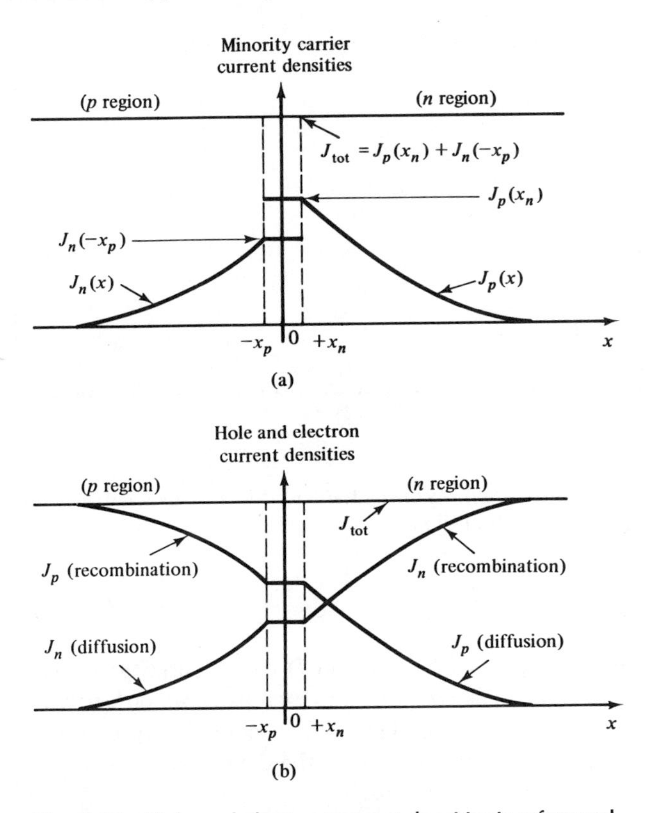

Fig. 6-16 Hole and electron current densities in a forward-biased *p-n* junction diode: (a) minority carrier current densities; (b) hole and electron current densities.

The total diode current is then determined by including the cross-sectional area of the *p-n* junction. Thus,

$$I = qA\left(\frac{D_p p_{n0}}{L_p} + \frac{D_n n_{p0}}{L_n}\right)\left[\exp\left(\frac{qV}{kT}\right) - 1\right] \tag{6-67}$$

Equation (6-67) represents the model that is conventionally used to characterize diode behavior. A plot of current vs. voltage for this model with an applied forward bias is shown in Fig. 6-17. The most important characteristic of this device is that it only permits current when sufficiently forward-biased. It is almost all-pervasive as a building block for electronic structures. Understanding diode behavior is a fundamental prerequisite to understanding the design, fabrication, or operation of most of the electronic structures in use today.

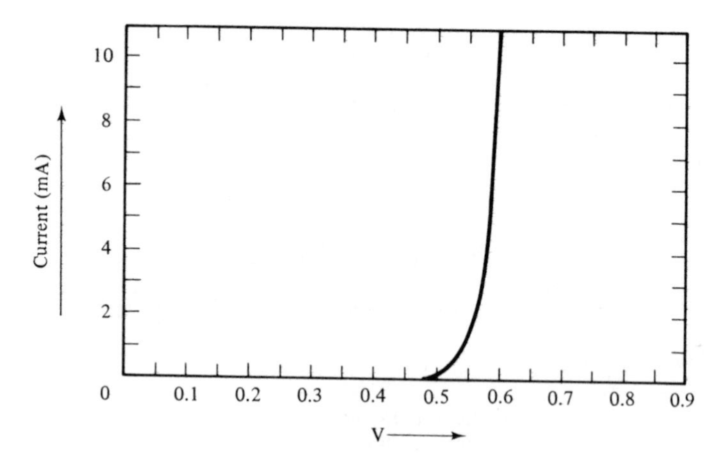

Fig. 6-17 Current–voltage characteristic for an ideal forward-biased *p-n* junction diode.

6-7-2 Reverse Bias

The model represented by Eq. (6-67) can be used to characterize diode behavior under reverse bias conditions as well as those of forward bias. Again, there are restrictions on the usefulness of the model because of the assumptions made in its development. In this section the development of Eq. (6-67) will be discussed for reverse bias conditions.

When an external voltage source is connected to a junction diode so as to raise the junction potential (reverse bias), equilibrium conditions are disturbed. The depletion layer is widened[3] and the electric field intensity in the depletion layer increases. Because of the nonequilibrium conditions imposed, available charge carriers will move in a direction that tends to return the system to equilibrium (tends to lower the junction potential). The externally connected voltage source that raised the junction potential in the first place will, of course, prevent it from being lowered, but the movement of charge carriers will occur nonetheless. Finding an expression for the diode current under conditions of reverse bias (i.e., the *reverse current*) then becomes a matter of determining what charge carriers are available. Any attempt to lower the junction potential must involve hole motion, from the n to the p side of the depletion layer, and electron motion from the p to the n side of the depletion layer, as shown in Fig. 6-18. Because the junction is reverse-biased, the only source of these mobile charge carriers is thermal generation, in the depletion layer and within a diffusion length of the depletion-layer edge. To be consistent with the forward-biased diode model developed in Section 6–7–1, we will *assume that the effects of carrier generation within the depletion layer are negligibly small.* For the time

[3] Variations in depletion layer width with applied voltage are discussed in Section 6–8.

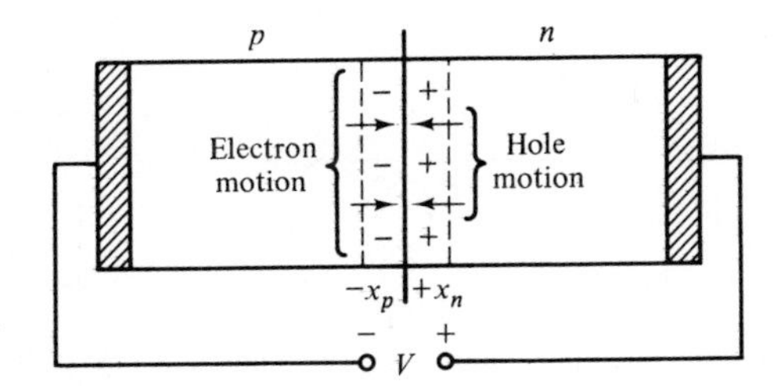

Fig. 6-18 Charge carrier motion in a reverse-biased diode.

being, then, we will consider only the effects of minority carrier generation in the p and n regions, within a diffusion length of the depletion-layer edges. After this model is developed, we will briefly discuss the effects of carrier generation within the depletion layer.

Energy-level diagrams for an unbiased diode and a reverse-biased diode are shown in Fig. 6-19. With the application of a reverse bias voltage V, the junction potential is increased to $V_B + V$ and the depletion layer is widened as shown in Fig. 6-19(b). The Fermi level is shifted across the depletion layer by an amount qV, where V denotes the applied reverse voltage. Again, in the vicinity of the depletion layer, the concept of Fermi levels is replaced by that of quasi-Fermi levels.

Carrier concentrations in the p and n regions of a junction diode are shown in Fig. 6-20. Figure 6-20(a) shows equilibrium concentrations and (b) shows carrier concentrations under reverse bias conditions. Since the electric field is assumed to exist only in the depletion layer, minority holes in the n region and electrons in the p region (those within a diffusion length of the depletion layer) *diffuse* to the edges of the depletion layer and are swept across by the electric field.

Referring to Fig. 6-20(b), one might suspect that positively charged holes in the n region, diffusing toward trapped positive ions in the depletion layer, would be repelled. This is not generally the case. The trapped ions are so widely

Fig. 6-19 Energy levels for an idealized *p-n* junction diode: (a) in equilibrium; (b) reverse-biased with an applied voltage V.

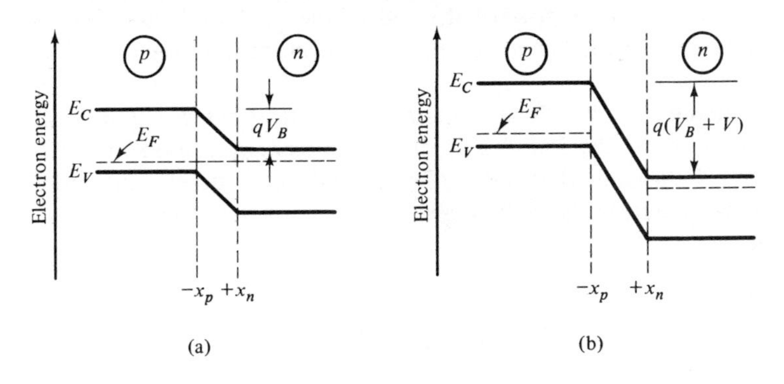

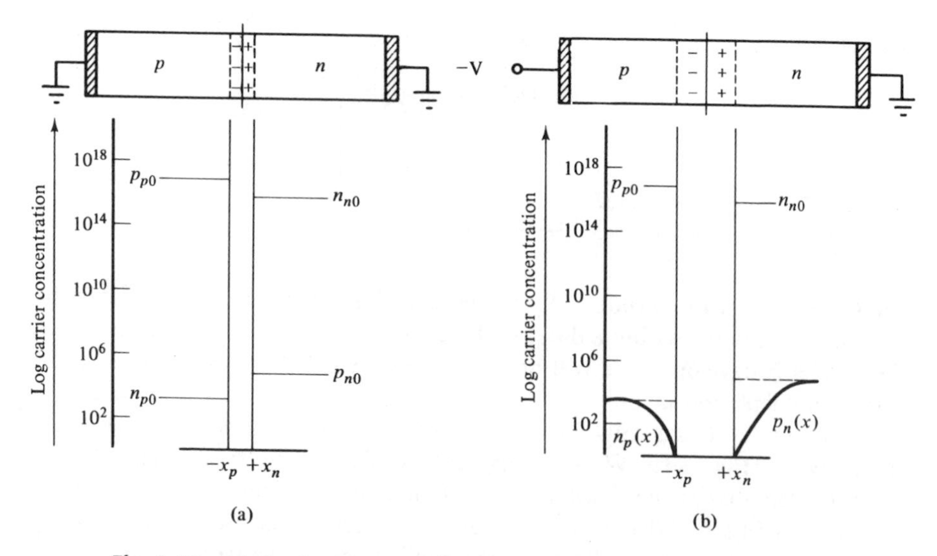

Fig. 6-20 (a) Carrier concentrations for a *p-n* junction diode in equilibrium. (b) Carrier concentrations for a reverse-biased *p-n* junction diode.

spaced that this effect is not significant. The bulk of the minority holes that reach the depletion layer edge are swept across.

Near the depletion region, minority carrier concentrations $p_n(x)$ and $n_p(x)$ are depressed below their equilibrium concentrations (denoted by p_{no} and n_{po}), as shown in Fig. 6-20(b). The motion of minority holes and electrons, generated within diffusion range of the depletion-layer edges, constitutes the reverse current that exists within this conceptual framework (or model). The reverse current is many orders of magnitude smaller than the forward current discussed in Section 6–7–1.

Holes generated in the *n* region, near the depletion layer, are swept across the depletion layer and into the *p* region. Other holes are generated within the *p* region, near the depletion-layer boundary. These majority holes diffuse away from the junction, toward the end of the semiconductor. Far from the depletion layer, practically all of the reverse current is seen as the flow of majority carriers.

The manner in which the minority concentrations $p_n(x)$ and $n_p(x)$ vary with distance x can be found with the same technique used in the forward-biased case. We assume that the equilibrium balance between drift and diffusion components of current is only slightly affected by the application of reverse bias. Then

$$qp\mu_p \frac{dV}{dx} \approx -qD_p \frac{dp}{dx} \tag{6-68}$$

Rearranging terms and integrating, we have that

$$V_B + V = -\frac{kT}{q} \ln \frac{p_n(x_n)}{p_{po}} \tag{6-69}$$

Solving Eq. (6-69) for $p_n(x_n)$, we find that

$$p_n(x_n) = p_{no} \exp\left(-\frac{qV}{kT}\right) \tag{6-70}$$

A similar treatment for electrons in the *p* region yields

$$n_p(-x_p) = n_{po} \exp\left(-\frac{qV}{kT}\right) \tag{6-71}$$

The development of expressions for minority carrier concentrations due to generation, recombination, and diffusion [Eqs. (6-48) through (6-59)] is valid whether the diode is forward- or reverse-biased. In the reverse-biased diode, however, $p_n(x) < p_{no}$ and $n_p(x) < n_{po}$. For reverse bias, Eqs. (6-58) and (6-59) can be written

$$p_n(x) - p_{no} = p_{no}\left[\exp\left(-\frac{qV}{kT}\right) - 1\right] \exp\left[\frac{(x_n - x)}{L_p}\right] \tag{6-72}$$

for $(x_n - x) < 0$ and

$$n_p(x) - n_{po} = n_{po}\left[\exp\left(-\frac{qV}{kT}\right) - 1\right] \exp\left[\frac{(x_p + x)}{L_n}\right] \tag{6-73}$$

for $(x_p + x) < 0$. Since $\exp\left(-\frac{qV}{kt}\right) << 1$ for $|V| > 0.026$ V., Eqs. (6-72) and (6-73) can be written

$$p_n(x) \approx p_{no}\left\{1 - \exp\left[\frac{(x_n - x)}{L_p}\right]\right\} \tag{6-74}$$

and

$$n_p(x) \approx n_{po}\left\{1 - \exp\left[\frac{(x_p + x)}{L_n}\right]\right\} \tag{6-75}$$

Equations (6-74) and (6-75) describe the exponential decrease in minority carrier concentrations as these carriers diffuse *toward* the depletion layer boundaries as shown in Fig. 6-20(b). As long as the diode is sufficiently reverse-biased, the minority carrier concentrations $p_n(x)$ and $n_p(x)$ in Eqs. (6-74) and (6-75) are independent of the magnitude of the reverse bias voltage. This is a direct consequence of the assumption that the electric field exists only in the depletion layer.

Minority carrier concentrations and diffusion currents are related by the expressions

$$J_p(x) = -qD_p \frac{dp_n(x)}{dx} \tag{6-76}$$

and

$$J_n(x) = -qD_n \frac{dn_p(x)}{dx} \tag{6-77}$$

Applying these expressions to Eqs. (6-72) and (6-73), we have

$$J_p(x) = \frac{qD_p p_{n0}}{L_p}\left[\exp\left(-\frac{qV}{kT}\right) - 1\right] \exp\left[\frac{(x_n - x)}{L_p}\right] \tag{6-78}$$

and

$$J_n(x) = \frac{qD_n n_{p0}}{L_n}\left[\exp\left(-\frac{qV}{kT}\right) - 1\right]\left[\exp \frac{(x_p + x)}{L_n}\right] \tag{6-79}$$

For values of reverse bias greater than 0.260 V, exp $(-qV/kT) \ll 1$. Then hole and electron current densities can be approximated by

$$J_p(x) \approx \frac{-qD_p p_{n0}}{L_p} \exp\left[\frac{(x_n - x)}{L_p}\right] \tag{6-80}$$

and

$$J_n(x) \approx \frac{-qD_n n_{p0}}{L_n} \exp\left[\frac{(x_p + x)}{L_n}\right] \tag{6-81}$$

These equations are plotted in Fig. 6-21(a).

Minority holes in the n region (within diffusion range of the depletion layer) diffuse to the depletion-layer boundary x_n and are swept across into the p region. Holes entering the p region combine with holes that are generated in the p region (near the depletion-layer edge). The direction of the resulting majority hole current is away from the depletion layer and toward the end of

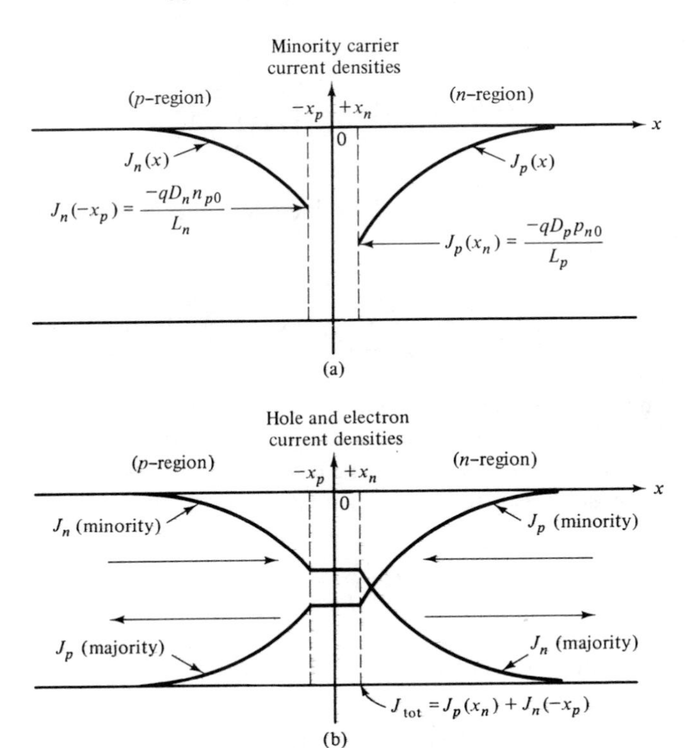

Fig. 6-21 Hole and electron current densities in a reverse-biased *p-n* junction diode: (a) minority carrier current densities; (b) hole and electron current densities.

the semiconductor. With an applied dc voltage, majority hole current and minority electron current sum to a constant value at every point in the *p* region, as shown in Fig. 6-21(b). A similar condition exists for majority electron current and minority hole current in the *n* region, as shown in the same figure. The total diode current density can then be found by summing the minority current densities at the edges of the depletion layer, where majority carrier generation is not a factor. Thus, from Eqs. (6-78) and (6-79),

$$J_p(x_n) = \frac{qD_p p_{n0}}{L_p}\left[\exp\left(-\frac{qV}{kT}\right) - 1\right] \tag{6-82}$$

and

$$J_n(-x_p) = \frac{qD_n n_{p0}}{L_n}\left[\exp\left(-\frac{qV}{kT}\right) - 1\right] \tag{6-83}$$

The total current density is given by

$$J_{\text{total}} = \left(\frac{qD_p p_{no}}{L_p} + \frac{qD_n n_{po}}{L_n} \right) \left[\exp\left(-\frac{qV}{kT} \right) - 1 \right] \tag{6-84}$$

and the total current is determined by including the cross-sectional area of the *p-n* junction. Thus,

$$I = qA \left(\frac{D_p p_{no}}{L_p} + \frac{D_n n_{po}}{L_n} \right) \left[\exp\left(-\frac{qV}{kT} \right) - 1 \right] \tag{6-85}$$

The only difference between Eqs. (6-67) and (6-85) is the negative sign on the applied voltage for the reverse bias condition. With this established, Eq. (6-67) can be used for both forward and reverse bias conditions. Since exp $(-qV/kT) << 1$ in most cases, the reverse current I is given by

$$I = -qA \left(\frac{D_p p_{no}}{L_p} + \frac{D_n n_{po}}{L_n} \right) \tag{6-86}$$

and is independent of the applied reverse bias voltage.

6-8 DEPLETION-LAYER WIDTH AND CAPACITANCE FOR REVERSE-BIASED JUNCTIONS

Figure 6-22(a) shows a sketch of an ideal step junction with no external voltage applied. Figure 6-22(b), (c), and (d) show the impurity density, the charge density, and the junction voltage, respectively.

Although most real junctions are much more heavily doped on one side than the other (one-sided step junctions), we will start with a simple situation in which the p and n dopings are comparable. Junctions are usually approximated to be abrupt or graded. An example of an abrupt junction would be a uniformly doped epi-layer on a uniformly doped substrate (with etch-back and diffusion effects assumed negligible). A p-type base region diffused into a uniformly doped n-type epi-layer would provide a graded junction. In the region of the junction it could be modeled as a linearly graded junction. Initially, we will consider abrupt junctions, as shown in Fig. 6-22.

Poisson's equation can be used to find the relationship between the width of the depletion layer and the contact potential. We can then introduce an external voltage and show how the depletion layer varies with applied voltage. Poisson's equation is

$$\nabla^2 \mathbf{V} = -\frac{\rho}{\epsilon} \tag{6-87}$$

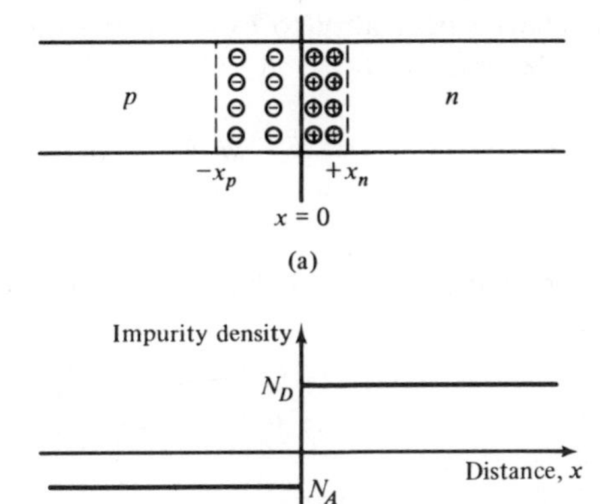

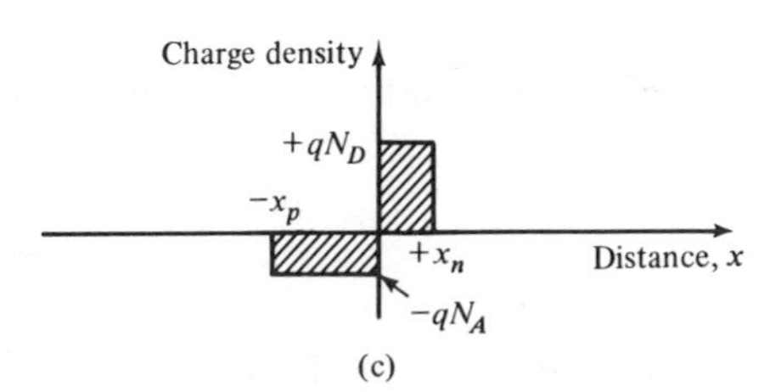

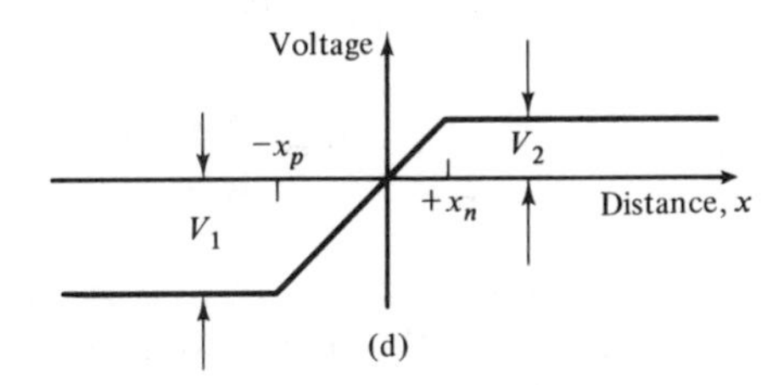

Fig. 6-22 Depletion layer of an idealized step junction (with $|N_D| = 2|N_A|$).

or in Cartesian coordinates,

$$\frac{\partial^2 \mathbf{V}}{\partial x^2} + \frac{\partial^2 \mathbf{V}}{\partial y^2} + \frac{\partial^2 \mathbf{V}}{\partial z^2} = -\frac{\rho}{\epsilon} \tag{6-88}$$

where ρ = volume charge density

$\epsilon = \epsilon_R \epsilon_0$ denotes the permittivity of the medium

Referring to Fig. 6-22(c), we have

$$N_D x_n = N_A x_p \tag{6-89}$$

for overall charge neutrality to exist. In the one-dimensional case of interest here, Eq. (6-88) reduces to

$$\frac{d^2V}{dx^2} = -\frac{\rho}{\epsilon} \tag{6-90}$$

For the space-charge region of the *p-type material,* $\rho = -qN_A$, so that

$$\frac{d^2V}{dx^2} = -\frac{\rho}{\epsilon} = \frac{qN_A}{\epsilon} \tag{6-91}$$

Integration of Eq. (6-91) gives

$$\frac{dV}{dx} = \frac{qN_A}{\epsilon}x + c \tag{6-92}$$

where c is the constant of integration. Integrating again, we have

$$V = \frac{qN_A}{2\epsilon}x^2 + cx + d \tag{6-93}$$

Equation (6-93) applies only in the space-charge region of the p-type material. Now we can apply boundary conditions to deal with the integration constants (i.e., at $x = 0$, $V = 0$), so that $d = 0$ or

$$V = \frac{qN_A}{2\epsilon}x^2 + cx \tag{6-94}$$

At $x = -x_p$, $dV/dx = 0$, since the potential is uniform in bulk p-type material. Applying this condition to Eq. (6-92), we have

$$0 = -\frac{qN_A}{\epsilon}x_p + c \qquad \text{or} \qquad c = \frac{qN_A}{\epsilon}x_p \tag{6-95}$$

Then

$$V = \frac{qN_A}{2\epsilon}x^2 + \frac{qN_A}{\epsilon}x_p x \tag{6-96}$$

If $V = V_1$ where $x = -x_p$, then

$$V_1 = \frac{qN_A}{2\epsilon}x_p^2 - \frac{qN_A}{\epsilon}x_p^2 = -\frac{qN_A}{2\epsilon}x_p^2 \tag{6-97}$$

Poisson's equation for the n-type depletion region is

$$\frac{d^2V}{dx^2} = -\frac{qN_D}{\epsilon} \tag{6-98}$$

The same techniques yield

$$V_2 = \frac{qN_D}{2\epsilon} x_n^2 \qquad \text{where } V = V_2 \text{ at } x = x_n \tag{6-99}$$

With no external voltage the contact potential can be expressed as

$$V_B = |V_1| + |V_2| = \frac{q}{2\epsilon}(N_A x_p^2 + N_D x_n^2) \tag{6-100}$$

From charge neutrality [Eq. (6-89)], $N_A x_p = N_D x_n$ or

$$x_n = \frac{N_A}{N_D} x_p \qquad \text{and} \qquad x_n^2 = \frac{N_A^2}{N_D^2} x_p^2 \tag{6-101}$$

Substituting Eq. (6-101) in (6-100) yields

$$V_B = \frac{qN_A x_p^2}{2\epsilon}\left(1 + \frac{N_A}{N_D}\right) \tag{6-102}$$

Solving this for x_p gives

$$x_p = \left[\frac{2\epsilon V_B}{qN_A(1 + N_A/N_D)}\right]^{\frac{1}{2}} \tag{6-103}$$

A similar treatment provides

$$x_n = \left[\frac{2\epsilon V_B}{qN_D(1 + N_D/N_A)}\right]^{\frac{1}{2}} \tag{6-104}$$

The total depletion-layer width is then

$$x_T = |x_p| + |x_n| \tag{6-105}$$

The presence of an applied voltage across the junction adds to or subtracts from the barrier voltage V_B and therefore also changes the depletion-layer width. The barrier voltage in the general case is given by V_T, where

$$V_T = V_B \mp V \tag{6-106}$$

If V_T is substituted for V_B in the equations for x_p and x_n, we obtain

$$x_p = \left[\frac{2\epsilon V_T}{qN_A(1 + N_A/N_D)}\right]^{\frac{1}{2}} \quad \text{and} \quad x_n = \left[\frac{2\epsilon V_T}{qN_D(1 + N_D/N_A)}\right]^{\frac{1}{2}} \tag{6-107}$$

Thus, with an abrupt junction, the depletion width is proportional to $\sqrt{V_T}$.

As discussed previously, most abrupt junctions are fabricated with one region much more heavily doped than the other. They are denoted as n^+-p or p^+-n junctions, depending on which region is more heavily doped, and are referred to as *one-sided step junctions.* For a junction of this type, x_p is orders of magnitude larger than x_n, or vice versa. Referring to Eq. (6-107), if $N_A >> N_D$, then $x_n >> x_p$, and the depletion-layer width *(W)* can be approximated as

$$W = x_n \approx \left(\frac{2\epsilon V_T}{qN_D}\right)^{\frac{1}{2}} \tag{6-108}$$

In general, the depletion-layer width of a one-sided step junction can be approximated by

$$W \approx \left(\frac{2\epsilon V_T}{qN_B}\right)^{\frac{1}{2}} \tag{6-109}$$

where N_B denotes the impurity concentration of the more lightly doped or background region.

The incremental junction capacitance of the reverse-biased p-n junction is given by

$$\frac{C_T}{A} = \frac{dQ}{dV} \tag{6-110}$$

where

$$Q = qN_A x_p = qN_D x_n \tag{6-111}$$

Then

$$Q = qN_A x_p = qN_A \left[\frac{2\epsilon V_T}{qN_A(1 + N_A/N_D)}\right]^{\frac{1}{2}} = \left[\frac{2\epsilon qN_A V_T}{1 + N_A/N_D}\right]^{\frac{1}{2}} \tag{6-112}$$

or

$$Q=\left(\frac{2q\epsilon N_A N_D V_T}{N_A+N_D}\right)^{\frac{1}{2}} \tag{6-113}$$

Since

$$\begin{aligned} V_T &= V_B + V \\ dV_T &= dV \end{aligned} \tag{6-114}$$

then

$$\frac{C_T}{A}=\frac{dQ}{dV}=\left[\frac{q\epsilon N_A N_D}{2(N_A+N_D)}\right]^{\frac{1}{2}} V_T^{-\frac{1}{2}} \tag{6-115}$$

In summary, the depletion width increases as the reverse voltage increases. The depletion-layer capacitance, or transition capacitance *(C_T)*, decreases with increase in reverse voltage. The application of a sinusoidal variation to the reverse voltage would cause a continuous variation in the thickness of the depletion layer.

Many diodes are fabricated by diffusing *p*-type impurities into uniformly doped *n* regions. The metallurgical junction occurs where the *p*- and *n*-type impurity concentrations are equal, as shown in Fig. 6-23(a). A small reverse bias voltage will result in a depletion layer width W_1, as shown in Fig. 6-23(a) and (b). Under these circumstances, we can use a straight-line approximation for the net impurity concentration [the dashed line in Fig. 6-23(a)]. In this case, the junction is referred to as a *graded* junction. Then,

$$|N_A - N_D| = ax \tag{6-116}$$

where a is the impurity concentration gradient *(dN/dx)* near the metallurgical junction. Using Poisson's equation, the width of the depletion layer *(W_1)* and the depletion-layer capacitance *(C/A)* can then be derived for the case where a is constant. The resulting equations are

$$W=\left(\frac{12\epsilon V_T}{qa}\right)^{\frac{1}{3}} \tag{6-117}$$

and

$$\frac{C}{A}=\left(\frac{qa\epsilon^2}{12V_T}\right)^{\frac{1}{3}} \tag{6-118}$$

If the reverse bias voltage is now increased to V_2, the depletion-layer width will increase to some value W_2, as shown in Fig. 6-23. The difference in impurity

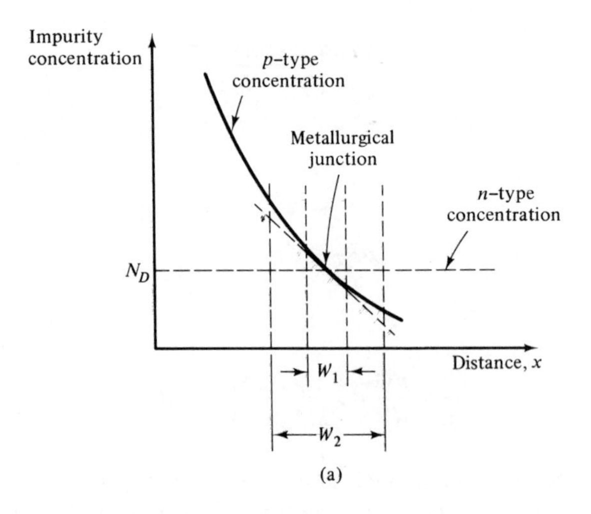

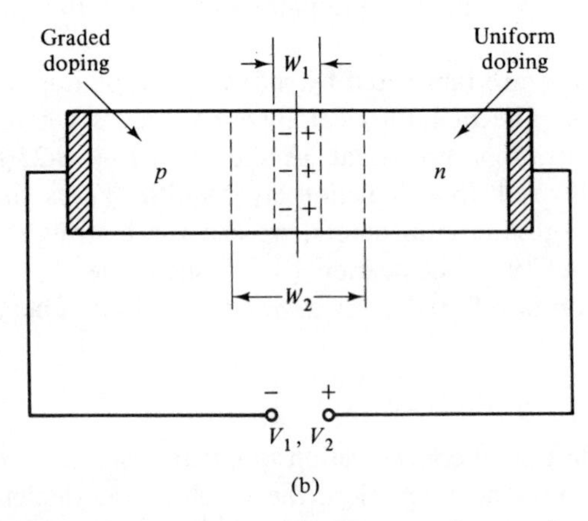

Fig. 6-23 (a) Impurity concentration vs. distance for a graded *p-n* junction. (b) Depletion width for a reverse biased *p-n* junction.

concentration on either side of the depletion layer becomes much greater, and the straight-line approximation used in the derivation of Eqs. (6-117) and (6-118) becomes considerably less accurate. Dependent on the magnitude of the reverse voltage (V_2), the abrupt junction model of Eqs. (6-107) and (6-115) may, in this case, provide more accurate estimates of depletion width and capacitance than the graded junction model. Data provided in reference [6] can be used to determine the relative accuracy of these two models for a useful range of impurity concentrations and reverse voltages.

6-9 VOLTAGE BREAKDOWN IN *p-n* JUNCTIONS

The magnitude of the electric field across the depletion layer of a reverse-biased diode is determined by the applied reverse voltage and the impurity concentrations in the p and n regions. If this electric field is sufficiently high, an appreciable reverse current will exist across the junction. The two mechanisms by which this voltage breakdown can take place are called *zener breakdown* and *avalanche breakdown.* Since these breakdown phenomena exceed the realm of our mathematical diode model, they must be treated separately.

Zener breakdown can occur when both the p and n regions of a junction are very heavily doped and a reverse voltage is applied. Figure 6-24 shows energy-level diagrams for a heavily doped p-n junction in equilibrium, and with an applied reverse bias voltage. As shown in Fig. 6-24(b), applying reverse bias to a p-n junction has the effect of aligning a large concentration of filled energy states in the p-type valence band just opposite a concentration of empty energy states in the n-type conduction band. If the energy barrier separating these full and empty energy states is sufficiently narrow, electrons can *tunnel* from the p-type valence band to the n-type conduction band. This quantum mechanical tunneling process constitutes a reverse current across the p-n junction.

As discussed in the preceding section, increasing acceptor and donor impurity concentrations decreases the width of the depletion layer, whereas increasing the applied reverse bias voltage increases it. With very heavy doping levels, and reverse bias voltages of a few volts (or less), the energy gap is narrow enough to permit tunneling. Electron tunneling causes a potentially large reverse current. The overall process is called the zener effect or zener breakdown.

Voltage breakdown with a few volts of applied reverse voltage is not usually a desired goal in junction devices. Either the p or the n region of most devices is lightly doped, so that breakdown voltages exceed the useful range of device operation. Under these circumstances, the mechanism by which voltage breakdown usually occurs is called avalanche breakdown.

Fig. 6-24 Zener breakdown due to electron tunneling: (a) heavily doped junction in equilibrium; (b) heavily doped junction under reverse bias.

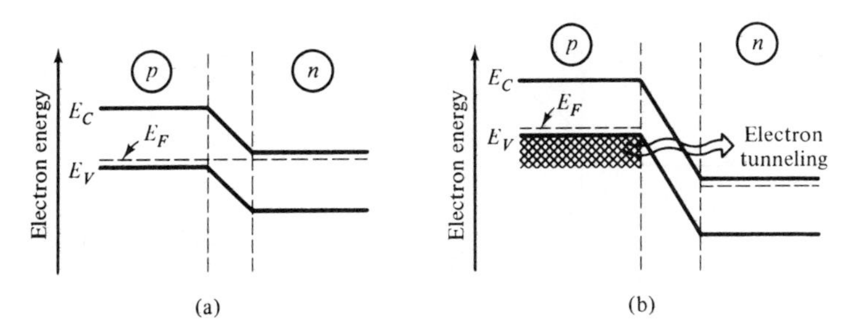

Avalanche breakdown is caused by collisions between charge carriers and valence electrons in reverse-biased depletion layers. As reverse bias voltage is increased, electrons and/or holes achieve sufficient kinetic energies to generate hole–electron pairs when they collide with valence electrons. The new holes and electrons are accelerated by the electric field, achieving sufficient energy to generate more hole–electron pairs through collisions and so on. The avalanche multiplication and breakdown processes are most probable in the lightly doped region of the depletion layer. In a one-sided step junction, for example, this region is orders of magnitude wider than the heavily doped region. There is a larger distance over which charge carrier acceleration and collision can take place. In the avalanche region, a small change in reverse voltage can now cause a very large change in reverse current, as shown in Fig. 6-25. The process is not necessarily destructive. The magnitude of the reverse current can be limited by an external resistor such that the current–voltage product does not exceed the rated power dissipation of the device. Diodes that are designed to work in the avalanche breakdown region are called zener diodes for historical reasons, and are widely used as voltage regulators. Impurity concentrations determine the breakdown voltage. Device area and heat sinking determine the rated power dissipation.

It is useful to have a way of determining the relationship between impurity concentrations and avalanche breakdown voltages. This relationship can be used as a parameter or boundary condition in the design of solid-state devices. Mathematical models have been developed that relate concentrations and breakdown voltages [7,8], and reasonably good correlations exist between these models and experimental data. Figure 6-26 shows a graph of avalanche breakdown voltage vs. impurity concentration (of the lightly doped region) for one-sided step junctions. This graph is based on direct experimental observations of n^+-p and p^+-n junctions by Miller [9]. It is a good approximation to the data points observed by Miller and indicates that a straight-line relationship exists between log V_{BD} and log N_B for values of V_{BD} between 300 and 10 V. The straight-line region of Fig. 6-26 can be modeled by the equation

$$V_{BD} = (2.3 \times 10^{12})(N_B)^{-0.66} \tag{6-119}$$

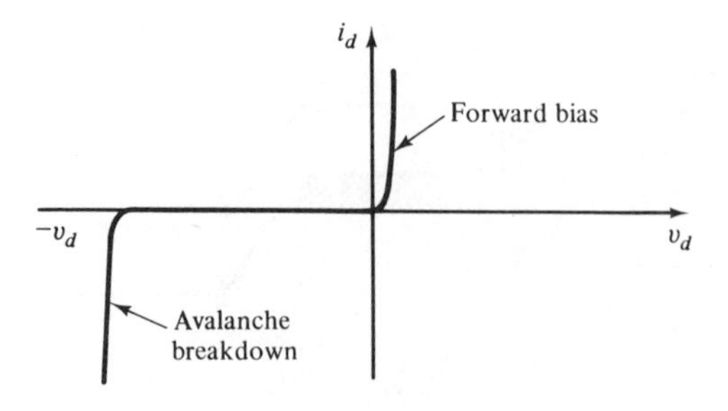

Fig. 6-25 Diode characteristic curve showing the region of avalanche breakdown.

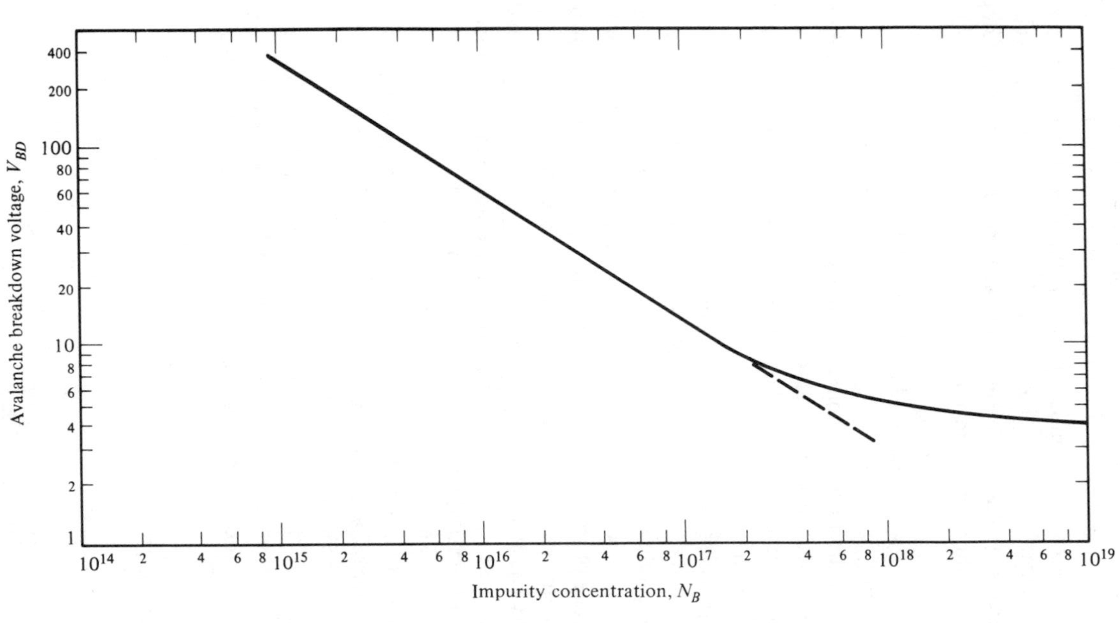

Fig. 6-26 Avalanche breakdown voltage as a function of the impurity concentration on the high-resistivity side of the junction.

When the impurity concentration exceeds 2×10^{17} atoms/cm³, the curve deviates from the straight-line approximation due to a "saturation" process. As the impurity concentration increases beyond this value, the electric field increases beyond the magnitude necessary to break covalent bonds. The curve levels off. Further increases in N_B no longer cause corresponding decreases in V_{BD}.

The relationship between impurity concentration and avalanche breakdown voltage has been established for linearly graded *p-n* junctions as well as one-sided step junctions. A mathematical model suggested by Sze and Gibbons [7] for linearly graded junctions is given by

$$V_{BD} = (9.31 \times 10^9) a^{-0.4} \tag{6-120}$$

where a represents the impurity concentration gradient (dn/dx atoms/cm⁴) in the vicinity of the metallurgical junction.

As a practical example of avalanche breakdown and impurity concentrations, consider the emitter–base diode of a bipolar *npn* transistor, as shown in Fig. 6-27. The *p*-type base region was formed by a Gaussian diffusion. With reverse bias, impurity concentrations are such that avalanche breakdown of the emitter–base diode occurs at about 6 V. The boron impurity concentration is highest at the device surface. Avalanche breakdown will therefore tend to occur at the surface of the emitter–base depletion region. The breakdown voltage is dependent on the surface boron concentration next to the n^+ emitter. This is one reason why a Gaussian diffusion was used. Had an error function diffusion been used for the base, the surface concentration would be much higher, causing zener or avalanche breakdown to occur at a lower voltage.

We must also consider the fact that Fig. 6-26, Eq. (6-119), and Eq. (6-120) are intended for use with *planar p-n* junctions (i.e., junctions formed between parallel slabs of *p* and *n* materials). When an impurity is diffused into silicon through an oxide window, the resulting junction is not actually planar; it has corners. When the *p-n* junction is reverse-biased, the electric field will be larger at these corners. Voltage breakdown will occur at lower voltages at the corners than in the planar regions. Actual breakdown voltages will thus

Fig. 6-27 Emitter–base of an *npn* bipolar transistor.

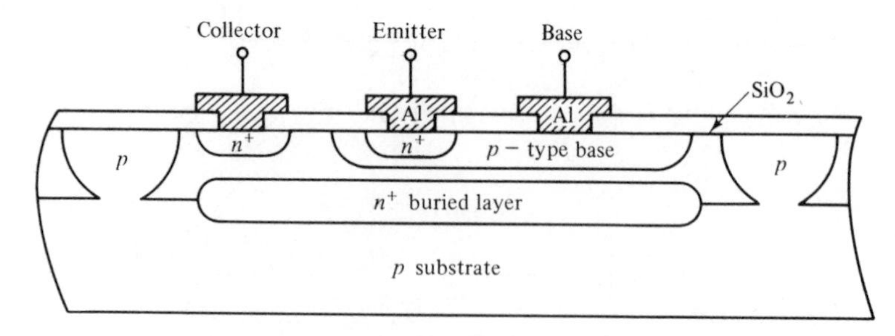

be lower than indicated through using Fig. 6-26, Eq. (6-119), or Eq. (6-120). Without detailed analysis, it is difficult to determine whether voltage breakdown will occur at the surface or the corners. If more than a rough estimate of avalanche breakdown is needed, the reader should consult Chapter 6 of reference [5] for more information on this subject.

6-10 OTHER DEPARTURES FROM THE IDEAL MODEL

The development of an idealized diode model culminated with a mathematical description that is valid for both forward and reverse bias conditions [Eq. (6-58)]. In the interest of developing a useful diode model that was neither too simple nor too complex, a number of mechanisms that affect the performance of junction devices were temporarily ignored. Junction device behavior is significantly affected by:

1. The generation and recombination of hole–electron pairs in the depletion layer.
2. Leakage currents at the surface of a reverse-biased *p-n* junction (caused by surface states at the silicon–silicon dioxide interface).
3. High-level injection conditions (large forward currents) that affect majority carrier concentrations at the depletion-layer edges and cause electric fields to exist in the *p* and *n* regions.

6-10-1 Generation and Recombination in the Depletion Layer

In the development of the diode model, we assumed that all hole–electron pair generation and recombination occurred in the electrically neutral *p* and *n* regions. In a real junction device, it is possible for most carrier generation and recombination to occur in the depletion region. Whether these mechanisms occur in the neutral or depletion regions depends on conditions of device processing, applied voltage, and operating temperature.

Let us first consider the magnitude of the reverse current caused by carrier *generation* in the depletion region. The rate of charge carrier generation per unit volume in this region is approximately n_i/T_0, where n_i represents the thermally generated charge carriers and T_0 denotes their effective lifetime. The magnitude of the current generated in this region is found by multiplying the rate of carrier generation per unit volume, the volume involved, and the charge per carrier. Thus,

$$I_{\text{gen}} \approx q\left(\frac{n_i}{T_0}\right)wA_j \tag{6-121}$$

where w denotes the depletion layer width and A_j denotes its cross-sectional area. Since the depletion width w varies with impurity concentration and applied voltage, the current generated in this region is dependent on these parameters. The current I_{gen} in Eq. (6-121) will add to the reverse current caused by minority carrier generation in the neutral regions [given by Eq. (6-86)].

Under forward bias conditions, some *recombination* of hole–electron pairs takes place in the depletion region. As with carrier generation in this region, the resultant current magnitude is dependent on the concentration of energy states near the center of the energy gap. With forward bias, however, many more charge carriers are available for recombination. The recombination rate in the depletion region is dependent on exp $(qV/2kT)$, where V denotes the applied voltage. The factor 2 occurs in the exponent because only one-half of the gap energy is involved in charge carrier transitions between the conduction and valence bands. The recombination current under forward bias conditions is approximately equal to I_{gen} [Eq. (6-121)] multiplied by the exponential factor exp $(qV/2kT)$. Thus,

$$I_{recomb} \approx q\left(\frac{n_i}{T_0}\right) wA_j \exp\left(\frac{qV}{2kT}\right) \tag{6-122}$$

Whether recombination current in the neutral regions [given by Eq. (6-67)] or recombination in the depletion region dominates junction behavior is determined by device temperature and the magnitude of the applied voltage. For silicon at room temperature with small forward voltages (a few tenths of a volt), depletion-region recombination current is usually larger. At higher voltages, neutral-region recombination current becomes larger. Thus, forward diode currents can be expressed as a function of $exp\ (qv/mkT)$, with m varying from 1 to 2. With small applied voltages, $m \approx 2$. With larger applied voltages, $m \approx 1$. At transitional values of applied voltage, where the magnitudes of depletion-region recombination and neutral-region recombination are comparable, the value of m is intermediate.

6-10-2 Leakage Currents Due to Surface States

At a thermally oxidized silicon surface, the highly ordered crystal lattice terminates in an interface with amorphous silicon dioxide. The major transition between silicon and silicon dioxide takes place over about 25 Å. The interface area provides numerous centers or traps for the generation and recombination of hole–electron pairs. From an energy-band point of view, the interface causes the existence of energy states near the center of the forbidden band. If an excess of mobile charge (with respect to an equilibrium charge density) exists near the interface, these *surface states* act primarily as recombination centers. If a deficit exists, they tend to act as generation centers.

Imagine, for the moment, that excess minority carriers are diffusing laterally through a region of bulk silicon, near the Si–SiO_2 interface. Recombination will be enhanced and carrier lifetimes will be shortened due to the presence of the interface. Charge carrier velocities can be divided into two components: a lateral velocity component, which acts as one might otherwise expect, and a vertical velocity component, which moves up to the interface and disappears. This vertical velocity component is referred to as the *surface recombination velocity.* It is denoted by the symbol s and has units of cm/s.

The symbol s and the term surface recombination velocity apply whether excess charge carriers are present and a cloud of mobile charge is moving toward the surface, or a charge deficit exists and the charge generated by the surface states is moving down into the bulk material.

Surface states affect device operation in areas where large excesses or deficits of mobile charge carriers exist, or where large electric fields exist near the surface. The symbol A_s is used to denote *effective area of surface recombination.* A distinction is thus made between A_s and the total surface area of the device. An approximate equation for the surface generation current in a reverse biased diode is given by

$$I_{\text{surface gen.}} \approx \tfrac{1}{2}\, q n_i s A_s \tag{6-123}$$

Surface recombination velocities are roughly on the order of 1 cm/s. For reverse-biased *p-n* junctions, surface recombination occurs predominantly near the surface of the depletion layer. Surface recombination currents are on the order of picoamperes and increase with increasing reverse bias. The leakage process also involves some ionic conduction due to the presence of mobile ions in the oxide layer and impurities at the Si–SiO_2 interface.

The integrated-circuit process technology has advanced to a point where surface states can be minimized to the extent that they do not pose a serious problem regarding device characteristics or circuit yields.

6-10-3 High-Level Injection Conditions

In the milliampere range of forward currents, the ideal diode model is reasonably accurate. As forward diode currents are increased, however, the assumptions made in developing this model become less and less accurate. We assumed, for example, that majority carrier concentrations are not significantly affected by minority carriers injected across the forward-biased depletion layer [see Fig. 6-14(b)]. As forward diode currents increase, this assumption becomes less accurate. We assumed that the electric fields and thus the drift currents in the "neutral" regions were negligibly small. Under conditions of high-level injection, the voltage drops in these "neutral" regions do affect the device I–V characteristics. The electric fields in the now quasi-neutral regions alter their conductivities.

The overall effects of high-level injection on device behavior are nonlinear. They cannot be modeled by simply taking the series resistance of the neutral regions into account. For specific values of high forward current, these effects are sometimes modeled with series resistances, offset by constant-voltage drops. In any case, the effects of high-level injection lead to lower forward diode current than the ideal diode model predicts.

BIBLIOGRAPHY

[1] RYDER, E. J., "Mobility of Holes and Electrons in High Electric Fields," *Physical Review,* Vol. 90, No. 5, pp. 766–769, June 1953.

[2] CONWELL, E. M., "Properties of Silicon and Germanium: II," *Proceedings of the IRE,* Vol. 46, pp. 1281–1300, 1958.

[3] GARTNER, W. W., *Transistors: Principles, Design and Applications,* p. 46. Princeton, N.J.: D. Van Nostrand Company, Inc., 1960.

[4] IRVIN, J. C., "Resistivity of Bulk Silicon and of Diffused Layers in Silicon," *The Bell System Technical Journal,* Vol. 41, pp. 387–410, March 1962.

[5] GROVE, A. S., *Physics and Technology of Semiconductor Devices,* p. 184. New York: John Wiley & Sons, Inc., 1967.

[6] LAWRENCE, A., AND R. M. WARNER, "Diffused Junction Depletion Layer Calculations," *The Bell System Technical Journal,* Vol. 39, p. 389, 1960.

[7] SZE, S. M., AND G. GIBBONS, "Avalanche Breakdown Voltages of Abrupt and Linearly Graded *P-N* Junctions in Ge, Si, GaAs, and GaP," *Applied Physical Letters,* Vol. 8, No. 5, pp. 111–113, March 1, 1966.

[8] MCKAY, K. G., "Avalanche Breakdown in Silicon," *Physical Review,* Vol. 94, No. 4, pp. 877–884, May 15, 1954.

[9] MILLER, S. L., "Ionization Rates for Holes and Electrons in Silicon," *Physical Review,* Vol. 105, No. 4, pp. 1246–1249, February 15, 1957.

[10] ADLER, R. B., A. C. SMITH, AND R. L. LONGINI, "Introduction to Semiconductor Physics," *SEEC,* Vol. 1. New York: John Wiley & Sons, Inc., 1964.

[11] GRAY, P. E., D. DEWITT, A. R. BOOTHROYD, AND J. F. GIBBONS, "Physical Electronics and Circuit Models of Transistors," *Semiconductor Electronics Education Committee,* Vol. 2. New York: John Wiley & Sons, Inc., 1964.

[12] WARNER, R. M., J. N. FORDEMWALT, ET AL. *Integrated Circuits—Design Principles and Fabrication.* New York: McGraw-Hill Book Company, 1965.

[13] CARR, W. N., AND J. P. MIZE, *MOS/LSI Design and Application.* New York: McGraw-Hill Book Company, 1972.

[14] LECROISSETTE, D., *Transistors.* Englewood Cliffs, N.J.: Prentice-Hall, Inc., 1963.

PROBLEMS

6-1 Given that the forbidden energy gap is 1.1 eV for silicon, find the number of valence 5 impurity atoms per unit volume needed to move the Fermi level to the edge of the conduction band at room temperature. ($T = 300$K.)

6-2 A p-type specimen of silicon having a resistivity of 0.5 Ω-cm is exposed to a flash of light that liberates an additional 2×10^{16} holes/cm³ and 2×10^{16} electrons/cm³. The temperature remains at 300K throughout the experiment. Find:
(a) The hole and electron concentrations of the original specimen.
(b) The maximum change in resistivity caused by the flash of light.

6-3 An n-type silicon crystal specimen 1.0-mm cube has a resistivity of 0.5 Ω-cm. Find the electron and hole currents which exist when the specimen is placed in a circuit with 2.0 V across the specimen.

6-4 An abrupt p-n junction device is made of silicon. The conductivities of the two sides are 1 mho/cm (p side) and 10 mhos/cm (n side). Compute the contact potential V_B at $T = 300$K.

6-5 A bar of n-type silicon has length, width, and depth dimensions of 100, 10, and 10 μm, respectively, as shown. The impurity concentration varies linearly along the length, from $N_D = 1 \times 10^{20}$ atoms/cm³ at the left end to $N_D = 1 \times 10^{14}$ atoms/cm³ at the right end. Does an electric field exist within the silicon bar? If so, determine its magnitude. If the silicon bar was shorted with an external wire, what value of dc current would flow?

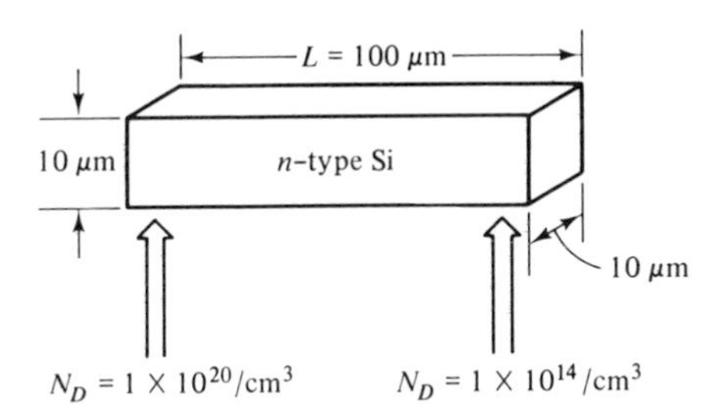

6-6 Find the current magnitude across a silicon p-n junction of area 10^{-6} m² at 300K when the applied voltage is 2 V in the reverse direction. The resistivities of the p and n regions are both 1 Ω-cm. Assume that the average diffusion length of minority holes L_p is 0.04 cm and the average diffusion length of minority electrons L_n is 0.1 cm.

6-7 Given a p-n junction diode with doping concentrations, carrier lifetimes, and cross-sectional area as listed:
(a) Find the reverse current i_s.
(b) Find the total diode current i_d if the diode is forward-biased by 0.7 V.

$$N_D = 1 \times 10^{19} \text{ atoms/cm}^3$$
$$N_A = 4 \times 10^{16} \text{ atoms/cm}^3$$
$$T_p = 1 \times 10^{-6} \text{ s}$$
$$T_n = 1 \times 10^{-7} \text{ s}$$
$$A = 1 \times 10^{-3} \text{ cm}^2$$

6-8 A solid-state device contains a step p-n junction with impurity concentrations of $N_D = 1 \times 10^{19}$ phosphorus atoms/cm³ and $N_A = 1 \times 10^{17}$ boron atoms/cm³.

Find the width of the depletion layer, the per unit area capacitance of the depletion layer, and the approximate avalanche breakdown voltage.

6-9 A solid-state device contains a graded *p-n* junction. The impurity concentration changes from $N_D = 10^{18}$ atoms/cm³ to $N_A = 10^{16}$ atoms/cm³ over a distance of 1×10^{-4} cm in the vicinity of the junction. Assuming that the concentration change is linear, the slope $a = \Delta N/\Delta X \approx 10^{22}$ cm^{-4}. Find the approximate width of the depletion layer, the per unit area capacitance of the depletion layer, and the approximate avalanche breakdown voltage. Assume that the contact potential $V_B = 0.7$ V.

6-10 All three regions of a narrow-base *npn* transistor are uniformly doped. We will assume that breakdown occurs due to punch-through rather than by avalanche effect. With no bias voltages applied, the base width (from the emitter–base metallurgical junction to the base–collector metallurgical junction) is 4×10^{-5} cm wide. The transistor is made of silicon, the collector doping is $N_{dc} = 5 \times 10^{14}$ atoms/cm³, the base doping is $N_{ab} = 1 \times 10^{16}$ atoms/cm³, and the emitter doping is $N_{de} = 1 \times 10^{20}$ atoms/cm³. Assume that the forward bias voltage across the emitter–base diode is held constant at 0.6 V and that the reverse bias voltage across the base–collector diode is increased until punch-through occurs. Find the collector–base voltage needed to cause punch-through.

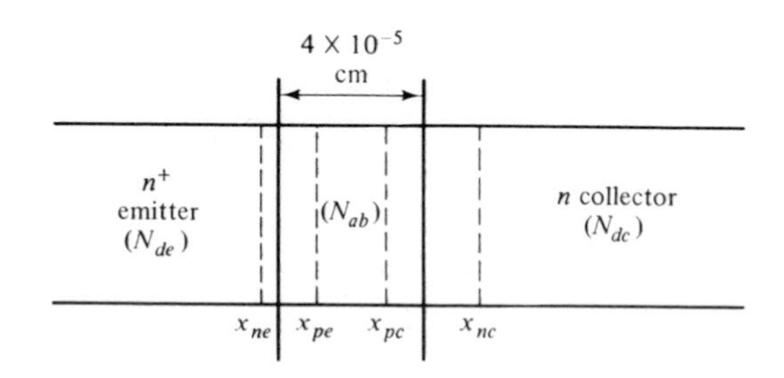

bipolar transistors

7

The most important property of a transistor is that it can amplify electrical signals. It is basically a three-terminal device that exhibits different impedance characteristics at the output terminals than at the input terminals. The word *transistor* is a contraction of *transfer resistor.*

Solid-state amplification is achieved by moving relatively large currents through a solid-state region that is very sensitive to a small externally applied voltage or current. In a bipolar transistor, the current to be controlled enters the *emitter* region, passes through the sensitive *base* region, and is collected in the *collector* region. Small changes in the base current, or the base-to-emitter voltage, can cause large changes in the current from emitter to collector.

The word *bipolar* denotes the fact that both holes and electrons are involved in the physical operation of the device. In contrast, field-effect transistors are often called *unipolar* because their operation involves the motion of one type of charge carrier, holes or electrons.

There are two types of bipolar transistors: *npn* and *pnp*. The letters denote the type of impurity in the emitter, base, and collector regions, respectively. In an *npn* transistor, minority electrons must diffuse through the *p*-type base region to get from the *n*-type emitter to the *n*-type collector. In *pnp* transistors, minority holes diffuse through the *n*-type base region.

In integrated-circuit structures, *npn* transistors are generally favored over *pnp* transistors for historical reasons. Electron mobilities are two to three times larger than hole mobilities. Charge carrier mobility is important to the signal amplification, the current handling ability, and the frequency response of a solid-state device. In the 1960s, transistor base widths were relatively large. The higher mobility of electrons favored the *npn* transistor. Advantages in processing also favored the *npn* transistor. With modern, narrow-base-width transistors, and with advanced processing techniques, however, the differences between *npn* and *pnp* transistors are no longer distinct; but with no significant advantages to the *pnp* device, the industry has stayed with bipolar structures which favor the *npn* transistor. However, *pnp* transistors do provide a very useful design tool. There are many design advantages to using *pnp* and *npn* transistors in the same circuit. Integrated injection logic and push-pull amplifiers are two examples of this.

Bipolar transistors are used over an extremely wide range of frequencies and power levels. They are used extensively in digital and nondigital circuits.

Transistor behavior is approximated by models. The major purposes of models are to make device behavior more understandable and predictable, to make possible the design of *devices* with predictable results, and to make possible the design of *circuits* with predictable results. A given model will best approximate device behavior over a certain range of operating parameters. When the operating parameters go beyond the useful range of a model, it is either modified or replaced.

Section 7-1 will provide a general description of bipolar transistor operation and the basic operating parameters.

In Section 7-2, a large signal static model of device operation, called the *Ebers–Moll model,* will be developed. This development serves two purposes: it provides a physical–mathematical explanation of transistor behavior, and it relates useful transistor parameters back to impurity concentrations and device geometries. The limitations and useful range of the model will be discussed, and terms will be added to improve its accuracy and extend its usefulness to a wider range of operating parameters. Although other models exist that more accurately depict large signal transistor behavior (the integral charge control model by Gummel and Poon, for example), they are considered beyond the scope of this text.

Bipolar transistors are used extensively for the amplification and processing of small ac signals. For this type of application, the transistor can be modeled as a linear bilateral device. In effect, it can be replaced by an equivalent circuit consisting of impedances and dependent voltage and current sources. Mesh and node analysis and the network theorems can then be used to predict small-signal *circuit* behavior. In Section 7-4, a small-signal model called the *hybrid-π* model will be presented and discussed for this purpose. The parameters of the model will be developed and then related to transistor frequency response.

Substrate and *lateral pnp* transistors, which are made with standard integrated-circuit process techniques, will be discussed in Section 7-5.

7-1 BASIC TRANSISTOR OPERATION

A planar silicon *npn* transistor is shown schematically in Fig. 7-1. When the transistor is used in a circuit, the *p*-type substrate is generally connected to the most negative potential available so that the collector–substrate diode

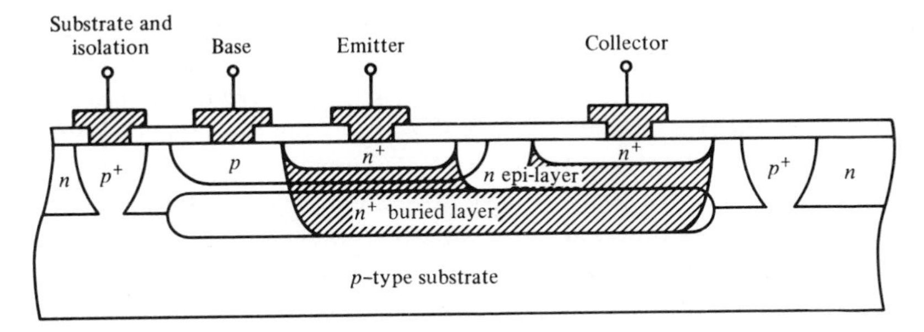

Fig. 7-1 Planar silicon *npn* transistor. The shaded area indicates the electron current path from emitter to collector.

is reverse-biased. In most operating modes, the most important region of device operation is the vertical region under the n^+ emitter.

Assume for the moment that the transistor is biased to conduct current from the emitter to the collector. Most of the injected emitter current then passes downward, through the *p*-type base region, through the *n*-type epi-layer, and into the n^+ buried layer. The buried layer forms a low-resistance current path between the epi-layer and the collector contact. If Fig. 7-1 could be conveniently drawn to scale, the fact that the shaded area forms a low-resistance path would become more obvious.

The vertical current path from the n^+ emitter to the n^+ buried layer can be represented by an idealized one-dimensional transistor model, as shown in Fig. 7-2(a). V_{BE} and V_{BC} represent external supply voltages that are connected to forward or reverse bias the base–emitter and base–collector diodes, respectively. Figure 7-2(b) shows the symbol for an *npn* transistor. The emitter arrow indicates the direction of *conventional* current.

By choosing polarities for V_{BE} and V_{BC}, we can put the transistor into four different modes or regions of operation. If we forward-bias both the base–emitter *(B-E)* junction and the base–collector *(B-C)* junction, the transistor

Fig. 7-2 (a) Idealized one-dimensional *npn* transistor model. (b) Symbol for an *npn* transistor.

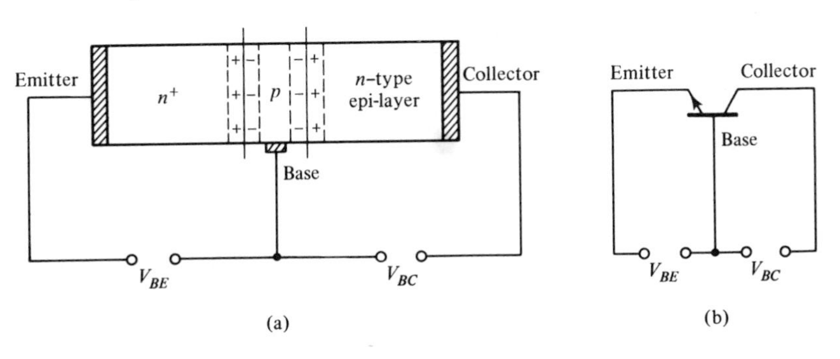

will operate in the *saturation region.* If we reverse-bias both junctions, the device is in the *cutoff region.* When the *B-E* junction is forward-biased and the *B-C* junction is reverse-biased, the device operates in the *forward-active region.* Alternatively, when the *B-E* junction is reverse-biased and the *B-C* junction is forward-biased, the device operates in the *inverse-active region.*

The forward-active region of device operation provides useful signal amplification. Voltage polarities and *electron* current directions for device operation in this region are shown in Fig. 7-3. We will assume that the applied voltages act to raise or lower the contact potentials and that the electric fields in the n and p regions are negligibly small. The *B-E* diode is forward-biased and the emitter region is much more heavily doped than the base region. Electrons *diffuse* across the forward-biased *B-E* junction and enter the p-type base region. At low to medium current levels, they are too far apart to mutually repel each other. They *diffuse* across the base region. The base width is much shorter than the average diffusion length of an electron in this region. (Typically, $W \approx 0.5\ \mu\text{m}$ and $L_{nb} \approx 10\ \mu\text{m}$.) The base width is also much shorter than the average distance from the active base region to the base contact. Most of the minority electrons that enter the base region therefore reach the *B-C* depletion layer, and are swept across by the electric field. As shown in Fig. 7-3,

$$I_E = I_B + I_C \tag{7-1}$$

Typically, I_B will be a very small fraction of I_E (less than one one-hundredth). Two major causes of base current are hole–electron recombination in the emitter–base space–charge region and majority hole current crossing the forward-biased *B-E* junction into the emitter region. The recombination rate in the base is kept small by keeping W much smaller than L_{nb}. The majority hole current is kept small by using a much smaller impurity concentration in the base region than in the emitter region.

In summary, a large current can cross the reverse-biased *B-C* junction because the forward-biased *B-E* junction is less than a diffusion length away. Hence, $I_C \approx I_E$, even though I_E crosses a low-resistance forward-biased diode and I_C crosses a high-resistance reverse-biased diode.

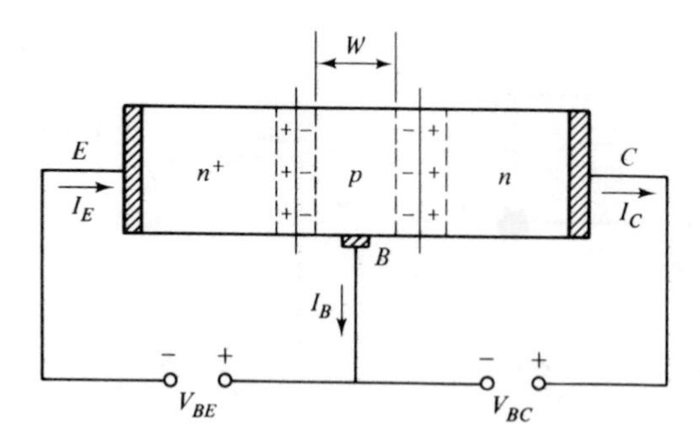

Fig. 7-3 *npn* transistor biased in the forward-active region.

In transistor circuits, one device terminal is normally used for the signal input and another for the signal output. The third terminal is common to input and output. Of the six possible circuit configurations that this allows, only three provide useful power gain. They are the common emitter, common base, and common collector circuits, as shown in Fig. 7-4. Each of these configurations has unique circuit advantages that depend upon the required values of input impedance, output impedance, current gain, and voltage gain.

Common-emitter, common-base, and common-collector circuits are used in digital applications by defining two fixed operating points (cutoff and saturation, for example), and arranging that the transistors operate at one point or the other. They are used for small-signal amplification by biasing the transistors in the forward active region and superimposing small ac signals on these biasing conditions.

In a common-emitter circuit, I_C and V_{CE} represent the output current and voltage, where, from Fig. 7-3,

$$V_{CE} = V_{CB} + V_{BE} \tag{7-2}$$

I_C and V_{CE}, in turn, are dependent on the input current I_B. Figure 7-5(a) shows a family of output characteristic curves for an *npn* transistor (common emitter) operating in the 5-V, 5-mA range. Figure 7-5(b) shows the characteristic curves for operation in the 50-V, 50-mA range. Ideally, these curves would be uniformly spaced horizontal straight lines for even increments of base current. Comparison of Fig. 7-5(a) and (b) shows that the degree of nonuniformity (and therefore nonlinearity) obtained depends on the range of output currents and voltages over which the device is operated. Figure 7-5(a) shows the saturation, cutoff, and forward active regions of device operation. An avalanche breakdown voltage for the device (BV_{CEO}) is shown in Fig. 7-5(b). A maximum power dissipation curve for a transistor designed to dissipate 1 W is shown as a dashed line in Fig. 7-5(b). Operation to the right of this line would cause the device to overheat and probably burn out.

Fig. 7-4 Useful transistor circuit configurations: (a) common emitter; (b) common base; (c) common collector.

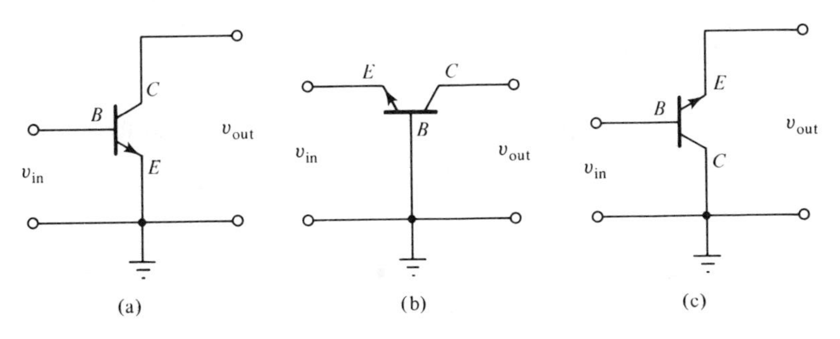

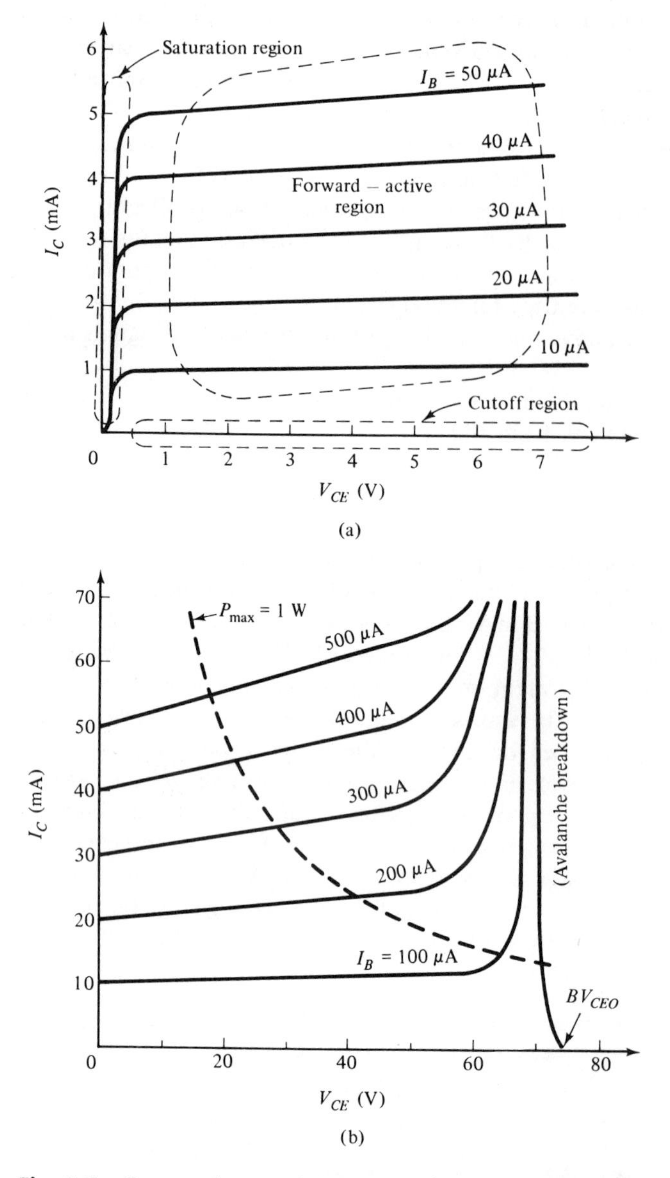

Fig. 7-5 Output characteristics curves for an *npn* transistor in the common-emitter configuration: (a) operating in the 5-V, 5-mA range; (b) operating in the 50-V, 50-mA range.

The *forward transistor alpha* (α_F) is an important parameter in transistor characterization. Using the notation of Fig. 7-3, it is the ratio of collector current to emitter current with the *B-C* diode shorted:

$$\alpha_F = \left.\frac{I_C}{I_E}\right|_{V_{BC}=0} \tag{7-3}$$

Defining α_F with $V_{BC} = 0$ makes it exclusively dependent on the emitter–base voltage V_{BE}.

The *forward transistor beta* (β_F) is defined as the ratio of collector current to base current with the *B-C* diode shorted:

$$\beta_F = \left.\frac{I_C}{I_B}\right|_{V_{BC}=0} \tag{7-4}$$

Combining Eq. (7-1) and Eq. (7-3), we have

$$\alpha_F = \frac{\beta_F}{\beta_F + 1} \tag{7-5}$$

which can be rewritten

$$\beta_F = \frac{\alpha_F}{1 - \alpha_F} \tag{7-6}$$

A few words about notation and terminology are necessary at this point. Both α_F and β_F are dc transistor parameters and are defined as ratios of dc currents with $V_{BC} = 0$. Small-signal ac values of α and β are described with lowercase subscripts (i.e., α_f and β_f). For current and voltage, we will follow the notation

$$i_E = I_E + i_e \tag{7-7}$$

where i_E = total current
I_E = dc current component
i_e = ac current component

Setting the small-signal component of collector–base voltage equal to zero is equivalent to holding the dc component constant. The small-signal parameters α_f and β_f will then be defined:

$$\alpha_f = \left.\frac{i_c}{i_e}\right|_{v_{bc}=0} = \left.\frac{i_c}{i_e}\right|_{V_{BC}=\text{const.}} \tag{7-8}$$

and

$$\beta_f = \left.\frac{i_c}{i_b}\right|_{v_{bc}=0} = \left.\frac{i_c}{i_e}\right|_{V_{BC}=\text{const.}} \tag{7–9}$$

Since

$$i_e = i_b + i_c \tag{7–10}$$

Eqs. (7–5) and (7–6) also apply to α_f and β_f.

Since α_f relates the input and output currents in the common-base configuration, it *denotes the current gain of the common-base circuit.* β_f *denotes the common-emitter current gain.*

Given a family of transistor characteristics as shown in Fig. 7–5(a), β_f can be graphically determined by taking the ratio $\Delta I_C/\Delta I_B$ with V_{CE} held constant. Because the characteristic curves are not quite linear, the magnitude of β_f will be somewhat dependent on where the measurement is made.

Both α_f and β_f are useful indicators of transistor behavior. The magnitude of α_f is generally between 0.990 and 0.998; β_f is therefore between 100 and 500 for a typical transistor.

On transistor specification sheets, three subscripts are often used to denote transistor breakdown voltages and leakage currents. For example, BV_{CEO} represents the *collector–emitter breakdown voltage* with the base open, and BV_{CBO} denotes the *collector–base breakdown voltage* with the emitter open. BV_{CES} denotes the collector–emitter breakdown voltage with the emitter–base diode shorted. I_{CBO} represents the reverse leakage current of the collector–base diode with the emitter open. I_{CES} denotes the collector–emitter reverse leakage current with the emitter–base diode shorted.

7-2 LARGE-SIGNAL OPERATION—THE EBERS–MOLL MODEL

We will use diode currents across the B-E and *B-C* junctions to develop a set of static equations that are useful for modeling the large-signal terminal behavior of bipolar transistors. Our simplifying assumptions will be similar to those used in Chapter 6. They are:

1. The model is one-dimensional.
2. Conditions of low-level injection exist (i.e., minority currents are not large enough to significantly affect majority concentrations).
3. Electric fields in the *p* and *n* regions are negligibly small. Externally applied junction voltages act only to raise or lower the junction potentials.

4. Generation and recombination occur only in the neutral p and n regions.
5. Impurity concentrations are constant in the p and n regions. (Although the model is valid with graded impurity distributions, this assumption simplifies the development.)
6. The emitter and collector regions are much longer than minority carrier diffusion lengths in these regions.
7. The effective cross-sectional area of the base-emitter diode is the same as that of the collector–base diode.
8. Variations in base width with applied junction voltage are neglected. Base width W is assumed to be constant.
9. I_E, I_B, and I_C are conventional currents that enter the transistor terminals.

The terminal currents, I_E, I_B and I_C, are treated in terms of four junction components as shown in Fig. 7–6. I_{ne} and I_{pe} represent the respective electron and hole currents crossing the *B-E* junction. I_{nc} and I_{pc} represent the electron and hole currents crossing the *B-C* junction. Hole current in the direction of conventional current entering a terminal is taken as positive. Thus,

$$I_E = -I_{ne} - I_{pe} \tag{7–11}$$

$$I_C = -I_{nc} - I_{pc} \tag{7–12}$$

and

$$I_B = -(I_E + I_C) \tag{7–13}$$

The *p-n* junction diode equation, Eq. (6–67), can be written

$$I_d = \frac{qAD_p p_{no}}{L_p}\left[\exp\left(\frac{qV}{kT}\right) - 1\right] + \frac{qAD_n n_{po}}{L_n}\left[\exp\left(\frac{qV}{kT}\right) - 1\right] \tag{7–14}$$

where the left-hand term represents the injected hole current at the n-type edge of the depletion layer and the right-hand term represents the injected

Fig. 7-6 Reference current directions for the Ebers–Moll model.

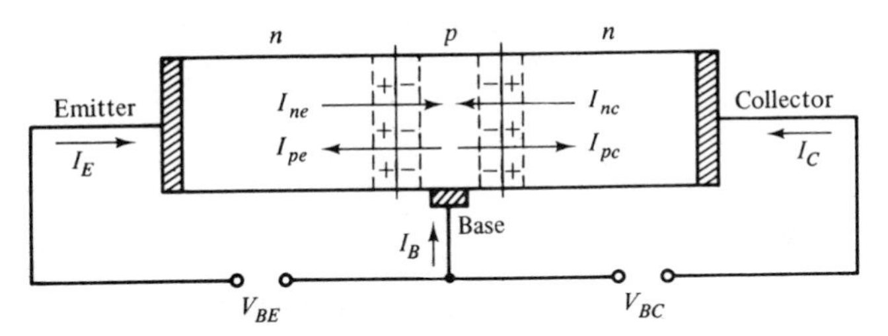

electron current at the p-type edge. Since the emitter and collector n regions are assumed to be long, the diode currents I_{pe} and I_{pc} in Eq. (7–11) and Eq. (7–12) can be directly written as

$$I_{pe} = \frac{qAD_{pe}p_{ne}}{L_{pe}} \left[\exp\left(\frac{qV_{BE}}{kT}\right) - 1 \right] \tag{7-15}$$

and

$$I_{pc} = \frac{qAD_{pc}p_{nc}}{L_{pc}} \left[\exp\left(\frac{qV_{BC}}{kT}\right) - 1 \right] \tag{7-16}$$

The equilibrium minority hole concentrations in the emitter and collector regions are represented by p_{ne} and p_{nc}, respectively. The second subscripts are needed on D_p and L_p because impurity concentrations are different in the emitter and collector regions.

Since the base width is narrow compared to the average electron diffusion length in the base region (L_{nb}), we cannot write I_{ne} and I_{nc} directly from the diode equation. They must be determined from the mechanisms governing electron motion and the boundary conditions that confine these mechanisms.

With the limiting assumptions that we made, the only processes occurring in the base region are generation, recombination, and diffusion. The time rate of change of electron concentration in a differential element (dx) of the p-type base region is given by

$$\frac{\partial n_p(x)}{\partial t} = \frac{n_{pb} - n_p(x)}{T_n} + D_{nb}\frac{\partial^2 n_p(x)}{\partial x^2} \tag{7–17}$$

Here n_{pb} denotes the equilibrium electron concentration in the base region, $n_p(x)$ denotes the nonequilibrium concentration as a function of distance, T_n denotes electron lifetime, and D_{nb} denotes the diffusivity of electrons in the base. The rate of change of electron concentration due to the difference between the generation and recombination rates is stated as $[n_{pb} - n_p(x)]/T_n$. The change in electron concentration due to diffusion is given by $D_{nb}[\partial^2 n_p(x)/\partial x^2]$.

If dc bias voltages are applied to the transistor and transients are disregarded, the electron concentration in a differential element of the base region will remain constant. Thus, $\partial n_p(x)/\partial t = 0$ in Eq. (7–17) and

$$\frac{d^2 n_p(x)}{dx^2} = \frac{n_p(x) - n_{pb}}{L_{nb}^2} \tag{7–18}$$

L_{nb} represents the average electron diffusion length in this region:

$$L_{nb} = \sqrt{D_{nb} T_n} \tag{7–19}$$

If we let $n'_p(x)$ represent the *excess* or *deficit* electron concentration,

$$n'_p(x) = n_p(x) - n_{pb} \tag{7–20}$$

then

$$\frac{d^2 n'_p(x)}{dx^2} = \frac{n'_p(x)}{L_{nb}^2} \tag{7-21}$$

Equation (7-21) has the general solution

$$n'_p(x) = A \exp\left(-\frac{x}{L_{nb}}\right) + B \exp\left(\frac{x}{L_{nb}}\right) \tag{7-22}$$

where the constants A and B must be determined from the boundary conditions imposed on the base region.

The base width (W) is defined as the distance between the B-E depletion-layer edge and the B-C depletion-layer edge, as shown in Fig. 7-7. Variations in base width with applied junction voltage are neglected in this development. Minority electron concentrations at the emitter and collector depletion-region edges are determined by the equilibrium concentrations and the applied junction voltages as developed in Eq. (6-59). They are given as

$$n'_p(x=0) = n'_p(0) = n_{pb}\left[\exp\left(\frac{qV_{BE}}{kT}\right) - 1\right] \tag{7-23}$$

and

$$n'_p(x=W) = n'_p(W) = n_{pb}\left[\exp\left(\frac{qV_{BC}}{kT}\right) - 1\right] \tag{7-24}$$

where V_{BE} and V_{BC} assume positive values when the B-E and B-C diodes are both forward-biased.

The constants A and B in Eq. (7-22) are found by applying the boundary conditions of Eq. (7-23) and Eq. (7-24). This provides us with an expression for the minority carrier concentration in the base region.

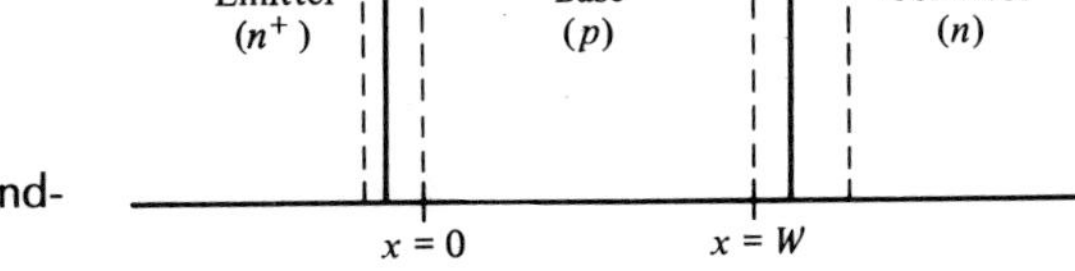

Fig. 7-7 Base region boundaries.

$$n_p'(x) = \frac{1}{\sinh\,(W/L_{nb})}\left[n_p'(0)\sinh\left(\frac{W-x}{L_{nb}}\right) + n_p'(W)\sinh\left(\frac{x}{L_{nb}}\right)\right] \tag{7-25}$$

where

$$\sinh y = \frac{e^y - e^{-y}}{2} = y + \frac{y^3}{3!} + \frac{y^5}{5!} + \ldots \tag{7-26}$$

Figure 7-8 shows $n_p'(x)$ plotted against distance for the four regions or modes of transistor operation. Both the magnitude and curvature of $n_p'(x)$ are exaggerated for the sake of clarity. When the transistor is operated in saturation, $+V_{BE}$ is normally larger than $+V_{BC}$. As a result, $n_p'(x)$ is larger at $x = 0$ than at $x = W$, as shown in Fig. 7-8(c). In the saturation region, the minority carrier distribution is determined by finding the distribution due to V_{BE} with $V_{BC} = 0$, and then due to V_{BC} with $V_{BE} = 0$. These distributions are then superimposed as shown.

Since the minority electron current through the base is a diffusion current,

$$I_{\text{base diffusion}} = -qAD_{nb}\frac{dn_p'(x)}{dx} \tag{7-27}$$

or, from Eq. (7-25),

Fig. 7-8 Minority carrier distributions in the base region: (a) forward-active region ($+V_{BE}$, $-V_{BC}$): (b) inverse-active region ($-V_{BE}$, $+V_{BC}$): (c) saturation ($+V_{BE}$, V_{BC}); (d) cutoff ($-V_{BE}$, $-V_{BC}$).

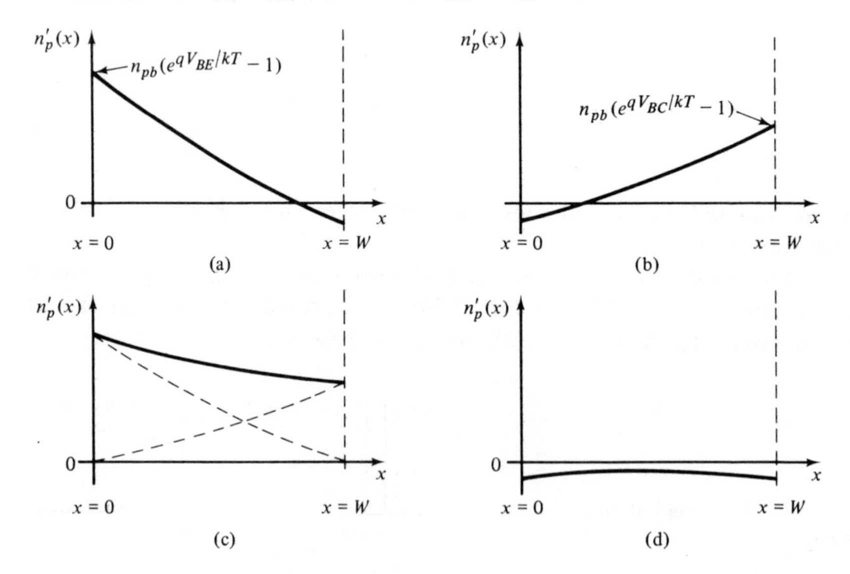

$$I_{\text{base diffusion}} = \frac{qAD_{nb}}{L_{nb}\sinh(W/L_{nb})}\left[n_p'(0)\cosh\left(\frac{W-x}{L_{nb}}\right) - n_p'(W)\cosh\left(\frac{x}{L_{nb}}\right)\right] \tag{7-28}$$

Cosh y in Eq. (7-28) can be written

$$\cosh y = \frac{e^y + e^{-y}}{2} = 1 + \frac{y^2}{2!} + \frac{y^4}{4!} + \ldots \tag{7-29}$$

With D_{nb} held constant (with a uniformly doped base region), Eq. (7-27) shows that the magnitude of the electron current diffusing across the base region is proportional to the slopes of the curves in Fig. 7-8.

To find the electron current crossing the *B-E* junction, Eq. (7-28) can be evaluated at $x = 0$. Thus,

$$I_{ne} = \frac{qAD_{nb}}{L_{nb}\sinh(W/L_{nb})}\left[n_p'(0)\cosh\left(\frac{W}{L_{nb}}\right) - n_p'(W)\right] \tag{7-30}$$

Substituting for $n_p'(0)$ and $n_p'(W)$ from Eqs. (7-23) and (7-24), we obtain

$$I_{ne} = \frac{qAD_{nb}n_{pb}}{L_{nb}\sinh(W/L_{nb})}\left\{\cosh\left(\frac{W}{L_{nb}}\right)\left[\exp\left(\frac{qV_{BE}}{kT}\right) - 1\right] - \left[\exp\left(\frac{qV_{BC}}{kT}\right) - 1\right]\right\} \tag{7-31}$$

The electron current crossing the *B-C* junction (I_{nc}) can be found by evaluating the electron base current of Eq. (7-28) at $x = W$. A sign change is needed here because $I_{\text{base diffusion}}$ was defined as crossing the base region from left to right, whereas I_{nc} was defined from right to left. Thus,

$$I_{nc} = -\frac{qAD_{nb}}{L_{nb}\sinh(W/L_{nb})}\left[n_p'(0) - n_p'(W)\cosh\left(\frac{W}{L_{nb}}\right)\right] \tag{7-32}$$

Substituting for $n_p'(0)$ and $n_p'(W)$ in Eq. (7-32), we have that

$$I_{nc} = -\frac{qAD_{nb}n_{pb}}{L_{nb}\sinh(W/L_{nb})}\left\{\left[\exp\left(\frac{qV_{BE}}{kT}\right) - 1\right] - \cosh\left(\frac{W}{L_{nb}}\right)\left[\exp\left(\frac{qV_{BC}}{kT}\right) - 1\right]\right\} \tag{7-33}$$

Our major goal in this section has been to find expressions for I_E and I_C as functions of V_{BE} and V_{BC}. Terminal currents I_E and I_C were related to the junction hole and electron currents by

$$I_E = -I_{ne} - I_{pe} \tag{7-11}$$

and

$$I_C = -I_{nc} - I_{pc} \tag{7-12}$$

I_{pe} and I_{pc} are expressed as functions of V_{BE} and V_{BC} in Eqs. (7-15) and (7-16). I_{ne} and I_{nc} are now expressed as functions of V_{BE} and V_{BC} in Eqs. (7-31) and (7-33). When the junction currents are summed as indicated in Eqs. (7-11) and (7-12), the terminal currents have the form

$$I_E = -I_{ES}\left[\exp\left(\frac{qV_{BE}}{kT}\right) - 1\right] + \alpha_R I_{CS}\left[\exp\left(\frac{qV_{BC}}{kT}\right) - 1\right] \tag{7-34}$$

and

$$I_C = \alpha_F I_{ES}\left[\exp\left(\frac{qV_{BE}}{kT}\right) - 1\right] - I_{CS}\left[\exp\left(\frac{qV_{BC}}{kT}\right) - 1\right] \tag{7-35}$$

Equations (7-34) and (7-35) are the Ebers–Moll equations for an *npn* transistor.

By writing out the complete expressions for I_E and I_C, and comparing them to the Ebers–Moll equations, we can evaluate I_{ES}, I_{CS}, α_F, and α_R. In other words, we can directly relate device parameters (the dc current gains and leakage currents of a transistor) to diffusion constants, diffusion lengths, equilibrium minority carrier concentrations, base width, and device area. Thus,

$$I_{ES} = qA\left[\frac{D_{pe}p_{ne}}{L_{pe}} + \frac{D_{nb}n_{pb}\cosh(W/L_{nb})}{L_{nb}\sinh(W/L_{nb})}\right] \tag{7-36}$$

$$I_{CS} = qA\left[\frac{D_{pc}p_{nc}}{L_{pc}} + \frac{D_{nb}n_{pb}\cosh(W/L_{nb})}{L_{nb}\sinh(W/L_{nb})}\right] \tag{7-37}$$

and

$$\alpha_R I_{CS} = \alpha_F I_{ES} = qA\left[\frac{D_{nb}n_{pb}}{L_{nb}\sinh(W/L_{nb})}\right] \tag{7-38}$$

I_{ES}, I_{CS}, α_F, and α_R can be defined from the Ebers–Moll equations through a process of elimination. If, for example, we short-circuit the *B-C* terminals, then Eqs. (7-34) and (7-35) can be written

$$I_E|_{V_{BC}=0} = -I_{ES}\left[\exp\left(\frac{qV_{BE}}{kT}\right) - 1\right] \tag{7-39}$$

and

$$I_C|_{V_{BC}=0} = \alpha_F I_{ES}\left[\exp\left(\frac{qV_{BE}}{kT}\right) - 1\right] \tag{7-40}$$

Combining Eqs. (7-39) and (7-40) and solving for α_F, we have that

$$\alpha_F = -\frac{I_C}{I_E}\bigg|_{V_{BC}=0} \tag{7-41}$$

This is the *forward transistor alpha* which was defined in Eq. (7-3). In Eq. (7-41), however, a minus sign is present, because α_F is defined here in terms of conventional currents entering the transistor terminals.

If the *B-E* diode is reverse-biased such that exp $(qV_{BE}/kT) \ll 1$ in Eq. (7-39), then $I_E|_{V_{BC}=0}$ is equal to I_{ES}. In other words, I_{ES} is the *reverse saturation current of the B-E diode with the B-C terminals shorted.*

With the *B-E* terminals shorted, Eqs. (7-34) and (7-35) can be written

$$I_E|_{V_{BE}=0} = \alpha_R I_{CS}\left[\exp\left(\frac{qV_{BC}}{kT}\right) - 1\right] \tag{7-42}$$

and

$$I_C|_{V_{BE}=0} = -I_{CS}\left[\exp\left(\frac{qV_{BC}}{kT}\right) - 1\right] \tag{7-43}$$

Combining Eqs. (7-42) and (7-43) and solving for α_R, we have

$$\alpha_R = -\frac{I_E}{I_C}\bigg|_{V_{BE}=0} \tag{7-44}$$

This is the *reverse transistor alpha.* With the *B-C* terminals forward-biased and the *B-E* terminals shorted, the collector injects electrons into the base region (it acts as an emitter). The ratio of the dc current crossing into the emitter region (I_E) to the current that was injected from the collector region (I_C) determines α_R. Since the transistor is not designed to be used backwards, α_R is generally much smaller than α_F.

If the *B-C* diode in Eq. (7-43) is reverse-biased such that exp $(qV_{BC}/kT) \ll 1$, then $I_C|_{V_{BE}=0}$ is equal to I_{CS}. Thus, I_{CS} is the *reverse saturation current of the B-C diode with the B-E terminals shorted.*

In terms of understanding device operation and being able to make devices with some degree of assurance, it is important to express α_F in terms of impurity concentrations, base width, and so on. The forward alpha can be determined from Eqs. (7-38) and (7-36) as follows:

$$\alpha_F = \frac{\alpha_F I_{ES}}{I_{ES}} = \frac{1}{\dfrac{D_p p_{ne}}{L_{pe}}\left[\dfrac{L_{nb}\sinh(W/L_{nb})}{D_{nb}n_{pb}}\right] + \cosh\left(\dfrac{W}{L_{nb}}\right)} \tag{7-45}$$

Since W is much smaller than L_{nb}, we can make the approximations

$$\sinh\left(\frac{W}{L_{nb}}\right) \approx \frac{W}{L_{nb}} \tag{7-46}$$

and

$$\cosh\left(\frac{W}{L_{nb}}\right) \approx 1 + \frac{1}{2}\left(\frac{W}{L_{nb}}\right)^2 \tag{7-47}$$

Then

$$\alpha_F \approx \frac{1}{1 + \dfrac{D_{pe}p_{ne}W}{D_{nb}n_{pb}L_{pe}} + \dfrac{1}{2}\left(\dfrac{W}{L_{nb}}\right)^2} \tag{7-48}$$

For small x, $1/(1 + x) \approx 1 - x$, so that Eq. (7-48) can be written

$$\alpha_F \approx 1 - \frac{D_{pe}p_{ne}W}{D_{nb}n_{pb}L_{pe}} - \frac{1}{2}\left(\frac{W}{L_{nb}}\right)^2 \tag{7-49}$$

The forward alpha is an important figure of merit for transistor operation. If fewer current losses occur in a transistor, a greater percentage of the emitter current will cross to the collector junction, and α_F will be closer to unity. Referring to Eq. (7-49), α_F will be close to unity if the ratios W/L_{pe}, W/L_{nb}, and $D_{pe}p_{ne}/D_{nb}n_{pb}$ are kept very small. This is accomplished by using a thin base region and doping the n-type emitter region much more heavily than the p-type base region.

The reverse alpha (α_R) deserves consideration here because situations occur in transistor circuits in which the *B-E* diode is reverse-biased and the *B-C* diode is forward-biased. It is usually desirable in this circumstance to keep α_R small so that the device does *not* operate efficiently. We can determine α_R by using the same methods and approximations that were used for α_F. Then

$$\alpha_R = \frac{\alpha_R I_{CS}}{I_{CS}} \approx 1 - \frac{D_{pc} p_{nc} W}{D_{nb} n_{pb} L_{pc}} - \frac{1}{2}\left(\frac{W}{L_{nb}}\right)^2 \tag{7-50}$$

Typically, α_R will be 0.4 or 0.5, and α_F is on the order of 0.99. The reverse alpha is small because the n-type collector region is more lightly doped than the p-type base region [i.e., the ratio $D_{pc}p_{nc}/D_{nb}n_{pb}$ in Eq. (7-50) is larger than unity]. There is also a practical consideration here that does not reveal itself through the Ebers–Moll model. Bipolar transistors are not constructed symmetrically. The fact that the effective emitter area is much smaller than the effective collector area tends to make α_R smaller than α_F.

When a transistor is operated in the forward-active region, the *B-C* junction is reverse-biased so that exp $(qV_{BC}/kT) << 1$. The magnitude of I_{CS} is negligibly small. The *B-E* junction is forward-biased such that exp $(qV_{BE}/kT) >> 1$. The Ebers–Moll equations, Eqs. (7-34) and (7-35), can then be written

$$I_E(\text{forward active}) \approx -I_{ES} \exp\left(\frac{qV_{BE}}{kT}\right) \tag{7-51}$$

and

$$I_C(\text{forward active}) \approx \alpha_F I_{ES} \exp\left(\frac{qV_{BE}}{kT}\right) \tag{7-52}$$

Thus,

$$I_C(\text{forward active}) \approx -\alpha_F I_E \tag{7-53}$$

Although α_F is strictly defined with the *B-C* terminals shorted, it is also a good measure of transistor current gain in the forward active region.

Another way to think of transistor operation in the forward-active region is in terms of the efficiency with which useful carriers (electrons in the *npn* case) travel from the emitter to the collector. This device efficiency can be broken into two components; emitter efficiency and base transport factor.

Emitter efficiency γ *is defined as the ratio of electron current to total current crossing the B-E junction:*

$$\text{emitter efficiency } \gamma = \frac{I_{ne}}{I_{ne} + I_{pe}} \tag{7-54}$$

Base transport factor α_T *is defined as the ratio of the electron current reaching the B-C depletion region to the electron current entering the base region:*

$$\text{base transport factor } \alpha_T = \left|\frac{I_{nc}}{I_{ne}}\right| \tag{7-55}$$

Expressions for the component currents I_{pe}, I_{ne}, and I_{nc} are given in Eqs. (7-15), (7-31), and (7-33), respectively. If the forward-active region approximations of Eqs. (7-51) and (7-52) are used with these component currents, then

$$\gamma \approx \frac{1}{1 + \left(\frac{D_{pe}p_{ne}L_{nb}}{D_{nb}n_{pb}L_{pe}}\right)\tanh\left(\frac{W}{L_{nb}}\right)} \tag{7-56}$$

If we make the approximations tanh $(W/L_{nb}) \approx (W/L_{nb})$ for small W, and $1/(1 + x) \approx (1 - x)$ for small x, then

$$\gamma = 1 - \frac{D_{pe}p_{ne}W}{D_{nb}n_{pb}L_{pe}} \tag{7-57}$$

Solving for the base transport factor in the same manner,

$$\alpha_T \approx \frac{1}{\cosh\,(W/L_{nb})} \approx 1 - \frac{1}{2}\left(\frac{W}{L_{nb}}\right)^2 \tag{7-58}$$

Taking the product $\gamma\alpha_T$ and neglecting the small cross-product term, we find that

$$\alpha_F \approx \gamma\alpha_T \tag{7-59}$$

Both α_F and $\gamma\alpha_T$ are measures of the efficiency with which useful current is transported from emitter to collector.

The forward transistor beta β_F was defined in Section 7-1 as

$$\beta_F = \left.\frac{I_C}{I_B}\right|_{V_{BC}=0} \tag{7-4}$$

It was related to α_F by

$$\alpha_F = \frac{\beta_F}{\beta_F + 1} \tag{7-5}$$

From Eq. (7-5),

$$1 - \alpha_F = 1 - \frac{\beta_F}{\beta_F + 1} = \frac{1}{\beta_F + 1} \approx \frac{1}{\beta_F} \tag{7-60}$$

since $\beta_F \geq 100$. Then, from Eq. (7-49), $1/\beta_F$ can be expressed as

$$\frac{1}{\beta_F} = \frac{D_{pe} p_{ne} W}{D_{nb} n_{pb} L_{pe}} + \frac{1}{2}\left(\frac{W}{L_{nb}}\right)^2 \tag{7-61}$$

where the left term denotes emitter efficiency and the right term denotes base transport factor. To maximize β_F, these terms should be small. This is accomplished by making W smaller than L_{pe} and L_{nb}, and by doping the emitter more heavily than the base region such that $p_{ne} << n_{pb}$.

7-3 LIMITATIONS AND MODIFICATIONS OF THE MODEL

Transistor performance is affected by many factors that cannot be directly accounted for in the Ebers–Moll model. Some of these factors can be dealt with by adding terms to the $1/\beta_F$ expression of Eq. (7-61). Others must be treated separately.

7-3-1 Surface States

Hole current from base to emitter, and hole–electron recombination in the base region, cause a base current to exist between the base contact and the active base region of a transistor. This base current has a lateral direction, under an Si–SiO_2 interface containing an abundance of surface states. The surface states can be characterized by a surface recombination velocity s, as discussed in Chapter 6. A vertical component of base current is lost through surface recombination. For a given value of I_C, an increase in I_B is needed to compensate this loss. As shown in Eq. (7-4), this causes a decrease in β_F. The β_F decrease can be accounted for by adding a term to Eq. (7-61). Thus,

$$\frac{1}{\beta_F} = \frac{D_{pe} p_{ne} W}{D_{nb} n_{pb} L_{pe}} + \frac{1}{2}\left(\frac{W}{L_{nb}}\right)^2 + \frac{s A_s W}{A_E D_{nb}} \tag{7-62}$$

where s denotes surface recombination velocity, A_s represents the effective area of surface recombination, and A_E denotes the emitter area.

7-3-2 Beta Variations with Collector Current

In Eq. (7-62), β_F is independent of transistor collector current I_C. When β_F is plotted for a wide range of collector currents, however, it decreases at both high and low values of current, as shown in Fig. 7–9. At low currents β_F decreases due to hole–electron recombination in the *B-E* depletion region. For large values of I_C, β_F decreases due to changes in the conductivity of the base region (conductivity modulation) and due to current crowding in the emitter

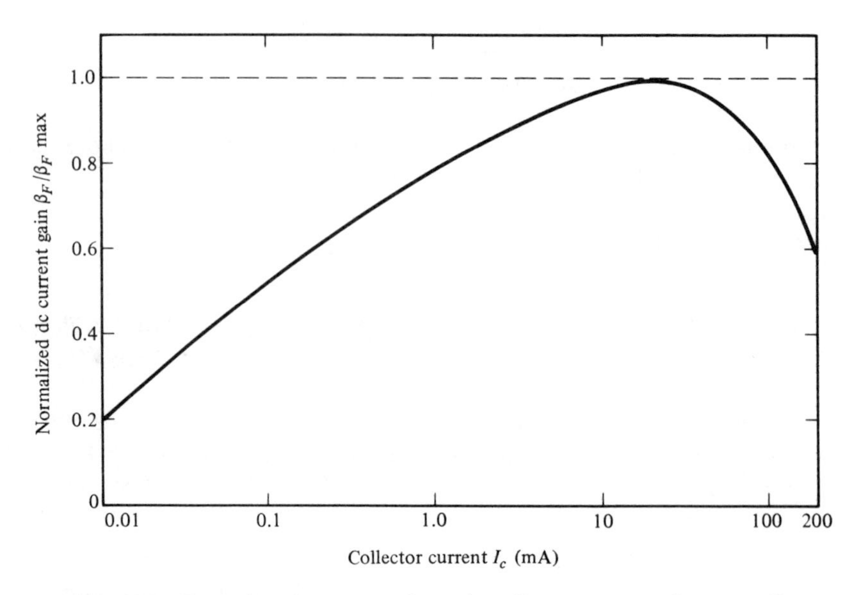

Fig. 7-9 Typical variations in β_F with collector current for a small-signal silicon transistor.

region (emitter crowding). These three mechanisms will now be discussed in more detail.

Recombination in the *B-E* Depletion Region. Hole–electron recombination in the depletion region of a forward-biased diode was discussed in Chapter 6. Equation (6-122) is repeated here for convenience.

$$I_{\text{recomb}} \approx q\left(\frac{n_i}{T_0}\right) WA_j \exp\left(\frac{qV}{2kT}\right) \tag{6-122}$$

Applied to the emitter–base diode of a transistor, W and A_j represent the width and cross-sectional area of the *B-E* depletion region, and V represents the forward bias voltage V_{BE}. I_{recomb} is dependent on the number of recombination centers in the *B-E* depletion region. Centers near the silicon surface (surface states) are dominant. This is a relatively constant current in the 100-nA range. Since it is independent of I_B, it does not contribute to the forward current gain β_F. It can be modeled by adding a term I_{RE}/I_E to the $1/\beta_F$ expression given in Eq. (7-62). Thus,

$$\frac{1}{\beta_F} = \frac{D_{pe}p_{ne}W}{D_{nb}n_{pb}L_{pe}} + \frac{1}{2}\left(\frac{W}{L_{nb}}\right)^2 + \frac{sA_sW}{A_ED_{nb}} + \frac{I_{RE}}{I_E} \tag{7-63}$$

where I_{RE} denotes the *B-E* recombination current and I_E represents the emitter current. If I_{RE} for a transistor is 100 nA and I_E is 1 mA, then I_{RE} has little effect on β_F. If this same transistor is operated with a collector current of 1 μA, however, the term I_{RE}/I_E will limit β_F to less than 10.

Conductivity Modulation. In developing the Ebers–Moll model for an *npn* transistor, we assumed conditions of *low-level injection.* In other words, we assumed that the electron concentration injected from the emitter into the base region was not sufficient to disturb the majority hole concentration in the base. As the current level and the injected electron concentration are increased, majority and minority concentrations in the base region become comparable. At still higher current levels, the injected electron concentration greatly exceeds the base doping concentration. By definition, conditions of *high-level injection* begin when the minority concentration in the base becomes comparable with the base doping concentration. For most low-power transistors, this corresponds to collector currents in the range 1 to 10 mA.

Regardless of current magnitudes, conditions of *near-charge neutrality* must exist in any localized volume within the base region. A local excess of electrons causes an E field, which causes hole motion, and so on. When $n_p(x)$ exceeds the doping concentration N_{AB}, an electric field exists across the base region, causing hole and electron concentrations to become approximately equal (i.e., $n \approx p$). Because near-charge neutrality must exist, local concentration gradients dn/dx and dp/dx are also approximately equal.

Hole and electron current densities in the base region can be written

$$J_n(\text{net}) = \overrightarrow{qD_n \frac{\partial n}{\partial x}} + \overrightarrow{qn\mu_n \frac{\partial v}{\partial x}} \tag{7-64}$$

and

$$J_p(\text{net}) = \overrightarrow{-qD_p \frac{\partial p}{\partial x}} + \overleftarrow{qp\mu_p \frac{\partial v}{\partial x}} \tag{7-65}$$

The arrows on the drift and diffusion components of current density correspond to the directions shown in Fig. 7-10.

Fig. 7-10 Base current density components under high-level injection conditions.

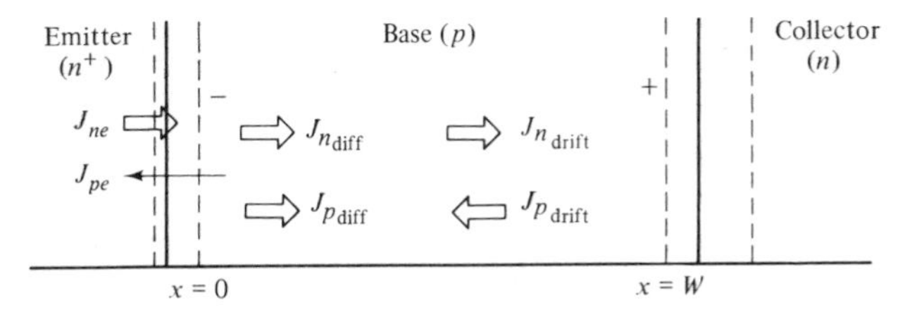

The electron current density injected into the base (J_{ne}) is much greater in magnitude than the hole current density injected into the emitter (J_{pe}). As a result, the hole currents that maintain charge neutrality within the base are much larger than J_{pe}. Within the base region, we can then make the approximation

$$J_p(\text{diff}) \approx J_p(\text{drift}) \tag{7-66}$$

along with the near-neutrality statements,

$$p \approx n \tag{7-67}$$

and

$$\frac{dp}{dx} \approx \frac{dn}{dx} \tag{7-68}$$

When Eq. (7-65) is solved for the voltage developed across the base region with the conditions given in Eq. (7-66) through Eq. (7-68), we find that

$$V_{\text{base}} = V_{BE} \tag{7-69}$$

The voltage across the base region is equal to the *B-E* junction voltage. In other words, half of the externally applied forward bias voltage appears across the *B-E* junction and the other half appears across the base region.

For a transistor operating in the forward-active region under conditions of low-level injection,

$$I_C \approx K \exp\left(\frac{qV_{BE}}{kT}\right) \tag{7-70}$$

For conditions of high-level injection,

$$I_C \approx K \exp\left(\frac{qV_{BE}}{2kT}\right) \tag{7-71}$$

The transition from Eq. (7-70) to Eq. (7-71) corresponds to a decrease in β_F.

If the conditions of Eqs. (7-66) through (7-68) are applied to Eq. (7-64), it is easily shown that the drift and diffusion components of electron current density are equal for high-level injection:

$$\left| qD_n \frac{\partial n}{\partial x} \right| = \left| qn\mu_n \frac{\partial v}{\partial x} \right| \tag{7-72}$$

Equation (7-64) can then be written

$$J_n(\text{net}) = q(2D_n)\frac{\partial n}{\partial x} \tag{7-73}$$

The increase in electron current density due to the electric field in the base region can be treated as an *apparent doubling of the diffusion constant.* Equation (7-62) is rewritten here for convenience:

$$\frac{1}{\beta_F} = \underbrace{\frac{D_{pe}p_{ne}W}{D_{nb}n_{pb}L_{pe}}}_{\substack{\text{emitter}\\\text{efficiency}}} + \underbrace{\frac{1}{2}\left(\frac{W}{L_{nb}}\right)^2}_{\substack{\text{base}\\\text{transport}\\\text{factor}}} + \underbrace{\frac{sA_sW}{A_ED_{nb}}}_{\substack{\text{surface}\\\text{states}}} \tag{7-62}$$

Under low-level conditions, β_F is related to emitter efficiency, base transport factor, and surface states as shown.

It is useful to show the effect on Eq. (7-62) of high-level injection. If we make the very rough approximations that $\mu_{nb} \approx \mu_{ne}$, and $\mu_{pe} \approx \mu_{pb}$, then emitter efficiency can be rewritten

$$\frac{D_{pe}p_{ne}}{D_{nb}n_{pb}}\left(\frac{W}{L_{pe}}\right) \approx \frac{\sigma_{pb}}{\sigma_{ne}}\left(\frac{W}{L_{pe}}\right) \tag{7-74}$$

Equation (7-62) can then be approximated by

$$\frac{1}{\beta_F} \approx \frac{\sigma_{pb}}{\sigma_{ne}}\left(\frac{W}{L_{pe}}\right) + \frac{1}{2}\left(\frac{W}{L_{nb}}\right)^2 + \frac{sA_sW}{A_ED_{nb}} \tag{7-75}$$

When high-level injection occurs, Eq. (7–75) is modified in two ways: (1) As the injected minority carrier concentration in the base region increases, majority concentration increases as well. This increases the conductivity of the base region to σ'_{pb} (thus the term "conductivity modulation"). (2) The electric field in the base region causes an apparent doubling of the diffusion constant.

The $1/\beta'_F$ expression for high-level injection is then given by

$$\frac{1}{\beta'_F} \approx \frac{\sigma'_{pb}}{\sigma_{ne}}\left(\frac{W}{L_{pe}}\right) + \frac{1}{4}\left(\frac{W}{L_{nb}}\right)^2 + \frac{sA_sW}{2A_ED_{nb}} \tag{7–76}$$

Comparison of Eq. (7–75] and Eq. (7–76) shows that the doubled diffusion constant has the apparent effect of improving the base transport factor and reducing losses due to surface states. Physically, however, these effects cannot be divorced from the increase in base conductivity (σ'_{pb}), which reduces emitter

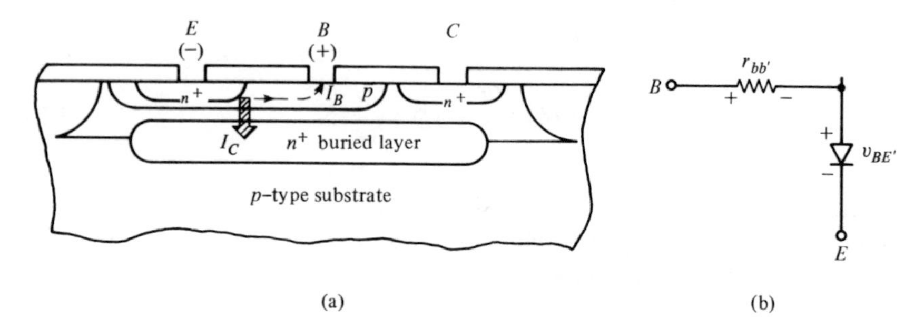

Fig. 7-11 Current crowding under conditions of high-level injection: (a) *npn* transistor cross section showing I_C and I_B; (b) schematic of the base–emitter circuit, showing the base spreading resistance $r_{bb'}$.

efficiency and causes β_F to decrease with increased collector current due to conductivity modulation.

Current Crowding. Base current exists in a transistor primarily due to minority injection from base to emitter and recombination in the base region. As shown in Fig. 7-11, the base current has a direction parallel to the device surface between the active transistor region and the base contact. The base resistance in this region is called the *base spreading resistance* and is denoted by $r_{bb'}$. As shown in Fig. 7-11(b),

$$v_{BE} = i_B r_{bb'} + v_{BE'} \tag{7–77}$$

where $v_{BE'}$ is the base–emitter voltage in the active transistor region. The $i_B r_{bb'}$ voltage drop reduces the base–emitter voltage in the active region. It varies with the path length between the base contact and the active transistor region, and with the magnitude of base current. This base voltage drop causes a lateral variation in the forward bias across the active emitter–base region, so that $v_{BE'}$ is larger near the base contact. As emitter and base currents are increased, the effect becomes more pronounced, until the current density is restricted to the emitter periphery near the base contact, as shown in Fig. 7-11(a). This effect is called *current crowding* or *emitter crowding.*

As a practical rule of thumb, a bipolar transistor can carry 0.8 to 1.2 mA of emitter current per 5 μm of *useful* emitter periphery (i.e., emitter periphery next to a base contact). Useful emitter periphery can be doubled by using two base stripes, as shown in Fig. 7-12. The two collector stripes shown in this figure will reduce the collector saturation resistance.

The current-handling capacity of a transistor can be further increased by using interdigitation, as shown in Fig. 7–13. In this transistor, three emitter stripes and three base stripes are used. For this geometry, the maximum beta will occur at a specific value of I_E. If a maximum beta is desired at a larger value of current, more emitter and base stripes can be used. The maximum

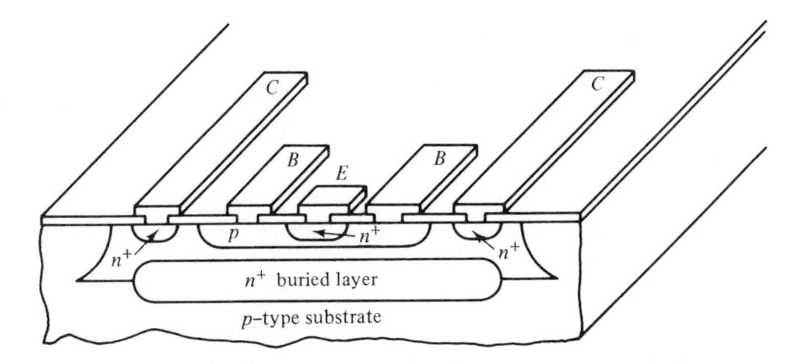

Fig. 7-12 *npn* Transistor with two base stripes to increase the useful emitter periphery.

current-handling capacity of a planar bipolar transistor is limited by the overall thermal resistance of the chip and its mounting and the resultant power rating of the packaged circuit. The thermal resistance of a dual-in-line plastic package, which is soldered to a printed circuit board and in still air, is about 145°C/*W*.

Fig. 7-13 Interdigitation to maximize the useful emiter periphery of an *npn* transistor.

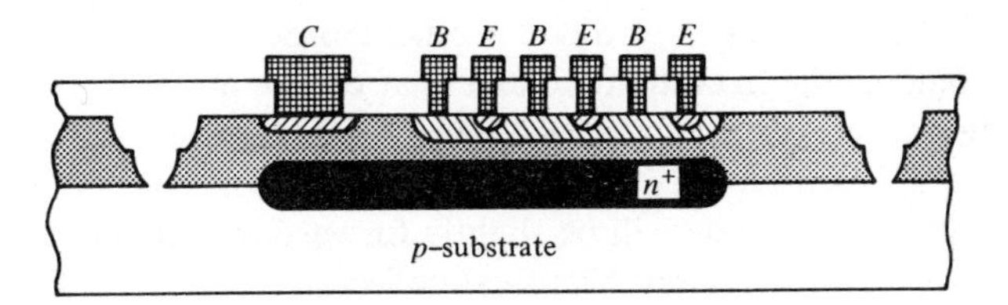

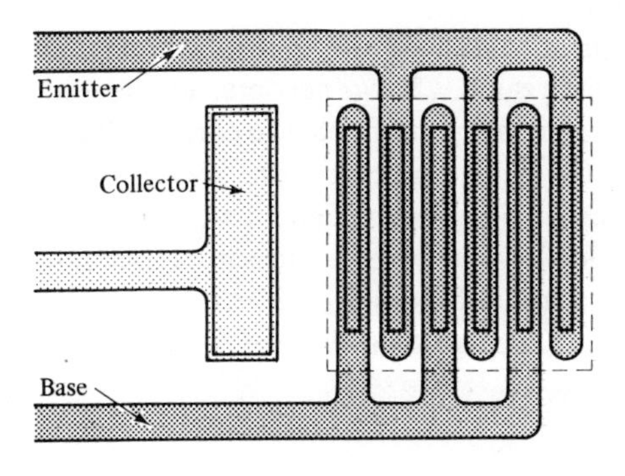

7-3-3 Voltage Limitations

Phenomena such as base width modulation, punch-through, and avalanche breakdown are not accounted for in the Ebers–Moll model.

As the magnitude of reverse bias across the collector–base diode of a transistor is increased, the depletion layer widens, decreasing the undepleted base width. This has the effect of increasing the transistor beta and is known as *base width modulation.*

In *npn* transistors with betas of 1000 to 3000, the undepleted base width may be on the order of tenths of micrometers with low-bias voltages. As the reverse collector–base voltage is increased, the depletion region moves across the base region until it touches the emitter–base depletion region at some point. In effect, the base is shorted and a large current exists between emitter and collector. The localized heating destroys the transistor. This breakdown phenomenon is referred to as *punch-through.*

In *npn* transistors with betas between 100 and 500 (which is more typical), the base is normally wide enough so that *avalanche breakdown* occurs at a reverse bias voltage that is lower than the punch-through voltage. It therefore predominates over punch-through as a boundary condition in device design. Avalanche breakdown occurs at a lower voltage in a common-emitter than in a common-base transistor configuration. The reason for this is discussed next.

The maximum voltage that can be applied to a transistor in the common-base configuration is BV_{CBO}, the collector–base (avalanche) breakdown voltage with the emitter lead open. It is determined by the impurity concentration in the collector region, as discussed in Chapter 6. The leakage current (I_{CBO}) across the reverse biased base–collector diode is on the order of nanoamperes.

BV_{CEO} denotes the maximum voltage that can be applied in the common-emitter configuration with the base lead open. With the transistor base floating, it will acquire a potential between the emitter and collector potentials. As a result, the emitter–base diode will be slightly forward-biased, and electrons will be injected into the base region. The total collector current will be the sum of the leakage current across the reverse-biased collector–base diode and the injected electron current that reaches the collector depletion region. It is denoted by I_{CEO}.

Avalanche breakdown is caused by the multiplication of charge carriers in the base–collector depletion region. Charge carriers, accelerated by the electric field, break covalent bonds, generating more charge carriers. The surplus charge carriers are accelerated by the field, breaking more covalent bonds, and so on. The magnitude of the reverse voltage at which this avalanche process occurs is sensitive to the magnitude of the reverse current. Since I_{CEO} is orders of magnitude larger than I_{CBO}, a transistor connected in a common-emitter configuration will break down at a lower voltage than one connected in a common-base configuration. Typically,

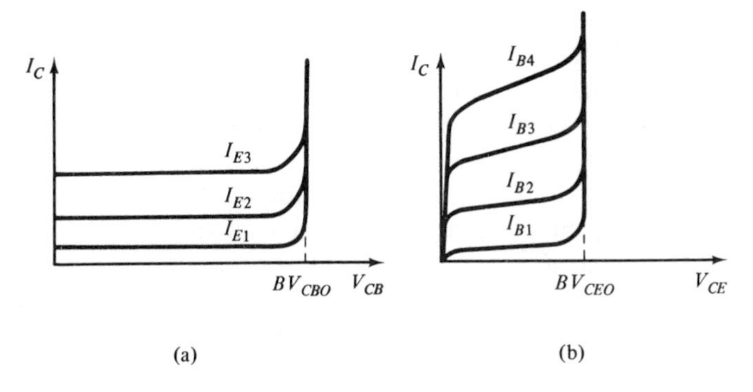

Fig. 7-14 Transistor characteristic curves, indicating avalanche breakdown and base width modulation: (a) common base; (b) common emitter.

$$BV_{CEO} = \frac{BV_{CBO}}{N\sqrt{\beta_F}} \tag{7-78}$$

where N varies between 4 and 5 and is determined empirically rather than derived.

Common-base and common-emitter characteristic curves are shown in Fig. 7-14. Avalanche breakdown voltages BV_{CBO} and BV_{CEO} are shown. The increase in the slopes of the curves as V_{CE} increases in Fig. 7-14(b) is caused by the base width modulation.

7-4 SMALL-SIGNAL TRANSISTOR OPERATION—THE HYBRID-π MODEL

In many applications, transistors are used to process small signals. A transistor is biased in the forward active mode, at a quiescent operating point (or Q point). Small ac signal variations are superimposed on the dc values of voltage and current. The transistor behaves linearly as far as these small signals are concerned. Although the nonlinear Ebers–Moll model is still valid, a linear model is more appropriate for small-signal analysis, because mesh and nodal techniques and the network theorems can be used to predict electronic circuit behavior. The transistors in a given circuit are simply replaced by the linear model. Standard ac circuit analysis techniques are then used to predict circuit parameters such as voltage gain, input and output resistance, and frequency response. Although many good small-signal models exist, the hybrid-π model has been particularly well accepted. This acceptance may be due to the fact that the model parameters are well related to physical device processes and can be measured with minimum difficulty.

We can use the Ebers–Moll model to define the parameters in the hybrid-π model. It is convenient, however, to make some changes in notation and reference current directions. In the development of the Ebers–Moll model, conventional currents into the transistor were assumed positive. In the hybrid-π model, we will treat I_E as the sum of I_B and I_C, as shown in Fig. 7-15. This will cause sign changes in the Ebers–Moll equations. Double-subscript notation will be used for the terminal voltages. Since the emitter–base diode is forward-biased, we will assume that $\exp(qV_{BE}/KT) >> 1$. With the base–collector diode reverse-biased, $\exp(qV_{BC}/kT) << 1$, and the saturation current I_{CS} is negligibly small. With these considerations in mind, we can write the Ebers–Moll equations as follows:

$$I_E = I_{ES} \exp\left(\frac{qV_{BE}}{kT}\right) \tag{7-79}$$

$$I_C = \alpha_F I_{ES} \exp\left(\frac{qV_{BE}}{kT}\right) = \alpha_F I_E \tag{7-80}$$

and

$$I_E = I_B + I_C \tag{7-81}$$

With the transistor at a fixed operating point in the forward-active mode, we can also write

$$I_B = (1 - \alpha_F) I_E \tag{7-82}$$

and

$$\beta_F = \frac{I_C}{I_B} = \frac{\alpha_F}{1 - \alpha_F} \tag{7-83}$$

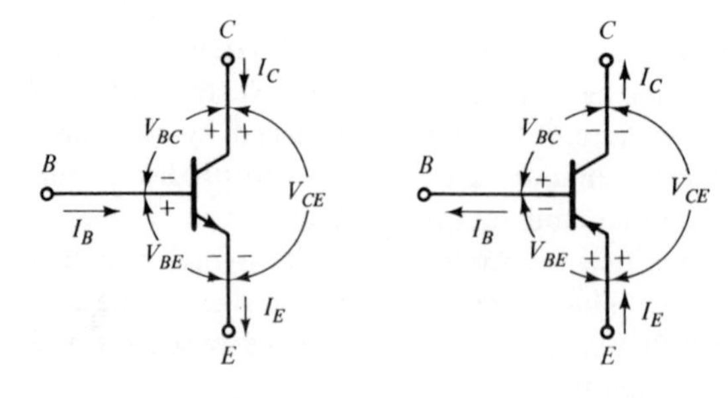

Fig. 7-15 Reference directions for (a) *npn* and (b) *pnp* transistors.

Figure 7-16 shows a low-frequency, first-order version of the hybrid-π model. A small change in the ac input voltage v_{be} will cause a change in the collector current i_c at the output of the transistor. The constant of proportionality between i_c and v_{be} is g_m, the *transconductance,* or transfer conductance of the transistor. Thus

$$i_c = g_m v_{be} \tag{7-84}$$

From Eq. (7-80),

$$g_m = \frac{\partial I_C}{\partial V_{BE}} = \frac{qI_C}{kT} \tag{7-85}$$

The small-signal base–emitter resistance of the transistor (looking into the base) is denoted by r_π. From Eqs. (7-82) and (7-79), the dc base current is

$$I_B = (1 - \alpha_F)I_{ES} \exp\left(\frac{qV_{BE}}{kT}\right)$$

Thus,

$$\frac{1}{r_\pi} = \frac{\partial I_B}{\partial V_{BE}} = \frac{qI_B}{kT} \tag{7-86}$$

or

$$r_\pi = \frac{kT}{qI_B} = \frac{kT}{q(I_C/\beta_F)} = \frac{\beta_F}{g_m} \tag{7-87}$$

The small-signal ac resistance of the transistor looking into the emitter is denoted by r_e, where from Eq. (7-79),

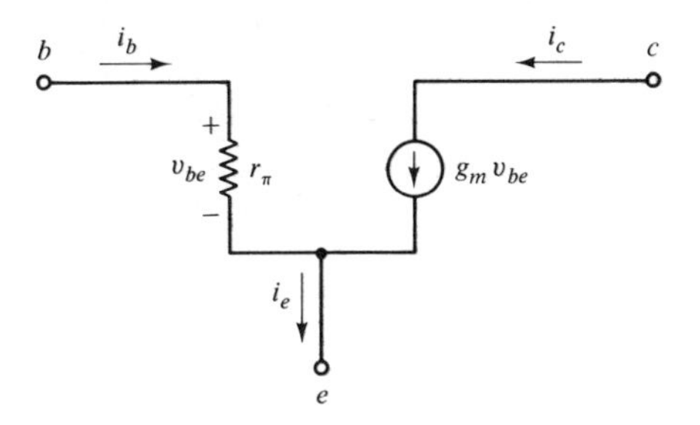

Fig. 7-16 Low-frequency first-order version of the hybrid-π model.

$$\frac{1}{r_e} = \frac{\partial I_E}{\partial V_{BE}} = \frac{qI_E}{kT}$$

Then

$$r_e = \frac{kT}{qI_E} \tag{7-88}$$

and

$$r_\pi = (\beta_F + 1)r_e \tag{7-89}$$

If $I_E = 1$ mA at room temperature ($T = 300$ K), and $\beta_F = 100$, r_e will be about 26 Ω and r_π will be about 2.626 kΩ.

Figure 7-17 shows one example of how the hybrid-π model can be applied. Since the internal resistance of the dc power supply ($+V_{CC}$) is negligibly small, V_{CC} is treated as ground in the ac equivalent circuit. Applying Kirchhoff's voltage law to Fig. 7-17(b), we find that the voltage gain A_v is given by

$$A_v = \frac{v_o}{v_{\text{in}}} = -\frac{\beta_F R_C}{(\beta_F + 1)(r_e + R_E)} \approx -\frac{R_C}{R_E} \tag{7-90}$$

The input resistance R_{in} is given by

$$R_{\text{in}} = \frac{v_{\text{in}}}{i_{\text{in}}} = R_1 || R_2 || (\beta_F + 1)(r_e + R_E) \tag{7-91}$$

where $R_1 || R_2$ denotes the parallel combination of R_1 and R_2.

Fig. 7-17 Common-emitter amplifier and its ac equivalent circuit.

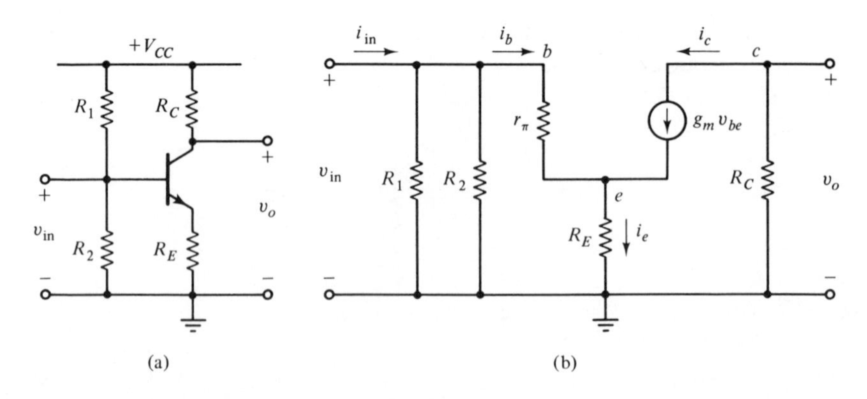

A high-frequency version of the hybrid-π model which takes base spreading resistance, base width modulation, and charge-storage effects into account is shown in Fig. 7-18.

Approximate values for the hybrid-π parameters of a small-signal transistor operating at $I_C = 1$ mA and at room temperature are as follows:

$$r_{bb'} \approx 50\text{–}100\ \Omega \qquad C_{jc} \approx 1\text{–}10\ \text{pF}$$
$$r_\pi \approx 2\text{–}3\ \text{k}\Omega \qquad r_{b'c} \approx 1\text{–}10\ \text{M}\Omega$$
$$g_m \approx 40\ \text{mA/V} \qquad r_{ce} \approx 50\text{–}250\ \text{k}\Omega$$
$$C_{je} \approx 1\text{–}10\ \text{pF} \qquad C_D \approx 50\text{–}100\ \text{pF}$$

The base spreading resistance $r_{bb'}$ is determined primarily by the device geometry and base region conductivity. The effects of $r_{bb'}$ are negligible at low current levels and low frequencies. Base spreading resistance is particularly important at high frequencies, however. The base currents that charge the device capacitances must pass through $r_{bb'}$.

C_{je} represents the capacitance of the forward-biased emitter base depletion layer. If the base and emitter impurity concentrations are known, C_{je} can be calculated from Eq. (6-115) or (6-118).

C_{jc} represents the capacitance of the reverse-biased base–collector depletion layer.

Resistors $r_{b'c}$ and r_{ce} represent the effects of base width modulation. They are second-order effects and are dependent on bias conditions, the impurity profile in the base region, base width, and the resistivity of the epi-layer.

C_D represents the diffusion capacitance of the transistor. At a given time, a finite concentration of minority electrons is diffusing across the p-type base region. If V_{BE} changes, the total electron concentration in the base region must also change. This does not happen instantaneously. The relation between stored base charge and base–emitter voltage can be modeled as diffusion capacitance C_D. Figure 7-19 shows the minority carrier distribution in the base region of

Fig. 7-18 High-frequency hybrid-π model. Second-order effects are included.

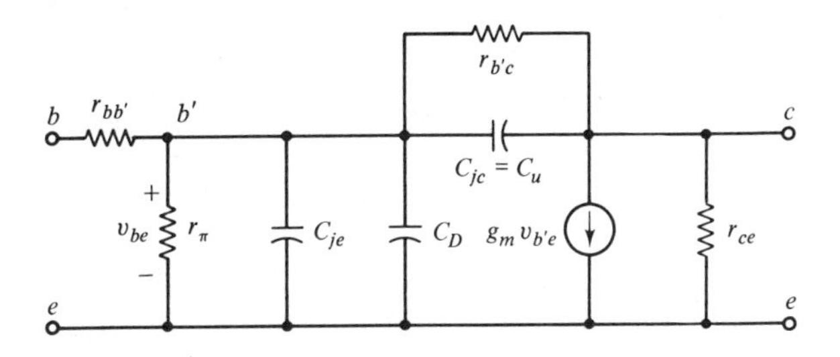

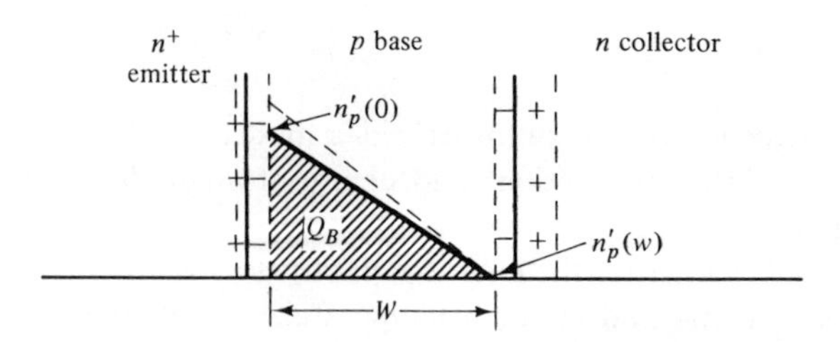

Fig. 7-19 Minority carrier distribution in the base region of an *npn* transistor.

an *npn* transistor. The electron concentration on the base side of the emitter–base depletion region is

$$n_p'(0) = n_{pb} \exp\left(\frac{qV_{BE}}{kT}\right) \tag{7-92}$$

The minority concentration at $n_p'(W)$ is approximately zero. The stored charge in the base region is represented by the shaded area in Fig. 7-19. It can be written

$$Q_B = \tfrac{1}{2} q n_{pb} WA \exp\left(\frac{qV_{BE}}{kT}\right) \tag{7-93}$$

where A is the cross-sectional area of the device and W is the base width. The diffusion capacitance is then given by

$$C_D = \frac{dQ_B}{dV_{BE}} = \tfrac{1}{2} q n_{pb} WA \left(\frac{q}{kT}\right) \exp\left(\frac{qV_{BE}}{kT}\right) \tag{7-94}$$

The emitter current I_E can be approximated as

$$I_E \approx \frac{qAD_{nb}n_{pb}}{W} \exp\left(\frac{qV_{BE}}{kT}\right) \tag{7-95}$$

Combining Eqs. (7-94) and (7-95), we find that

$$C_D = \frac{W^2}{2D_{nb}} \frac{q}{kT} I_E \tag{7-96}$$

Since $I_E \approx I_C$ and $g_m = qI_C/kT$, Eq. (7-96) can be written

$$C_D = \frac{W^2}{2D_{nb}} g_m \tag{7-97}$$

7-4-3 Common-Emitter Short-Circuit Current Gain

We would like to determine the high-frequency response of a common-emitter amplifier in terms of the hybrid-π device parameters. This is accomplished by driving the amplifier with a sinusoidal constant-current source and creating an ac short between collector and emitter. For these conditions, we will write an expression for the short-circuit current gain. From this expression we can find the high half-power frequency f_β and the frequency f_t at which the amplifier gain becomes unity. The frequency response of a common-emitter amplifier decreases as the value of load resistance is increased. By determining the short-circuit current gain, we obtain frequency data that are dependent on the hybrid-π device parameters rather than the magnitude of the load resistance. For simplicity, we will neglect the effects of $r_{bb'}$ and $r_{b'c}$ in this development, as shown in Fig. 7-20. The base modulation resistance r_{ce} disappears because it is in parallel with the short circuit. C_π represents the sum of the diffusion capacitance C_D and the emitter–base depletion layer capacitance C_{je}. C_u represents the collector–base depletion capacitance C_{jc}.

Using Kirchhoff's current law and standard Laplace notation, we can write expressions for $I_b(s)$ and $I_c(s)$ as follows:

$$I_b(s) = \left(\frac{1}{r_\pi} + C_\pi s + C_u s\right) V_{be}(s) \tag{7-98}$$

and

$$I_c(s) = g_m V_{be}(s) - C_u s V_{be}(s) \tag{7-99}$$

or

$$I_c(s) = \left(\frac{\beta}{r_\pi} - C_u s\right) V_{be}(s) \tag{7-100}$$

Combining terms and taking the ratio $I_c(s)/I_b(s)$, we have

$$\frac{I_c(s)}{I_b(s)} = \beta \left[\frac{1 - \dfrac{r_\pi C_u}{\beta} s}{1 + r_\pi (C_\pi + C_u) s}\right] \tag{7-101}$$

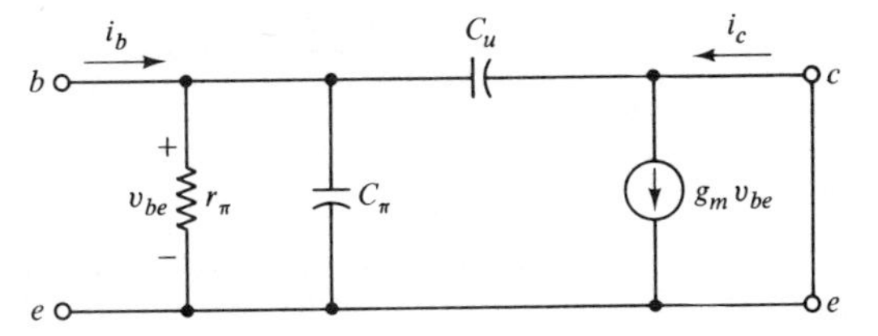

Fig. 7-20 High-frequency hybrid-π model for the calculation of short-circuit common-emitter current gain.

If $\beta = 100$, $r_\pi = 2\ \text{k}\Omega$, $C_u = 5$ pF, and $C_\pi = 50$ pF, then $r_\pi C_u/\beta = 1 \times 10^{-10}$ s, and $r_\pi(C_\pi + C_u) = 1.1 \times 10^{-7}$ s. We can therefore neglect $r_\pi C_u/\beta$ without introducing much error. Also, for sinusoidal excitation, $s = j\omega$, so that Eq. (7-101) can be written

$$\frac{I_c(j\omega)}{I_b(j\omega)} = \frac{\beta}{1 + j\omega r_\pi(C_\pi + C_u)} = \frac{\beta}{1 + j\dfrac{f}{\left[\dfrac{1}{2\pi r_\pi(C_\pi + C_u)}\right]}} \tag{7-102}$$

or

$$\frac{I_c(j\omega)}{I_b(j\omega)} = \frac{\beta}{1 + jf/f_\beta} \tag{7-103}$$

where

$$f_\beta = \frac{1}{2\pi r_\pi(C_u + C_\pi)} \tag{7-104}$$

If we again use the values $\beta = 100$, $r_\pi = 2\ \text{k}\Omega$, $C_u = 5$ pF, and $C_\pi = 50$ pF, then the *beta cutoff frequency* $f_\beta = 1.47$ MHz. At $f = f_\beta$, the current gain $|A_i|$ is equal to $1/\sqrt{2} = 0.707$ times the low-frequency short-circuit common-emitter current gain β. Since power gain is proportional to $|A_i|^2$, f_β represents the high half-power frequency of the common-emitter circuit.

The frequency at which the common-emitter short-circuit current gain drops to unity is called f_t. It can be determined as follows:

$$|A_i| = \left|\frac{I_c(j\omega)}{I_b(j\omega)}\right| = \left|\frac{\beta}{1 + jf_t/f_\beta}\right| = 1 \tag{7-105}$$

Fig. 7-21 Common-emitter short circuit current gain vs. frequency.

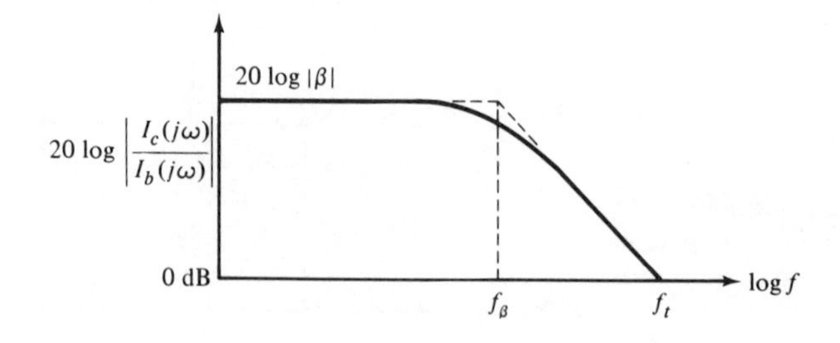

Since $\beta >> 1$, $|A_i| = 1$ at

$$f_t = \beta f_\beta \tag{7-106}$$

Thus, f_t represents the *gain–bandwidth product* of the short-circuited common-emitter amplifier. Usually, f_t and β are specified by the device manufacturer. Using our hybrid-π parameter values again, $f_t = 100 f_\beta = 147$ MHz. Figure 7-21 shows the variation in common-emitter short-circuit current gain with frequency.

7-5 *pnp* TRANSISTORS

Two types of *pnp* transistors can be made using techniques compatible with standard IC fabrication methods. They are called *substrate and lateral pnp* transistors. Because hole mobilities are smaller than electron mobilities and the process techniques are not designed to optimize *pnp* transistor behavior, they generally have lower betas and poorer frequency response than do *npn* transistors.

7-5-1 Substrate *pnp* Transistors

These devices already exist in conventional integrated circuits. The *p*-type base diffusion becomes the *pnp* emitter, the epi-layer becomes the base, and the *p*-type substrate becomes the collector, as shown in Fig. 7-22. The substrate must be connected to the negative power supply voltage in order to reverse-bias the substrate–epitaxial layer junctions of other devices and circuits on the same IC chip. Thus, the substrate *pnp* transistor can be used only as an emitter follower (common collector).

The *n*-type epi-layer between the *p*-diffusion and the substrate determines the base width, as shown in Fig. 7-22. If we assume that the epi-layer is 8 μm thick and the *p* diffusion is 2 μm deep, the base width is approximately 6 μm. However, because the epi-layer has not been doped with an n^+ buried

Fig. 7-22 Cross section of a substrate *pnp* transistor.

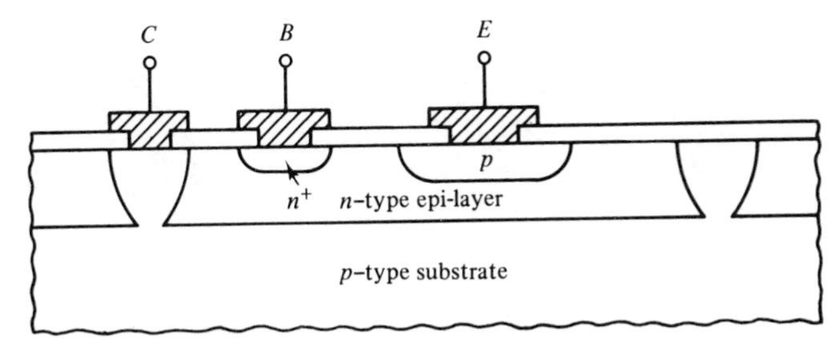

layer in the active base region, it will allow minority hole diffusion lengths (L_{pb}) on the order of 25 μm. Considering only emitter efficiency and base transport factor, the beta expression is

$$\frac{1}{\beta} = \frac{D_{ne} n_{pe} W}{D_{pb} p_{nb} L_{ne}} + \frac{1}{2}\left(\frac{W}{L_{pb}}\right)^2 \tag{7-107}$$

We will assume that $N_{Db} = 1 \times 10^{15}$ atoms/cm³ and $L_{pb} = 25$ μm for the *pnp* base region (the *n*-type epi-layer). We will further assume that $N_{Ae} = 1 \times 10^{17}$ atoms/cm³ and $L_{ne} = 10$ μm for the emitter region of the *pnp* transistor. Then

$$\frac{1}{\beta} = \frac{(23.4)(2.25 \times 10^3)(6 \times 10^{-4})}{(12.2)(2.25 \times 10^5)(1 \times 10^{-3})} + \frac{1}{2}\left(\frac{6 \times 10^{-4}}{25 \times 10^{-4}}\right)^2 = 0.04$$

and $\beta = 25$. Actual betas for substrate *pnp* transistors generally range between 10 and 100.

The behavior of substrate *pnp* transistors is also influenced by other factors. The high resistivity of the epi-layer causes the high-level injection effects to occur at a lower value of I_C than in an *npn* transistor of comparable size. The collector–base diode is formed at the epitaxial layer–substrate interface. The large area involved gives rise to a large value of collector–base junction capacitance C_u. The combination of large collector capacitance and large base resistance degrades the frequency behavior of the substrate *pnp* transistor.

7-5-2 Lateral *pnp* Transistors

This is the most commonly used *pnp* structure in monolithic ICs. The standard *npn* base diffusion is used to form both the emitter and collector of the lateral *pnp* device, as shown in Fig. 7-23. The *p*-type collector ring completely

Fig. 7-23 Lateral *pnp* transistor.

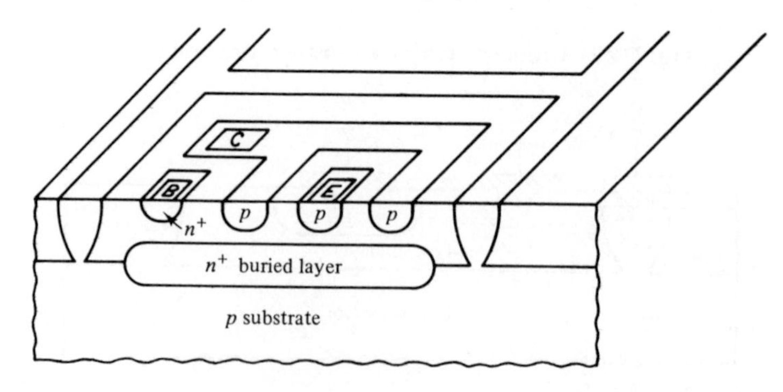

surrounds the emitter. The *n*-type epi-layer serves as the transistor base. It is contacted through an n^+ region outside the collector ring, as shown.

Device operation depends on holes, injected from the side walls of the *p*-type emitter diffusion. These minority holes diffuse laterally across the *n*-type base region until they approach the sidewalls of the *p*-type collector ring and are swept across the reverse-biased base–collector depletion layer. The base width depends on the lateral distance between emitter and collector sidewalls. This distance, in turn, is determined by the minimum line width allowed by the lithography, and by the degree to which lateral diffusion can be controlled. In practice, this base width is large compared to the *npn* base width, leading to poorer frequency response and lower gain.

In constructing a lateral *pnp* transistor, two parasitic vertical *pnp* transistors were created. Referring to Fig. 7-23, one vertical *pnp* transistor exists between collector and substrate. Another exists between emitter and substrate. When the lateral *pnp* device is biased in the forward active region, both junctions of the collector-to-substrate transistor are reverse-biased, so that it is in cutoff. But the vertical *pnp* transistor between emitter and substrate is biased in the forward active region and this presents a problem. The n^+ buried layer is used to ruin the beta of the emitter–substrate transistor. In effect, the buried layer heavily dopes part of the base region, decreasing the emitter efficiency and base transport factor. Nevertheless, the vertical component of emitter current is significant with respect to the useful sidewall component. The parasitic emitter–substrate transistor thus degrades the beta of the lateral *pnp* transistor. The beta is also lower than that of an *npn* transistor, because the lateral base width cannot be as well controlled. The betas of lateral *pnp* transistors range between 5 and 100.

Typical gain–bandwidth products (f_t) for lateral *pnp* transistors are on the order of several megahertz. This compares with f_t values of hundreds of megahertz for *npn* transistors. The poor frequency response of lateral *pnp* transistors is caused primarily by the large base width and by the ratio of vertical to sidewall emitter current. The epi-layer under the emitter provides a large volume for minority carrier storage, which does not contribute to collector current.

Lateral *pnp* transistors can be used in the common-emitter, common-base, or common-collector configuration because the collector does not have to be tied to the negative power supply (as it does in substrate *pnp* devices).

The base–emitter diode of a conventional *npn* transistor will go into avalanche breakdown at approximately 6 to 8 V. The breakdown voltage is dependent on the base doping. Since the lightly doped epi-layer forms the base of a lateral *pnp* transistor, avalanche breakdown of the base–emitter diode occurs at much higher voltages. Lateral *pnp* transistors are sometimes used in amplifier input circuits for this reason.

The collector of a lateral *pnp* transistor can be split as shown in Fig. 7-24. Collector currents I_{C1} and I_{C2} are proportional to collector areas A_{C1} and A_{C2}. If the base terminal of Fig. 7-24(b) is connected to a constant-current

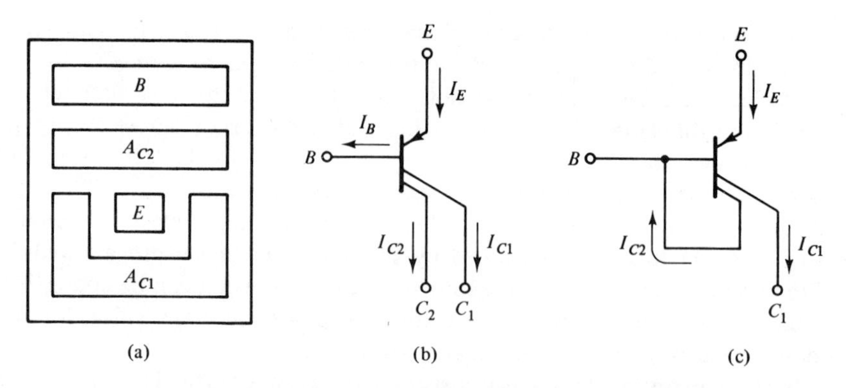

Fig. 7-24 Multiple collector lateral *pnp* transistors.

source, then I_{C1} and I_{C2} will act as constant-current sources. The two collector current magnitudes will be directly proportional to collector areas A_{C1} and A_{C2}. Beta can be tightly controlled by tying one of the collectors back to the base, as shown in Fig. 7-24(c). This serves to reduce beta and g_m and to improve frequency response. Beta becomes approximately equal to I_{C1}/I_{C2}, or correspondingly, A_{C1}/A_{C2}.

BIBLIOGRAPHY

[1] GUMMEL, H. K., AND H. C. POON, "An Integrated Charge Control Model of Bipolar Transistors," *The Bell System Technical Journal,* Vol. 49, pp. 827–851, May–June 1970.

[2] GLASER, A. B., AND G. E. SUBAK-SHARPE, *Integrated Circuit Engineering: Design, Fabrication and Applications.* Reading, Mass.: Addison-Wesley Publishing Company, Inc., 1977.

[3] GROVE, A. S., *Physics and Technology of Semiconductor Devices.* New York: John Wiley & Sons, Inc., 1967.

[4] HAMILTON, D. J., AND W. G. HOWARD, *Basic Integrated Circuit Engineering.* New York: McGraw-Hill Book Company, 1975.

[5] MILLMAN, J., AND C. C. HALKIAS, *Integrated Electronics: Analog and Digital Circuits and Systems.* New York: McGraw-Hill Book Company, 1972.

[6] STREETMAN, B. G., *Solid State Electronic Devices.* Englewood Cliffs, N.J.: Prentice-Hall, Inc., 1972.

PROBLEMS

7-1 Draw energy-level diagrams for an *npn* transistor biased in the
(a) Forward-active region.
(b) Inverse-active region.

7-2 Draw energy-level diagrams for a *pnp* transistor biased in the
(a) Saturation region.
(b) Cutoff region.

7-3 An *npn* transistor is connected as a diode as shown. The diode is forward-biased by $V_D = 0.7$ V. If $I_{ES} = 1 \times 10^{-14}$ A, $\alpha_F = 0.98$, $\alpha_R = 0.4$, and $kT/q = 0.026$ V, find the base current I_B.

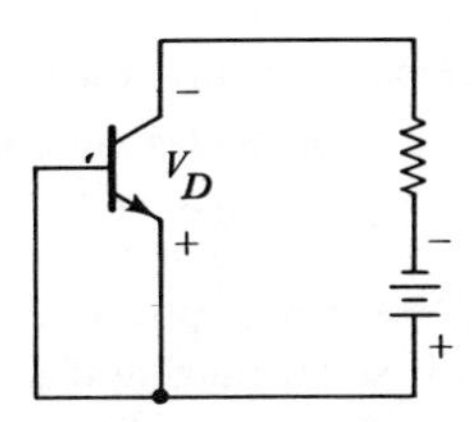

7-4 An *npn* transistor is connected as a diode as shown. The diode is forward-biased by $V_{BE} = 0.65$ V. If $I_{ES} = 1 \times 10^{-14}$ A and $kT/q = 0.026$ V, use the Ebers–Moll equations to find the total diode current.

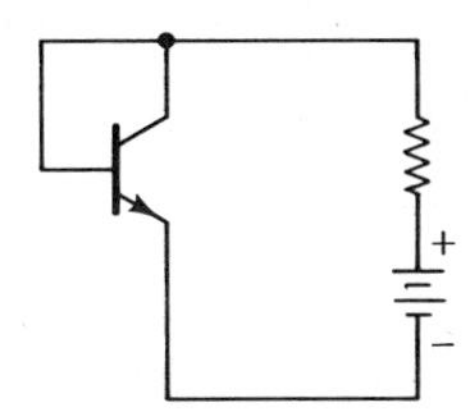

7-5 A bipolar transistor is biased in the forward-active mode with $|I_{ES}| = 1 \times 10^{-12}$ A, $|I_{CS}| = 1.98 \times 10^{-12}$ A, $\alpha_F = 0.99$, and $\alpha_R = 0.50$. If $|V_{BE}| = 0.65$ V and $|V_{BC}| = 4.35$ V, find the values of I_C, I_B, and I_E.

7-6 An *npn* transistor is biased in the saturation region with $|V_{BE}| = 0.65$ V and $|V_{CE}| = 0.2$ V. Find I_E, I_B, and I_C if $I_{ES} = 5 \times 10^{-15}$ A, $\alpha_F = 0.995$, and $\alpha_R = 0.50$.

7-7 An *npn* transistor is biased in the forward-active region with:

Active base width $W = 1.5 \times 10^{-4}$ cm
Average base doping $N_{Ab} = 1 \times 10^{17}$ atoms/cm^3
Average emitter doping $N_{De} = 1 \times 10^{19}$ atoms/cm^3
Hole diffusion length $L_{pe} = 1.5 \times 10^{-3}$ cm
Electron diffusion length $L_{nb} = 2 \times 10^{-3}$ cm
Minority electron lifetime $T_n = 2 \times 10^{-7}$ s
Minority hole lifetime $T_p = 1 \times 10^{-6}$ s

Find the forward transistor beta. Neglect surface states and beta variations with current.

7-8 (a) Rewrite the Ebers–Moll equations (7-34) and (7-35) to express V_{BE} and V_{BC} as the dependent variables [i.e., $V_{BE} = f(I_E, I_C)$ and $V_{BC} = f(I_E, I_C)$].
(b) Express $V_{CE} = V_{CB} - V_{BE}$ as a function of I_E and I_C.

(c) Show that if $I_C = 0$, $V_{CE} \approx (kT/q)|\ln \alpha_R|$. This indicates that when the base–emitter diode is forward-biased and the collector lead is open, V_{CE} has a finite nonzero value.

7-9 Given a transistor with $\beta_F = 200$, $g_m = 66.7$ mA/V, $C_u = 7$ pF, and $C_\pi = 60$ pF, find the beta cutoff frequency f_β and the gain–bandwidth product f_t.

7-10 A transistor with $\beta_F = 370$ and $r_\pi = 2.6$ kΩ has a gain–bandwidth product of 250 MHz. Find the diffusion capacitance C_D by assuming that the depletion-layer capacitances C_{je} and C_{jc} are much smaller than C_D.

7-11 The alpha cutoff frequency of a common-base amplifier can be determined by driving the emitter with a constant-current source and creating an ac short from collector to base. The hybrid-π model for the common-base transistor configuration is then as shown. The alpha cutoff frequency f_α can be found by writing an expression for current gain $I_c(j\omega)/I_e(j\omega)$ and arranging it in the form $K/(1 + jf/f_\alpha)$. Derive an expression for f_α using the hybrid-π model as shown.

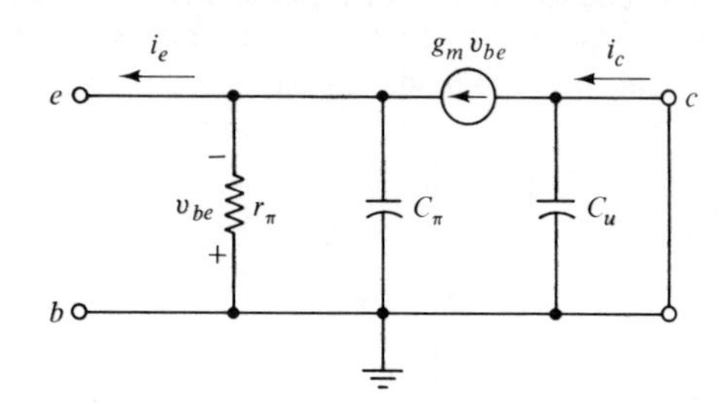

systems of metal, oxide, and semiconductors

8

Because of the ease with which silica (SiO_2) films can be grown on Si, SiO_2 is used for passivation (protection from ambient atmosphere) and diffusant masking in the manufacture of Si integrated circuits. To overcome some of the less desirable features of SiO_2 films, particularly metal ion diffusion, silicon nitride (Si_3N_4) films are used in special applications. Emphasis will be placed on the properties of SiO_2 and Si_3N_4 and some discussion of the techniques for producing such films will be presented when necessary.

The electrical properties of bipolar transistors and field-effect devices depend strongly on the condition of the semiconductor surface and the insulating layer used for passivation. Hence, the peculiarities of semiconductor surfaces and oxide layers that affect semiconductor device properties will be considered.

8-1 PROPERTIES OF SILICA

Pure silica glass[1] is thermodynamically unstable below 1710°C, which means that it tends to devitrify (crystallize) into crystalline quartz. However, below 1000°C, the rate of devitrification is so slow that silica glass is stable, for all intents and purposes. When the temperature is raised much above 1000°C, the glass will crystallize upon cooling. Although an amorphous state, pure silica glass does have short-range structure. The molecular formula is SiO_2 but the structural formula is SiO_4^{4-}, which means it is an ion with a valence of −4. The Si atom, with a valence of 4, is located at the center of a regular tetrahedron

[1] Glass is basically a supercooled liquid state lacking the long-range order present in crystalline materials.

with oxygen ions (O^{2-}) at each corner. Each Si atom shares one electron with each of four oxygen atoms. This leaves each oxygen atom with one unpaired electron. Bonding between tetrahedra occurs via the sharing of oxygen atoms. The shared oxygen atoms are called bridging atoms. There are always some nonbridging atoms (oxygen atoms not shared by two tetrahedra). The greater the percentage of bridging oxygens, the greater the cohesiveness of the glass. In the pure crystalline state all the oxygen atoms are bridging atoms. Figure 8-1 schematically depicts bridging and nonbridging oxygen atoms in silica glass. The movement of Si in silica glass requires breaking four Si—O bonds, so Si vacancies are rare. However, only two Si—O bonds must be broken to free a bridging oxygen atom, or one Si—O must be broken to free a nonbridging oxygen atom. Consequently, oxygen vacancies in silica are fairly common and act as positively charged defects.

Diffusion in silica obeys the same laws as in crystalline materials (see Chapter 4). The diffusion coefficient D of a given species (atom, ion, or molecule) is given by

$$D = D_0 \exp\left(\frac{-E_d}{kT}\right) \qquad (8\text{-}1)$$

where D_0 is a constant and E_d is the activation energy. For oxygen in SiO_2, $D_0 = 1.5 \times 10^{-2}$ cm^2/s, $E_d = 3.09$ eV and for H_2O, $D_0 = 10^{-6}$ cm^2/s, $E_d = 0.794$ eV. The motion of ionic species due to applied electric fields may be present as well as diffusion.

There are numerous impurities that play important roles in the modification of silica glass properties. Elements from group III of the periodic table, such as boron, B, and elements from group V, such as phosphorus, are called network formers. These elements replace the Si atoms in the network and as such become part of the normal framework. Atoms such as phosphorus cause excess nonbridging because charge neutrality can be achieved with one nonbridging oxygen in each affected tetrahedron since P can share two electrons with one of the four oxygens. Group III elements can supply only three electrons, so it is more likely that the tetrahedrally coordinated oxygens will be shared by another tetrahedron to achieve charge neutrality. Localized charge neutrality is not always accomplished, thus leaving charged defects; in this case, acceptor or donor-like defects for B and P, respectively, since B will tend to pick up an electron and P will tend to lose one to achieve a minimum energy condition.

Interstitial impurities, such as Na^+, K^+, Pb^{2+}, and Ba^{2+}, are called network modifiers because, rather than becoming a part of the framework, they disrupt it. Of particular concern in microcircuit fabrication is Na^+. Large metal ions such as these tend to weaken the silica structure and make it more porous. They can also seriously affect the electronic properties of the circuits directly because of their charge. If the sodium enters silica as an oxide, NaO_2, the

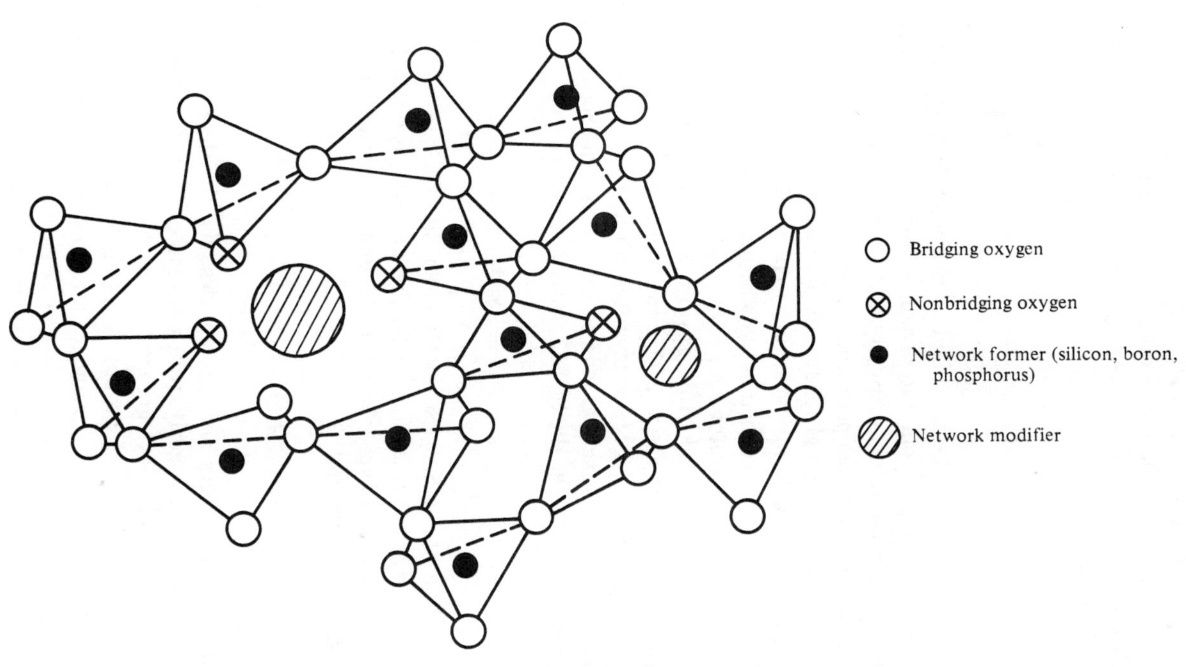

Fig. 8-1 SiO_4^4 Tetrahedra in silica glass.

excess oxygen tends to cause more nonbridging, hence further weakening the silica structure.

Water vapor also has a weakening effect on the silica structure and makes it more porous, much like the metal cations. The water molecule dissociates, forming hydroxyl groups (OH^-). These hydroxyl groups replace O^{2-} and are nonbridging since they have a valence of -1.

8-2 SILICA GROWTH AND DEPOSITION

An oxide of silicon can be deposited by pyrolytic decomposition of oxysilane compounds or grown from the silicon substrate by a thermal process. Considerable use is being made of pyrolytic deposition to avoid the exposure of silicon wafers to the high temperature, around 1100°C, required for thermal oxide growth. However, thermal oxides are smoother and generally of higher quality and are required in some cases of selective oxidation for isolation purposes. Pyrolytic deposition is carried out in a furnace somewhat like an epitaxial reactor, but at a lower temperature.

In the process of thermal oxide growth, Si is consumed. The thickness of the grown oxide layer is about 2.27 times the thickness of the consumed Si. For example, growth of a 2270-Å layer of SiO_2 consumes about 1000 Å of Si. The basic chemistry of oxide growth is simple, as evidenced by Eq. (8-2), which gives the interaction at the Si–SiO_2 interface:

$$Si + O_2 \rightleftharpoons SiO_2 \tag{8-2}$$

Growth takes place at the SiO_2–Si interface, so that O_2 must diffuse through the existing oxide layer. Permeation[2] of O_2 through SiO_2 is negligible.

When wet oxygen (oxygen bubbled through water maintained at about 95°C) is used, the reaction is

$$Si + 2H_2O \rightleftharpoons SiO_2 + 2H_2 \tag{8-3}$$

This equation, however, does not actually describe the details of the overall process. As pointed out previously, water molecules tend to dissociate in SiO_2 and form nonbridging hydroxyl groups. At the Si surface, the hydroxyl ion reacts with Si to form Si–O–Si groups releasing the hydrogen. The escaping H_2 can react with bridging oxygen ions to form nonbridging hydroxyl groups, tending to weaken the silica structure and making it more porous.

The rate of oxide growth in wet oxygen exceeds the rate of growth in dry oxygen by about a factor of 10. This is partly due to the larger diffusion coefficient (D) of H_2O and partly due to the greater solid solubility of H_2O in

[2] Permeation is the flow of gas through a material due to a pressure differential.

SiO_2 (3×10^{19} molecules/cm^3 for H_2O; 5.2×10^{16} molecules/cm^3 for O_2 both at 1000°C and 1 atm). It is the solid solubility that controls the surface concentration of the diffusant and, in combination with the diffusion constant, determines the rate of diffusion of the species through the oxide. The rate of oxide growth is initially limited by the rate of chemical reaction at the surface of the Si. For large thicknesses, however, the rate is diffusion-limited.

At the gas-oxide interface, oxygen forms a "superoxide ion," O_2^-, and a hole. The hole, having a higher mobility than the O_2^- ion, precedes it into the oxide layer. The resulting space-charge region increases the diffusion rate of the O_2^-. (It is actually a drift effect.) The thickness of the region in which this takes place is of the order of a Debye length[3] which is inversely proportional to the square root of the diffusant concentration. Since H_2O has such a large solid solubility, the Debye length is only a few angstroms. For dry oxygen, however, it is nearly 200 Å. Hence, an extremely rapid initial growth occurs with dry oxygen but not with wet oxygen. Since gate oxides in MOSFETs are only a few hundred angstroms thick, this is a very important phenomenon.

Silica layers have excellent masking capability for network formers such as B and P. The diffusion constant for these dopants in SiO_2 is two to three orders of magnitude smaller than in Si. Gallium and Al have higher diffusion constants in SiO_2 than in Si, so SiO_2 cannot be effectively used to mask these impurities.

8-3 SILICA QUALITY

To some extent the quality of a layer of silica can be determined visually. It can be inspected microscopically for crystallites, blemishes, or discoloration. Such defects may be due to improper oxide growth technique or incomplete cleaning or photoresist removal. A scanning electron microscope or sensitive profilometer is required to observe or measure the contour variations generated in fabricating a microcircuit. Different thicknesses of oxide may be grown on various parts of the circuit. Figure 8-2 is an scanning electron micrograph that shows several oxide steps on a bipolar IC circuit.

The dielectric breakdown strength of an oxide is measured by evaporating aluminum over the entire oxide layer to form a capacitor and then measuring the capacitor's breakdown voltage. Typically, thermal oxide breakdown strength is about 600 V/μm. Breakdown results in physical damage, usually leading to a short, through the oxide. In MOSFETs, where the tendency is to produce a very thin oxide over the gate, breakdown may occur at a fairly low voltage. This puts a fundamental lower limit on useful oxide thickness. An oxide thickness of 500 Å would result in a breakdown around 30 V.

Oxide thickness can be determined accurately by a lapping technique or

[3] The Debye length is defined as the effective range of the field of a charge, such as an ion or a hole, in a material.

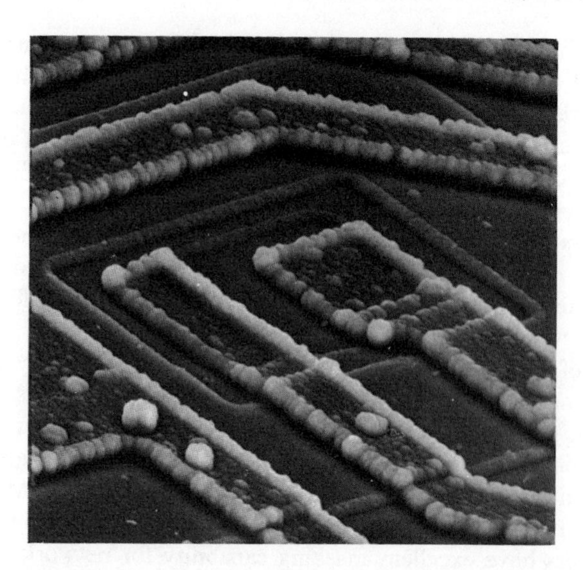

Fig. 8-2 SEM micrograph of oxide steps in a bipolar IC (40×).

an interferometric technique. A fair estimate of the oxide thickness can be made from its color in normally incident white light by use of the theory of thin-film interference if a rough estimate of film thickness can be made from a knowledge of growth conditions. For very thin films, 100 Å and less, an ellipsometer must be used. The ellipsometer is used to determine film thickness and optical properties of thin films based on the polarization of light reflected from films.

8-4 SILICON NITRIDE

Silicon nitride, Si_3N_4, is deposited by decomposition of $SiCl_4$ in ammonia at about 850°C. It is used to provide a barrier to heavy metal ions, such as Na^+, and to diffusants normally used in microelectronics fabrication. Gate oxides of MOS devices have been built with a sandwich structure of thermally grown SiO_2 to preserve the charge characteristics at the Si–SiO_2 interface, and a layer of Si_3N_4 for the additional passivating capability. Si_3N_4 has a higher dielectric constant than SiO_2, so a thicker layer can be used, for mechanical strength, while obtaining the same electrical effect as with SiO_2 alone.

Si_3N_4 is also used as a mask on Si to achieve selective SiO_2 growth. The Si_3N_4 is then removed by etching in hot phosphoric acid. Figure 8-3 illustrates the process of selective oxide growth, which is used for isolation in integrated circuits.

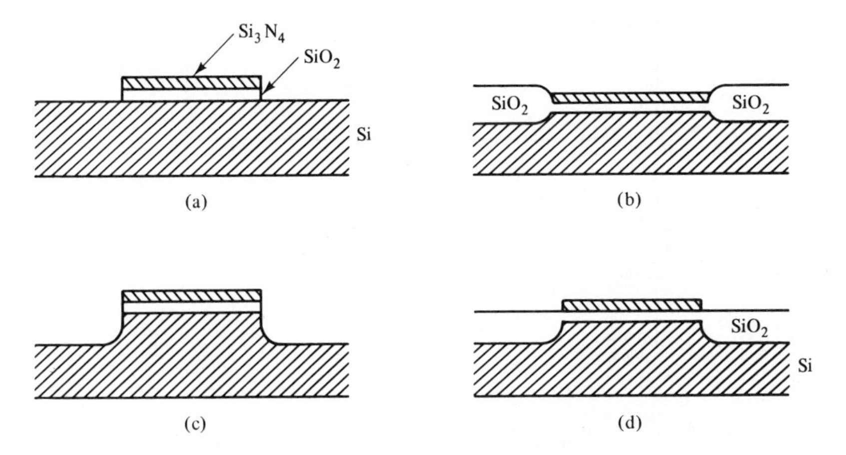

Fig. 8-3 Selective oxidation using Si_3N_4.

8-5 SEMICONDUCTOR SURFACES

A chemically polished or newly cleaved semiconductor surface has incomplete covalent bonds (dangling bonds) at the surface. There are about 10^{15} cm^{-2} such incomplete bonds present at an ideally clean surface which behave as acceptors. Consequently, the surface of an ideally clean semiconductor will acquire a negative surface charge density. Such an ideal surface seldom occurs with Si because an oxide layer is immediately formed unless the wafer is kept in vacuum or an inert atmosphere. However, a brief discussion of the ideal case may help in understanding the actual situation.

For charge neutrality to be maintained at an ideal surface, a positive space-charge layer must accumulate in the bulk of the semiconductor near the surface. This causes distortion of the energy bands at the surface as depicted in Fig. 8-4. The space-charge region consists of positive holes, which tend to make this region *p*-type. The depth of the space-charge region can be as great as 1 μm, depending on the doping concentration. This depth is a sizable fraction of the depth of the entire circuit in a monolithic IC.

In the case of thermally grown oxide layers on Si, the dangling bonds are donor-like instead of acceptor-like, which is the case for a clean surface. Owing to the disruption of the crystal structure, states are introduced deep in the energy gap. Because many of the Si bonds are tied up in SiO_2 tetrahedra, the number of states is reduced to 10^{11} to 10^{12} cm^{-2} depending on crystal orientation. These states occur within about the first 25 Å of the Si–SiO_2 interface and are referred to as Tamm states. The next 100 Å or so into the SiO_2 is a transition region. Excess Si ions in the transition region create a positive space-charge region of about 10^{11} ions/cm^2. Figure 8-5 depicts these regions at the Si–SiO_2 interface.

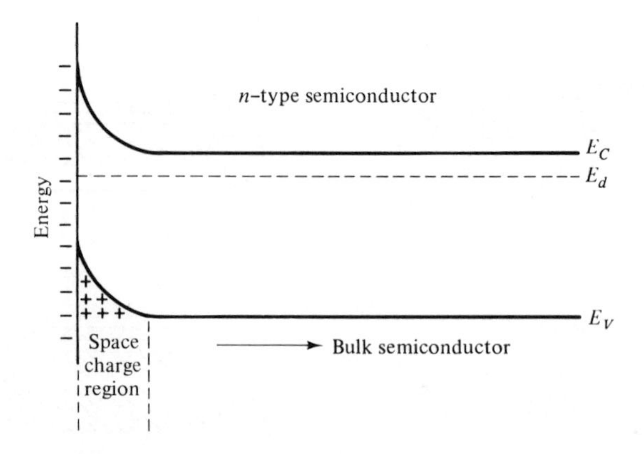

Fig. 8-4 Energy bands at an ideally clean semiconductor surface.

The Tamm states are states that can undergo transitions quite rapidly. Their transition times are on the order of microseconds. Consequently, these states are referred to as "fast" states. The ionic states in the Si–SiO_2 transition region may remain unchanged for minutes to months and are correspondingly referred to as "slow" states. The "fast" states affect such properties of *p-n* junctions as the reverse leakage current, reverse breakdown voltage and, in transistors, the common-emitter current gain (β). Slow ionic states can cause long-term drift in device parameters and apparent changes in the impurity concentrations if the ions are mobile. The excess Si ions in the transition region are fixed (immobile), whereas defects such as Na ions are mobile. Figures 8-6 and 8-7 contain illustrations of what may happen to the energy bands at the surfaces of thermally oxidized *n*-type and *p*-type semiconductors, respectively.

It is clear from the diagrams that the presence of positive surface states causes an *n*-type crystal to become more *n*-type near the surface (enhancement) and *p*-type crystal to become less *p*-type (depletion), and may even cause a *p*-type crystal to become inverted (changed to *n*-type) near the surface.

The phenomena of depletion, enhancement, and inversion occur in an

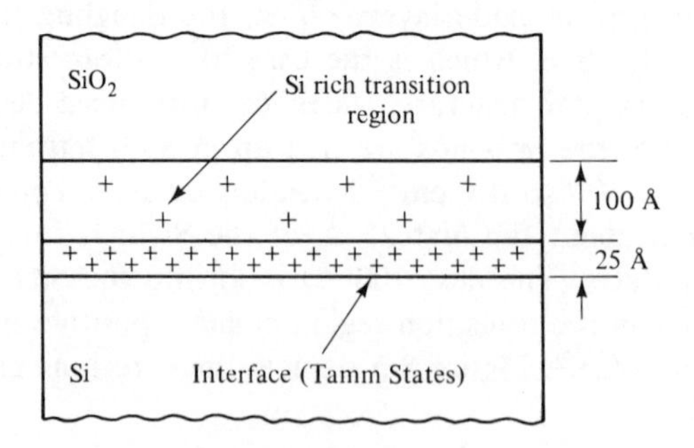

Fig. 8-5 Transition from Si to SiO_2.

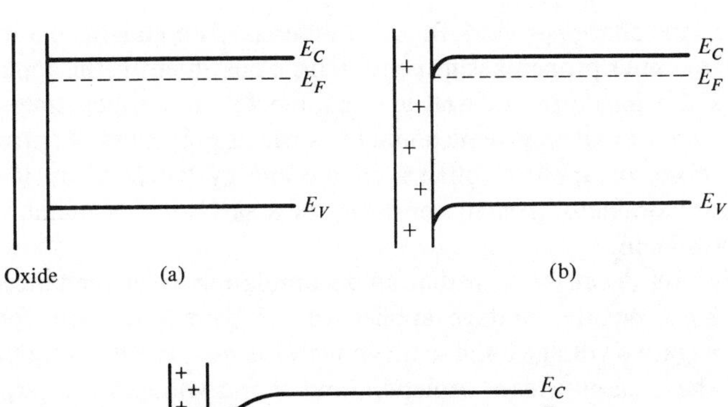

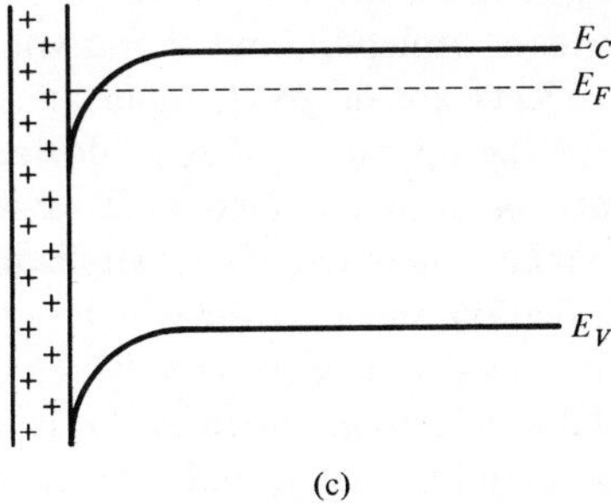

Fig. 8-6 Thermally oxidized Si *n*-type surface.

Fig. 8-7 Thermally oxidized Si *p*-type surface.

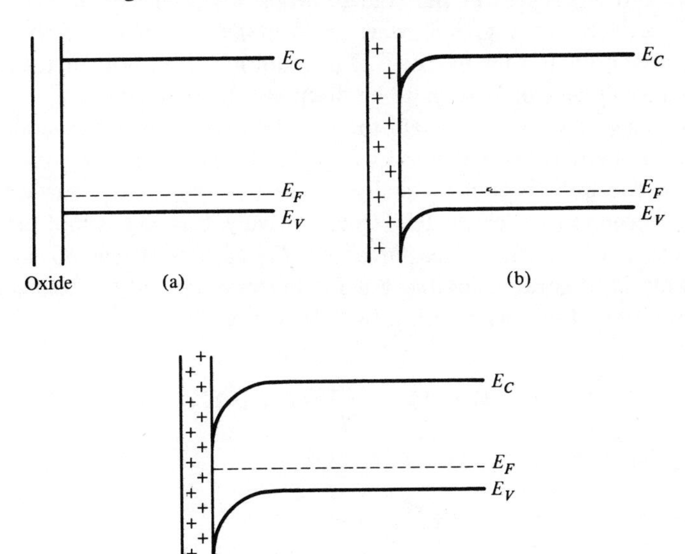

ideal (no surface charge or work-function difference[4]) metal–insulator–semiconductor system under proper biasing conditions. Consequently, the applied voltage drop across the insulator, as well as that due to the surface states, must be considered when analyzing semiconductor surface properties. Figure 8-8 illustrates the effects of applied voltages on the energy bands of an ideal metal–insulator–semiconductor system consisting of a sandwich of metal, oxide, and *n*- or *p*-type silicon.

In Fig. 8-8 it can be seen that an accumulation layer (enhancement) can be caused by a positive voltage applied to the gate if the semiconductor is *n*-type or a negative voltage if the semiconductor is *p*-type. Reversing the voltages in each of these cases causes depletion, and if the voltages are large enough, causes inversion. These effects are of great importance in MOS devices and, in fact, form the basis for the operation of such devices. In bipolar ICs (*npn* transistors), the metalizations are maintained as far away from the emitters as possible except where making contact. The metalization is separated from the base, collector, and isolation regions, over which it passes by a considerable thickness of SiO_2. A thick insulating layer greatly reduces the electric field at the semiconductor–insulator interface, which is the cause of phenomena such as depletion, accumulation, and inversion under ideal conditions.

The voltage applied to the gate metal with respect to a grounded substrate is referred to as V_G. This voltage is the sum of the voltage drops across the oxide and across the depletion layer (if there is one) beneath the oxide. Thus,

$$V_G = V_{ox} + V_{Si} \tag{8-4}$$

where V_{ox} and V_{Si} represent the voltage drops across the oxide and depletion layer, respectively. In Fig. 8-8, when no voltage is applied to the gate, the flat-band condition is said to exist. This condition does not, in reality, occur at zero gate voltage, for reasons to be discussed subsequently.

The onset of inversion is said to exist (by definition) when sufficient gate voltage is applied to bend the band edge to the point that a *p*-type (*n*-type) material in the bulk becomes equally *n*-type (*p*-type) at the surface. This is called the "strong inversion model." Quantitatively, this means that the extrinsic Fermi level at the surface is as far above E_{F_i} as it is below E_{F_i} in the bulk silicon. This is illustrated in Fig. 8-8 for inversion in *n*-type semiconductors as well as *p*-type. For this case Eq. (8-4) becomes

$$V_G = V_{ox} + \frac{2\phi_f}{q} = V_{ox} + 2V_f \tag{8-5}$$

[4] Work function is the energy required to lift an electron from the Fermi level to free space. The work-function difference is just the difference between the work functions of two contiguous materials.

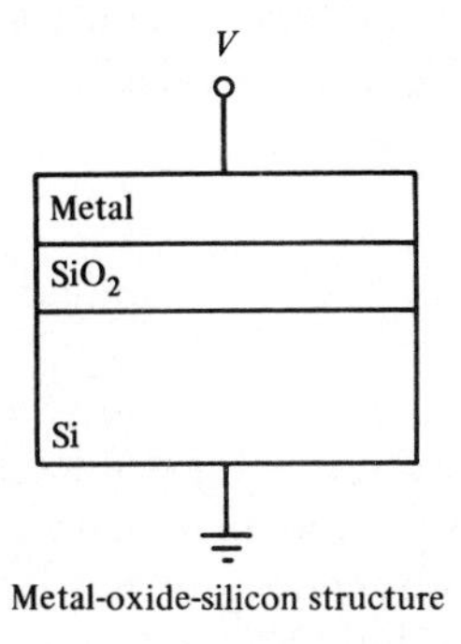

Metal-oxide-silicon structure

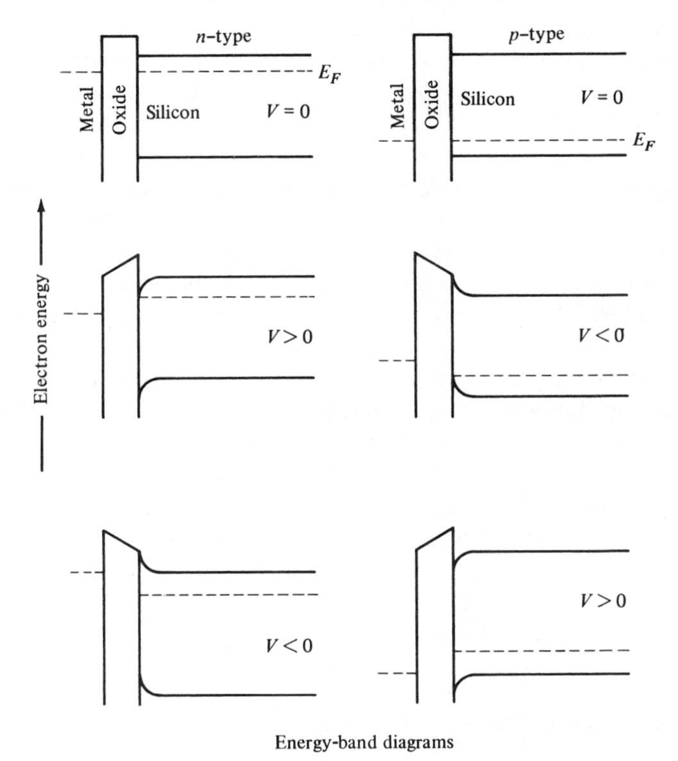

Energy-band diagrams

Fig. 8-8 Energy-band scheme for an ideal metal–oxide–silicon sandwich structure is depicted for various applied voltages. The slope of the curve across the oxide indicates an energy rise or decrease for electrons.

where ϕ_f is the difference between the actual Fermi level and the intrinsic Fermi level, and $V_f = \phi_f/q$. In Chapter 6, ϕ_f was referred to as ΔE, or $\Delta E'$. It will be shown later that V_{ox} can be calculated from the oxide capacitance and the charge in the depletion layer.

8-6 REAL METAL–INSULATOR–SEMICONDUCTOR SYSTEMS

In Section 8-5, ideal metal–insulator–semiconductor systems were considered. Two important factors were ignored in that simplified interpretation: specifically, the difference in work functions of the materials and the presence of positive charge at the SiO_2–Si interface. To simplify the development, we will first consider a system with only a work-function difference and then the effect of the oxide surface charge will be added to this. The type of device studied here will be one in which the bulk material is *p*-type and a positive gate voltage is applied to produce an *n*-type layer beneath the oxide. This case is illustrated in Fig. 8-9.

The definition of work function, repeated here for convenience, is that energy required to move an electron from the Fermi level in that material to free space. At the interface of two different materials with different work functions, the work-function difference is that energy required to move an electron from the Fermi level in the material of lower work function to the Fermi level of the material with higher work function. It has been experimentally established that 3.2 eV is required to move an electron from Al into SiO_2 and 3.25 eV is required to move an electron from the conduction band of Si into SiO_2. This is illustrated by the energy-level diagrams of Fig. 8-9(a). The work-function difference between the metal–oxide and the silicon–oxide systems, the energy required to move an electron from the Si Fermi level into Al in a MOS structure, can be determined if the semiconductor energy gap, E_g, and the distance of E_F from E_{F_i} are known. Thus,

$$W_{\text{Si-O}} = 3.25\ \text{eV} + \frac{E_g}{2} + \phi_f \tag{8-6}$$

and the work function difference $W_{\text{M-Si}}$ is

$$\begin{aligned} W_{\text{M-Si}} &= W_{\text{M-O}} - W_{\text{Si-O}} \\ &= 3.2\ \text{eV} - (3.25\ \text{eV} + 0.55\ \text{eV} + \phi_f) \\ &= -0.6\ \text{eV} - \phi_f \end{aligned} \tag{8-7}$$

Note that ϕ_f is positive for a *p*-type material and negative for an *n*-type material, so that this expression is general. For a hypothetical *p*-type Si with $\phi_f = 0.3$

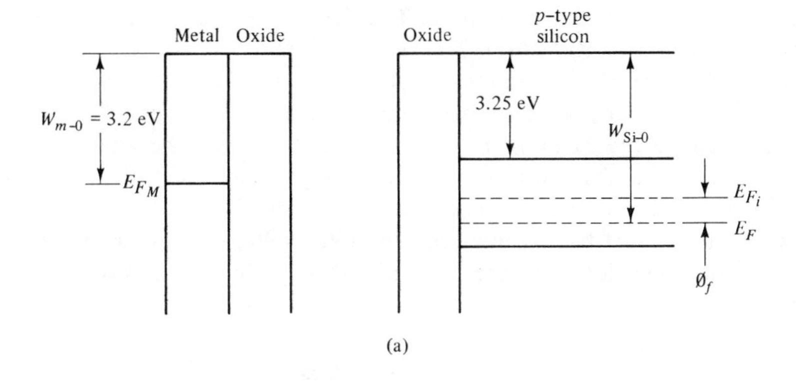

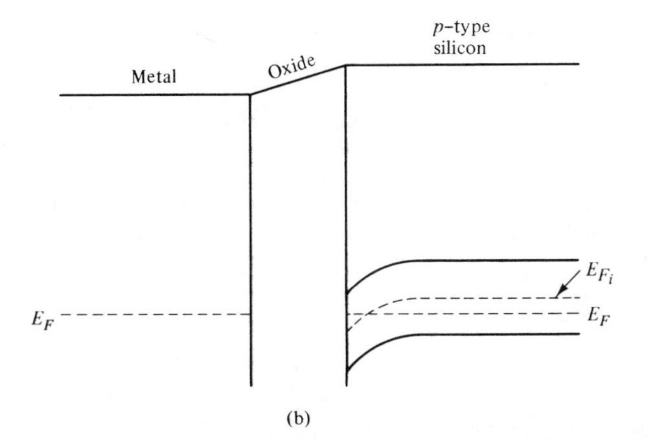

Fig. 8-9 Effect of work function differences on a metal–insulator–semiconductor system.

eV, $W_{M\text{-}Si} = -0.9$ eV. This means that a voltage of $V_G = -0.9$ V would have to be applied to the gate to achieve the flat-band condition; or, in general,[5]

$$V_{fb} = \frac{-0.6\text{ eV} - \phi_f}{q} = -0.6\text{ V} - V_f \tag{8-8}$$

The final item to be considered to complete the interpretation of the inversion process in a metal–insulator–semiconductor system is the surface charge Q_{SS} located at the Si–SiO_2 interface. This charge is always positive and is a function

[5] In Eq. (8-8), q is always positive and simply converts electron volts to volts. ϕ_f is the only term that can change sign. V_{fb} is always negative since in Si the maximum value of ϕ_f is about ±0.55 eV.

of crystal orientation and processing. The density of charges at the interface is approximately 5×10^{11} cm^{-2}, 2×10^{11} cm^{-2}, and 9×10^{10} cm^{-2} for (111), (110), and (100) crystal orientations, respectively. This charge causes band bending at the surface because of the attraction of electrons and repulsion of holes, tending to make the Si more n-type at the surface (i.e., the bands are bent downward). In the absence of a work-function difference, a voltage of $-Q_{SS}/C_{ox}$ (where C_{ox} is the oxide capacitance per unit area) must be applied to the gate to achieve a flat-band condition. The total voltage required to achieve a flat-band condition due to surface charge and the difference in work functions is then

$$V_{fb} = \frac{W_{\text{M-Si}}}{q} - \frac{Q_{SS}}{C_{ox}} \tag{8-9}$$

The oxide capacitance per unit area can be calculated from

$$C_{ox} = \frac{\epsilon_0 \epsilon_{ox}}{t_{ox}} \tag{8-10}$$

where ϵ_0 is the permittivity of free space, ϵ_{ox} the dielectric constant of SiO_2, and t_{ox} the thickness of the oxide.

Superimposing Eqs. (8-9) and (8-4) yields the complete expression for the gate voltage:

$$V_G = V_{ox} + V_{Si} + V_{WF} - V_{SS} \tag{8-11}$$

where $V_{WF} = W_{\text{M-Si}}/q$ and $V_{SS} = Q_{SS}/C_{ox}$.

The first two terms can change sign, but the second two terms cannot change sign. The voltage drop across the oxide for the ideal case ($W_{\text{M-Si}}$ and Q_{SS} both zero) is

$$V_{ox} = \frac{Q_s}{C_{ox}} \tag{8-12}$$

where Q_s is the charge in the Si surface region. Q_s is composed of the bound charge Q_b in the depleted region and the mobile charge Q_m that is pulled to the Si–SiO_2 interface by the applied voltage.

To simplify the calculation of the gate voltage required to bring about the onset of inversion, $V_G = V_T$, called the threshold voltage, it will be assumed that $Q_s = Q_b$ and $Q_m = 0$ until the onset of inversion. For voltages larger than V_T, any additional charge developed in the Si surface region will be assumed to be mobile charge. Hence, based on the results of Chapter 6 for a one-sided step junction, the thickness of the depletion layer can be estimated to be, for a p-type material,

$$W = \sqrt{\frac{2\epsilon_{\mathrm{Si}}\epsilon_0 |2V_f|}{qN_A}} \tag{8-13}$$

and the bound charge in the depletion layer is

$$Q_b = qN_A W \tag{8-14}$$

The threshold voltage for an enhancement-mode device such as shown in Fig. 8-10 is

$$V_T = \frac{\sqrt{2\epsilon_{\mathrm{SI}}\epsilon_0 q |2V_f| N_A}}{C_{\mathrm{ox}}} + 2V_f + V_{WF} - V_{SS} \tag{8-15}$$

The sign on the first term is positive for a p-type silicon substrate and negative for an n-type substrate. For voltages above the threshold voltage, V_G is given by

$$V_G = V_T + \frac{Q_m}{C_{\mathrm{ox}}}$$

The threshold-voltage concept will be utilized extensively in Chapter 9, where MOS devices are discussed in detail. A simplified illustration of an MOS device is given in Fig. 8-10.

The threshold voltage is defined such that when $V_G = V_T$, the region just under the gate oxide becomes as much n-type as the bulk of the Si is p-type. For gate voltages larger than V_T, the device is considered to be "on"; that is, conduction can take place between the source and drain when a voltage V_{DS} is applied between them. As mentioned before, this is called the "strong inversion" model.

Fig. 8-10 Simplified diagram of an MOS device.

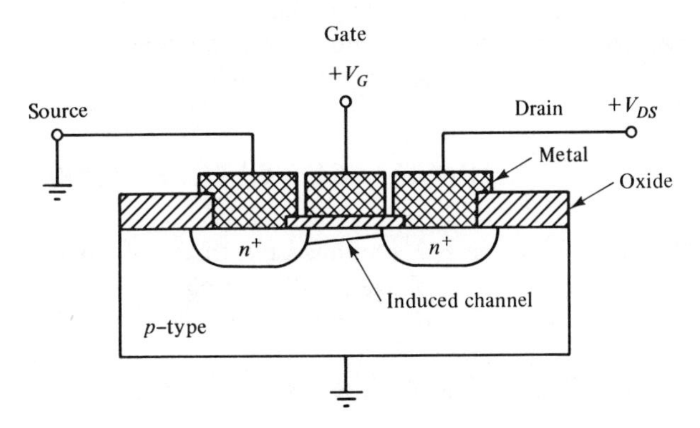

The following example will illustrate the calculation of the threshold voltage for the device in Fig. 8-10. Given that $N_A = 2 \times 10^{16}\ \text{cm}^{-3}$, $t_{ox} = 1000$ Å, (100) crystal orientation with $Q_{SS} = (1.6 \times 10^{-19}\ \text{C})\ (9 \times 10^{10}\ \text{cm}^{-2}) = 1.4 \times 10^{-8}\ \text{C/cm}^2$, and an Al gate,

$$V_{\text{Si}} = 2V_f = 2\frac{kT}{q} \ln \frac{N_A}{n_i} = 0.73\ \text{V}$$

This equation for V_{Si} is developed in Chapter 6 for a one-sided step junction.

$$C_{ox} = \frac{\epsilon_0 \epsilon_{ox}}{t_{ox}} = \frac{4(8.85 \times 10^{-14})}{1 \times 10^{-5}} = 3.54 \times 10^{-8}\ \text{F/cm}^2$$

$$Q_b = (2\epsilon_0 \epsilon_{\text{Si}} q N_A |2V_f|)^{1/2}$$
$$= 7.04 \times 10^{-8}\ \text{C/cm}^2$$

where $\epsilon_{\text{Si}} = 12$,

$$V_{ox} = \frac{Q_b}{C_{ox}} = 1.99\ \text{V}$$

$$V_{SS} = \frac{Q_{SS}}{C_{ox}} = 0.4\ \text{V}$$

$$V_{WF} = \frac{W_{\text{M-Si}}}{q} = \frac{-0.6\ \text{eV} - 0.37\ \text{eV}}{q} = -0.97\ \text{V}$$

$$V_T = V_{\text{Si}} + V_{ox} - V_{SS} + V_{WF} = (0.73 + 1.99 - 0.4 - 0.97)\ \text{V} = 1.35\ \text{V}$$

A positive gate voltage of 1.35 V is required to turn this device on. If N_A were $1 \times 10^{15}\ \text{cm}^{-3}$, the threshold voltage would be negative. For light doping, inversion occurs due to the work function and surface charge; thus, the device is on for voltage above some negative value. To avoid this problem when lightly doped *p*-type substrates are used, an adjustment of the near-surface dopant concentration is made by means of an ion implantation. That is, the concentration is adjusted upward in the gate region and the other *p*-type regions just beneath the oxide, to prevent the occurrence of undesirable inversions.

8-7 CAPACITANCE–VOLTAGE CHARACTERISTICS

The following discussion is a simplified treatment of capacitance–voltage characteristics and capacitance–voltage *(C–V)* testing of metal–insulator–semiconductor systems. For a more rigorous treatment, see one of the references listed at the end of the chapter, such as Richman [2] or Grove [1]. The treatment given here generally follows that of Grove, but with much less rigor.

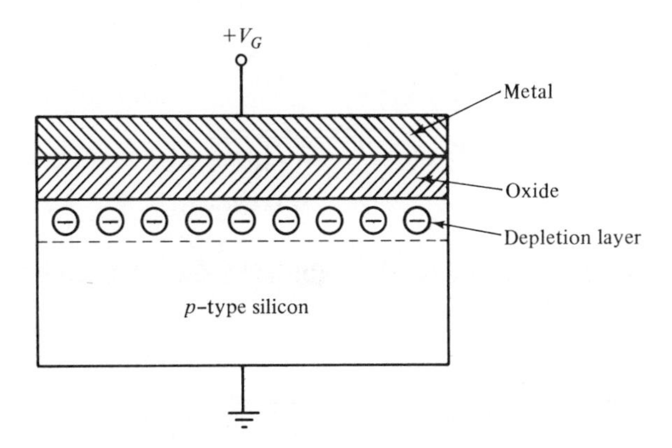

Fig. 8-11 MIS system with a p-type substrate and positive gate voltage.

The capacitance of an MIS system, such as that depicted in Fig. 8-11, can be expressed as

$$\frac{1}{C_T} = \frac{1}{C_{\text{ox}}} + \frac{1}{C_s} \tag{8-16}$$

This expresses the total capacitance as a combination of the oxide capacitance and the depletion-layer capacitance connected in series. Recall that the capacitances $C_{\text{ox}} = \epsilon_{\text{ox}}\epsilon_0/t_{\text{ox}}$ and $C_s = \epsilon_{\text{Si}}\epsilon_0/W$, where

$$W = \sqrt{\frac{2\epsilon_{\text{Si}}\epsilon_0|V_{\text{Si}}|}{qN_A}}$$

In the absence of an oxide–Si interface charge Q_{SS} and a work-function difference, recall that $V_G = (Q_S/C_{\text{ox}}) + V_{\text{Si}}$. It will be assumed that this condition holds and thus $V_{FB} = 0$. To obtain an expression for C_T/C_{ox}, an equation for V_G in terms of W (with V_{Si} and Q_s eliminated) is obtained:

$$V_G = V_{\text{Si}} + \frac{Q_s}{C_{\text{ox}}} = \frac{qN_A W^2}{2\epsilon_{\text{Si}}\epsilon_0} + \frac{qN_A W}{C_{\text{ox}}} \tag{8-17}$$

If W is expressed in terms of C_s, the resulting quadratic equation in C_{ox}/C_s can be solved. Retaining the physically meaningful root, the result is

$$\frac{C_{\text{ox}}}{C_s} = -1 + \sqrt{1 + (2C_{\text{ox}}^2|V_G|)/(qN_A\epsilon_{\text{Si}}\epsilon_0)} \tag{8-18}$$

Combining Eqs. (8-16) and (8-18) gives

$$\frac{C_T}{C_{ox}} = \frac{1}{\sqrt{1 + \dfrac{2C_{ox}^2|V_G|}{qN_A\epsilon_{Si}\epsilon_0}}} \tag{8-19}$$

If C_{ox} is replaced by $\epsilon_{ox}\epsilon_0/t_{ox}$ on the right-hand side of Eq. (8-19), the result is

$$\frac{C_T}{C_{ox}} = \frac{1}{\sqrt{1 + [2(\epsilon_0\epsilon_{ox})^2|V_G|]/(qN_A\epsilon_{Si}\epsilon_0(t_{ox}^2)}} \tag{8-20}$$

The absolute value of V_G is used in order that Eq. (8-20) is applicable to either p-type or n-type enhancement devices. When V_G is negative for a p-type substrate, holes are drawn to the oxide and the system acts like a capacitor (C_{ox}) in series with a resistor (the Si substrate). The same interpretation holds for a positive gate voltage with an n-type substrate. As the gate voltage is made more positive in Fig. 8-11, C_T/C_{ox} decreases inversely as the square root of V_G until the onset of inversion, that is, until $V_G = V_T$. It must be realized that we are dealing with low-frequency incremental capacitance and that if the frequency is too high for the charge to respond (the period small compared with the minority carrier lifetime), the capacitance term will not contribute to the total capacitance. This is indeed the case for the inversion charge of a structure such as the one depicted in Fig. 8-11. Once inversion starts, fixed charge will no longer accrue to the depletion layer. It would be expected that this would tend to cause C_TC_{ox} to increase with further increases in V_G by providing terminal charges beneath the oxide for the electric field, thus dropping the entire gate voltage across the oxide. This is true for frequencies of a few hertz (see Fig. 8-12). However, the inversion-layer charges are minority carriers

Fig. 8-12 Capacitance vs. gate voltage for a p-type substrate, MIS system.

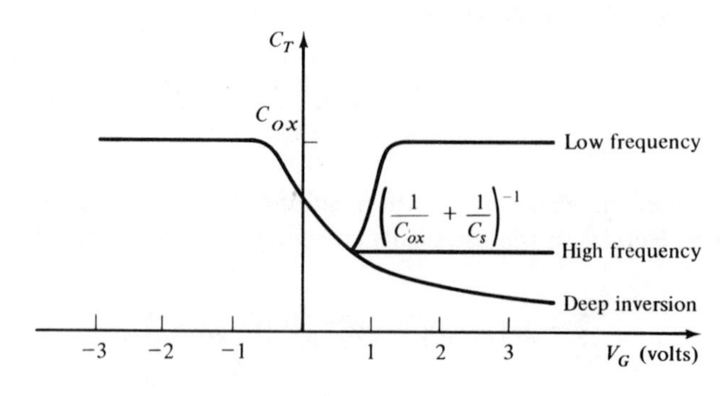

that take about one one-hundredth of a second (minority carrier lifetime) to reach the oxide. For frequencies above a few hertz, the inversion-layer charges do not change rapidly enough to affect the capacitance; therefore, C_T approaches a constant $(1/C_{ox} + 1/C_s)^{-1}$ with increasing V_G. At low frequencies, C_T approaches C_{ox} for $V_G > V_T$. Both cases are depicted in Fig. 8-12.

Deep inversion occurs when large voltage pulses are applied to the unbiased gate. If the pulse width is short compared with the minority carrier lifetime, the carriers do not have sufficient time to create an inversion layer and, consequently, the depletion layer width increases beyond the width at threshold. The series capacitance of the oxide and depletion layer decreases correspondingly. The deep inversion case is also depicted in Fig. 8-12.

The effect of a finite flat-band voltage on the MOS capacitance characteristics is to shift the curves to the left by an amount equal to the flat-band voltage. V_{fb} can be determined by a comparison of actual C–V curves with the ideal curve. Since the work function difference is known, the contribution of Q_{SS} can be determined from Eq. (8-9) and the value of Q_{SS} can be calculated. To separate the contribution of the mobile ionic charge from that of the bound surface charge, a large negative voltage is applied to the gate at a temperature of around 200°C to pull the mobile ionic charges to the gate–oxide interface, where they are neutralized. Now the surface-charge contribution to the flat-band voltage is due only to the bound charge and can be ascertained from the shift in the C–V curve from the ideal (zero flat-band voltage) curve. The ideal and actual C–V curves are depicted in Fig. 8-13 for a p-type substrate.

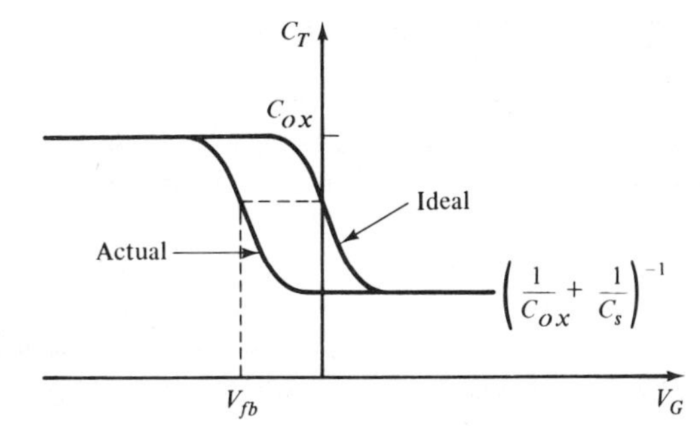

Fig. 8-13 Plot of ideal and actual C–V curves to illustrate the effect of a finite flat-band voltage.

With the aid of computers, more precise theoretical ideal curves can be obtained for comparison with actual curves. Numerous examples of such curves can be found in reference [5].

BIBLIOGRAPHY

[1] GROVE, A. S., *Physics and Technology of Semiconductor Devices.* New York: John Wiley & Sons, Inc., 1967.

[2] RICHMAN, D., *MOS Field-Effect Transistors and Integrated Circuits.* New York: John Wiley & Sons, Inc., 1973.

[3] PHILOFSKY, E., "Intermetallic Formation in Gold–Aluminum Systems," *Solid State Electronics,* Vol. 13, pp. 1391–1399, 1970.

[4] CARR, W. N., AND J. P. MIZE, *MOS/LSI Design and Application.* New York: McGraw-Hill Book Company, Inc., 1972.

[5] GOETZBERGER, A., "Ideal MOS Curves for Silicon," *Bell System Tech. J.,* Vol. 45, pp. 1097–1122, 1966.

PROBLEMS

8-1 Calculate the diffusion constants for O_2 and H_2O in silica at 300 K and 1300 K.

8-2 Explain what is meant by enhancement, depletion, and inversion.

8-3 Calculate the flat-band voltage for an Al–SiO_2–Si system for *p*-type silicon with a hole concentration of 5.0×10^{15} cm^{-3} at 300 K.

8-4 The dielectric constants of SiO_2 and Si_3N_4 are 3.9 and 7.5, respectively. Calculate the capacitance of a 0.1-μm layer of each.

8-5 For an Al–SiO_2–Si MOS system, calculate the bound charge per cubic centimeter if $|\phi_f| = 0.25$ eV for *p*-type Si. The dielectric constant of Si is 12.

8-6 Change N_A to 1×10^{15} cm^{-3} in the example of Fig. 8-10 and recalculate the threshold voltage.

8-7 In the example of Fig. 8-10, the Al gate is replaced by a heavily doped polysilicon gate, $N_D = 1 \times 10^{20}$ cm^{-3}. (*Note:* At this level of doping Si is degenerate, which means that the Fermi level lies in the conduction band; however, ϕ_f can still be reasonably estimated by the usual method.) Calculate the new work-function difference for the poly-Si–SiO_2–Si MOS system and the new threshold voltage.

8-8 Given a *p*-type substrate with $N_A = 1 \times 10^{15}$ cm^{-3}, an oxide thickness of 1.2 μm, a (111) crystal orientation, and an aluminum gate, find the threshold voltage. What implication does this have for metalizations over thick-field oxides (areas between devices) in MOS ICs if the regions underneath are lightly doped *p*-type?

8-9 Change the doping level to $N_A = 3 \times 10^{17}$ cm^{-3} in Problem 8-8 and calculate the threshold voltage. Regions of this sort are called "guard rings" or "channel stoppers" and are used to prevent inversion in isolation regions under metalization.

8-10 Given an *n*-type substrate with $N_D = 1 \times 10^{15}$ cm^{-3}, a gate oxide thickness of 0.1 μm, a (100) crystal orientation, and an aluminum gate, find the threshold voltage of a *p*-channel MOS device.

8-11 The measured flat-band voltage shift for an MOS capacitor with (aluminum metalization) a *p*-type substrate, $N_A = 1 \times 10^{15}$ cm^{-3}, and an oxide thickness of 0.1 μm is -2.3 V. Calculate the surface-charge density Q_{SS}. A negative voltage of several hundred volts is applied to the metalization of this capacitor while it is at an elevated temperature. The voltage shift is reduced to -1.3 V. What are the respective densities of the bound and mobile surface states?

field-effect transistors

9

A large percentage of the world electronics market today involves devices that use electric fields to control the charges or currents in silicon channels. These *field-effect devices* are often interconnected by the tens of thousands and buried in integrated-circuit chips 1 cm or so on a side. The chips, in turn, are buried in computer systems, microprocessors, pocket calculators, digital watches, and so on. Even the computers and microprocessors often disappear into communications systems, transportation systems, production machinery, instrumentation, and appliances.

The general category of silicon field-effect transistors includes Metal–Oxide–Semiconductor Field-Effect Transistors, or MOSFETs,[1] and Junction Field-Effect Transistors (JFETs). In this chapter we discuss how the basic devices work, first descriptively and then analytically. Some models that can be used for device design and interconnection are developed. Some basic circuits are discussed which show how the devices are used. Field-effect device fabrication and processing techniques are discussed in Chapter 11. The basic devices and circuits introduced in this chapter will be used as building blocks for some of the digital ICs discussed in Chapter 14.

The basis for bipolar transistor action is the *diffusion* of *minority carriers* across the *base* region. Betas are limited because some majority carriers move from base to emitter and because some hole–electron recombination occurs in the base region. We call them bipolar transistors because two current-carrying mechanisms exist (i.e., holes and electrons).

[1] MOSFETs are also known as Metal–Insulator–Semiconductor FETs (MISFETs) and Insulated Gate FETs (IGFETs).

Conversely, the basis for field-effect transistor action is the *drift* of *majority carriers* across the *channel.* The bulk of the current is carried by electrons in n-channel devices, or by holes in p-channel devices, but not *both.* Thus, FETs are sometimes called *unipolar* devices. In comparison to most bipolar transistors, FETs exhibit higher values of input impedance, they are easier to fabricate, and they occupy less space. FETs are less noisy, but they exhibit smaller gain–bandwidth products.

9-1 METAL–OXIDE–SEMICONDUCTOR FETS (MOSFETS)

There are four basic types of MOSFETs, as shown in Fig. 9-1. Each device consists of a substrate, two heavily diffused regions of opposite polarity from the substrate called the *source* and the *drain,* and a metal *gate* which sits above a thin (~1000Å) oxide layer. The gate voltage controls the movement of charge in the *channel* between source and drain. These devices are classified as p-channel or n-channel, and as normally ON or normally OFF. The normally ON mode of operation is called the *depletion mode.* Depletion-mode devices have channels with the same type of charge carrier as the source and drain regions [Fig. 9-1(b) and (d)]. In the normally OFF or *enhancement-mode* devices, the channel doping is opposite to the source and drain regions [Fig. 9-1(a)

Fig. 9-1 Idealized MOSFET structures: (a) n-channel enhancement; (b) n-channel depletion; (c) p-channel enhancement; (d) p-channel depletion.

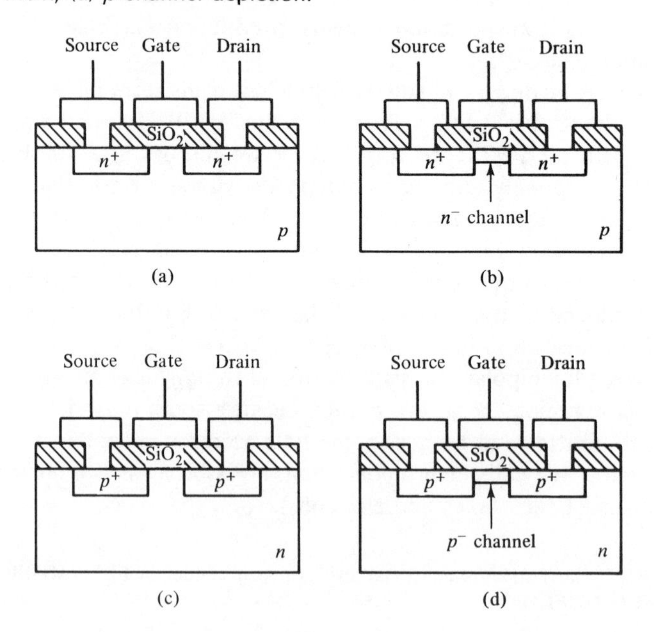

and (c)]. The gate–substrate voltage is applied to create an inversion layer, which turns them on.

As a digital IC element, the MOSFET provides an attractive alternative to the bipolar technologies. Because of the relatively low number of process steps, and the resulting high process yields obtained, an impressive array of digital large-scale integrated (LSI) circuits are commercially available. The major reasons for this are as follows:

1. *No resistors* are needed in MOSFET circuits. The channels function as miniature resistors with sheet resistances on the order of 10 kΩ/□.
2. Because of the insulating layer between gate and channel, the dc *input impedance* of a MOSFET is on the order of 10^{15} Ω. A minimal gate current is needed to charge the gate capacitance. This results in a high fan-out capability.
3. Enhancement-mode devices are *normally off.* A finite level of gate–substrate voltage (~ 1 V) is required to turn them on. This is a very desirable feature in the design of logic and memory elements.
4. The high input impedance and finite gate capacitance of a MOSFET leads to its utilization as a *memory element.* An enhancement-mode device can be held on with the charge on the gate capacitance.
5. MOSFETs have *interchangeable source and drain* contacts. This factor provides a unique advantage in the design of some types of switching circuits.

9-1-1 Description of Operation

Figure 9-2 shows the *n*-channel enhancement-mode MOSFET of Fig. 9-1(a) under various bias conditions. The *n*-type inversion layer is shaded in Figs. 9-2(b) through (d). Although the MOSFET can be operated as a four-terminal device, for now we will connect both the substrate and the source to ground. We are then allowing for voltages $(+V_G)$ and $(+V_D)$ to be present with respect to ground. For now, let us make $V_D = 0$ and look at the effects of V_G acting alone. With $V_G = 0$, our idealized[2] *n*-channel enhancement-mode MOSFET has a *p*-channel as indicated in Fig. 9-2(a).

As V_G is made positive, holes will be repelled from the Si–SiO_2 interface, leaving trapped negatively charged boron ions. These ions begin to form a depletion layer under the oxide interface. The *E* field, originating on the gate, terminates on these trapped negative ions. The ion density is not sufficient to allow an equilibrium condition, however. As V_G is further increased, the depletion layer widens. Mobile electrons, which are available in the n^+ source and drain

[2] With reference to the threshold voltage discussion in Chapter 8, we are temporarily ignoring surface states and the work-function voltage.

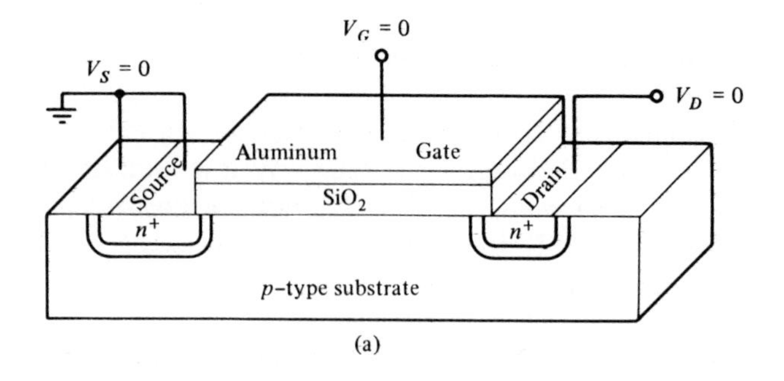

(a)

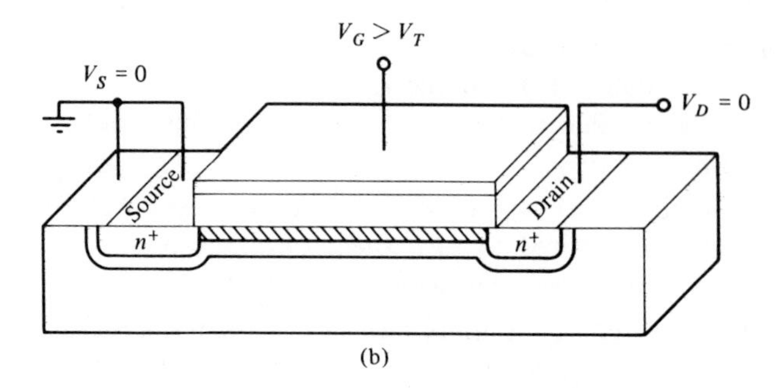

(b)

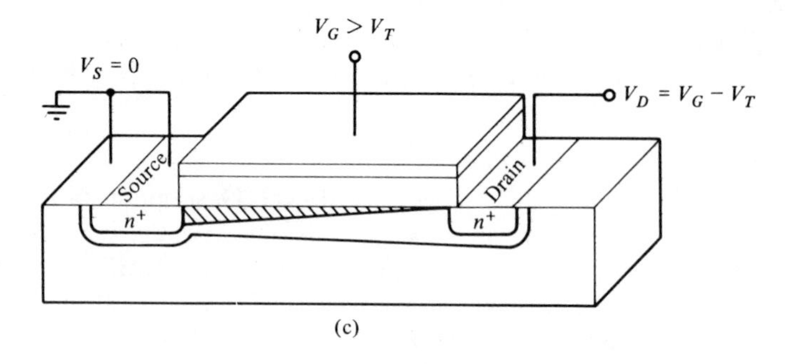

(c)

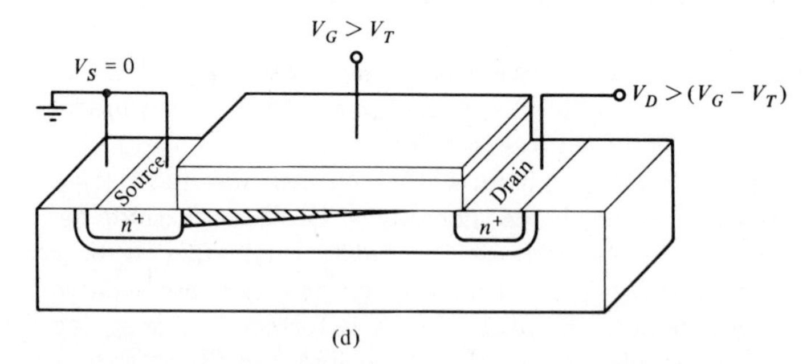

(d)

Fig. 9-2 *n*-Channel enhancement-mode MOSFET under various bias conditions.

regions, are injected into the gate region and attracted to the Si–SiO_2 interface. When the mobile electron layer begins to form, the depletion-layer width remains constant as V_G increases. The mobile negative charge in the channel now changes when V_G changes. This situation is shown in Fig. 9-2(b).

The threshold voltage V_T is defined as that gate voltage which will make the electron concentration in the channel (at the Si–SiO_2 interface) the same as the hole concentration in the *p*-type substrate (see Chapter 8). In an *n*-channel enhancement-mode MOSFET, the threshold of channel formation occurs when V_G equals V_T. With V_G less than V_T, the channel does not exist. With V_G greater than V_T, the channel is present. This approximation is the basis of the "strong inversion model," which will be used to develop the current–voltage characteristics of the MOSFET later in this section.

When a gate voltage just sufficient to cause strong inversion $(V_G = V_T)$ is applied to the device, a portion of this voltage is dropped across the oxide layer (V_{ox}), and a portion is dropped across the silicon depletion layer (V_{Si}). Positive *surface* states exist at the Si–SiO_2 interface, causing positive surface charge Q_{SS}. A negative component of V_T is needed to compensate for this $(-V_{SS})$. The difference in the *work function* of the gate and semiconductor materials causes bending of the energy bands. A negative component of V_T is needed to compensate for this (V_{WF}). Thus, the four factors involved in determining V_T for an *n*-channel enhancement-mode MOSFET are

$$V_T = V_{ox} + V_{Si} - V_{SS} + V_{WF} \tag{9-1}$$

V_T is thus dependent on silicon impurity concentrations, the properties of the insulating layer, whether (001) or (111) silicon is used (this affects Q_{SS}), and the type of gate material used (aluminum or polycrystalline silicon). Channel impurity concentrations are often altered through ion implantation to adjust V_T to the desired value.

With V_G greater than V_T, as shown in Fig. 9-2(b), trapped negative ions exist in the depletion layer and a considerable supply of mobile electrons is available in the channel. The depletion layer is shown not shaded. A uniform channel (shaded) exists between the depletion layer and the SiO_2 layer.

We can now make use of the mobile charge in the channel by applying a positive drain voltage to the device. This causes an electron current from source to drain. The gate voltage controls the magnitude of this current. Our thinking about the device must be two-dimensional. We have a vertical electric field due to the gate to substrate voltage and a horizontal electric field due to the drain to source voltage. The electric fields interact, causing the depletion layer to get wider and the channel to get thinner as we move from left to right, approaching the drain region. This is shown in Fig. 9-2(c). Think of the drain voltage reverse-biasing the drain substrate diode, causing the depletion layer to be wider there. The drain–source voltage appears as a lateral *IR* drop across the channel, decreasing the gate–substrate voltage and narrowing the channel near the drain region. Figure 9-2(c) really shows the MOSFET with

$$V_D = V_G - V_T \tag{9-2}$$

In this situation, the gate–channel voltage near the drain contact is

$$V_G - V_D = V_G - (V_G - V_T) = V_T \tag{9-3}$$

The gate voltage is just sufficient to maintain strong inversion near the drain region.

If V_D is increased again, a section of the channel near the n^+ drain region will "pinch off" and disappear because the gate–substrate voltage in this region is no longer sufficient to cause strong inversion. A widened depletion layer occurs in this region because of the increased reverse bias. A horizontal electric field exists in the depletion layer between the pinch-off point and the n^+ drain region. The mobile carriers from the channel are swept into the n^+ drain region by this electric field. This situation is shown in Fig. 9-2(d). We will refer to the region of device operation in which a conductive channel exists from source to drain as the *linear* region and the region beyond channel pinch-off as the *saturation* region. In the linear region, an increase in V_D causes the channel current to increase. In the saturation region, the channel current remains fairly constant regardless of increases in V_D. The increase in V_D shows up across the depletion region. The remaining channel determines the magnitude of the current. The channel length does decrease slightly as V_D increases because the drain–substrate depletion layer widens in all directions. This shortens the channel and causes a slight increase in I_D. This is referred to as *channel modulation.*[3] It causes a finite value of output resistance in MOS devices.

Figure 9-3 shows a family of characteristic curves for an n-channel enhancement-mode MOSFET. The dashed line represents the boundary between the linear and saturation regions of operation. Since this is determined by whether a condition of strong inversion or pinch-off exists, it is defined by the equation

$$V_D = V_G - V_T$$

It will be shown later that a square-law relationship exists between drain current and gate voltage in the saturation region:

$$I_D = K(V_G - V_T)^2 \tag{9-4}$$

[3] One can imagine an ac signal applied to the drain, causing a relatively high frequency modulation of the channel length.

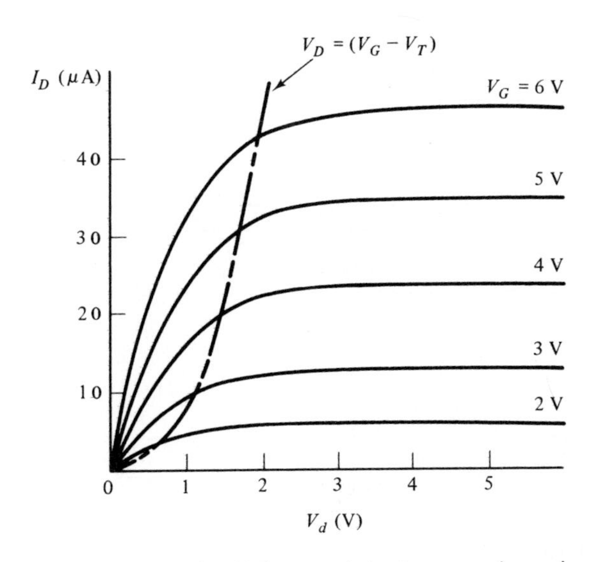

Fig. 9-3 Typical I–V characteristics for an n-channel enhancement-mode MOSFET.

where K is a constant. This is graphically demonstrated by the change in the spacing of the characteristic curves. In other words, $\Delta I_D/\Delta V_G$ is not uniform with V_D constant.

Figure 9-4(a) shows a more detailed drawing of the *n-channel depletion mode* device of Fig. 9-1(b). The n-type conducting channel is shown shaded with $V_S = V_G = V_D = 0$. The conducting channel exists as shown from source to drain.

If V_G is made positive under these circumstances, mobile electrons will be injected into the channel from the source and drain regions, and it will become more conductive.

If, on the other hand, V_G is made negative, electrons are repelled from the channel. The channel becomes narrower until, with a sufficiently negative gate voltage, the channel will pinch off. We will again define the threshold voltage V_T as that gate voltage which will make the electron concentration in the channel the same as the hole concentration in the p-type substrate. In this case, however, $V_G = -V_T$, and the threshold voltage defines the condition in which conduction is *terminated.* Remember that we are discussing a situation in which $V_S = V_D = 0$. Since there is no IR drop in the channel, it is symmetrical from source to drain. When $V_G = -V_T$ in this case, the channel closes.

Now we will apply a positive drain voltage to the device, so that mobile electrons flow from source to drain, and an IR drop exists across the channel. With a sufficiently positive V_D and/or a sufficiently negative V_G, the channel will pinch off. Mobile electrons will be swept from the pinch-off point, across the depletion layer, to the n^+ drain region. The drain current will become rela-

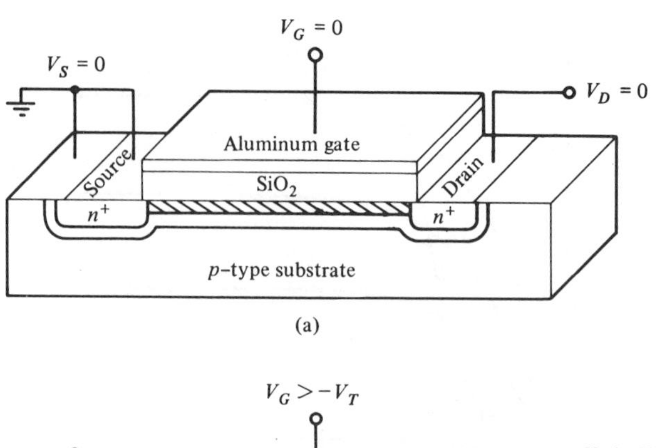

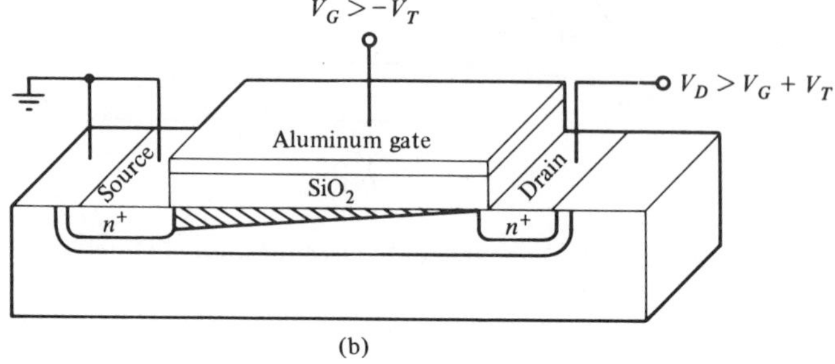

Fig. 9-4 *n*-Channel depletion-mode MOSFET: (a) with no applied voltages; (b) operating in the saturation region.

tively constant with changes in drain voltage, and the device will be operating in the *saturation* region. This mode of operation is sketched in Fig. 9-4(b). Pinch-off occurs when

$$V_D = V_G + V_T$$

but in this case, V_T is a negative voltage. If $V_G = -V_T$, pinch-off occurs with $V_D = 0$. If $V_G = -\frac{1}{2}V_T$, then pinch-off occurs with $V_D = +\frac{1}{2}V_T$.

In enhancement-mode devices, the channel is *field-induced,* whereas in depletion-mode devices, the channel is *built in.*

The characteristic curves for an *n*-channel depletion-mode MOSFET are shown in Fig. 9-5. The square-law relationship of Eq. (9-4) again exists between drain current and gate voltage in the saturation region, but V_T is a negative voltage. The dashed line in Fig. 9-5 represents the boundary between the linear and saturation regions of operation for the strong inversion model. It is defined by Eq. (9-2).

n-Channel depletion-mode devices are used in place of resistive loads in conjunction with *n*-channel enhancement-mode "drivers" in NMOS circuits.

The operation of *p*-channel enhancement and depletion-mode MOSFETs

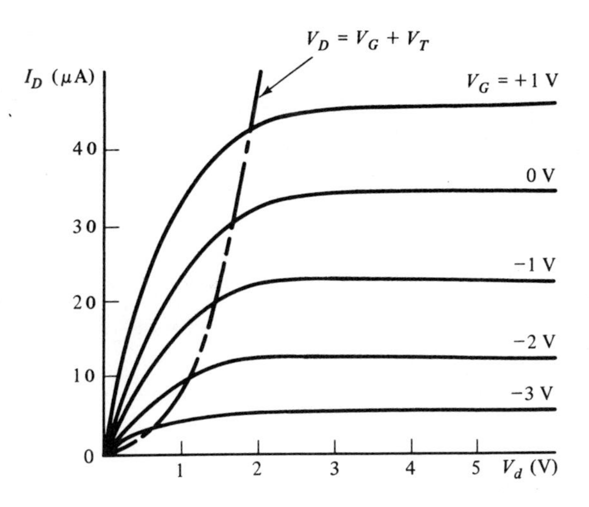

Fig. 9-5 Typical I–V characteristics for an n-channel depletion-mode MOSFET.

is very similar to that of n-channel devices, as shown in Fig. 9-1. They are not used as extensively as n-channel devices because holes are considerably less mobile than electrons. The mobility difference allows n-channel devices to be made smaller and faster. In CMOS (complementary MOS) circuits, p-channel and n-channel enhancement mode devices are used in series.

9-1-2 Current–Voltage Characteristics of n-Channel Enhancement-Mode Devices

Operation in the Linear Region. An n-channel enhancement-mode MOSFET, operating in the linear region, with a field-induced inversion layer is shown in Fig. 9-6. The source and substrate are connected to ground and a positive gate–source voltage $(V_G > V_T)$ is applied so that a conducting inversion layer exists between source and drain. A small drain–source voltage (V_D) is applied so that mobile electrons move from source to drain.

A vertical electric field exists between the gate and the channel. A lateral electric field exists across the channel between the drain and source. Since these fields interact, deriving I–V characteristics becomes a two-dimensional problem. The derivation can be simplified, however, by assuming that the inversion layer is extremely shallow.[4] The derivation can then be dealt with as two one-dimensional problems. We can deal with a vertical electric field across the oxide layer to find the resulting mobile charge density in a section of the channel.

[4] This is known as the "shallow channel" approximation.

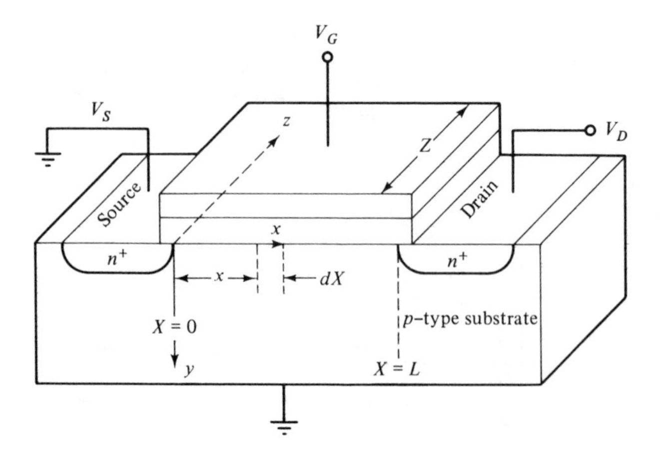

Fig. 9-6 *n*-Channel enhancement-mode MOSFET.

This mobile charge density then leads to an expression for the lateral source to drain current across the channel as a function of applied voltages V_D and V_G.

Referring to Fig. 9-6, let the channel voltage a distance *(x)* from the source be $V(x)$. The vertical potential drop across the oxide layer must exceed the threshold voltage V_T for channel inversion to occur. Thus,

$$E_{ox} = \frac{V_G - V(x) - V_T}{t_{ox}} \tag{9-5}$$

where E_{ox} is the electric field across the oxide layer and t_{ox} is the thickness (in centimeters) of the oxide layer under the gate.

The charge induced in the semiconductor consists of bound negative ions Q_b and mobile electrons Q_m. In equilibrium, this charge is equal to the charge on the gate. Gauss' law tells us that the total flux leaving the gate normal to the surface is numerically equal to the surface-charge density. Thus,

$$\epsilon_{ox}\epsilon_0 E_{ox} = \rho_s(x) \tag{9-6}$$

where $\rho_s(x)$ = surface charge density in C/cm². Combining Eqs. (9-5) and (9-6), we have

$$\rho_s(x) = \frac{\epsilon_{ox}\epsilon_0}{t_{ox}}[V_G - V(x) - V_T] \qquad \text{C/cm}^2 \tag{9-7}$$

Referring to Fig. 9-6, we now assume a uniform sheet of charge of thickness y_0. The volume charge density $\rho_v(x)$ can now be expressed as

$$\rho_v(x) = \frac{\rho_s(x)}{y_0} \qquad \text{C/cm}^3 \tag{9-8}$$

$\rho_v(x)$ is caused by trapped negative ions and mobile electrons. We will assume that the density of trapped ions is small with respect to the mobile electron density. Although this approximation leads to some error in practical devices, it provides a much less complex mathematical model. The conductivity of the channel at point (x) can then be expressed as

$$\sigma_n(x) = \rho_v(x)\mu_{ns} = \frac{\rho_s(x)\mu_{ns}}{y_0} \qquad \text{mhos/cm} \tag{9-9}$$

where μ_{ns} is the surface mobility of electrons in the channel. (Surface mobility values are about half the magnitude of bulk mobility values. The difference is caused by additional scattering mechanisms associated with the nearness of the surface. We will assume that surface mobility values are constant across the channel.)

Combining Eqs. (9-7) and (9-9), we have

$$\sigma_n(x) = \frac{\mu_{ns}\epsilon_0\epsilon_{\text{ox}}}{y_0 t_{\text{ox}}}[V_G - V(x) - V_T] \qquad \text{mhos/cm} \tag{9-10}$$

Ohm's law at a point relates current density to conductivity and the magnitude of the electric field. Thus,

$$J_n(x) = \sigma_n(x)E(x) = \sigma_n(x)\frac{dV(x)}{dx} \qquad \text{A/cm}^2 \tag{9-11}$$

We are now assuming a variation in channel voltage only in the (x) direction due to the IR drop in the channel. Then

$$I_n(x) = J_n(x)A \tag{9-12}$$

Referring to Fig. 9-6, the cross-sectional area of the channel is

$$A = Zy_0 \tag{9-13}$$

The drain current through the channel must be constant at every point (x), so that

$$I_n(x) = I_D \tag{9-14}$$

Then

$$I_D = Zy_0\sigma_n(x)\frac{dV(x)}{dx} \qquad \text{A} \tag{9-15}$$

or

$$I_D dx = Zy_0\sigma_n(x)\, dV(x) \tag{9-16}$$

Substituting Eq. (9-10) in Eq. (9-16), we have

$$I_D\, dx = \frac{Z\mu_{ns}\epsilon_0\epsilon_{ox}}{t_{ox}}[V_G - V(x) - V_T]\, dV(x) \tag{9-17}$$

or

$$I_D\int_0^L dx = \frac{Z\mu_{ns}\epsilon_0\epsilon_{ox}}{t_{ox}}\int_0^{V_D}[V_G - V(x) - V_T]\, dV(x) \tag{9-18}$$

Then

$$I_D = \frac{Z\mu_{ns}\epsilon_0\epsilon_{ox}}{Lt_{ox}}[(V_G - V_T)V_D - \tfrac{1}{2}V_D^2] \tag{9-19}$$

Since

$$C_{ox} = \frac{\epsilon_0\epsilon_{ox}}{t_{ox}} \qquad \text{F/cm}^2 \tag{9-20}$$

Eq. (9-19) can be written

$$I_D = \frac{Z}{L}\mu_{ns}C_{ox}[(V_G - V_T)V_D - \tfrac{1}{2}V_D^2] \tag{9-21}$$

This is the characteristic equation for device operation in the linear region. It is valid only for the region of operation in which an inversion layer exists from source to drain. In other words, it is valid for the region in which

$$V_G > V_T \qquad \text{and} \qquad (V_G - V_T) > V_D \tag{9-22}$$

Note that when $V_D = 0$, $I_D = 0$ in Eq. (9-21). For very small values of V_D^2, that is, for

$$\tfrac{1}{2}V_D^2 << (V_G - V_T)\, V_D, \tag{9-23}$$

Eq. (9-21) can be approximated as

$$I_D \approx \frac{Z}{L} \mu_{ns} C_{ox}(V_G - V_T)V_D \tag{9-24}$$

With V_G constant, I_D is a linear function of V_D. When

$$V_D = V_G - V_T \tag{9-25}$$

the channel disappears near the drain, since the gate–channel voltage is not sufficient to maintain strong inversion in that region.

Operation in the Saturation Region. For $V_D > V_G - V_T$, the MOSFET is operating in the saturation region with the channel pinched off. Mobile charge is swept across the depletion region by the lateral electric field, from the point where the channel is pinched off to the n^+ drain region. The drain voltage at pinch-off is $V_{D\text{sat}}$. Then

$$V_{D\text{sat}} = (V_G - V_T) \tag{9-26}$$

If we substitute this in Eq. (9-21), we obtain

$$I_{D\text{sat}} = \frac{Z}{L} \mu_{ns} C_{ox} \frac{(V_G - V_T)^2}{2} \tag{9-27}$$

This is the characteristic equation for device operation in the saturation region. It tells us that with V_G constant, I_D will be constant. I_D is no longer dependent on V_D. It also tells us that a square-law relation exists between I_D and $(V_G - V_T)$.

The current–voltage characteristics for an n-channel enhancement-mode MOSFET are shown in Fig. 9-7. The dashed line represents the transition between operation in the linear and saturation regions. The square-law relation is seen in the spacing between constant V_G lines.

Channel conductance g_D is defined by

$$g_D = \left.\frac{\partial I_D}{\partial V_D}\right|_{V_G = \text{const.}} \tag{9-28}$$

It is a measure of the slope of the I–V characteristic curves with constant gate voltage. It is also the reciprocal of the channel resistance r of the device. To find g_D in the linear region, we apply Eq. (9-28) to the equation for I_D in the linear region (9-21). Then

$$g_D = \frac{Z}{L} \mu_{ns} C_{ox}(V_G - V_T - V_D) \qquad \text{(linear region)} \tag{9-29}$$

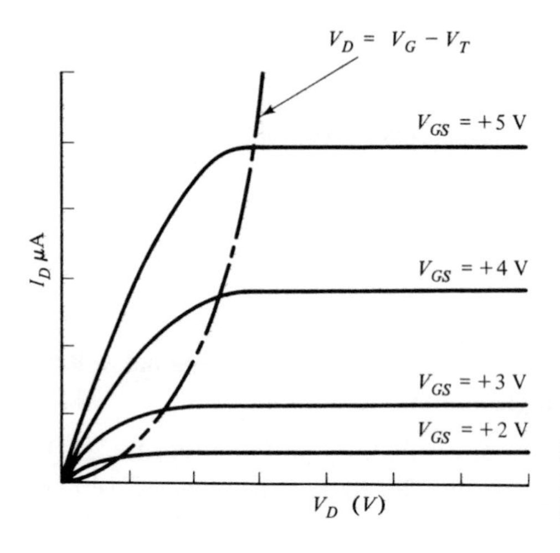

Fig. 9-7 Current–voltage characteristics for an n-channel enhancement-mode MOSFET.

Since our model specifies that the I–V curves are flat in the saturation region, application of Eq. (9-28) to the equation for $I_{D\mathrm{sat}}$ (9-27) will yield $g_D = 0$. Departing from our simple model, however, we find that as V_D is increased in the saturation region, the depletion region around the n^+ drain becomes wider. The channel length L therefore decreases. This causes $I_{D\mathrm{sat}}$ to increase with increasing V_D and leads to a nonzero value of $g_{D\mathrm{sat}}$. Its reciprocal, $r_{D\mathrm{sat}}$, is therefore finite.

Transconductance g_m is defined by

$$g_m = \left.\frac{\partial I_D}{\partial V_G}\right|_{V_D = \mathrm{const.}} \tag{9-30}$$

Referring to the I–V curves of Fig. 9-7, g_m is a measure of the spacing between constant V_G lines for fixed values of V_D. It is used as a measure of gain for field-effect transistors. Applying Eq. (9-30) to the equation for I_D in the linear region (9-21), we obtain

$$g_m = \frac{Z}{L}\mu_{ns} C_{\mathrm{ox}} V_D \qquad \text{(linear region)} \tag{9-31}$$

Applying Eq. (9-30) to the equation for I_D in the saturation region (9-27), we obtain

$$g_{m\,\mathrm{sat}} = \frac{Z}{L}\mu_{ns} C_{\mathrm{ox}}(V_G - V_T) \tag{9-32}$$

9-1-3 MOSFET Symbols and Small-Signal Models

Figure 9-8 shows the IEEE standard symbols for MOSFETs. In each case, the gate is shown separated from the rest of the transistor, indicating that a high impedance exists between the gate and the rest of the device. The arrow indicates the doping of the substrate or bulk material (sometimes denoted with a B for body). An arrow into the device indicates a p-type substrate, which corresponds to an n-channel device; and so on. Enhancement-mode MOSFETs are shown with a dashed vertical line to indicate normally "off" behavior with zero gate voltage. A solid vertical line is used for depletion-mode MOSFETs to indicate that a current path exists between drain and source with zero gate voltage.

MOSFETs can usually be obtained from manufacturers as four-terminal devices, with separate substrate connections available, or as three-terminal devices, with substrates internally connected to sources. Since these devices are usually constructed symmetrically, the source and drain contacts are interchangeable.

A small-signal common-source equivalent circuit, which is useful for analyzing linear MOSFET circuits, is shown in Fig. 9-9. C_{gs} represents the gate-to-source capacitance. Its magnitude is roughly two-thirds that of the gate-to-oxide capacitance (C_{ox}). C_{gd} represents the gate-to-drain capacitance which is normally smaller than C_{gs}. In some MOS devices, the oxide layer and gate metalization overlap the source and drain regions. This tends to increase C_{gs} and C_{gd} to larger values than would occur due to fringing fields with no overlap present. C_{ds} represents the drain-to-source capacitance. If the substrate is not connected to the source, C_{ds} is the series capacitance of the reverse-biased drain–

Fig. 9-8 MOSFET symbols.

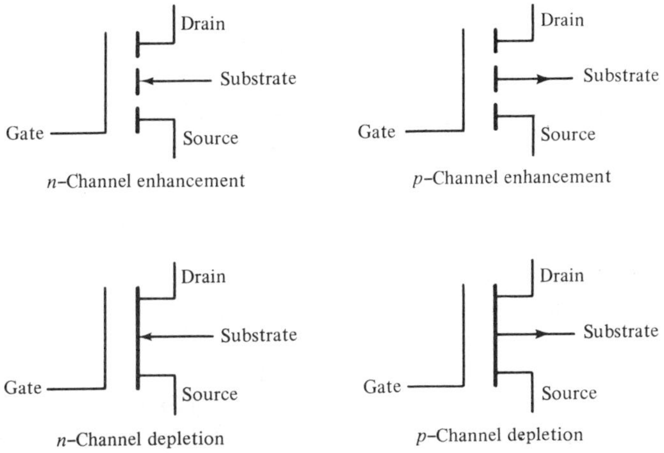

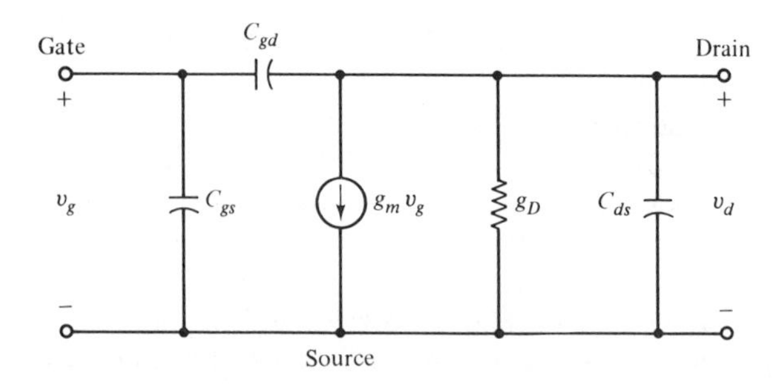

Fig. 9-9 Small-signal equivalent circuit for a MOSFET.

substrate diode and source–substrate diode. If the source–substrate connection is made, then C_{ds} is just the capacitance of the drain–substrate diode. g_m represents the device transconductance and g_D represents the channel conductance. Because of the nonlinear nature of the MOSFET, all these circuit elements are differential and their magnitudes are dependent on the device operating point.

The maximum operating frequency of the device modeled in Fig. 9-9 can be defined as the frequency at which the current through C_{gs} is equal to the current through the dependent current source $g_m V_g$. Then

$$2\pi f_m C_{gs} V_g = g_m V_g \tag{9-33}$$

and

$$f_m = \frac{g_m}{2\pi C_{gs}} \tag{9-34}$$

Now we will assume that

$$C_{gs} \approx C_{ox} Z L \tag{9-35}$$

Substituting Eq. (9-32) for $g_{m\,\text{sat}}$ and Eq. (9-35) in Eq. (9-34), we obtain

$$f_m \approx \frac{\mu_{ns}(V_G - V_T)}{2\pi L^2} \tag{9-36}$$

The present frequency limit of MOS devices is much lower than that indicated by Eq. (9-36). The present limit is determined by the speed with which MOS devices can charge and discharge load capacitances.

Figure 9-10 shows a small-signal low-frequency equivalent circuit for a MOSFET. This model is useful for deriving equations for the voltage gains and output resistances of the various small-signal amplifier configurations.

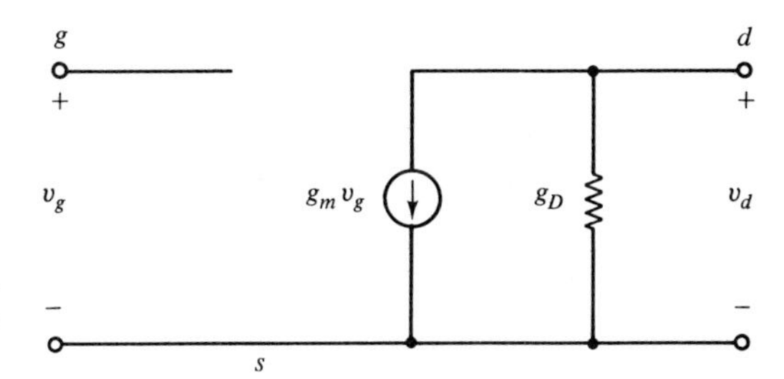

Fig. 9-10 Small-signal low-frequency equivalent circuit for a MOSFET.

9-1-4 Basic Digital Inverters

The most basic digital logic circuit is an inverter, which performs the logical NOT function. We will restrict our discussion to the two types of MOSFET inverters that are most frequently used: the *n*-channel *enhancement driver–depletion load* configuration, and the complementary MOS or *CMOS inverter,* which consists of a *p*-channel enhancement-mode MOSFET in series with an *n*-channel enhancement-mode MOSFET. An inverter consists of an active device or a "driver," which provides gain, and a passive or "load" device, which, in effect, replaces the load resistor.

We will use the word "high" to denote a voltage level close to the supply voltage ($+V_{DD}$). "Low" will denote a voltage level close to zero volts or ground. Input and output voltages can only be high or low or be making a transition between these two levels. If the input voltage to an inverter is high, the output is low, and vice versa.

The two inverters discussed in this section are the basic building blocks for whole families of logic functions, including gates, flip-flops, arithmetic logic units (ALUs), random-access memories (RAMs), and read-only memories (ROMs).

***n*-Channel Enhancement Driver/Depletion Load Inverter.** Ion implantation made it economically feasible to process enhancement- and depletion-mode devices on the same chip. Threshold voltages can be optimized for either or both types of device.

Figure 9-11 shows two inverters. Each has an enhancement driver: Fig. 9-11(a) has an enhancement load and (b) has a depletion load. C_L represents stray capacitance and the input capacitance to succeeding stages. Charging and discharging C_L limits the speed of MOS circuits.

If we prepared an I_D vs. V_{DS} graph for the *driver* MOSFETs, and then constructed load lines for the *load* devices in Fig. 9-11(a) and (b), they would be nonlinear and would appear as shown in Fig. 9-12. The load line for a conventional resistive load is shown for comparison. The depletion load in Fig. 9-12 is seen to be much more attractive because (1) it intercepts the voltage axis at $+V_{DD}$, whereas the enhancement load does not and (2) while making the transition from point *A* to point *B,* it acts primarily as a current source.

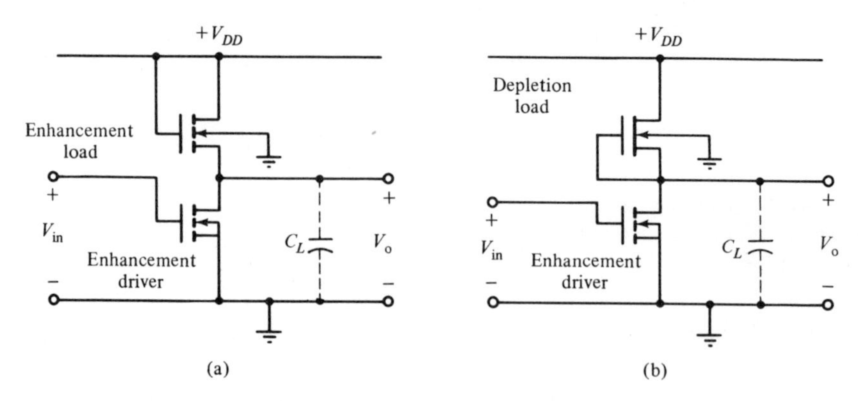

Fig. 9-11 Digital inverter configurations: (a) enhancement driver, enhancement load; (b) enhancement driver, depletion load.

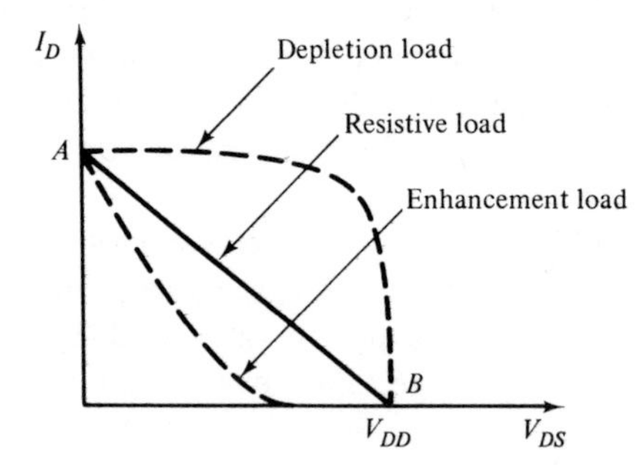

Fig. 9-12 Load lines for enhancement, depletion, and resistive loads.

This maximizes the speed at which the external load capacitance is charged.

For convenience, Fig. 9-11(b) is redrawn in Fig. 9-13, together with the depletion load line and the I–V characteristics of the enhancement driver. When V_{in} is high, the driver unit is on. Its characteristic curve is shown as I_D(ON)

Fig. 9-13 (a) Enhancement driver/depletion load inverter. (b) I–V characteristics and load lines.

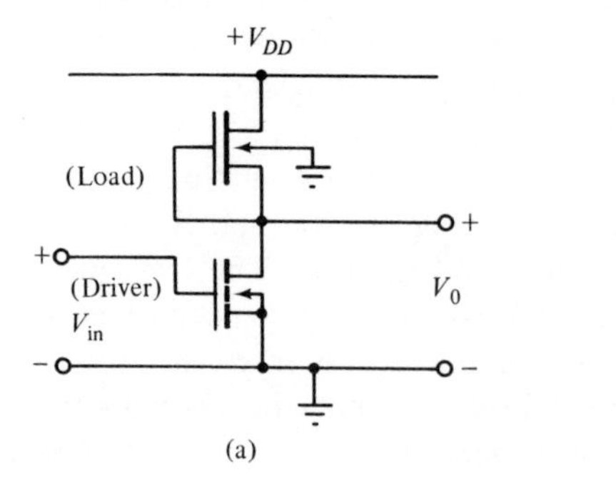

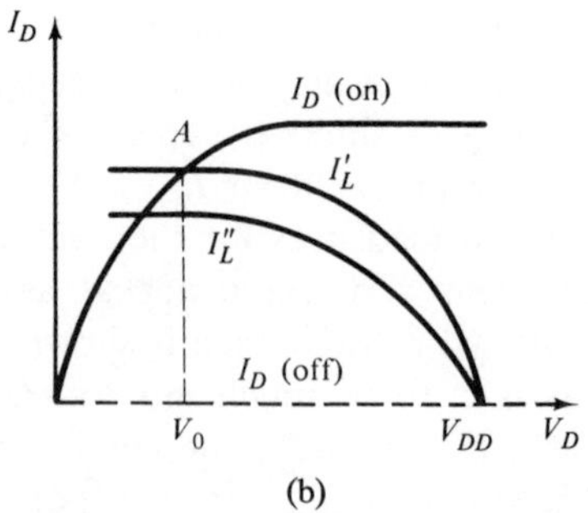

in Fig. 9-13(b). The load unit current is I'_L. The operating point occurs at A, where I_D(ON) and I'_L meet. The output V_0 is low.

When V_{in} is low, the driver unit is off. Its characteristic curve is shown as I_D(OFF). The load device has a current I''_L. The operating point occurs where I_D(OFF) and I''_L meet, at V_{DD}. Thus V_0 is high.

I''_L is smaller than I'_L because when V_0 is high, the source of the load device is at $+V_{DD}$ with respect to the grounded substrate. In other words, the substrate is reverse-biased with respect to the source. This alters the threshold voltage and thus the load characteristics of the device.

We can use the device equations developed earlier in this chapter to find an expression for V_0 when V_{in} is high. With V_{in} high, the load device is saturated. Then, from Eq. (9-27),

$$I'_L = I_{DL} = \frac{Z}{L} \mu_{ns} C_{ox} \frac{(V_G - V_{TL})^2}{2} \tag{9-37}$$

where V_{TL} is the threshold voltage of the depletion load device. Let

$$\beta_L = \frac{Z}{L} \mu_{ns} C_{ox} \tag{9-38}$$

for the load device. Since the gate and source are connected, $V_G = 0$. Then

$$I'_L = \frac{\beta_L}{2} V_{TL}^2 \tag{9-39}$$

The enhancement driver is operating in the linear region. Then, from Eq. (9-21),

$$I_D(\text{ON}) = \frac{Z}{L} \mu_{ns} C_{ox} [(V_G - V_{TD}) V_D - \tfrac{1}{2} V_D^2] \tag{9-40}$$

where V_{TD} is the driver threshold voltage. In this situation, $V_G = +V_{DD}$, and $V_D = V_0$. Let

$$\beta_D = \frac{Z}{L} \mu_{ns} C_{ox} \tag{9-41}$$

for the driver. Then

$$I_D(\text{ON}) = \beta_D [(V_{DD} - V_{TD}) V_0 - \tfrac{1}{2} V_0^2] \tag{9-42}$$

The two devices are in series, and the operating point occurs where

$$I'_L = I_D(\text{ON}) \tag{9-43}$$

Then

$$\frac{\beta_L}{2} V_{TL}^2 = \beta_D[(V_{DD} - V_{TD})V_0 - \tfrac{1}{2}V_0^2] \tag{9-44}$$

Rearranging terms, we have

$$V_0^2 - 2(V_{DD} - V_{TD})V_0 + \frac{\beta_L}{\beta_D} V_{TL}^2 = 0 \tag{9-45}$$

Using the quadratic equation to find V_0 and taking the useful solution, we have that

$$V_0 = V_{DD} - V_{TD} - \left[(V_{DD} - V_{TD})^2 - \frac{\beta_L}{\beta_D} V_{TL}^2 \right]^{\frac{1}{2}} \tag{9-46}$$

To get an approximate figure for V_0 when V_{in} is high, let $V_{DD} = +5$ V, $|V_{TD}| = |V_{TL}| = 1$ V, and $\beta_L = \beta_D$. Then

$$V_0 = (5 - 1) - [(5 - 1)^2 - (1)^2]^{\frac{1}{2}} = 0.127 \text{ V}$$

This is a reasonably "low" level of output voltage for use in digital applications.

Complementary MOS (CMOS) Inverter. In complementary MOS or CMOS circuits, *p*- and *n*-channel enhancement-mode MOSFETs are arranged in series combinations between a power supply and ground. Figure 9-14 shows a circuit diagram for a digital CMOS inverter. This type of structure has many attractive features, such as low standby power dissipation (in the nanowatt range), high noise immunity, a wide range of operating voltages (typically 3 to 15 V), and a wide range of operating temperatures (−55 to 125°C).

Under static conditions, with V_{in} high or low, one device is on and the

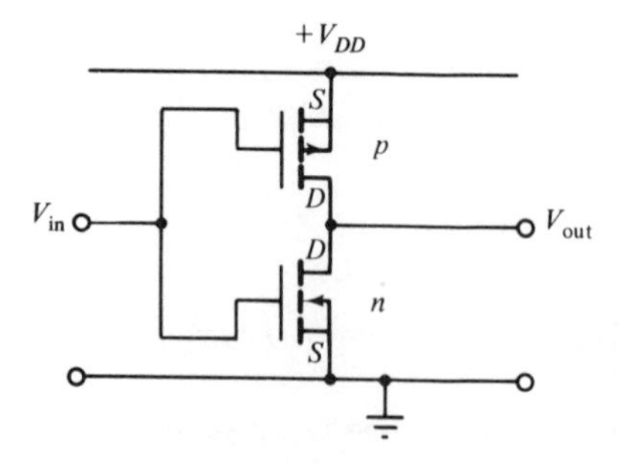

Fig. 9-14 CMOS inverter.

other device is off. While V_{out} is connected through a low-impedance path to either the supply voltage or ground, the path from the power supply to ground always contains an off device, so that only leakage currents exist.

Figure 9-15 shows a cross section of a simplified CMOS structure. The starting material is an *n*-type substrate with $N_D \approx 1 \times 10^{15}$ cm^{-3}. A *p*-type well with $N_A \approx 3 \times 10^{16}$ cm^{-3} is diffused into the substrate. The *n*-type substrate accommodates *p*-channel devices, and the *p*-well accommodates *n*-channel devices.

As shown in Fig. 9-15, the source and substrate of the *p*-channel MOSFET are connected to $+V_{DD}$. When the gate voltage is low, the gate will be sufficiently negative *with respect to the channel* to turn the unit on. The unit will be off with a positive voltage applied to the gate.

Since the source and substrate of the *n*-channel MOSFET are connected to ground, the unit will be off for a low gate voltage and on for a high gate voltage.

With digital CMOS circuits, it is useful to remember that a low gate voltage turns a *p* unit on and a high gate voltage turns a *n* unit on. When V_{in} is low in Fig. 9-14, the *p* unit is on and the *n* unit is off. Since a conductive path exists between V_{DD} and V_{out}, V_{out} is high. When V_{in} is high, the *p* unit is off and the *n* unit is on. A conductive path exists between V_{out} and ground, and V_{out} is low.

Under static conditions, CMOS devices operate in a linear or a cutoff region. The value of transconductance (g_m) for an *n*-channel device is given by Eq. (9-31) and is repeated here for convenience.

$$g_m(n\text{-channel}) = \frac{Z_1}{L_1} \mu_{ns} C_{ox} V_D \tag{9-31}$$

Since g_m must have a positive value, absolute values are used for voltages applied to *p*-channel enhancement-mode devices. Thus,

Fig. 9-15 Cross section of a CMOS inverter.

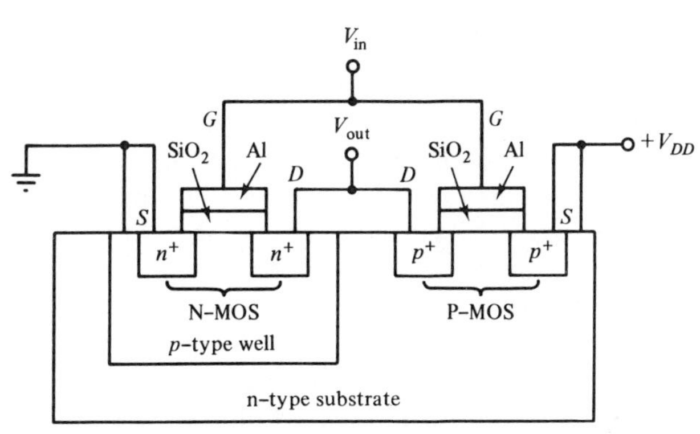

$$g_m(p\text{-channel}) = \frac{Z_2}{L_2}\mu_{ps}C_{ox}|V_D| \tag{9-47}$$

Electron surface mobility μ_{ns} is approximately twice as large as hole surface mobility μ_{ps}. Therefore, if the channel lengths are equal ($L_1 = L_2$) and we want the g_m's to be the same, the p unit must be twice as wide as the n unit ($Z_2 = 2Z_1$).

For static operation, power dissipation in a CMOS inverter is determined by the product of leakage current and supply voltage. It is negligibly small. During transient operation at high frequencies, however, the inevitable load capacitance must be charged and discharged. The energy stored in the load capacitance during each half cycle of the input is $\frac{1}{2}C_L V_{DD}^2$. The average power dissipated in the inverter is the energy stored during a cycle divided by the period of the waveform, or

$$P_{diss} = \frac{C_L V_{DD}^2}{T} = C_L V_{DD}^2 f \tag{9-48}$$

where f is the input frequency. Thus, power dissipation is proportional to frequency.

9-2 JUNCTION FIELD-EFFECT TRANSISTORS (JFETs)

The JFET is a voltage-controlled device whose input impedance is in the range between bipolar transistors and MOSFETs. It has nonlinear I–V characteristics and zero offset voltage. These properties make it a valuable device in electronic circuits. Also, design innovations can provide JFET devices with characteristics not presently available with bipolar or MOSFET technologies. Two examples of this are high-current vertical JFETs with uniquely low values of saturation resistance, and gallium arsenide JFETs, which operate in frequency ranges beyond the capabilities of MOSFET or bipolar devices.

9-2-1 Description of Operation

The width of the depletion layer across a reverse-biased p-n junction can be controlled by varying the doping of the p and n regions and by adjusting the applied voltage across the junction. The width of a channel, sandwiched between two depletion layers, can thus be controlled by doping and applied voltage.

An idealized n-channel JFET is shown in Fig. 9-16. The reverse biased p^+-n junctions ensure that the bulk of the depletion regions will exist in the lightly doped n-channel. The depletion layer thickness, and thus the channel

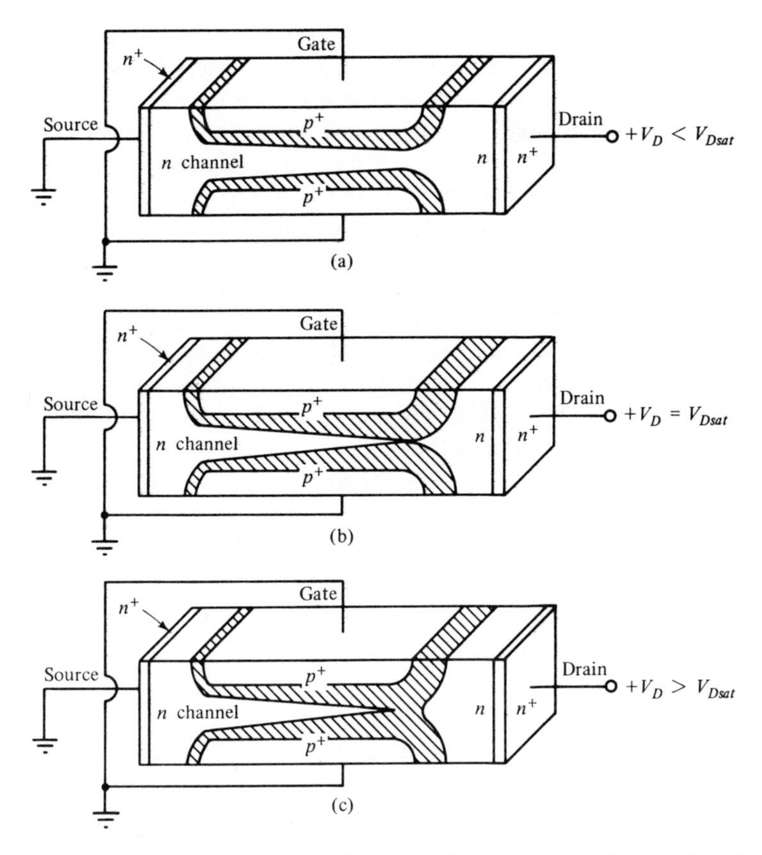

Fig. 9-16 *n*-Channel JFET with gate and source grounded. (a) $V_D < V_{D\,\mathrm{sat}}$—a resistive channel exists between source and drain. (b) $V_D = V_{D\,\mathrm{sat}}$—the channel is just pinched off; channel current is swept across the depletion layer. (c) $V_D > V_{D\,\mathrm{sat}}$—the device operates in the saturation region; the drain current is constant.

thickness, can then be controlled by adjusting the gate–source voltage or the drain–source voltage. In Fig. 9-16 the gate and source are connected to ground and a positive voltage is applied to the drain. If V_D is small, as in Fig. 9-16(a), the *n*-channel acts as a simple resistor. If we increase V_D, a linear increase in I_D will result. The device is operating in the *linear region.* As we move from source to drain, the *IR* drop across the channel increases and the depletion width across the reverse-biased channel gate diode increases. If V_D is made more positive, the channel becomes thinner and the channel resistance increases. Thus, the positive slope of the *I–V* characteristic decreases. When $V_D = V_{D\mathrm{sat}}$, the channel pinches off, as shown in Fig. 9-16(b). In the vicinity of pinch-off, the relationship between the horizontal and vertical electric fields and the current density is very complex. In the interest of brevity, we will assume here that the depletion regions join at the pinch-off point, isolating the source from the drain. We will further assume that the channel current is swept across the depletion region in much the same way that charge carriers are swept across the base–collector depletion layer in a bipolar transistor. We will refer to the drain current in this situation as I_{Dsat}. If the drain voltage is further increased, the depletion layer will become wider, as shown in Fig. 9-16(c), but the channel current will remain near the value I_{Dsat} because the bulk of the voltage increase will develop across the depletion layer. The JFET is now operating in the *saturation region.*

To observe the effect of the gate–source voltage V_G acting alone, let us hold $V_D = 0$ and $I_D = 0$ while we vary V_G. Because there is no *IR* drop in the channel, the depletion layer is uniform, as shown in Fig. 9-17. As V_G is made more negative, the depletion regions get wider. By setting $V_G = V_p$, it is possible to pinch the channel off completely.

If we now apply V_D and V_G at the same time, we can control the *I–V* characteristics of the device as shown in Fig. 9-18. For low values of V_D, the *I–V* characteristic is linear. As V_D becomes larger, the channel pinches off and

Fig. 9-17 Effect of V_G acting alone on an *n*-channel JFET.

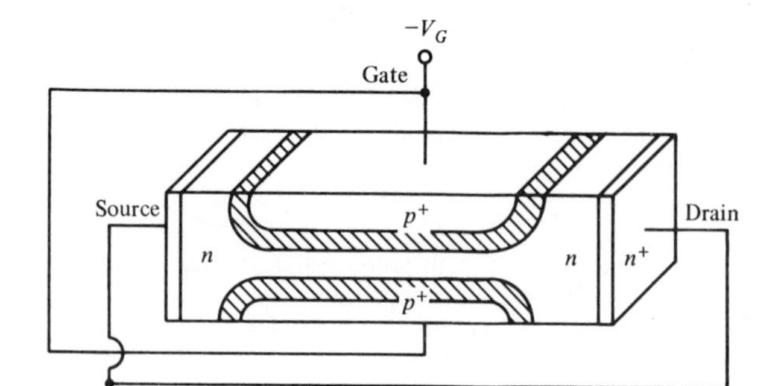

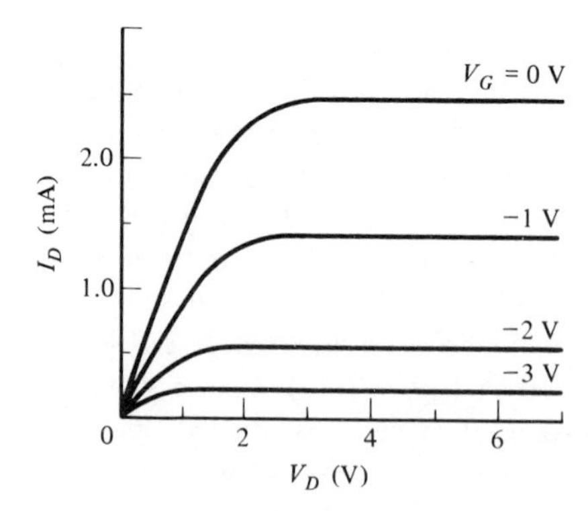

Fig. 9-18 *I–V* characteristics for an *n*-channel JFET.

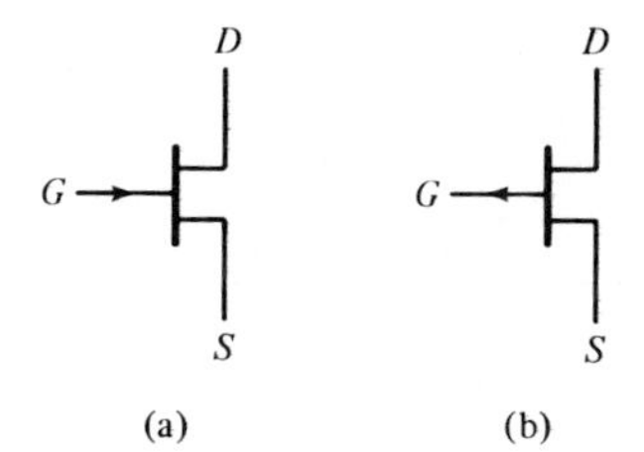

Fig. 9-19 Circuit symbols for JFETs: (a) *n*-channel JFET; (b) *p*-channel JFET.

we move into the constant-current or saturation region of operation. As V_G is made more negative, the channel current decreases. A sufficiently negative gate voltage ($V_G \approx -5$ V) will reduce the drain current to zero. As with the MOSFET, a nonlinear relationship exists between I_D and V_G.

Circuit symbols for *n*- and *p*-channel JFETs are shown in Fig. 9-19. The arrow "in" on Fig. 9-19(a) indicates a *p*-type gate and therefore an *n*-type channel. The arrow "out" on Fig. 9-19(b) indicates a *p*-channel device. As with MOSFETs, *n*-channel devices are generally preferred because of the higher electron mobilities.

9-2-2 *I–V* Characteristics of an Idealized JFET[5]

We want to develop a mathematical model of the *I–V* characteristics for an *n*-channel JFET. We need expressions for I_D as a function of V_D and V_G for the linear and saturation regions of operation. We start with a uniformly doped, abrupt-junction JFET as shown in Fig. 9-20. Although JFETs are not usually uniformly doped or constructed in this manner, this provides a good starting point for a model.

The resistance of the channel shown in Fig. 9-20 is given by

$$R = \frac{L}{\sigma A} = \frac{L}{q\mu_n N_D Z(d - 2W)} \tag{9-49}$$

where $q\mu_n N_D$ is the channel conductance; L, Z, and d represent the channel length, width, and thickness; and W represents the width of the depletion layer.

The depletion width of an abrupt p^+-n junction is given by Eq. (6-108), which is repeated here for convenience.

$$W \approx \left(\frac{2\epsilon_0\epsilon_{Si} V}{qN_D}\right)^{\frac{1}{2}} \tag{9-50}$$

where V represents the sum of the contact potential V_B and the applied gate–channel voltage. As the drain voltage is increased with $V_G = 0$, the depletion

[5] This development closely follows that of Grove [1].

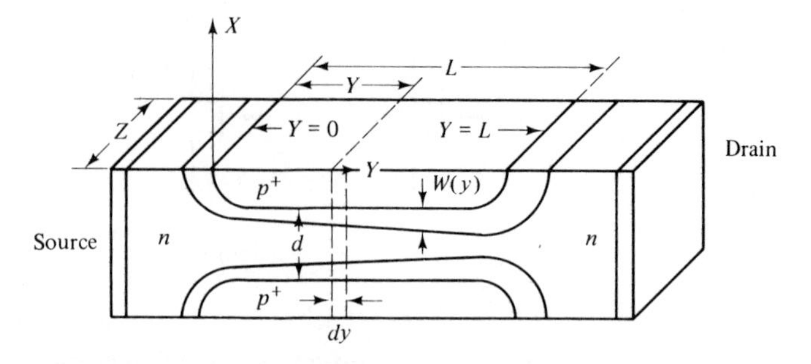

Fig. 9-20 Notation for a mathematical I–V model.

widths *(W)* increase until they touch near the drain when $V_D = V_{Dsat}$. At this point the depletion width W is equal to $d/2$. Equation (9-50) can then be rewritten

$$\frac{d}{2} = \left[\frac{2\epsilon_0\epsilon_{Si}(V_{Dsat} + V_B)}{qN_D}\right]^{\frac{1}{2}} \tag{9-51}$$

Solving for V_{Dsat}, we have

$$V_{Dsat} = \left(\frac{qN_D d^2}{8\epsilon_0\epsilon_{Si}} - V_B\right) \qquad \text{for } V_G = 0 \tag{9-52}$$

If a negative gate voltage (V_G) is applied to increase the reverse bias of the gate–channel junction, it will further increase the depletion-layer width. A smaller value of V_{Dsat} will then be needed to cause pinch-off. Then

$$V_{Dsat} = \frac{qN_D d^2}{8\epsilon_0\epsilon_{Si}} - V_B + V_G \tag{9-53}$$

where V_G is negative for an n-channel device.

Suppose, for example, that

$$N_D = 3 \times 10^{15} \text{ atoms/cm}^3$$
$$N_A = 1 \times 10^{18} \text{ atoms/cm}^3$$
$$d = 2\ \mu\text{m} = 2 \times 10^{-4} \text{ cm}$$

Then

$$V_B = \frac{kT}{q} \ln \frac{N_A N_D}{n_i^2} \approx 0.8 \text{ V}$$

$$\frac{qN_D d^2}{8\epsilon_0\epsilon_{Si}} = 2.26 \text{ V}$$

Then $V_{Dsat} = 2.26 - 0.8 + V_G$ volts. If $V_G = 0\text{V}$, $V_{Dsat} = 1.46$ V; if $V_G = -1\text{V}$, $V_{Dsat} = 0.46$ V; and if $V_G = -1.46\text{V}$, $V_{Dsat} = 0$ V. In the latter case, the gate is sufficiently negative to pinch off the channel with no drain voltage applied. This is called the gate pinch-off voltage V_p. From Eq. (9-53), with $V_{Dsat} = 0\text{V}$, the channel is pinched off when

$$V_G = V_p = -\frac{qN_D d^2}{8\epsilon_0\epsilon_{Si}} + V_B \tag{9-54}$$

For convenience, we will let

$$V_p' = V_p - V_B = -\frac{qN_D d^2}{8\epsilon_0\epsilon_{Si}} \tag{9-55}$$

From Eq. (9-49) and Fig. 9-20, the incremental voltage across a length of channel dy can be written

$$dV = I_D dR = \frac{I_D dy}{q\mu_n N_D Z[d - 2W(y)]} \tag{9-56}$$

or

$$I_D dy = q\mu_n N_D Z[d - 2W(y)]dV \tag{9-57}$$

The net voltage drop across the depletion layer a distance (y) from the source is given by

$$V = V(y) + V_B - V_G$$

From Eq. (9-50), the width of the depletion layer a distance (y) from the source is then

$$W(y) = \left\{\frac{2\epsilon_0\epsilon_{Si}[V(y) + V_B - V_G]}{qN_D}\right\}^{\frac{1}{2}} \tag{9-58}$$

Substituting Eq. (9-58) in Eq. (9-57), rearranging terms, and solving for I_D, we then have

$$I_D = G_0\left\{V_D + \frac{2}{3}V_p'\left[\left(\frac{V_G - V_B - V_D}{V_p'}\right)^{\frac{3}{2}} - \left(\frac{V_G - V_B}{V_p'}\right)^{\frac{3}{2}}\right]\right\} \tag{9-59}$$

where

$$G_0 = \frac{Z}{L} q\mu_n N_D d \qquad \text{mhos} \tag{9-60}$$

G_0 represents the conductance of the *n*-type channel with no depletion layers present. Equation (9-59) represents the abrupt junction model that we have been seeking. Since it is still somewhat cumbersome, we can simplify it for two cases: the linear region of device operation, in which V_D is small, and the saturation region $V_{D\text{sat}}$, in which I_D is relatively constant.

In the *linear region,* with $V_D << V_B - V_G$, a series expansion of Eq. (9-59) can be used to arrive at a simpler expression. Then

$$I_D \approx G_0\left[1 - \left(\frac{V_G - V_B}{V'_p}\right)^{\frac{1}{2}}\right] V_D \tag{9-61}$$

In the *saturation region,* Eq. (9-53) can be substituted in Eq. (9-59) to yield

$$I_{D\text{sat}} = I_{DSS}\left[1 - 3\left(\frac{V_G - V_B}{V_p - V_B}\right) + 2\left(\frac{V_G - V_B}{V_p - V_B}\right)^{\frac{3}{2}}\right] \tag{9-62}$$

where

$$I_{DSS} = -\tfrac{1}{3} G_0 (V_p - V_B) \tag{9-63}$$

or

$$I_{DSS} = I_D \qquad \text{with } (V_G - V_B) = 0 \tag{9-64}$$

A widely used engineering approximation for the saturation drain current of a JFET is given by

$$I_{D\text{sat}} \approx I_{DSS}\left[1 - \left(\frac{V_G - V_B}{V_p - V_B}\right)\right]^2 \tag{9-65}$$

where I_{DSS} is again given by Eqs. (9-63) and (9-64).

Let us refer back to our earlier example, with $N_D = 3 \times 10^{15}$ atoms/cm^3, $N_A = 1 \times 10^{18}$ atoms/cm^3, and $d = 2$ μm. We found that with these values, $V_B \approx 0.8$ V, $V'_p = -2.26$ V, and $V_p = -1.46$ V. The Z/L ratio can be chosen to set the magnitude of the drain current. Let $Z/L = 100$. Then, from Eq. (9-60),

$$G_0 = \left(\frac{Z}{L}\right) q\mu_n N_D d = 1.3 \times 10^{-2} \text{ mho}$$

and from Eq. (9-63),

$$I_{DSS} = -\tfrac{1}{3} G_0 V_p' = 9.79 \text{ mA}$$

From Eq. (9-62), $I_{D\text{sat}} = 3.52$ mA, and from Eq. (9-65), $I_{D\text{sat}} = 4.09$ mA. A comparison with experimental data would be required to determine which $I_{D\text{sat}}$ value is more realistic.

The conductance of the drain–source channel is defined as

$$g_D = \left.\frac{\partial I_D}{\partial V_D}\right|_{V_G=\text{const.}} \tag{9-66}$$

Applying this definition to Eq. (9-61), the channel conductance in the linear region is

$$g_D = G_0\left[1 - \left(\frac{V_G - V_B}{V_p'}\right)^{\frac{1}{2}}\right] \tag{9-67}$$

where V_G and V_p' are negative for an *n*-channel device. As V_G becomes more negative, g_D decreases until, when $V_G = V_p$, the channel is completely depleted and g_D goes to zero.

The transconductance is defined by

$$g_m = \left.\frac{\partial I_D}{\partial V_G}\right|_{V_D=\text{const.}} \tag{9-30}$$

Applying this definition to Eq. (9-61), the JFET transconductance in the *linear region* is given by

$$g_m = -\frac{G_0 V_D}{2V_p'}\left(\frac{V_p'}{V_G - V_B}\right)^{\frac{1}{2}} \tag{9-68}$$

From Eq. (9-62), the transconductance in the *saturation region* is

$$g_{m\ \text{sat}} = G_0\left[1 - \left(\frac{V_G - V_B}{V_p'}\right)^{\frac{1}{2}}\right] \tag{9-69}$$

Notice that the channel conductance in the linear region is equal to the transconductance in the saturation region.

Small-signal JFET models for low-and high-frequency operation are shown in Fig. 9-21(a) and (b), respectively. We are assuming here that the reverse leakage current of the gate diode is negligibly small, so that the input impedance

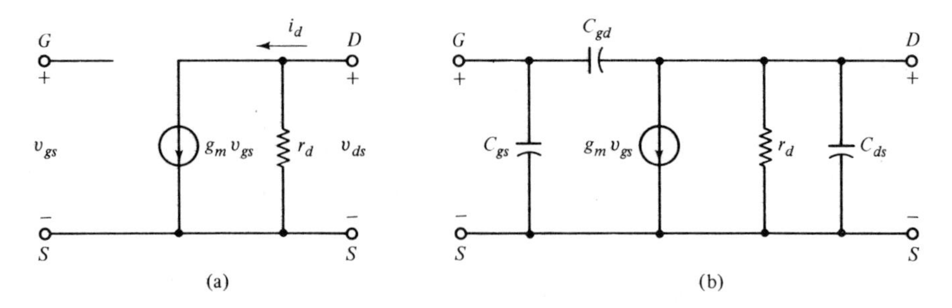

Fig. 9-21 (a) Low-frequency small-signal JFET model. (b) High-frequency small-signal model.

of the low-frequency model is infinite. The drain resistance r_d is the reciprocal of the channel conductance g_D. If the drain voltage is increased with the device operating in the saturation region, the depletion region widens, causing a slight decrease in channel length and a corresponding increase in drain current. This causes the I–V characteristic to have an upward slope and r_d to have a finite value. If this channel-shortening effect is ignored, however, r_d will be infinite in the saturation region. In Fig. 9-21(b), C_{gs} represents the gate–source capacitance, C_{gd} represents the gate–drain capacitance, and C_{ds} represents the drain–source capacitance. In the design of high-frequency JFETs, it is important to minimize C_{gd} because it has considerable influence on the gate input capacitance due to the Miller effect.

BIBLIOGRAPHY

[1] GROVE, A. S., *Physics and Technology of Semiconductor Devices.* New York: John Wiley & Sons, Inc., 1967.

[2] RICHMAN, P., *MOS Field-Effect Transistors and Integrated Circuits.* New York: John Wiley & Sons, Inc., 1973.

[3] CARR, W. N., AND J. P. MIZE, *MOS/LSI Design and Application.* New York: McGraw-Hill Book Company, 1972.

[4] WARNER, R. N., AND J. N. FORDEMWALT, *Integrated Circuits: Design Principles and Fabrication.* New York: McGraw-Hill Book Company, 1965.

[5] WALLMARK, J. T., AND H. JOHNSON, *Field Effect Transistors: Physics, Technology and Applications.* Englewood Cliffs, N.J.: Prentice-Hall, Inc., 1966.

[6] SAH, C. T., *Characteristics of the Metal–Oxide–Semiconductor Transistors. IEEE Transactions on Electron Devices,* Vol. ED-11, No. 7, July 1964, pp. 324–345.

PROBLEMS

9-1 Given that $E_{ox} = [V_G - V_T - V(x)]/t_{ox}$ for an n-channel enhancement-mode MOSFET, outline the major steps in the derivation of the equation

$$I_D = \frac{Z}{L}\mu_{ns}C_{ox}[(V_G - V_T)V_D - \tfrac{1}{2}V_D^2]$$

Briefly indicate the technique used to get from one equation to the next.

9-2 An n-channel enhancement-mode MOSFET is constructed with the following parameters:

$$Z = 100\ \mu\text{m} \qquad \epsilon_{ox} = 4$$
$$L = 10\ \mu\text{m} \qquad \epsilon_0 = 8.85 \times 10^{-14}\ \text{F/cm}$$
$$t_{ox} = 0.1\ \mu\text{m} \qquad \mu_{ns} = 450\ \text{cm}^2/\text{V-s}$$
$$V_T = +1.0\ \text{V}$$

Find $I_{D\text{sat}}$ and $g_{m\ \text{sat}}$ when the device is operating with $V_G = V_D = +5$ V (source and substrate grounded).

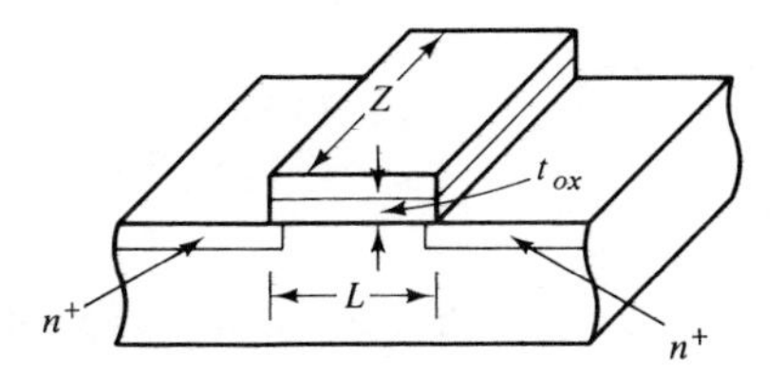

9-3 An n-channel depletion-mode MOSFET is connected between +5 V and ground as shown. Given that

$$Z = 200\ \mu\text{m} \qquad \epsilon_{ox} = 4$$
$$L = 10\ \mu\text{m} \qquad \epsilon_0 = 8.85 \times 10^{-14}\ \text{F/cm}$$
$$t_{ox} = 0.1\ \mu\text{m} \qquad \mu_{ns} = 600\ \text{cm}^2/\text{V-s}$$
$$V_T = -1.0\ \text{V}$$

Find the drain current I_D through the device.

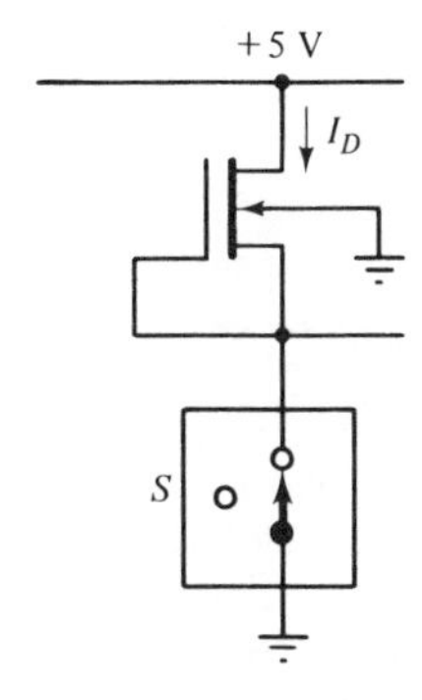

9-4 Switch S in Problem 9-3 is suddenly opened, and drain current I_D must charge a load capacitance C_L of 35 pF to +5.0 V. Assuming as a first approximation that I_D remains constant, how much time is required for C_L to charge to +5 V?

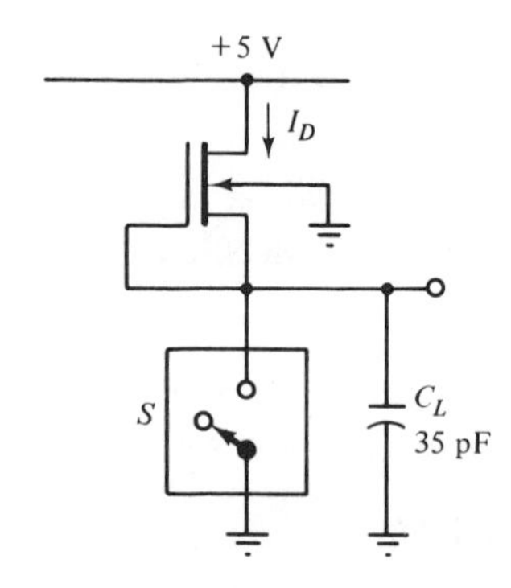

9-5 A *p*-channel enhancement-mode MOSFET is constructed with the following parameters:

$Z = 50\ \mu m$ $\quad \epsilon_{ox} = 4$
$L = 5\ \mu m$ $\quad \epsilon_0 = 8.85 \times 10^{-14}$ F/cm
$t_{ox} = 0.1\ \mu m$ $\quad \mu_{ps} = 190$ cm²/V-s
$V_T = -1.0$ V

(a) Find I_D, $r_D = 1/g_D$, and g_m when the device is operating in the linear region with $V_G = -3.0$ V and $V_D = -0.1$ V.

(b) Find I_{Dsat} and $g_{m\ sat}$ when the device is operating with $V_G = -4.0$ V and $V_D = -5.0$ V.

9-6 Given the low-frequency small-signal amplifier shown, with $V_T = 1.0$ V, $Z/L = 10$, $\mu_{ns} = 600$ cm²/V-s, and $C_{ox} = 3.54 \times 10^{-8}$ F/cm², find the voltage gain of the amplifier V_o/V_{in}. Assume that the transistor is operating in the saturation region.

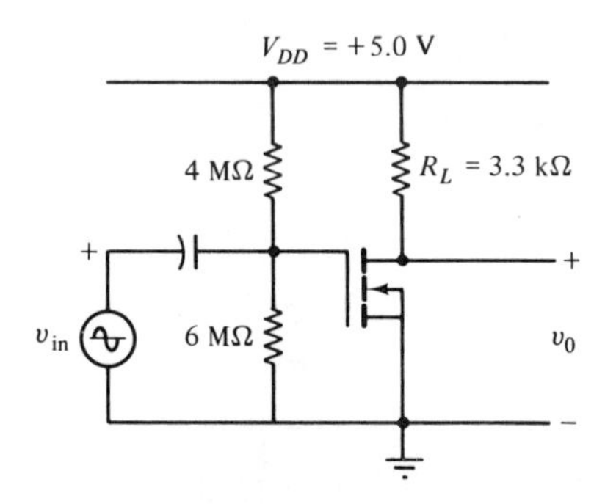

9-7 The parameters for this enhancement driver depletion–load digital inverter are listed below. Find V_o when V_{in} is high (5.0 V).

$\epsilon_0 = 8.85 \times 10^{-14}$ F/cm $\quad V_T(\text{load}) = -2.0$ V
$\epsilon_{ox} = 4$ $\quad V_{DD} = +5.0$ V
$t_{ox} = 0.1\ \mu m$

For both devices, $Z = L = 10\ \mu m$, N_A (substrate) $= 5 \times 10^{14}/\text{cm}^3$, μ_{ns}(driver) $= 450$ cm²/V-s, μ_{ns}(load) $= 600$ cm²/V-s, and V_T(driver) $= +0.8$ V.

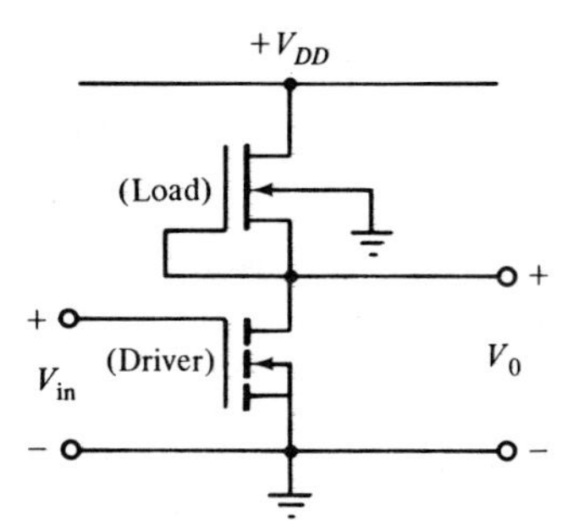

9-8 A CMOS inverter is driven by a square wave voltage source with a period of 1 μs. The power supply voltage $V_{DD} = +12$ V and the load capacitance C_L is 20 pF. Find the power dissipated by the inverter.

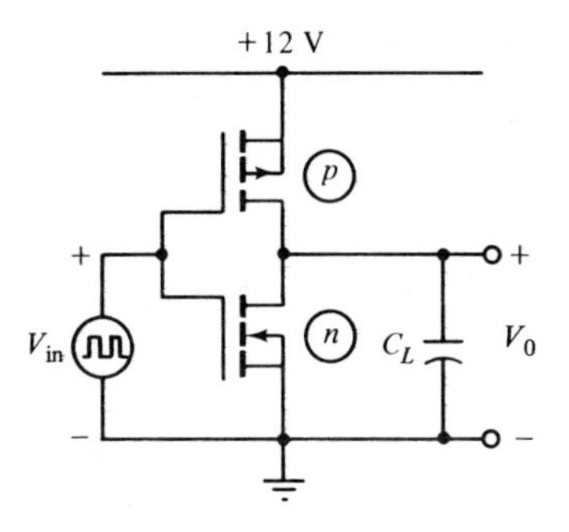

9-9 An n-channel JFET is constructed as shown. The design parameters are $N_D = 4 \times 10^{15}/\text{cm}^3$, $N_A = 1 \times 10^{18}/\text{cm}^3$, $d = 2\ \mu$m, $L = 10\ \mu$m, and $Z = 400\ \mu$m.

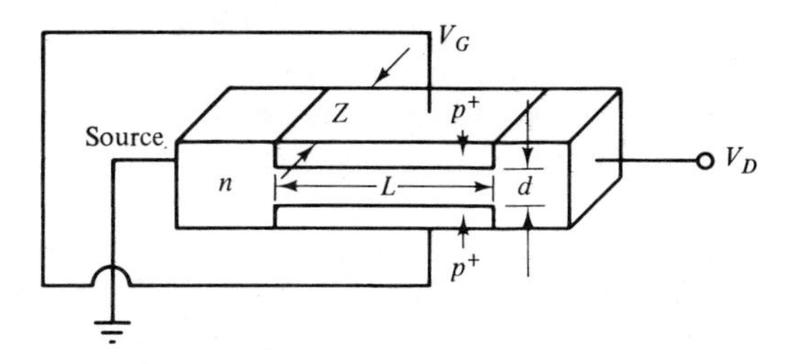

(a) Find the actual thickness of the channel $(d - 2w)$ in micrometers with the source, drain, and gate all connected to ground.
(b) Find the channel resistance under the circumstances of part (a).
(c) Find the gate pinch-off voltage of the device.
(d) If $V_G = -1.0$ V with the source grounded, find the drain voltage at which the device goes into saturation ($V_{D\text{sat}}$).
(e) If $V_G = -1.0$ V and $V_D = +3$ V, use Eqs. (9-62) and (9-65) to find $I_{D\text{sat}}$. Compare the results obtained.

9-10 Derive Eq. (9-59) from Eqs. (9-57) and (9-58).

9-11 Derive Eq. (9-62) from Eqs. (9-59) and (9-53).

lithography

10

In this chapter the process steps involved in progressing from a circuit diagram to a precisely defined pattern in the surface layer of a silicon wafer will be discussed. By the process of lithography, windows can be opened in layers of SiO_2 and fine-line patterns can be defined in polycrystalline silicon and aluminum layers. After completing a discussion of lithography in this chapter, the fabrication of integrated circuits will be described in Chapter 11.

Starting with a circuit diagram, it is necessary to lay out the artwork for a set of individual photomasks, corresponding, for example, to buried layers, isolation regions, transistor bases and resistors, emitters, contact windows, metalization patterns, and so on. Typically, a mask is required for each structurally defined step in the fabrication of an integrated circuit. The mask layout process results in a set of scaled drawings, typically 200 to 1000 times actual size, on Mylar-base graph paper.

Using a computer-operated digitizer, the mask pattern information is recorded and stored in computer memory. The layout information can then be plotted or displayed on an oscilloscope. Interactive graphic displays are often used so that the designer can make layout changes with a keyboard or a light pen. Completed mask layouts can then be stored digitally on magnetic tapes. Libraries of standard circuit layouts or cells are developed in this manner. Standard cell information from the "library" can be used for new IC layouts. Interactive graphics can be used to modify input and output circuits and interconnection patterns. The usefulness of mask libraries has been limited, however, by the rapid evolution of IC technology.

Radiation-sensitive emulsions called *resists* play a key role in the transfer of circuit patterns from the mask to the wafer. Typically, a thin layer (0.5 to

1.5 μm) of resist is deposited on the wafer surface. A mask, which is patterned with opaque and transparent surface areas, is placed between the wafer and a source of radiation. Resist areas under transparent mask regions are exposed through the mask to the radiation source, while the areas under opaque mask regions remain unexposed. The new mask patterns must be accurately aligned with previous patterns in the wafer surface. The alignment and exposure processes take place in a machine called a *mask aligner.* As a result of being exposed and then immersed in develop and rinse solutions, an image of the mask is preserved as a resist layer. In addition to being radiation sensitive, resists have a second important property. The resist used for a given process step is *chemically resistant* to the etchant used to remove the surface layer of the wafer. Thus, the surface resist patterns act as barriers against the etchant. When the unprotected surface areas have been etched away, the remaining resist is removed. The mask patterns are preserved in the surface layer. The wafer is now ready for a fabrication step such as an impurity diffusion. Between 7 and 14 masks are typically used to process a batch of IC wafers. The resist processes described above are repeated for every mask.

Both optical systems and electron-beam systems are used to make masks from the mask pattern information stored in computer memory. Some of the ways in which these systems are used are shown in Fig. 10-1.

Photolithography involves the use of photographic reductions to make masks. The standard photoreduction process actually involves two reductions. For the first reduction, a computer-driven light source is used to expose a photographic plate and generate a 5× or 10× size photoplate or reticle mask. This mask contains patterns for a single circuit. Two methods are presently used between the reticle mask and the mask aligner, as shown in Fig. 10-1.

The more modern technique uses the reticle mask in a direct-step mask aligner. Sections of the wafer are exposed, sequentially, through the mask. The final reduction to 1× size takes place with the mask image through the mask aligner optics. The present disadvantage of this technique is the cost of the direct-step mask aligner.

With the older, more standard technique, the first reduction is followed by a second step-and-repeat reduction. During this second reduction, the single circuit pattern is repeated sequentially, in rows and columns, to produce a 1× size master mask. This technique is shown in the second column of Fig. 10-1. Working mask copies are made from the master masks. They are used in the mask aligner to transfer patterns into the resist on the silicon wafer surfaces.

Electron-beam lithography (EBL) systems are available which are specifically designed to interface with digital computers and produce patterns from digitally stored mask information. Two types of EBL systems are in use: those designed to produce master masks and those designed to process wafers directly. Both types of systems are indicated in Fig. 10-1 and are described below.

For the production of master masks, an EBL system is used in conjunction with an electron-sensitive resist. The optically flat glass masks are coated with

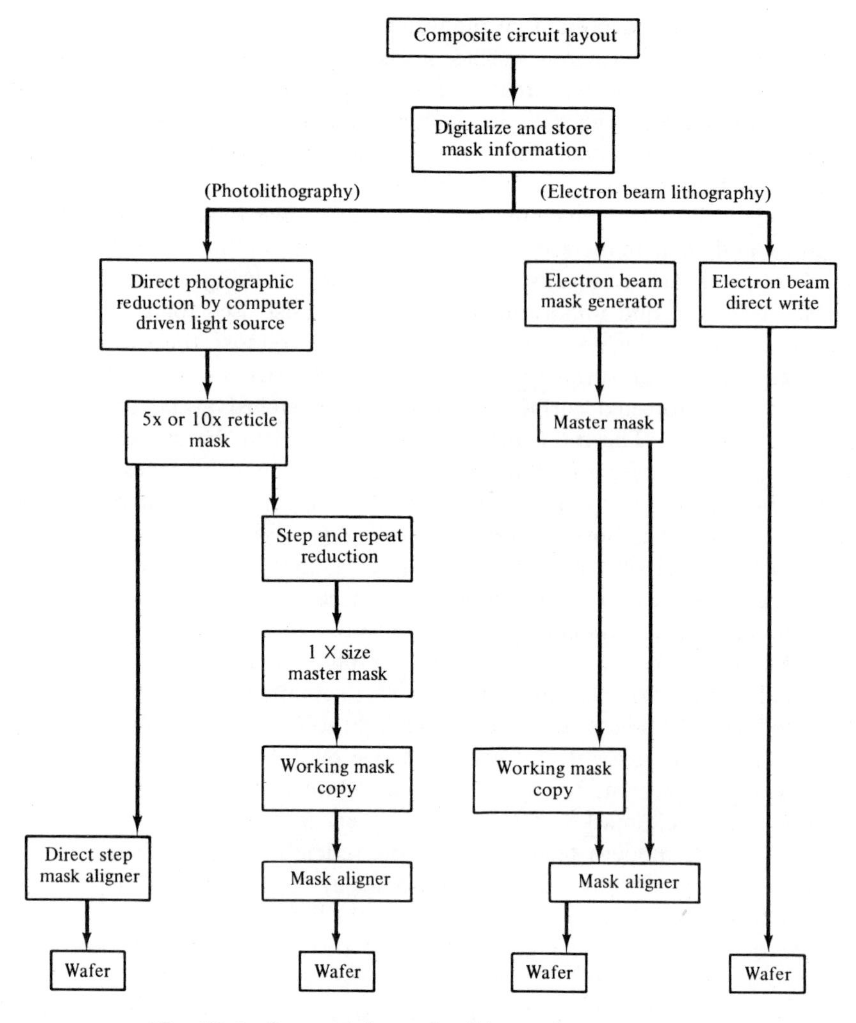

Fig. 10-1 Some of the major lithography systems.

a metal, such as chromium, and then coated again with the electron-sensitive resist. The electron beam, controlled by pattern information stored in the computer, is turned on and off and moved about on the resist surface. Thus, it exposes patterns in the resist. The EBL system can only expose small areas of the mask at a given time. The mask sits on an x-y table, which indexes to repeat a given circuit pattern until the entire mask is exposed. After exposure, the mask is immersed in develop and rinse solutions, and an image of the mask patterns is preserved in the remaining resist layer. The mask is then treated with an etchant, which removes the metal layer in areas not protected by the resist. The remaining resist is then removed in a stripping solution and the mask is chemically cleaned. For some applications, the electron-beam-produced mask is used as a working photomask. For other applications, copies of the master mask are produced photographically and used as working photomasks. To transfer the patterns from a working photomask to a wafer, the silicon wafer is coated with a resist that is sensitive to ultraviolet light or X-rays. The wafer is then exposed, through the photomask, to the radiation source. When the wafer is developed and rinsed, the resist adheres only in regions that were exposed.

EBL systems are available that can be used to process wafers directly. In this type of application, the silicon wafer is coated with an electron-sensitive resist and exposed to the electron beam. As with the previous EBL system, the wafer is positioned on an *x-y* table, which indexes to repeat a given circuit pattern, until the entire wafer is exposed. After exposure, the wafer is developed and rinsed, and subsequently etched. Because the EBL system can expose only a section of the wafer at a given time, and 7 to 14 masks may be needed to complete the wafer processing, this is a slow, expensive technique at present. It is in limited use for wafer production. Its advantages lie in the fact that the process is direct, line widths on the order of 0.5 μm can be obtained, and fast implementation of new designs is possible.

Now that an overview of the major lithography processes has been presented, the remainder of this chapter will be used to:

1. Develop the mask layout techniques generally used for bipolar circuits.
2. Present a more thorough explanation of electron-beam lithography systems.
3. Discuss resists in more detail.
4. Discuss mask alignment systems.
5. Develop the photo-processing techniques used to go from coating a wafer with resist to wafer etching and resist removal.

10-1 BIPOLAR MASK LAYOUT

In this section we discuss the methods through which we can progress from a bipolar circuit diagram to a set of masks. As an illustrative example, we will "walk through" the design of a mask set for an emitter-coupled-logic (ECL) gate, as shown in Fig. 10-2.

One of the first considerations to be resolved in laying out a mask set is to determine the number of isolation regions required. Since the isolation diffusion is the deepest, space must be allocated around the isolation mask patterns to allow for lateral diffusion. To minimize the chip or die size, we begin by minimizing the number of isolation regions used. This is decided by the number of epitaxial regions in the circuit that must assume different voltage levels. For example, transistors Q_1 and Q_2 can be put in one epitaxial region because the collectors are connected. Transistor Q_3 is put in a separate epitaxial region

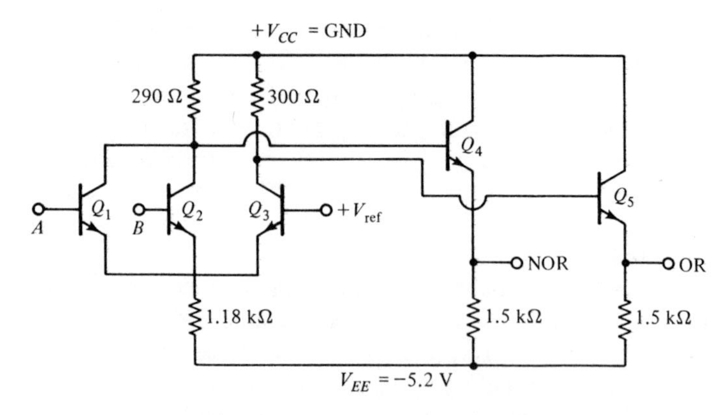

Fig. 10-2 Emitter-coupled logic (ECL) gate.

because its collector voltage is unique. Transistors Q_4 and Q_5 and all the resistors can be placed in a single epitaxial region. The region containing the resistors, and transistors Q_4 and Q_5, is tied to the most positive potential in the circuit. The resistors are isolated from each other because they are at more negative potentials than the *n*-epi region that they are seated in (i.e., they have reverse-biased *p-n* junctions between them). Thus, we should be able to build the circuit of Fig. 10-2 with three isolation regions.

This design example was developed so that the masks could be laid out on 3M's No. 341-M millimeter graph paper ($8\frac{1}{2} \times 11$ in.). The minimum etched window will be 5×5 μm, and the minimum separation between *p* and *n* regions will be 5 μm. By making 5 μm correspond to 2 mm, or two divisions on 341-M graph paper, we will be laying the circuit out at 400 times actual size. In this way, students can gain experience in mask layout without unnecessary expense. Typically, 400 or 500 times reductions are used with conventional photolithography. Minimum line widths are in the vicinity of 2 to 5 μm because of the wavelength of the ultraviolet light source used for mask alignment and mask alignment accuracy. Although the dimensions used here are somewhat large, they were chosen for convenience in laying out masks on millimeter graph paper.

Emitter and base geometries for an *npn* transistor that can be used for the ECL gate of Fig. 10-2 are shown in Fig. 10-3. The sizes of transistors used to conduct currents in the range 1 to 5 mA are usually minimized. Transistor layout usually begins with the emitter window and progresses to the emitter, base, and n^+ collector regions. If the collector voltage is unique, *p*-type isolation regions must surround the *n*-epi region that forms the collector. In Fig. 10-3 the emitter contact window is 5×5 μm. A 5-μm spacing has been used from the emitter contact to the emitter edge, and also from the emitter edge to the base edge. We want the metalizations to overlap the contact windows by 5 μm. We will use a minimum metalization width and spacing of 10 μm. To

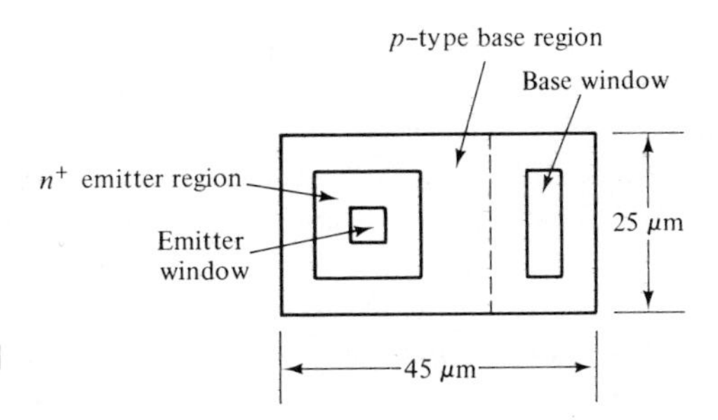

Fig. 10-3 Transistor base and emitter regions.

hold these dimensions, a spacing of 15 μm is needed between the n^+ emitter edge and the edge of the base window in Fig. 10-3. The dashed line represents the 5 μm by which the base contact metalization overlaps the base window. As shown, the base dimensions are then set at 25 × 45 μm.

A single isolated *npn* transistor is shown in Fig. 10-4. The *p*-type isolation

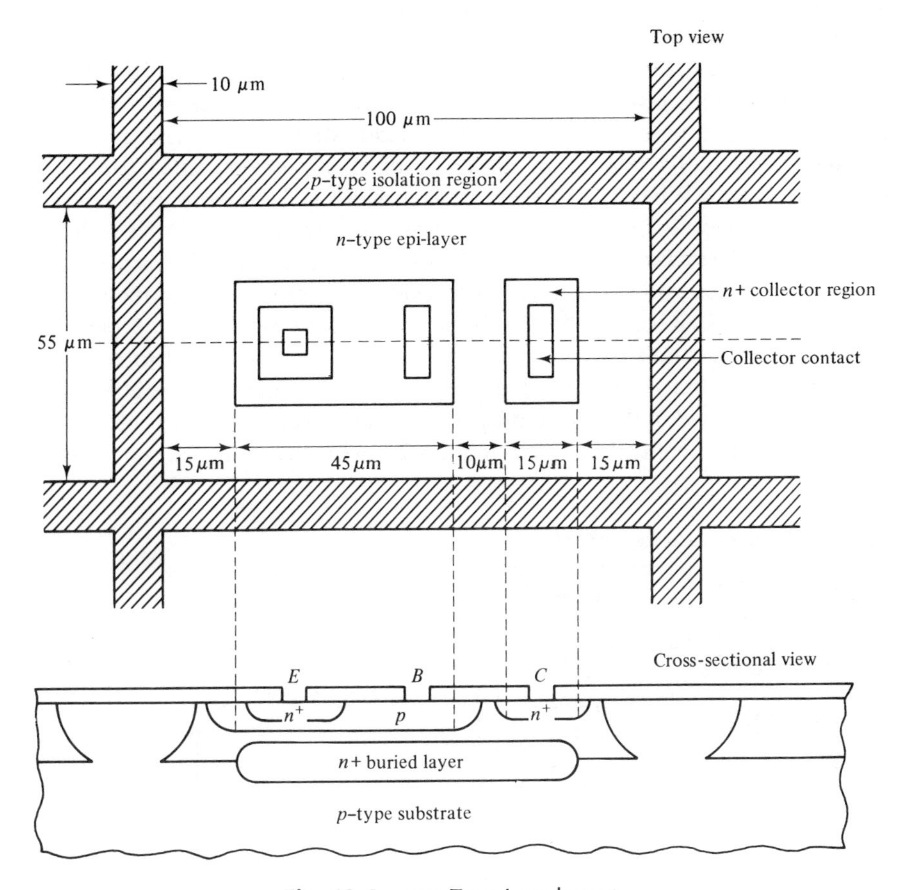

Fig. 10-4 *npn* Transistor layout.

region is 10 μm wide. A 15-μm space is used between the base region and the isolation region and also between the collector n^+ region and the isolation region. There is a good reason for this. *p*-type boron is used to define the isolation region. As it diffuses down, through the *n*-type epi-layer to the substrate, it also diffuses laterally. If the epi-layer is 10 μm deep, the boron will also diffuse laterally for about 10 μm. We allow for this in the mask layout by keeping everything 15 μm away from the isolation regions. As shown in Fig. 10-4, the transistor dimensions are 55 × 100 μm. The n^+ buried layer provides a low-resistance path for collector current. It sits under the base and collector n^+ regions, horizontally 15 μm away from the isolation region.

One way in which transistors Q_1, Q_2, and Q_3 of the ECL gate could be laid out is shown in Fig. 10-5. It is important to the operation of this circuit that these transistors be matched. They should have the same dimensions and be placed next to each other on the integrated-circuit chip. Metalizations *A*, *B*, and $+V_{REF}$ can be connected to pads for wire bonding. Metalizations $C_{1,2}$, *E*, and C_3 can be connected to resistors and transistors in the third isolation region below transistors Q_1 to Q_3.

The base diffusion is normally used to form most resistors in bipolar circuits. Alternatively, resistors can be ion implanted, made from emitter diffusions, or the epi-layer can be patterned by the isolation diffusion to form large, high-voltage resistors. Tolerances on the order of ±20% are obtained from base-diffused resistors. When variations in resistor width and temperature coefficients are added to these tolerances, resistor variations may be on the order of ±50% from design center. By designing circuits whose important characteristics depend on resistor ratios rather than absolute values, tolerances better than ±10% can be obtained. Resistors in the range from 10 to 50 Ω can be patterned from the emitter diffusion. Epitaxial-layer resistors can be biased to behave as current sources, but they have wide tolerances and are quite temperature sensitive. Ion-implanted resistors commonly provide sheet resistances of 1 to 10 kΩ/□ and can be controlled much more accurately than diffused resistors. Al-

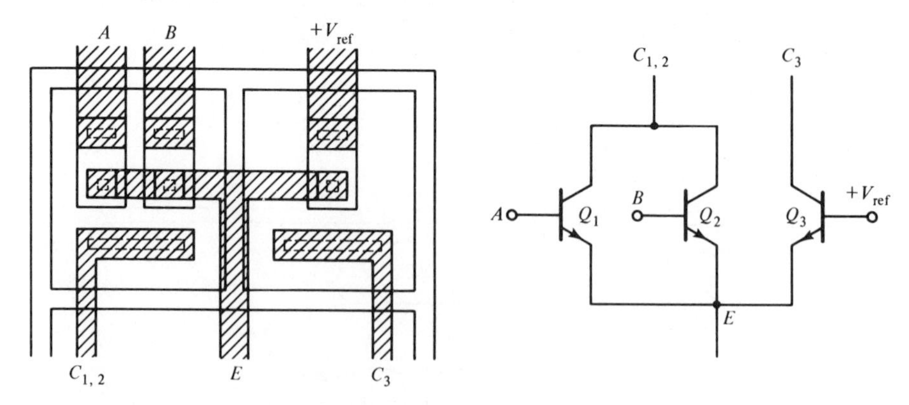

Fig. 10-5 Possible layout for ECL transistors Q_1, Q_2, and Q_3.

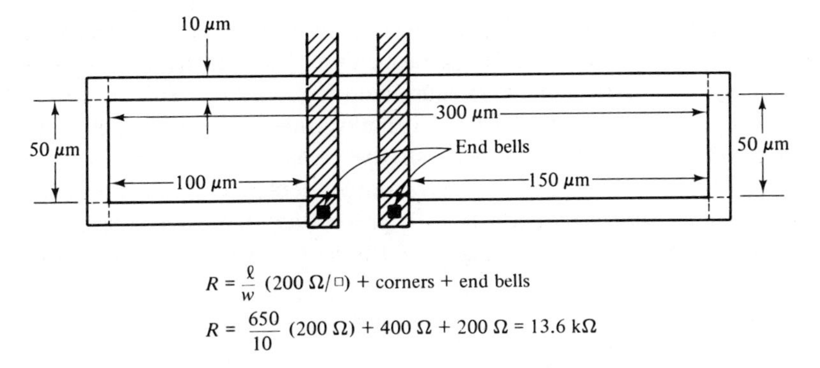

$$R = \frac{\ell}{w}\ (200\ \Omega/\square) + \text{corners} + \text{end bells}$$

$$R = \frac{650}{10}\ (200\ \Omega) + 400\ \Omega + 200\ \Omega = 13.6\ \text{k}\Omega$$

Fig. 10-6 Resistor layout.

though they require an additional masking step, they are replacing diffused resistors in many applications. Where large-value resistors are needed, designers generally prefer to use current sources or ion-implanted resistors because base-diffused resistors require too much chip area.

Base-diffused resistors typially have sheet resistances of approximately 200 Ω/□. They are placed in *n*-type epi-layers. Applied voltages ensure that the resistor–epi-layer diodes are reverse-biased. As shown in Fig. 10-6, end bells and corner squares are normally counted as half-squares, or as 100 Ω each for 200-Ω/□ sheet resistance. Resistance values are calculated as shown in Fig. 10-6.

When one end of a resistor is to be connected to the positive power supply voltage ($+V_{CC}$), the epitaxial layer should also be connected to $+V_{CC}$ at that point. This is done by diffusing an n^+ region adjacent to one side of the *p*-type resistor as shown in Fig. 10-7. This step avoids the possibility of building in an epitaxial resistor that is not wanted.

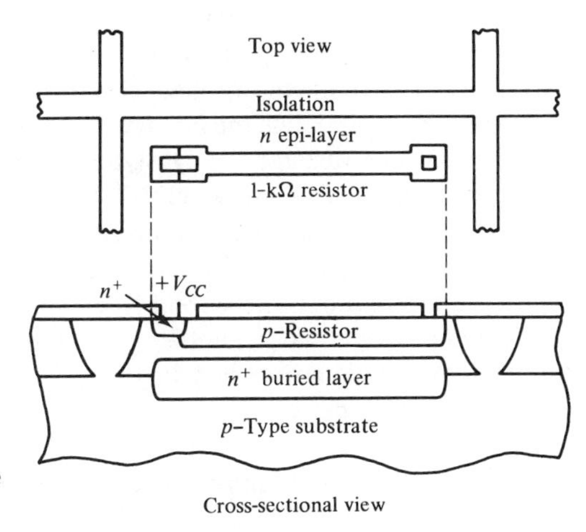

Fig. 10-7 Resistor with one side connected to $+V_{CC}$.

If a circuit cannot be made planar, it is necessary to make one conductor pass under the other without the two touching. This can be done with an n^+ region carrying a current under an aluminum metalization, with an oxide layer in between, as shown in Fig. 10-8. Normally, the cross-under will be placed in the n epi-region, which contains all the resistors and which is biased at $+V_{CC}$. Referring to Fig. 10-8, the cross-under looks very much like an *npn* transistor whose collector is at $+V_{CC}$. To prevent any parasitic transistor action here, it is advisable to tie the p-region to the most negative potential available.

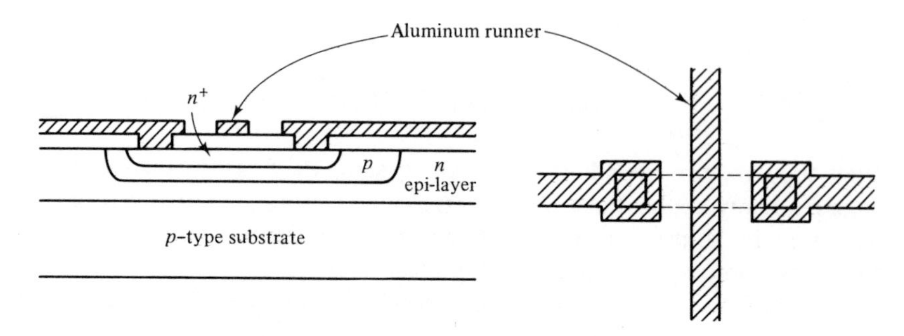

Fig. 10-8 Cross section and top view of a crossunder.

A potential problem with the cross-under is that a buildup of static charge between the aluminum runner and the n^+ region could break down the oxide layer. For this reason, none of the aluminum regions involved in a cross-under are brought to outside pins.

One of the smaller currents in a circuit might well be chosen to use the cross-under since the n^+ region has a finite resistivity and we do not want a large IR drop here. Since the emitter and base diffusions are normally used to form the cross-under, no additional processing steps are needed.

Where many, many cross-unders would be needed [e.g., in bipolar random-access memories (RAMs)], it is desirable to use two levels of metalization, with a layer of low-temperature glass deposited in between. Here, normal IC processing is used to complete the first level of metalization. Then a low-temperature glass is deposited through the pyrolytic decomposition of silane in an epitaxial reactor at temperatures below 300°C. (Because of the low temperature, no damage is done.) The standard photoresist processing steps are used to open windows in the pyrolytic glass where interconnects between the two levels of aluminum are desired, and the second layer of aluminum is deposited and processed. To complete the wafer, another protective layer of low-temperature glass is deposited. Windows are etched in this glass at the pad locations, so that wire bonds can be made.

A summary of typical layout dimensions to be used for the ECL gate and problem assignments is shown in Table 10-1.

TABLE 10-1 A Summary of Typical Bipolar Layout Dimensions (Micrometers)

Minimum etched window	5 × 5
Minimum line width	5
Minimum resistor width	10
Emitter to base edge	5
Emitter contact to emitter edge	5
Emitter to emitter separation	10
Base to base separation	10
Base contact to base edge	5
Base contact to emitter region	10
Base to isolation region	15
Collector contact to base region	10
Collector contact to collector N^+ edge	5
Collector n^+ to base region	10
Collector n^+ to isolation region	15
Buried layer to isolation region	15
Metal width	10
Metal separation	10
Metal overlap on contact window	5
Metal edge to scribe grid	40
Bonding pad dimensions	100 × 100
Pad to metal separation	40
Pad to pad separation	50
Width of isolation region	10
Width of base windows over isolation region	15
Width of scribe grid (total)	75

Some general rules for laying out this design example and the homework problems on mask layout at the end of this chapter are summarized as follows:

1. The mask layout is 400 times actual size. Buried layer, isolation, base, emitter, contact, and metalization masks are required on metric graph paper. Also, a composite graph with overall dimensions is needed.
2. Metalizations that go to pads are brought to the edge of the layout, but conductor pads and spaces between pads will not be shown on the mask layouts. The area required for seven 100 × 100 μm pads with 50-μm spacing would be on the order of 10^5 μm square. The chip area for a single ECL gate is on the order of 5.4×10^4 μm square. It would not be economically feasible to put only one ECL gate on a chip.
3. All matched components are placed as close together as possible.
4. The isolation diffusion is reinforced with the base diffusion.
5. The substrate is connected to the most negative potential available.

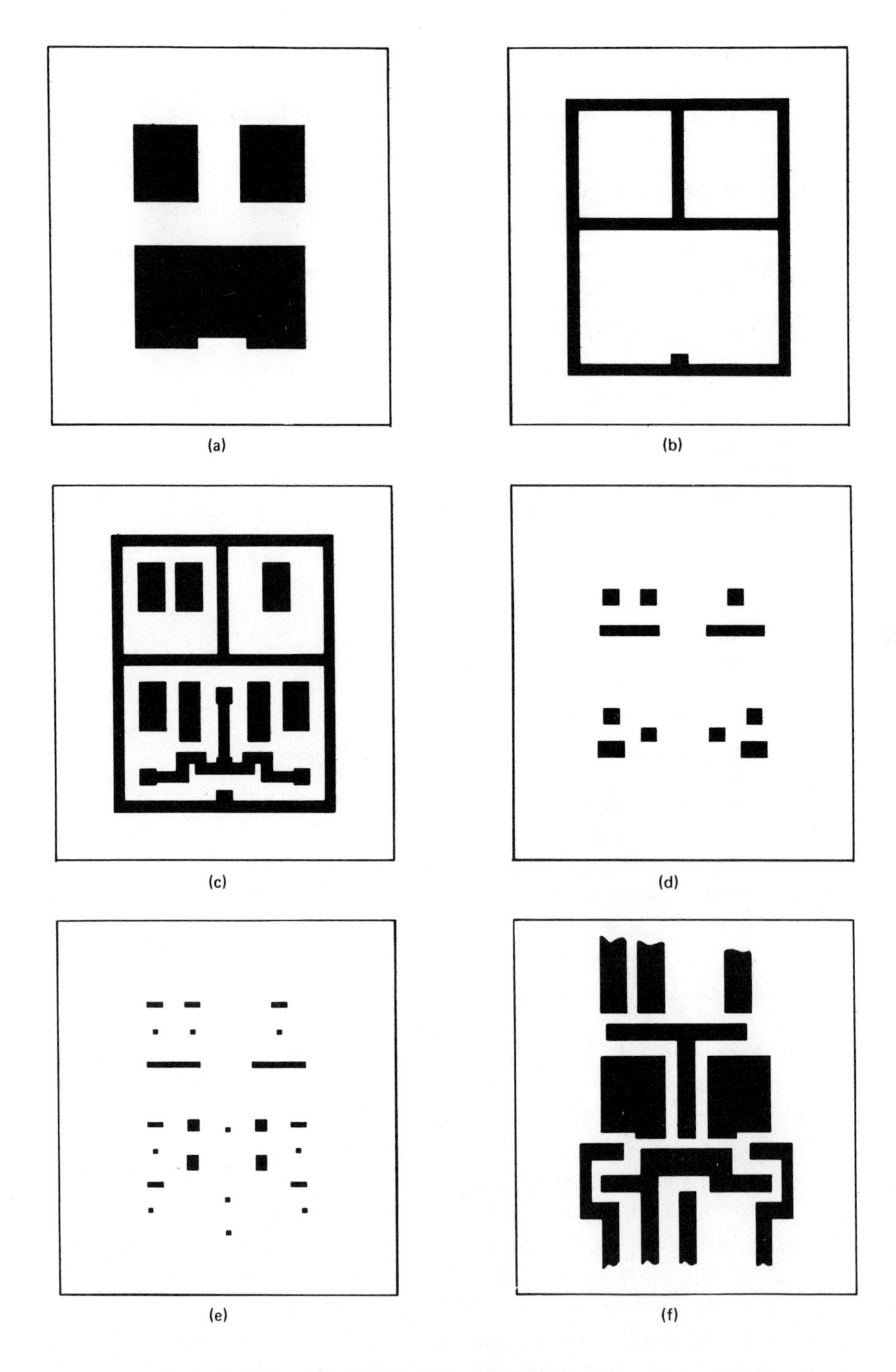

Fig. 10-9 Mask set for an emitter-coupled logic gate: (a) buried layer mask; (b) isolation mask: (c) base mask; (d) emitter mask; (e) contact mask; (f) metalization mask.

6. The final layout may be rectangular or square, but should be as nearly square as possible to facilitate chip handling.
7. Metalizations should not run over any n^+ regions that they do not make contact with.
8. Buried layers are placed under all transistors, diodes, and resistors.
9. You cannot run two metalizations into different locations in the epi-layer or the substrate and expect them to be at the same potential. In order for two points to be at the same potential, a metalization must exist between them.

Using these rules and the components shown in Figs. 10-4 through 10-7, a mask set for the ECL gate is shown in Fig. 10-9.

10-2 ELECTRON-BEAM LITHOGRAPHY

Shorter line widths can be obtained in silicon wafers with electron-beam systems than with photolithography systems. When a photolithography system is used with a 1×-size mask, diffraction effects can occur during wafer resist exposure. The diffraction pattern occurs where the edges of the opaque mask regions are imaged on the wafer. The severity of the problem is related to the wavelength of the light source used for the exposure. A shorter-wavelength exposure source reduces the diffraction effect. Some of the techniques used to overcome the diffraction problem involve the use of:

1. Shorter-wavelength ultraviolet light sources.
2. Direct step mask aligners, in which 1×-size masks are not involved.
3. X-ray exposure sources used in conjunction with X-ray-sensitive resists.
4. EBL systems to produce fine-line masks in conjunction with item 2 or 3.
5. EBL systems that directly expose resists on wafer surfaces.

It has been pointed out before that a typical EBL system can scan only a portion of a master mask (or wafer) at a given time. The mask (or wafer) is seated on an *x-y* table, which indexes to repeat a given circuit pattern until the entire mask (or wafer) is exposed. The positional accuracy of the *x-y* table is critical in this application. A laser interferometer is often used in a position control system to maximize the accuracy of the *x-y* table.

Since about 1960, scanning electron microscopes (SEM) have been commercially available. In an SEM, electrons emitted by a tungsten filament (or a lanthanum hexaboride, LaB_6, field emission source) are accelerated through a potential difference of a few kilovolts to 50 kV in vacuum. The electrons are then passed through magnetic focusing coils (electrostatic focusing is also used) and deflection coils to provide a raster scan pattern on a sample (see Fig.

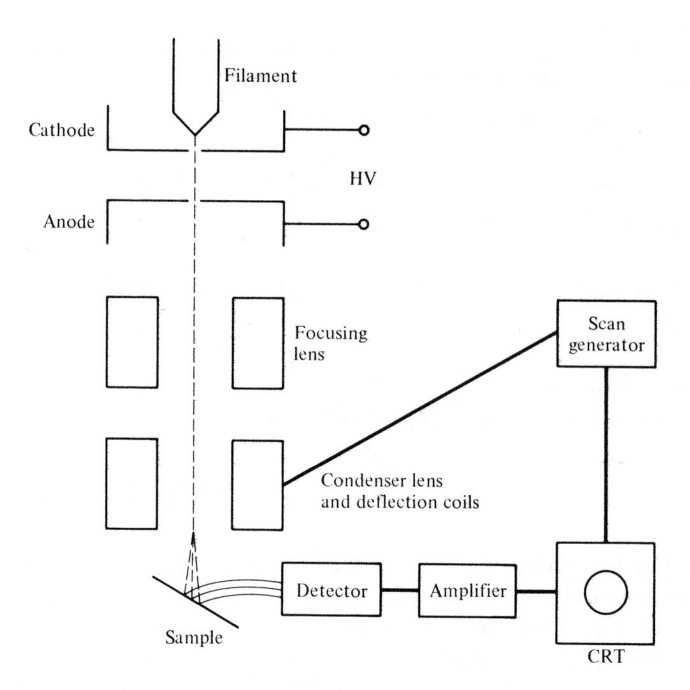

Fig. 10-10 Block diagram of SEM.

10-10 for a block diagram). The electron beam serves as a very small probe, easily less than 100 Å in diameter. Various imaging techniques are possible. The most common is secondary electron emission. Secondary electrons are ejected from the sample by the impinging primary electrons with energies typically less than 50 eV. These secondary electrons are detected and the quantity emitted at a particular point on the sample provides a level of contrast in a cathode-ray-tube (CRT) or television (TV) image. The CRT or TV scan is synchronized with the scan of the electron beam. Variations in the quantity of secondary electrons emitted occur both as a result of sample topography and material differences. Other imaging techniques used are cathodoluminescence, induced current, X-rays, and backscattering of primary electrons. Backscattered electrons are detected by allowing them to create electron–hole pairs, thus generating a current in a silicon detector. Various electron diffraction phenomena are also useful in characterizing materials. Some aspects of the use of the SEM in microelectronics are discussed in Chapter 13. Presently our concern is with the use of an electron beam to write patterns in appropriate electron resists at 1× size, either to define a pattern directly on a silicon wafer, or to produce a mask that serves as a master for replica masks that are to be used in conventional photolithography or X-ray lithography to produce the desired patterns on the Si wafer.

It has been pointed out before that the resolution capability of a lithographic process using ultraviolet light is limited by the wavelength of the light. This practical limit on line width is around 1 to 3 μm in microelectronic circuits, because of the necessity of maintaining clearly defined line edges. The wavelength of electrons at the voltages used in SEMs is of the order of 1 Å, so the limiting criterion is not wavelength, but the diameter of the electron beam. Electron-beam diameters of 100 Å are routine in commercial SEMs. Beam diameters of 20 Å and less have been attained using high-brightness-field emission electron guns. This, however, requires a very high vacuum and consequently a very clean system. Pumpdown time is, consequently, substantially longer. In any case, it is clear that the possibilities for the use of electron-beam fabrication of microelectronic circuits are extensive. The limit on line widths and packing density is not set by the electron-beam writing process as much as it is by the resolution of the resist and the capability of aligning mask patterns on the Si wafer.

Even when photolithographic techniques are used to align masks and expose resists on the wafer, production of masks by electron-beam lithography (EBL) has distinct advantages. Superior line definition can be achieved in EBL on the master mask, thus providing a higher-quality mask. Turnaround time from the digitizing of a mask design to the computer-controlled EBL production of the mask can be substantially less than when photographic reduction is used. Additionally, a change in design simply requires a modification of the computer program. Masks can be checked on the SEM after development of the resist by secondary electron imaging, or the computer can be used to print out overlays so that alignment of successive masks can be checked.

Commercial EBL systems are available and existing SEMs can be adapted to perform EBL work. In most cases, the electron beam in existing machines is probably not sufficiently stable to achieve the highest resolution or bright enough to achieve sufficiently rapid exposure for production purposes, but considerable improvement over photolithography can be made, and such adapted machines provide a useful tool for experimental circuit work and EBL development, such as in the area of electron-beam resists. Many adapted machines make use of a flying spot scanner for controlling the position of the electron beam. In this arrangement, an image tube (a photosensitive tube such as a TV camera tube) is masked with a positive or negative mask of the pattern to be defined in the electron resist. A CRT scans the mask-covered image tube and the output of the image tube is then utilized to turn the electron beam on and off at the appropriate locations in its scan.

In other systems, the electron beam is controlled by a computer to write the pattern directly in the resist. Work has been done on multiple-beam systems and imaging systems to achieve higher processing rates.

If the pattern is not defined directly on the wafer, as it is in such applications as bubble memories and charge-coupled devices, a mask must be used to do the exposure on the wafer. To take full advantage of the small line widths achievable with EBL, the masks and patterns must be aligned with even greater accuracy. Standard optical techniques will not suffice for alignment or exposure. Exposure must be with shortwave ultraviolet, X-rays, or electron beams to achieve adequate line definition at the wafer image. Alignment must be done automatically using one of the short-wavelength types of radiation mentioned above or by means of a sophisticated interferometric technique using visible or near-visible ultraviolet light.

10-3 RESISTS

Resists are radiation-sensitive emulsions used to transfer patterned images to masks or wafers. Where the resist has been removed from a mask or wafer, the surface layer is etched. Where present, the resist protects the surface layer from the etchant.

Resists are liquid formulations of resins rather than colloidal materials such as photographic emulsions. The size of the pattern in a mask or wafer is limited by the length of the polymer chain within the resist as well as by the wavelength of the radiation source used to expose the resist and the mask alignment tolerances to which patterns can be aligned.

Resists are classified by the type of radiation they are sensitive to. There are *photoresists* (sensitive to ultraviolet radiation), *electron resists*, and *X-ray resists*. Traditionally, photoresists have been used in conjunction with ultraviolet light sources with wavelengths in the range 3600 to 4600 Å. Alignment tolerances and diffraction around the edges of mask patterns have limited these systems to line widths on the order of 2 μm. To achieve 2-μm line widths, the mask and wafer must be in contact. This causes mask and wafer damage, and thus yield problems. Laser interferometry is being used to improve alignment tolerances, and *contact* mask aligners are being replaced by *projection aligners*, in which the mask image is projected on the wafer. The resolution problem caused by the diffraction of light around pattern edges is being improved by using shorter-wavelength light sources. Deep ultraviolet light sources are being used in some systems, with wavelengths in the range 2000 to 3000 Å. Deep ultraviolet sources require that masks be made of fused silica rather than glass, however, and fused silica masks are more fragile and expensive.

A system to mass-produce VLSI wafers with submicron line widths might well consist of SEM-generated masks, X-ray resists, and projection mask aligners using X-ray radiation sources. X-ray mask aligners with wavelengths in the 5-Å region are under development, but there are many problems to be solved before they can become commercially available. Organic X-ray resists, such as *copolymer*, are available. The resist is exposed through gold-patterned masks, on substrates such as Mylar or silicon.

Resists are also classified as negative working or positive working. A *negative-working resist* cross-links or polymerizes where it is exposed to radiation. The exposed regions are then insoluble in the developing solution. The unexposed, soluble portions wash away. A tough, chemically resistant image is left in the exposed areas. With a *positive-working resist,* on the other hand, the radiated energy breaks up the polymer so that the exposed portions wash away in the developing solution. Generally, the positive-working resists have shorter polymer lengths than the negative-working resists. This is one of the reasons that positive resists are used in the fabrication of modern large-scale integrated circuits. In summary, resists are categorized as being sensitive to ultraviolet light, or X-ray radiation, or electron beams. Each of these categories contains negative-working and positive-working resists.

A number of positive- and negative-working resists are in existence for use with scanning electron microscopes. Polymethylmethacrylate (PMMA) is used as a positive-working resist, and the tongue-twisting polymer chain that goes by the name "copolymer" or COP is widely used as a negative-working resist. With electron exposure, pattern resolution is limited to about 0.15 μm because the resist scatters the electron beam to some extent, causing it to spread out as it enters the resist.

10-4 MASK ALIGNERS

There are three basic types of optical mask aligners: the contact, off-contact or proximity, and the projection mask aligner. In contact mask alignment, the mask is aligned manually or automatically with the pattern on the wafer in a separated position; then the wafer is vacuum-clamped tightly against the mask. Alignment is rechecked and, if it is satisfactory, the exposure is carried out. The chief disadvantage of this method is that emulsion masks are easily damaged and can be used only a few times. Chrome and iron oxide masks have superior life, but even they can be damaged by protuberances from the oxide on the wafer.

Off-contact mask aligners operate with a slight separation, 5 to 30 μm, between the mask and wafer. The chief advantage of this technique is the lack of contact (and hence mask wear) between the wafer and mask. However, because of problems with bowing, warpage, and taper of wafers, great care must be taken in the instrument design to ensure that the wafer is adequately flattened and that close parallelism between the wafer surface and mask is achieved.

The projection mask alignment approach has the same advantage as the off-contact aligner as far as mask wear is concerned. In this type of alignment, an image of the mask is literally projected, using either transmitting or totally reflecting optics, over a considerable distance onto the wafer surface. In the event that transmitting optics are used, a monochromatic light source is required to minimize chromatic aberration. This results in longer exposure times and interference layers in negative photoresists. Interference layers occur due to the fact that light reflected from the top of the oxide layer and the Si surface will interfere constructively if the path-length difference is an integral multiple

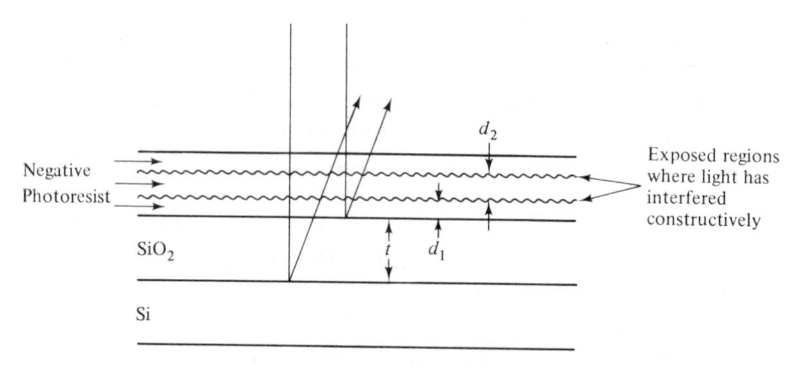

Fig. 10-11 Interference effect in negative photoresist.

of the wavelength, and destructively if the path-length difference is an odd integral multiple of a half-wavelength. Figure 10-11 is an illustration of this phenomenon. This interference effect results in layering of the photoresist and reduces edge definition since the develop-and-rinse operation will tend to remove some of the unexposed photoresist at an edge, causing a vertical scalloping effect. The use of purely reflective optics avoids the problem of chromatic aberration entirely so that a relatively broadband light source can be used.

The interference effect is particularly troublesome for oxide thicknesses of less than 3000 Å. A similar problem arises from the creation of standing waves due to reflection from the SiO_2 and resist surface. Both of these phenomena create what is referred to as reticulation or the "orange peel effect," due to

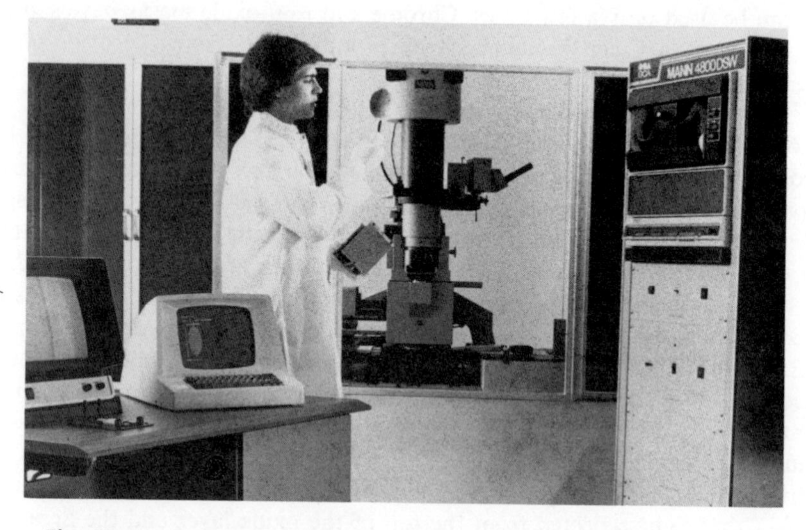

Fig. 10-12 Direct on-wafer step-and-repeat system. (Courtesy of GCA Corporation.)

solvent-caused swelling and shrinking of the undeveloped and developed planes of the resist.

Still another problem associated with negative resists is a photo-oxidation reaction that causes a reduction in the thickness of the resist layer. This problem is more severe for thinner resist films, where up to 50% of the thickness can be lost in 1-μm films. In proximity mask alignment, this problem can be minimized by vacuum clamping or nitrogen flushing to reduce the oxygen present during exposure. In projection alignment, nitrogen flushing is used.

An alternative to mask alignment is a system employing direct step-and-repeat on the wafer. In this process, an image of a 5× or 10× size pattern of a single circuit (or small group of circuits) is reduced and directly imaged onto the wafer and stepped and repeated. In this way, each individual circuit (or group) can be aligned, rather than aligning a whole mask relative to a pattern in the wafer. This technique can overcome, to some extent, the problems created by changes in wafer bowing and warpage between masking steps. Figure 10-12 is a picture of a direct step-and-repeat system.

10-5 PHOTOPROCESSING

Photoprocessing consists of transferring patterns from masks to material layers on wafer surfaces. This is generally accomplished by:

1. Cleaning the wafer and then baking it to remove surface moisture.
2. Coating the wafer surface with a uniform layer of resist.
3. Baking the wafer to remove solvents from the resist (called the *soft bake* or *prebake*).
4. Aligning the existing wafer patterns with the new mask patterns, and then exposing the resist, through the mask, to a radiation source.
5. Developing and rinsing the resist, leaving resist patterns where the surface layer is to be protected.
6. Baking the wafer to increase film resistance. This *hard bake* or *postbake* is important with negative-working resists because it increases the polymerization or cross-linking of the resist.
7. Etching the surface layer such that it is removed where not protected by the resist. *Wet* and *dry* processing are both used for surface-layer removal. *Wet processing* refers to liquid chemical etching, and *dry processing* refers to vacuum-controlled plasma processes such as plasma etching and sputter etching.
8. Removing the resist from the wafer by using wet or dry processing.

The photoprocessing steps listed above are summarized in Fig. 10-13.

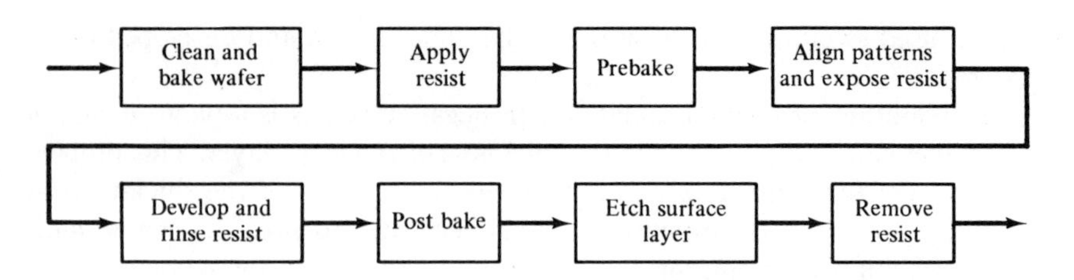

Fig. 10-13 Summary of photoprocessing steps.

10-5-1 Wet Chemical Etching

Integrated circuits can be produced using only negative-working photoresist or positive-working photoresist. To illustrate both systems, we will discuss the use of negative-working photoresist to selectively remove SiO_2, and positive-working resist to selectively remove aluminum.

Figure 10-14 indicates the process steps that will selectively remove SiO_2 using negative-working photoresist. Figure 10-14(a) shows a section of a silicon

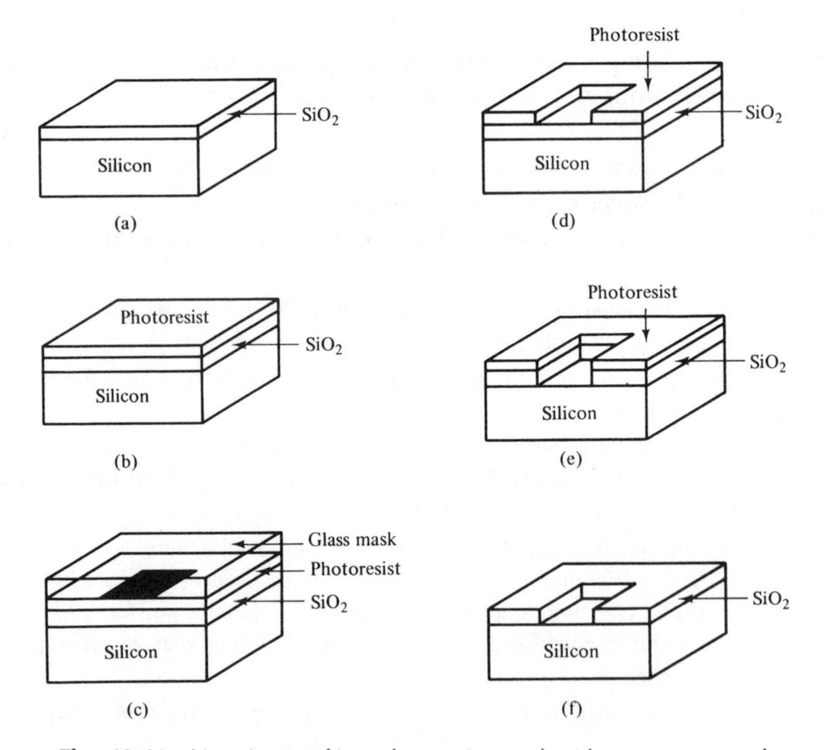

Fig. 10-14 Negative-working photoresist used with a contact mask to selectively remove SiO_2 from silicon.

wafer with a surface layer of SiO_2. In Fig. 10-14(b), the wafer is coated with negative photoresist, and in Fig. 10-14(c) a patterned glass mask is brought into contact with the photoresist surface. The wafer–mask system is then exposed to an ultraviolet light source in a mask aligner. The negative-working resist hardens where it is exposed to light. When the photoresist is developed and rinsed, the unexposed regions are washed away, as shown in Fig. 10-14(d). The wafer is then immersed in a hydrofluoric acid solution, which etches the exposed SiO_2 regions but does not significantly affect the photoresist or silicon. The wafer at this point is as shown in Fig. 10-14(e). A stripping solution is used to remove the photoresist and a cleaning procedure is used to remove contaminants. The wafer section is shown in Fig. 10-14(f).

Figure 10-15 indicates positive-working photoresist used with a mask to selectively remove aluminum from a silicon wafer section. Figure 10-15(a), (b), and (c) show a layer of photoresist and a mask being used in conjunction with an aluminum-coated silicon wafer. The positive-working photoresist is ex-

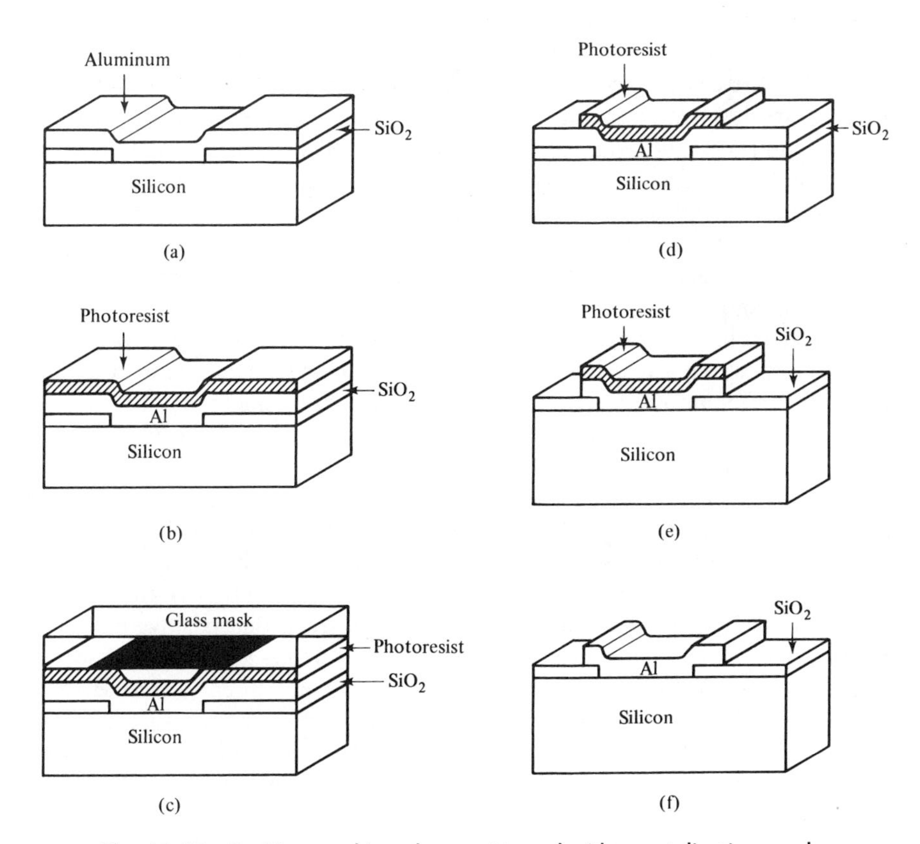

Fig. 10-15 Positive-working photoresist used with a metalization mask to selectively remove aluminum from a silicon wafer.

posed, through the mask, to the ultraviolet light source. The light energy breaks up the polymer chain. The exposed photoresist will be washed away during the develop and rinse steps. The unexposed resist adheres to the wafer, as shown in Fig. 10-15(d). When the wafer is immersed in a phosphoric acid solution, the exposed aluminum is leached away, as shown in Fig. 10-15(e). The SiO_2 and photoresist surfaces are relatively unaffected. A solvent, typically acetone, is used to remove the photoresist, and a cleaning procedure is used to remove contaminants. The completed wafer section is shown in Fig. 10-15(f).

10-5-2 Dry Plasma Etching

With wet chemical etching, photoresist patterns are used to mask surface layers on silicon wafers as shown in Figs. 10-14(d) and 10-15(d). When the wafers are immersed in chemical etchants, the surface layers that are not protected by photoresist are etched away. A wet chemical called photoresist stripper is subsequently used to remove the photoresist.

With plasma processing, photoresist patterns are again used to mask surface layers. In this case, however, the wafers are placed in a vacuum system. A gas mixture is introduced into the vacuum system at low pressures (0.01 to 20 torr), and a radio-frequency (RF) source is used to ionize some fraction of the gas molecules, creating a plasma. The plasma contains positive ions, negative ions, electrons, and neutral gas atoms, such that a condition of near charge neutrality exists. It may also contain *free radicals*—atoms or molecules that are extremely reactive. Depending on the surface layer to be etched and the desired outcome of the etching process, the exposed silicon surface may be chemically etched by the free radicals (plasma etching), or it may be physically etched by positive ion bombardment (sputter etching). There is also a third alternative; vacuum pressure, RF energy, and wafer placement may be arranged so that both chemical etching and ion bombardment take place. This third alternative is called *reactive ion etching.*

Plasma etching, reactive ion etching, and sputter etching are approximately related to gas pressure and RF energy, as shown in Fig. 10-16. For *sputter etching,* RF energy is used to ionize a nonreactive gas. The plasma is used to create positive ions. Relatively low gas pressures are used (0.01 to 0.1 torr). An electric field is created in the vacuum chamber such that positive ions bombard the silicon wafer surfaces. Ion bombardment breaks loose exposed surface atoms. In the case of *plasma etching,* higher gas pressures (0.1 to 20 torr) and lower values of RF energy are used. The plasma is used to create free radicals, which either diffuse or are directed to the wafer surface. For *reactive ion etching,* the plasma is used to create a combination of free radicals and positive ions so that a combination of chemical and physical etching takes place at the wafer surface.

Natural plasma processes (such as lightning) are transient phenomena.

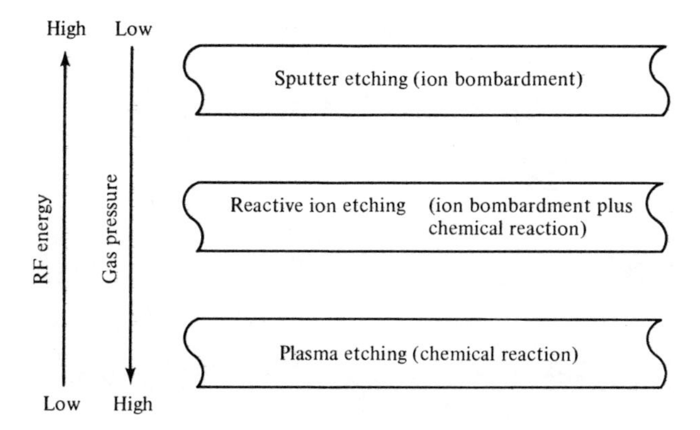

Fig. 10-16 Techniques for plasma processing.

In plasma processing, RF energy and a gas mixture are supplied continuously to maintain the plasma. Products from the etching reactions are continuously removed by the vacuum pump. Gas temperatures normally used in plasma processing are low enough so that conventional resists can be used to mask wafer surfaces. When the etching processes are completed, the reactive gas is pumped out of the system and replaced by oxygen. An RF field is then used to create an oxygen plasma, which strips away the resist.

Single-crystal silicon, polycrystalline silicon, thermally grown and deposited SiO_2, silicon nitride, and aluminum have all been etched by plasma processes. In the case of plasma etching, free radicals such as atomic fluorine, CF_3, and SiF_3 react strongly with silicon compounds. The gases that are typically used to create these radicals are carbon tetrafluoride (Freon-14),[1] hexafluoroethylene (Freon-116), trifluoromethane (Freon-23), and silicon tetrafluoride (SiF_4). All of these gases have applications for etching various silicon compounds. Carbon tetrafluoride (CF_4) has been the most widely used gas for plasma etching.

The creation of free radicals (atomic fluorine and CF_3) in a CF_4 plasma, and the reaction of these free radicals with silicon wafer surfaces, are roughly depicted in Fig. 10-17. In a plasma, only certain radicals have sufficient lifetime to diffuse to the silicon surfaces. In a plasma etching system, RF energy, gas flow, gas pressure, and temperature can be maintained so that over 90% of the particles are free fluorine radicals. The ion population can be kept very small. Silicon etches rather slowly in a pure CF_4 plasma. As oxygen is added to the plasma, however, the etch rate increases, goes through a maxima, and then decreases. The maxima occurs with approximately 90% CF_4 and 10% O_2.

Plasma etching processes can be classified according to *etch rate,* etching *selectivity, anisotropy,* and *loading effect. Etch rates* are generally between 100 and 10,000 Å/min.

[1] Freon is a registered trademark of the DuPont Company.

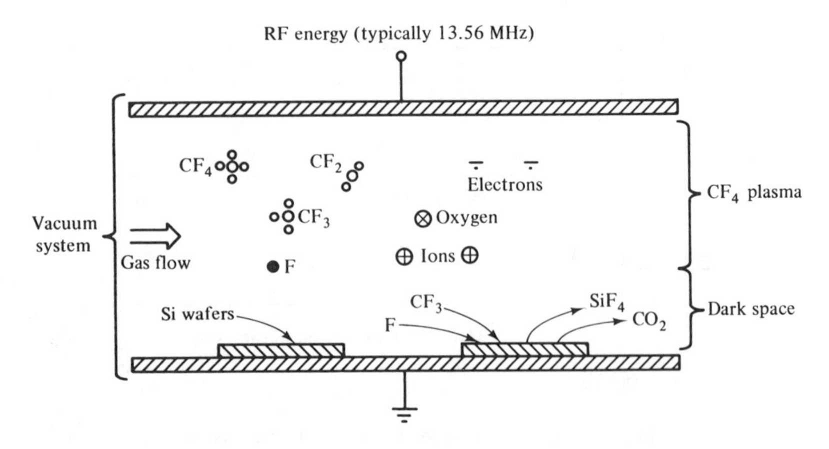

Fig. 10-17 Plasma etching with a CF_4/O_2 gas mixture.

Selectivity refers to the ratio of etch rates between two different materi immersed in the same plasma. For example, if windows are being etched in SiO_2, it is desirable that the etch rate decrease when the silicon surface is reached. Selectivity is more difficult to control with dry plasma processing than with wet chemical processing. Fluorine radicals usually etch silicon and silicon nitride much more rapidly than SiO_2. Depending on the plasma parameters, selectivities or etch ratios on the order of 20:1 might be obtained for Si:SiO_2 layers. However, the addition of H_2 to a CF_4 plasma suppresses the etching of silicon but not that of SiO_2, and a C_2F_6 plasma etches SiO_2 more rapidly than silicon. By judicious choice of gas mixture, flow rates, RF energy, and so on, a plasma system can be engineered to optimize the selectivity of a specific silicon compound.

To say that an etching process is *isotropic* means that the etch rate is the same in all directions. Figure 10-18 shows an isotropic and a fully *anisotropic* etch. Referring to Fig. 10-18(a), the anisotropy of an etching system can be defined by

$$A = \frac{d}{\mu}$$

where A represents the degree of anisotropy, d the vertical etch depth, and μ the amount of undercut. Thus, for an isotropic etch, $d = \mu$ and $A = 1$. For a fully anisotropic etch, as shown in Fig. 10-18(b), $A = \infty$ since $\mu = 0$.

Wet chemical etching is generally isotropic. With plasma processing, it is possible to get values of $A >> 1$ under certain conditions. For example, sputter etching (and, to some extent, reactive ion etching) relies on positive ion bombardment for material removal. Silicon target wafers are generally held at a negative dc potential with respect to the plasma. Because of the electric

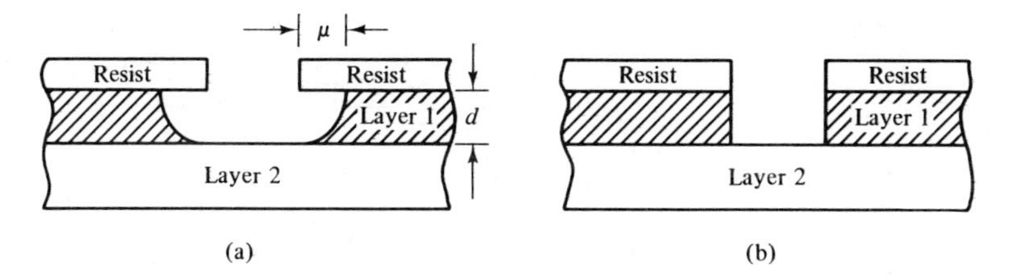

Fig. 10-18 (a) Isotropic etch. The undercut (μ) is equal to the etch depth *(d)*. (b) Fully anisotropic etch with zero undercut.

field, positive ions bombard the wafers perpendicular to the wafer surface and etch anisotropically. In general, sputter etching techniques provide poor selectivity with silicon compounds, however, and cause surface damage due to high particle energies. With plasma etching processes that are ion-assisted, both anisotropy ($A >> 1$) and selectivity can be obtained. Such processes include reactive ion etching and, under certain circumstances, plasma etching. As integrated-circuit dimensions shrink, anisotropy becomes a very important parameter. For example, it is not possible to isotropically etch a 1-μm-wide aluminum runner that is 1 μm thick. Very little aluminum would survive. Thus, to etch fine-line structures, anisotropy is necessary.

It was stated earlier that plasma etching processes could be classified according to etch rate, selectivity, anisotropy, and loading effect. *Loading effect* refers to the fact that the etch rate generally decreases when the number of wafers in the system (the wafer load) is increased. During the plasma etching process, wafers "consume" fluorine radicals. When the wafer load is increased, the rate of consumption of fluorine radicals increases. This decreases the steady-state concentration of radicals in the plasma, slowing down the reaction and decreasing the overall etch rate.

With plasma etching, the number of free radicals consumed is comparable to the number of silicon atoms removed. Products of the etching reactions are removed continuously by the vacuum pump. With wet chemical etching, on the other hand, chemicals etch the wafers and the products of the etching reactions stay in solution. Thus, much larger quantities of hazardous wet chemicals must be stored and used. The primary reasons that the integrated-circuit industry is moving toward plasma processing are that it is cheaper and safer. As dimensions shrink, the fact that ion-assisted anisotropic etching can provide finer etch patterns is certainly an added incentive.

Two types of plasma etching systems are in common usage: *barrel reactors,* which are volume-loaded, as shown in Fig. 10-19(a), and *planar reactors,* which are surface-loaded, as shown in Fig. 10-19(b). In a barrel reactor, the wafers are stacked vertically in quartz boats and surrounded by a cylindrical column of plasma. Because of the geometry of the wafer–plasma system, barrel reactors provide an isotropic etch and cannot be used for structures in the range 1 to

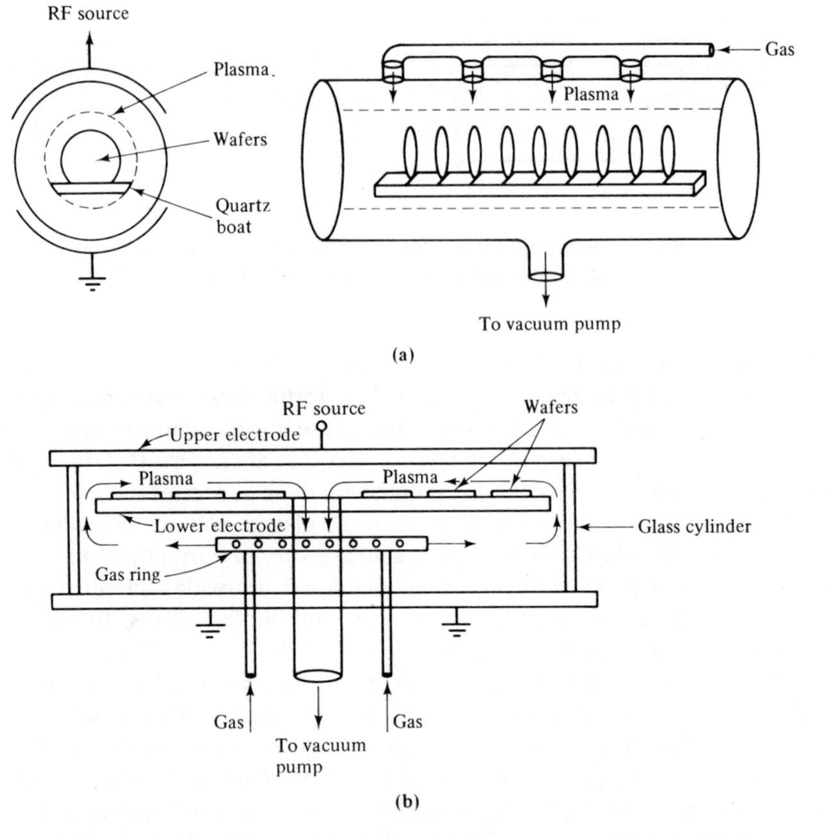

Fig. 10-19 Equipment for plasma etching: (a) A barrel reactor. The wafers are stacked vertically in a quartz boat and surrounded by a cylindrical column of plasma; (b) A planar reactor. The wafers lay flat on the lower electrode with the plasma directly above them.

2 μm. In planar reactors, the wafers lay flat on the lower electrode with the plasma directly above them. The geometry is such that the chemical reactions can receive a directional ion assist, so that fine-line anisotropic etches can be achieved.

BIBLIOGRAPHY

[1] *Techniques of Microphotography,* A Kodak Data Book. Eastman Kodak Company, Rochester, N.Y. 14650.

[2] GRAF, J. M., AND W. CONVERSE, EDS., "Technology Advances in Micro and Submicro Photofabrication Imagery," *Proceedings of Photo-optical Instrumentation Engineers,* Vol. 55.

[3] "Mask Making and Generation," *Solid State Technology,* June 1974.

[4] FLAMM, D. L., "Measurements and Mechanisms of Etchant Production during the Plasma Oxidation of CF_4 and C_2F_6," *Solid State Technology,* April 1979.

[5] COBURN, J. W., AND E. KAY, "Some Chemical Aspects of the Fluorocarbon Plasma Etching of Silicon and Its Compounds," *Solid State Technology,* April 1979.

[6] PARRY, P. D., AND A. F. RODDE, "Anisotropic Plasma Etching of Semiconductor Materials," *Solid State Technology,* April, 1979.

[7] BOYD, H., AND M. S. TANG, "Applications for Silicon Tetrafluoride in Plasma Etching," *Solid State Technology,* April 1979.

[8] MOGAB, C. J., AND W. R. HARSHBARGER, "Plasma Processes Set to Etch Finer Lines with Less Undercutting," *Electronics,* August 31, 1978.

[9] International Plasma Corporation, *Fundamentals of Plasma Etching.* 1976. Hayward, Calif. 94544.

PROBLEMS

10-1 Assume that a mask set is to be laid out for the transistor–transistor logic (T^2L) gate shown. Indicate how many isolation regions are needed (minimum), and what components should be put in each isolation region.

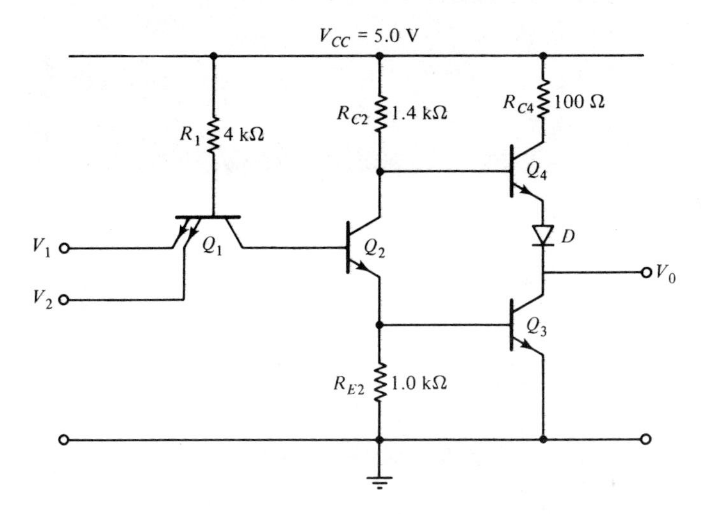

10-2 Design a mask set for the two-input T^2L gate shown in Problem 10-1. Masks should be laid out on No. 341-M millimeter graph paper ($8\frac{1}{2} \times 11$ in.). By making 5 μm correspond to 2 mm (two divisions on 341-M graph paper), the circuit will be laid out at 400 times actual size. *Buried layer, isolation, base, emitter, contact,* and *metalization* masks are needed. Also, a *composite mask* with overall dimensions is needed. Resistors will be formed during the base diffusion (200 Ω/□). Metalizations that go to bonding pads should be brought to the edge of the layout, but bonding pads and spaces between pads will not be shown in the mask layouts. The instructions and rules discussed in Section 10-1 should be followed. The mask set should be rectangular (nearly square). Since chip area is related to cost, the silicon area required for the circuit should be minimized.

10-3 A mask layout for a three-input digital logic gate is shown in parts (a) and (b) of the accompanying figure. Given the mask set, draw the circuit diagram and explain the circuit operation. Assuming that $R_s = 200\ \Omega/\square$, determine resistor sizes.

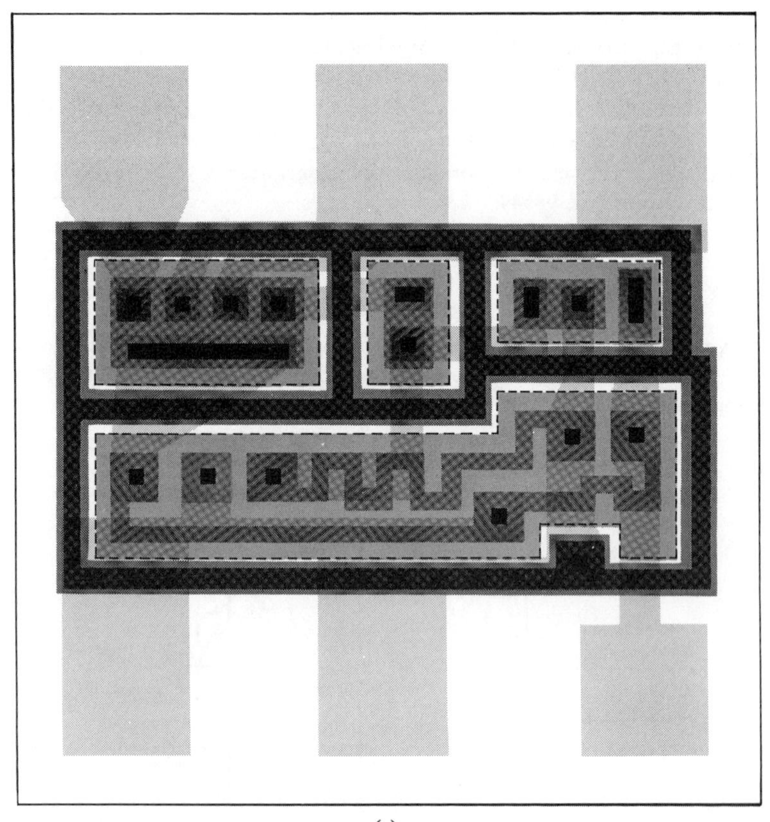

(a)

10-4 An integrated circuit is to be designed with a minimum line width of 1.0 μm. Would you recommend
(a) Wet or dry etching processes?
(b) SEM lithography or photolithography?

10-5 Explain the differences between negative- and positive-working photoresists.

10-6 What is meant by the terms *selectivity* and *anisotropy* in the plasma etching process?

10-7 Explain the differences between sputter etching, reactive ion etching, and plasma etching.

10-8 Briefly describe three types of optical mask aligners, and list their main advantages and disadvantages.

10-9 Describe three different techniques for producing IC master masks.

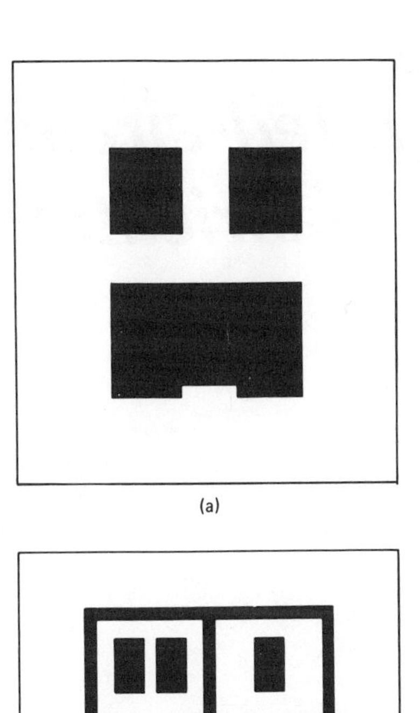

(a)

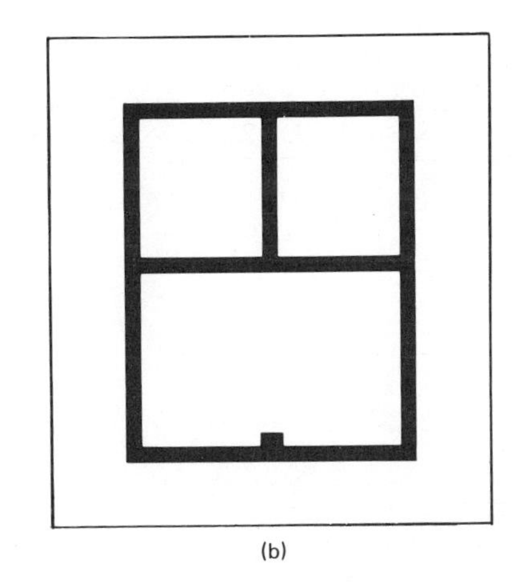

(b)

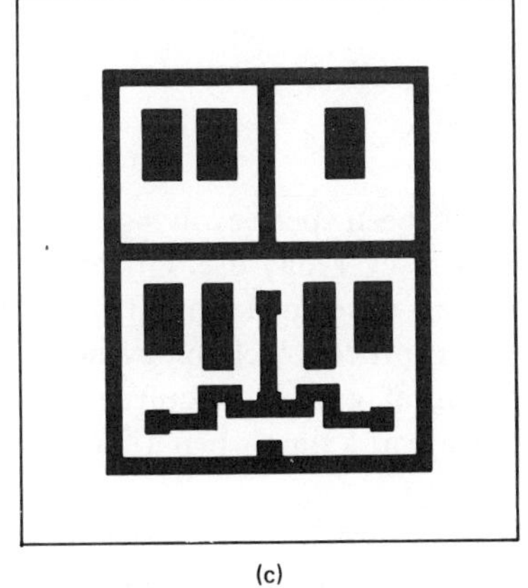

(c)

(d)

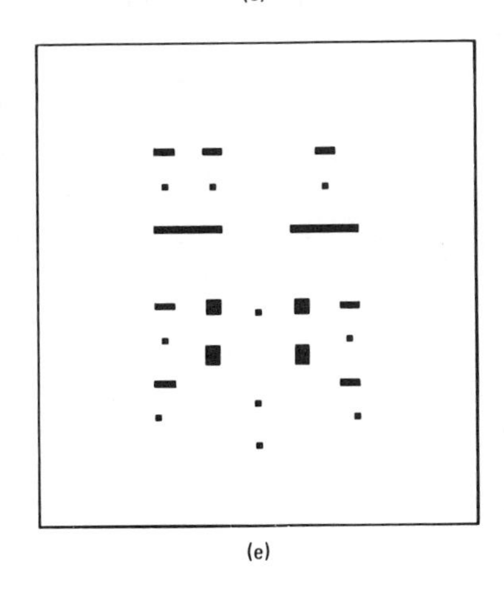

(e)

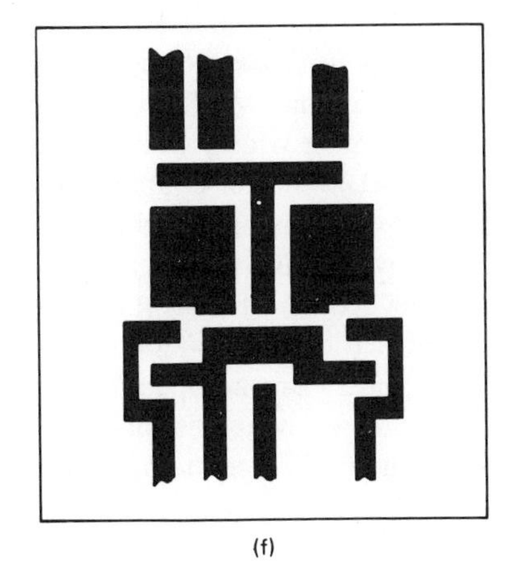

(f)

(b)

integrated-circuit fabrication

11

Hundreds of thousands of person-years have been spent evolving the integrated-circuit process technologies that are now in everyday use. Of the many fabrication and circuit techniques that have been developed since the 1960s, only the fittest have survived. To survive, a given process technology not only has to compete with proven processes in terms of product requirements, it also has to provide available hardware and software at a time when a "window" into the marketplace exists.

Fabrication and circuit techniques cannot be clearly distinguished, since both are dictated by the application. Applications can be roughly divided into digital and nondigital circuits. Then, depending on criteria such as packing density, power dissipation, switching speed, and noise immunity, choices between the bipolar and MOS technologies can be made. For example, in the digital area the most widely used bipolar circuit technologies are transistor–transistor logic (T^2L), emitter-coupled logic (ECL), and integrated-injection logic (I^2L). The most widely used digital MOS technologies are PMOS, NMOS, and CMOS. At a given point in time, each of these bipolar and MOS technologies has a well-defined set of advantages and disadvantages. Thus, a best fit can often be established between a technology (or a set of technologies) and a product. Since IC suppliers are competing for similar product lines, however, the optimum technology–product fit changes with time. Recent literature must be used to track the major technologies so that optimal choices can be made.

Recently, processes and circuit techniques that have appeared and disappeared seem to have been variations on existing technologies rather than major technological revisions. A solid technology base appears to be forming. It involves sets of MOS and bipolar process technologies and sets of circuit techniques.

In previous chapters, we described crystal properties, diffusion, epitaxial growth, silicon device behavior, lithography, and so on. Given a mask set, the ability to transfer mask patterns onto silicon surface layers, and the ability to diffuse impurities into silicon, we can now describe the fabrication of integrated circuits.

Section 11-1 emphasizes the basic bipolar processing technology that has been dominant since the 1960s. In Section 11-2 some of the process variations in use for digital bipolar circuits are described. Section 11-3 discusses the MOS process technologies.

11-1 BASIC BIPOLAR PROCESSING

In Sections 11-1-1 through 11-1-8, we use the fabrication of a planar *npn* transistor to describe the basic bipolar processing steps. This is followed by brief discussions of ion implantation in Section 11-1-9 and *pnp* transistors in Section 11-1-10.

11-1-1 The Slice Clean

We begin by realizing that our wafer surfaces are badly contaminated. *Molecular contaminants* such as waxes, resins, and oils are present after mechanical polishing steps. Greasy surface films may exist from handling and storing slices in plastic containers. Photoresist residues and organic solvent residues can also be present on silicon surfaces. *Ionic contaminants* are caused by etching wafers in hydrofluoric acid solutions or caustic solutions. They continue to adhere even after extensive rinsing in deionized water. Alkali ions are particularly harmful ionic contaminants, since they can move under the influence of electric fields, causing inversion layers, leakage currents, device parameter drift, and so on. *Atomic contaminants* include heavy metals such as gold, silver, and copper. They are usually plated out as metallic deposits when acid etchants are used with silicon. The development of a "slice clean," which is effective in removing the bulk of these contaminants, has been a major breakthrough in bipolar and MOS IC processing. Without an effective slice clean, high-beta, low-leakage transistors and the NMOS and CMOS technologies would not exist today. This slice clean is shown between many of the process steps in Fig. 11-1. Each slice clean takes about 1 h and involves immersing a group of silicon wafers in (or spraying them with) various chemical solutions and rinsing in deionized water.

11-1-2 The Buried Layer

Process steps P1 through P4 involve selectively diffusing n^+ buried layers into the *p*-type substrate. There are a number of ways that this can be done, only one of which will be discussed here. With the method shown (P1–P4), a

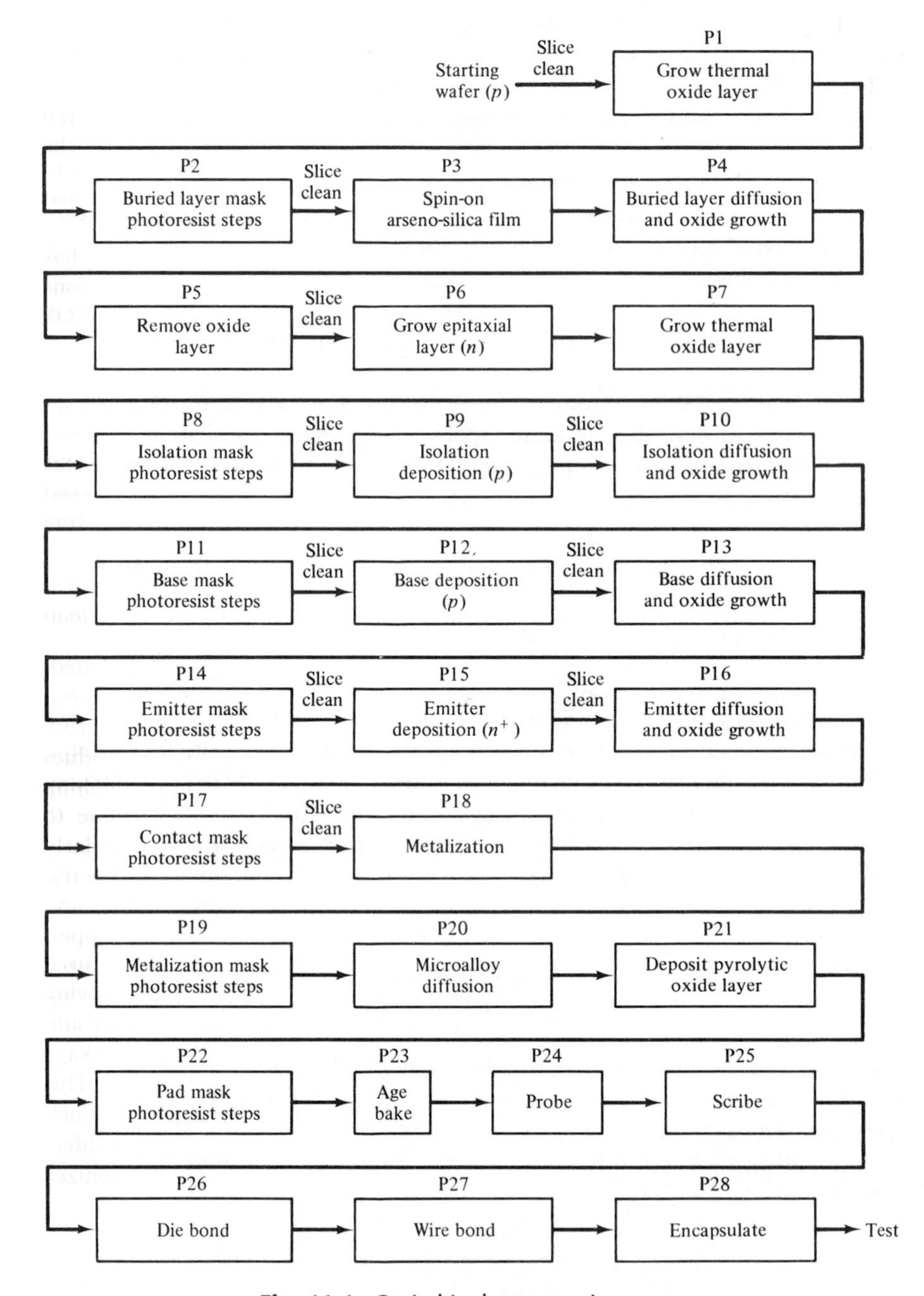

Fig. 11-1 Basic bipolar processing steps.

thermal oxide layer is grown by passing oxygen over the wafer surface in a furnace (1000 to 1200°C). Photoresist steps are used to open windows where the buried layers are needed [Fig. 11-2(b)]. Then a liquid source of arsenic or antimony is "painted"[1] on the wafer surface by spraying or spinning. If a spinner is used, film thickness can be varied by controlling the viscosity of the liquid and the spin speed. After a short heat treatment (not shown in Fig. 11-2), the wafers are placed in a diffusion furnace. Arsenic and antimony diffuse through glass much more slowly than through silicon. Thus, diffusion occurs only where

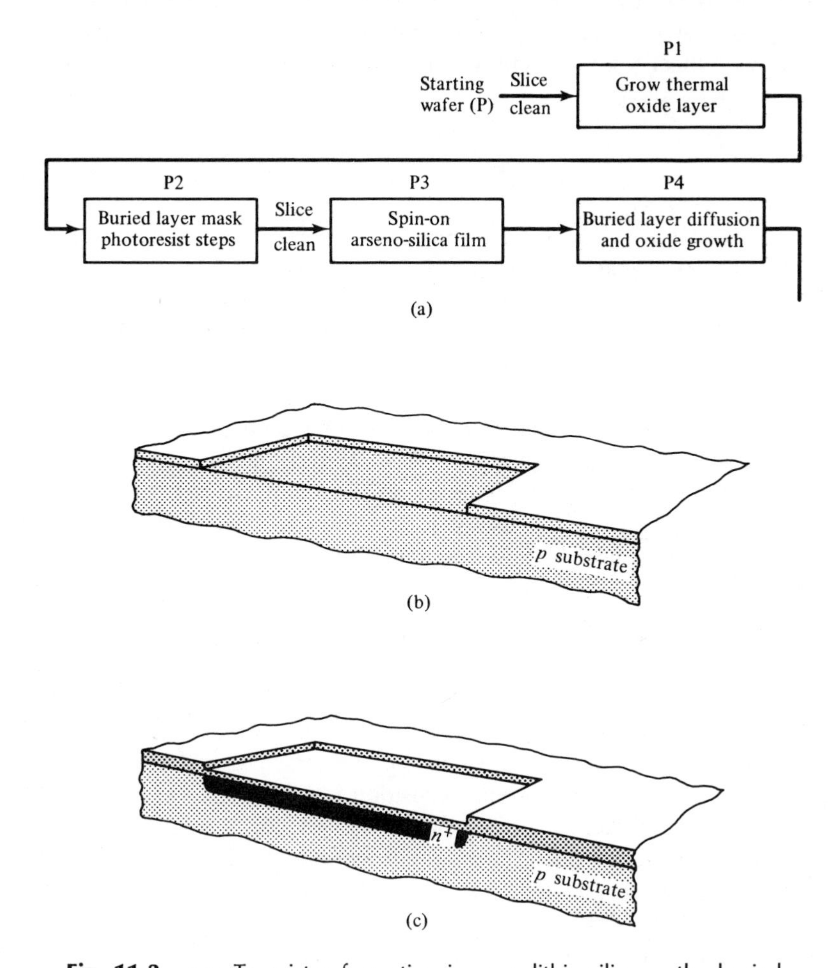

Fig. 11-2 *npn* Transistor formation in monolithic silicon—the buried layer: (a) buried layer process steps; (b) oxide windows are opened for n^+ buried layer deposition; (c) n^+ buried layer is diffused into the substrate. The oxide layer is regrown.

[1] This is traditionally known as a "paint-on" process.

windows have been opened in the glass. During the diffusion of arsenic or antimony, a new oxide layer is grown as shown in Fig. 11-2(c). Arsenic or antimony is used rather than phosphorus because they are slow diffusers. We want to minimize buried layer up-diffusion into the epitaxial layer while we do the other diffusions. Arsenic is normally introduced into the silicon lattice through the reaction of arsenic trioxide (As_2O_3) and silicon. Antimony is normally introduced through the reaction of the trioxide (Sb_2O_3) or the tetraoxide (Sb_2O_4) and silicon. In both cases, diffusion occurs from a glassy layer as a result of the surface reaction with silicon. A combination of nitrogen and oxygen is usually used as the ambient gas during diffusion. Buried-layer diffusions normally provide sheet resistivities in the range 10 to 30 $\Omega/\Box$ with layer thicknesses of 2 to 5 μm.

11-1-3 The Epitaxial Layer

The oxide layer grown during the buried-layer diffusion is removed in a dilute hydrofluoric acid etch, as shown in Fig. 11-3(b). Then a slice clean is performed and an *n*-type epitaxial layer is grown (P6). As discussed in Chapter 5, the epitaxial layer is independent of the substrate doping (i.e., it is not the result of a counterdoping process). New silicon, with appropriate dopant atoms, is deposited above the substrate. The crystal lattice is continuous. Arsenic or antimony is again used for *n*-type doping because each is a slow diffuser. An epitaxial reactor is as shown in Fig. 11-4. Typical *n*-type layers for bipolar ICs range from 3 to 25 μm in thickness depending on device application. Typical resistivities measure from 0.1 to 5.0 Ω-cm. We can discuss resistivity rather than sheet resistance here because the doping is uniform.

11-1-4 Isolation Deposition and Diffusion

Another thermal oxide layer is grown in a diffusion furnace (1000 to 1200°C) for masking purposes as shown in P7 and Fig. 11-3(c). The isolation mask is then used to open windows in the new oxide layer [see Fig. 11-5(b)], the slices are cleaned, and a *p*-type deposition is performed.

From the periodic table of elements, the valence 3 atoms that might be candidates for *p*-doping silicon are indium, gallium, aluminum, and boron. It takes 0.16 eV of energy to ionize indium atoms in silicon (compared to 0.045 eV to ionize boron). This amount of thermal energy is not available at normal operating temperatures, thus ruling out indium. Gallium diffuses through SiO_2 much faster than through silicon. Thus, oxide masking does not work, ruling out gallium. Aluminum is a very fast diffuser in both silicon and SiO_2. It is too difficult to control to be used regularly as a dopant. Boron, on the other hand, has a high value of solid solubility and diffuses much more slowly in SiO_2 than it does in silicon. It is the only logical choice for a *p* dopant.

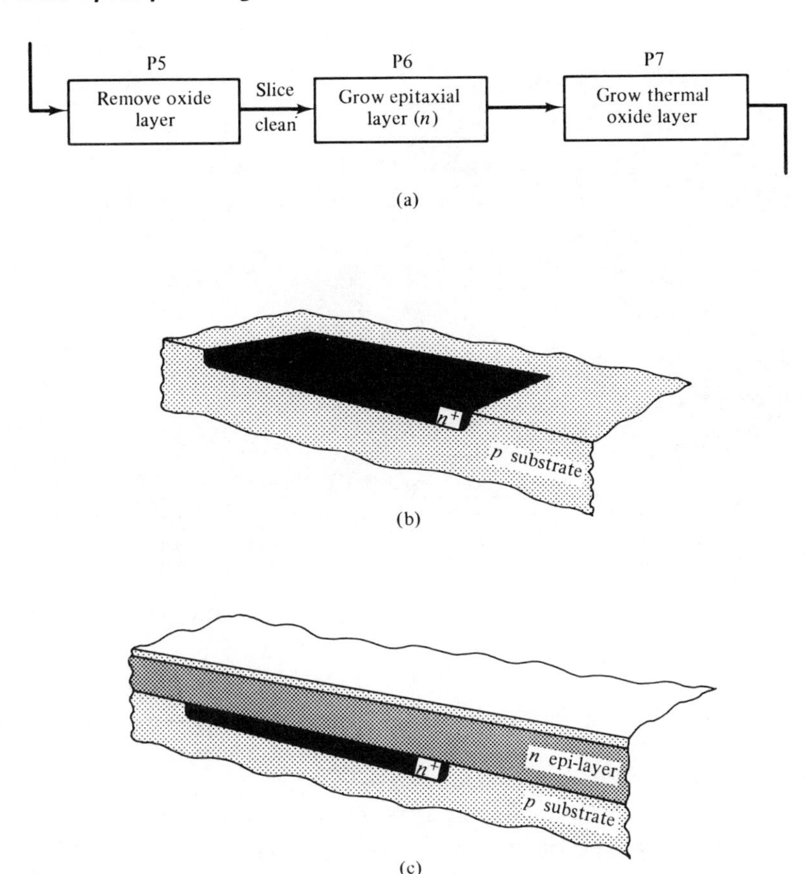

Fig. 11-3 *npn* Transistor formation in monolithic silicon—the epitaxial layer: (a) epitaxial layer process steps; (b) the oxide layer is removed in preparation for epitaxial growth; (c) *n*-type epitaxial layer and oxide layer are grown.

Solid, liquid, and gaseous sources of boron are available. Whatever source is used, however, its purpose is to react with oxygen to produce boron trioxide, B_2O_3. The surface reaction between boron trioxide and silicon is given by

$$2B_2O_3 + 3Si \rightleftharpoons 4B + 3SiO_2 \tag{11-1}$$

Boron trioxide is a liquid at normal deposition temperatures. It mixes with the glass on the silicon wafers (there is always a small amount of glass on silicon wafers) to form a borosilicate glass. Diffusion into silicon then takes place from this glass.

The *solid source,* boron nitride (BN), is available as inert nontoxic wafers. After purchase, the wafers are activated by placing in a diffusion furnace with

Fig. 11-4 Epitaxial reactors with wafers. (Courtesy of Delco Electronics Division, GMC.)

an oxygen atmosphere. This provides a surface layer of boron trioxide, which acts as the diffusion source. These wafers are then placed in a diffusion boat, close to and in parallel with silicon host wafers. One preoxidized boron nitride wafer is placed between two silicon wafers. The deposition furnace tube is loaded with wafers whose surfaces are perpendicular to the gas flow. The BN wafers react with the hydrogen and oxygen present to form an intermediate HBO_2 gas. Boron trioxide is formed in a second reaction. The boron trioxide then reacts with the silicon wafers to form a boron deposit where windows were opened.

Boron tribromide is available as a *liquid source* for boron deposits. An inert gas such as nitrogen is passed over the liquid dopant surface and mixed with oxygen in the furnace tube. The reaction is given by

$$4BBr_3 + 3O_2 \rightarrow 2B_2O_3 + 6Br_2 \qquad (11\text{-}2)$$

Provision must be made for venting the system because of the bromine.

Two *gaseous sources* of boron are boron trichloride (BCl_3) and diborane (B_2H_6). Boron trichloride is a gas at room temperature, but it can be liquified under pressure to provide a large supply in a small volume. It is mixed with oxygen and nitrogen in the furnace tube. Of the reactions that take place, the significant one is

$$4BCl_3 + 30_2 \rightarrow 2B_2O_3 + 6Cl_2 \qquad (11\text{-}3)$$

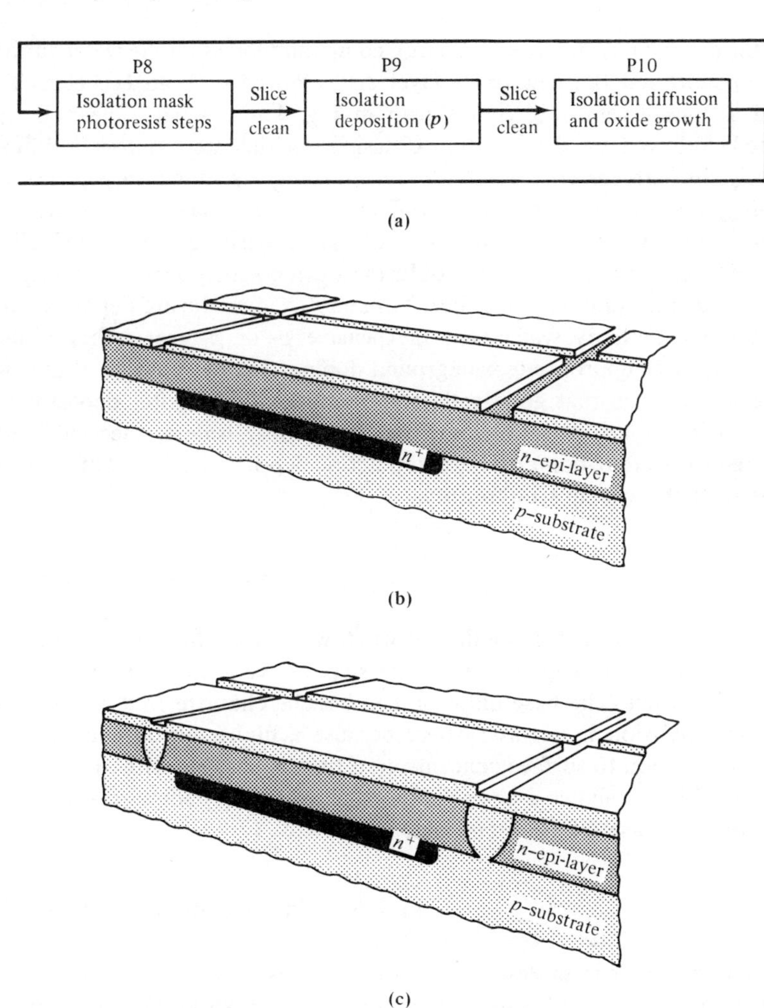

Fig. 11-5 *npn* Transistor formation in monolithic silicon—the isolation diffusion: (a) process steps for isolation diffusion; (b) oxide windows are opened for *p*-type isolation deposition and diffusion; (c) *p*-type isolation diffusion is completed and oxide layer is regrown.

Diborane is a highly poisonous and explosive gas. It is usually purchased in a gas cylinder, 99.9% diluted by argon before use. This mixture smells like a combination of cheap perfume and aroma of men's locker room. Again, mixed with oxygen and nitrogen in a furnace tube, the significant reaction products are

$$\begin{aligned} B_2H_6 + 3O_2 &\rightarrow 2HBO_2 + 2H_2O \\ 4HBO_2 &\longrightarrow 2B_2O_3 + 2H_2O \end{aligned} \qquad (11\text{-}4)$$

During the deposition, an undesired boron compound forms on the wafer surface. This compound and the new oxide layer are removed with an HF etch. This is called a "phase removal" (not shown in Fig. 11-1). A slice clean is then performed, followed by the isolation diffusion. In this step, boron is diffused completely through the epitaxial layer (or epi-layer), as shown in Fig. 11-5, to isolate *n*-type epi-regions. During the diffusion, the oxide layer is regrown. The new oxide layer prevents out-diffusion, protects the easily contaminated silicon surface, and acts as a mask against diffusion during subsequent processing.

It is common practice to include "dummy" wafers during deposition and diffusion steps. Dummy wafers are inexpensive *p*- or *n*-type wafers without epi-layers but with appropriate background doping. They are placed at the ends of diffusion boats, so that good wafers are between them. After deposition or diffusion, oxide layers are removed from the dummy wafers, sheet resistance is measured with a four-point probe, and junction depth is checked. This provides a good test of the system's uniformity.

11-1-5 Base Deposition and Diffusion

Figure 11-6 shows the opening of windows in glass for resistors and *npn* transistor bases. Boron is again used. These steps are very similar to the isolation steps just described. The base diffusion can best be characterized as Gaussian. A Gaussian diffusion is necessary here because a higher boron concentration is needed in relation to surface concentration than is possible with an erfc diffusion. Figure 11-6(b) shows a base window opened and (c) shows a wafer cross section after the base diffusion and oxide regrowth.

11-1-6 Emitter Deposition and Diffusion

Later in our process description, aluminum will be used to interconnect integrated components and connect ICs to the outside world. Where aluminum contacts the *n*-type collector regions, however, metal–semiconductor diodes known as Schottky diodes will be formed unless something is done to prevent them. We do not want diodes here. If the region in which the aluminum makes contact with the silicon is heavily doped n^+, it will cause the Schottky diode (the aluminum–n^+ silicon diode) to have a very low or degenerate barrier. Schottky diodes with low or degenerate barriers behave much like ohmic contacts. This is why an open window is shown for n^+ deposition into the *n*-type collector region in Fig. 11-7(b). When n^+ transistor emitters are put in, n^+ regions are added for all collector contacts at the same time.

Process step P14 indicates the emitter mask photoresist steps, which are followed by a slice clean and the n^+ emitter deposition. The candidates for n^+ doping (from the periodic table of valence 5 elements) are antimony, arsenic, and phosphorus. During the diffusion, while the emitter–base junction is moving

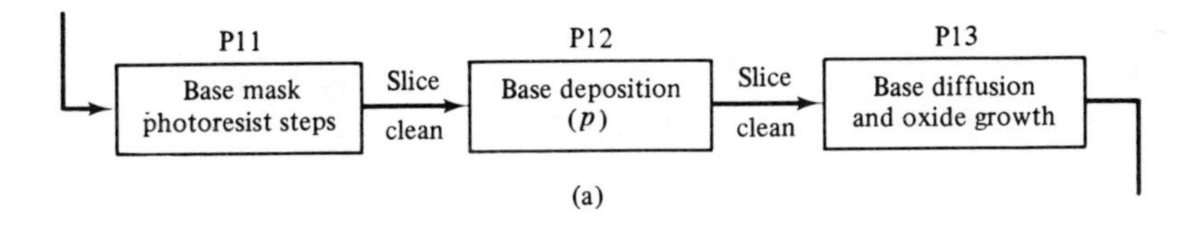

(a)

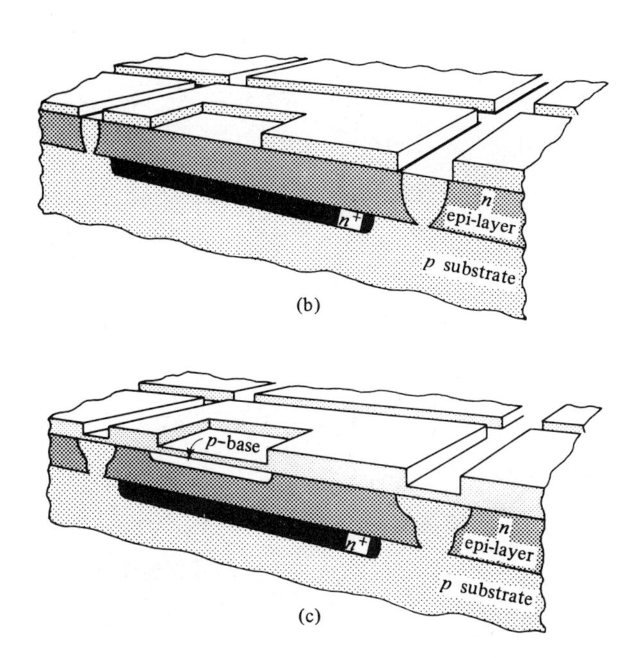

(b)

(c)

Fig. 11-6 *npn* Transistor formation in monolithic silicon—the base diffusion: (a) process steps for base diffusion; (b) oxide windows are opened for *p*-type base deposition and diffusion; (c) base diffusion is completed and oxide layer is regrown.

downward into the silicon, the base–collector junction will also be moving downward. To narrow the base width, the emitter–base junction has to move faster. It will, provided that the impurity diffusion coefficient is large enough and that the concentration gradient is larger. Referring to Fig. 4-2, boron and phosphorus have about the same diffusion coefficients, whereas those of arsenic and antimony are smaller. Figure 4-3 shows that you can put more boron in silicon than you can arsenic or antimony. It also shows that you can put more phosphorus in silicon than you can boron. The most logical choice, therefore, is phosphorus. In some shallow device structures, however, arsenic is used at high concentrations because it provides a steeper impurity profile.

Solid, liquid, and gaseous sources of phosphorus are all commonly used. Carborundum Company has a solid phosphorus source available in wafer form.

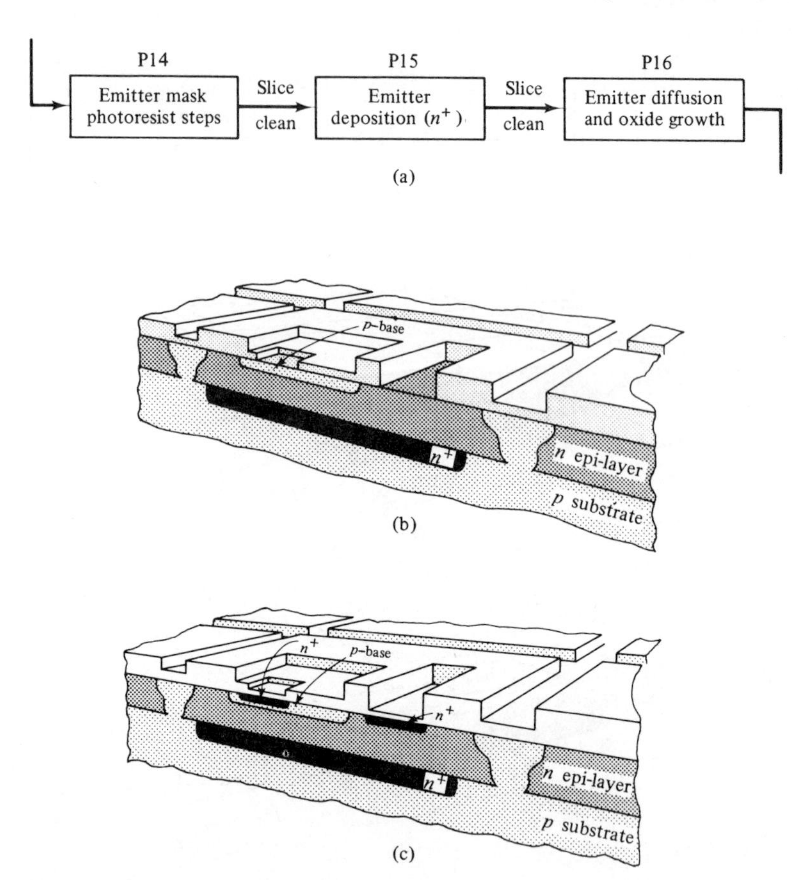

Fig. 11-7 *npn* Transistor formation in monolithic silicon—the emitter diffusion: (a) process steps for the emitter diffusion; (b) oxide windows are opened for n^+ emitter deposition and diffusion; (c) n^+ emitter diffusion is completed and oxide layer is regrown.

All of these sources provide phosphorus pentoxide (P_2O_5). The reaction is given by

$$2P_2O_5 + 5Si \longrightarrow 4P + 5SiO_2 \tag{11-5}$$

The actual source of the dopant usually reacts with oxygen to form P_2O_5 gas. At the surface of the wafers, the P_2O_5 combines with SiO_2 to form a phosphosilicate glass. Diffusion proceeds from this glassy layer.

Phosphorus oxychloride ($POCl_3$) is available as a *liquid source* for forming P_2O_5. It is kept in a constant-temperature bath between 0 and 45°C. Although it is a clear liquid like water, it reacts explosively with water, so it is kept in an ethylene glycol (antifreeze) bath. An inert gas (argon) is bubbled through the $POCl_3$ and then metered into the furnace tube. The preliminary reaction with oxygen is

$$4POCl_3 + 3O_2 \rightarrow 2P_2O_5 + 6Cl_2 \quad (11\text{-}6)$$

Surface concentration can be controlled by adjusting the bubbler temperature, gas concentrations, and deposition time. Two other liquid sources that are also used for phosphorus depositions are phosphorus tribromide (PBr_3) and phosphorus trichloride (PCl_3). "Paint-on" sources of phosphorus are also available.

Phosphine (PH_3) is the *gaseous source* of phosphorus. Although highly toxic and explosive in concentrated form, it is usually diluted with 99.9% nitrogen or argon, which makes it relatively safe to handle. A small amount of oxygen is added to the dilute mixture of phosphine in the furnace tube. The preliminary reaction is

$$2PH_3 + 4O_2 \rightarrow P_2O_5 + 3H_2O \quad (11\text{-}7)$$

Diluting the gas provides a convenient way of controlling the P_2O_5 concentration. One source of control problems with both liquid and gaseous sources is that the silica glass furnace tube absorbs a quantity of P_2O_5 and then acts as a second source. The problem can be alleviated somewhat by steam-cleaning the furnace tube and by making trial runs with dummy wafers.

Process steps P15 and P16 in Fig. 11-7 indicate the emitter deposition, emitter diffusion, and oxide growth. Figure 11-7(c) shows the wafer after the emitter diffusion and oxide growth. This diffusion closely approximates an erfc diffusion. The doping is sufficiently heavy to move the Fermi level into the conduction band. Typically, using $POCl_3$ as a dopant source, an emitter deposition might take 20 min at 1100°C, and an emitter diffusion might take 60 min at 1000°C. The time and temperature of the emitter diffusion will determine the base widths and betas of *npn* transistors in the integrated-circuit wafer. It is fortunate for us that the last doping is with phosphorus. The phosphorus "getters" sodium atoms and thus has a stabilizing influence on glass. Devices protected by phosphorus-doped SiO_2 appear to be less sensitive to temperature and bias conditions than do devices protected by non-phosphorus-doped oxides.

11-1-7 Contact Windows and the Metalization Layer

Process step P17 in Fig. 11-8(a) indicates the contact mask photoresist steps and Fig. 11-8(b) shows windows opened in the oxide layer so that contact can be made with aluminum. When the contact windows are opened, a slice clean is performed and the wafers are inserted in a vacuum chamber. The vacuum techniques generally used to deposit aluminum are evaporation from a tungsten filament, sputtering, and electron-beam deposition. These techniques are dis-

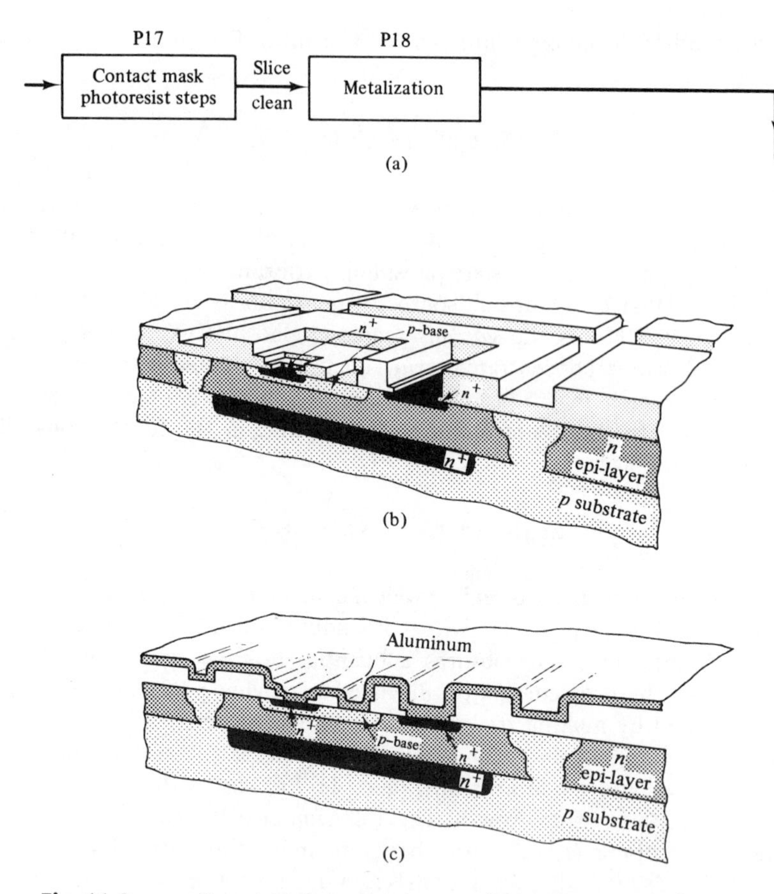

Fig. 11-8 *npn* Transistor formation in monolithic silicon—metalization: (a) process steps for aluminum metalization; (b) oxide windows are opened for contact with aluminum; (c) aluminum is deposited on the wafer.

cussed in detail in Chapter 12. Of these techniques, electron-beam deposition is most frequently used because it results in a very pure layer of aluminum. Sputtering is gaining in importance for the growing number of applications requiring binary or ternary alloys such as Al/Cu/Si. Typically, 1.0 to 1.5 μm of aluminum is deposited. Figure 11-8(c) shows the aluminum deposition step. The metalization mask and associated photoresist steps are used to remove aluminum as necessary. This is shown in Fig. 11-9. The microalloy diffusion (P20) is used to provide good ohmic contact between aluminum and silicon. Although the eutectic temperature for a silicon aluminum alloy is 577°C (Fig. 3-8), localized melting occurs at points of contact at considerably lower temperatures. It is advisable not to go higher in temperature than is necessary here because of the high diffusion rate of aluminum in silicon (Fig. 4-2). Given a

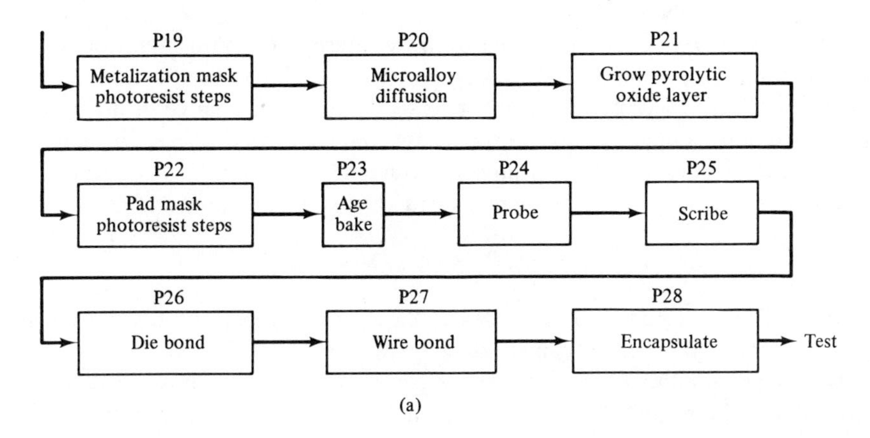

(a)

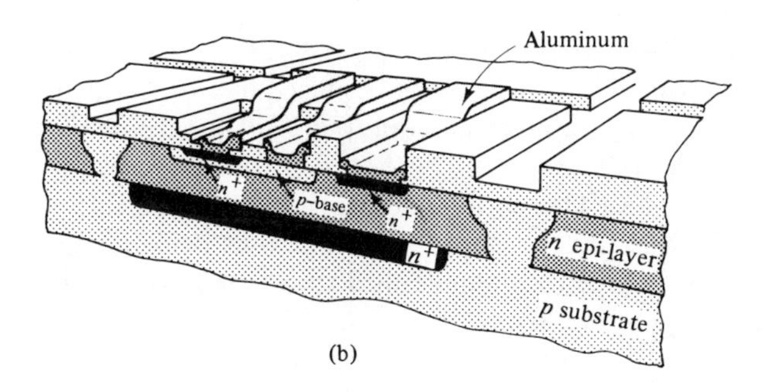

(b)

Fig. 11-9 Completed *npn* transistor: (a) process steps to complete *npn* transistor fabrication and assembly; (b) aluminum is removed as needed to complete the circuit (P19).

chance, the aluminum would diffuse right through the emitters. The microalloy diffusion therefore consists of placing the wafers in a diffusion furnace at 450 to 500°C for 5 to 60 min.

11-1-8 Pyrolytic Oxide Layer

Once the IC wafers are scribed into individual chips, the chips are often stored together in vials. Through handling and rubbing against each other, yields can be lowered due to surface scratches. Lowered yields can be avoided here by coating the wafers with a low-temperature glass. This glass layer also protects the chip from contamination and other failure modes. Windows are opened in this glass only over the pads, so that wire bonds can be made. In the deposition of low-temperature glass, the silicon substrate does not participate in the oxide-forming reaction. It merely acts as a substrate for deposition. Pyrolytically deposited oxides are prepared from the thermal decomposition of various silicon compounds. Silane (SiH_4) is commonly used for pyrolytic oxide depositions. The word "pyrolytic" simply refers to a chemical change brought about by the action of heat. Silane, oxygen, and an inert carrier gas are admitted into a reactor containing the heated substrate. The reaction between silane and oxygen, which occurs in the vicinity of the hot substrate surfaces, is

$$SiH_4 + 2O_2 \rightleftharpoons SiO_2 + 2H_2O \tag{11-8}$$

Although no external heat is needed to catalyze the reaction, less porous and more uniform films are obtained at temperatures above 300°C. Temperatures between 300 and 450°C are used. The upper limit is set by the possible continued diffusion of aluminum. In comparison with thermal oxide layers, pyrolytic oxides are more porous, have more pinholes, and have a higher defect density. They also etch more rapidly in HF. But they protect the chip surface, which is their purpose.

The photoresist steps [P22 in Fig. 11-9(a)] are used to open windows to the aluminum pads. Then the wafers are placed in a bake-out oven for approximately 4 h at 400°C in a nitrogen atmosphere. The purpose of this "age bake" (P23) is to getter the mobile ions in the oxide layers and to reduce the density of fast surface states. This is intended to minimize the drift of IC electrical parameters which might be caused by mobile ions distributing themselves and to reduce excess surface recombination. Process steps P24 through P28 in Fig. 11-9(a) are discussed in Chapter 13.

11-1-9 Ion Implantation

The basic bipolar process techniques were developed before ion implantation systems became commercially available. The NMOS and CMOS technologies were developed much later, and ion implantation quickly became a standard MOS process step. It is used primarily to adjust device threshold voltages and has led to a major improvement in MOS yields. Most commercial IC facilities are not designed to produce bipolar *or* MOS circuits. They can produce *both.* Thus, ion implantation systems are available for bipolar processing where the cost of changing to or adding this process step can be justified. Ion implantation is being used in bipolar circuits to make high-value boron resistors. It is also being used for the emitter and base deposition steps in situations where shallow bipolar devices are built in thin epitaxial layers (2 to 6 μm).

11-1-10 *pnp* Transistors

Substrate and lateral *pnp* transistors are discussed in Section 7-5. Figures 7-22 and 7-23 show cross sections of substrate and lateral *pnp* transistors, respectively. These particular device structures are used because the processing is fully compatible with the basic fabrication techniques used to make planar *npn* structures.

11-2 BIPOLAR PROCESSING FOR DIGITAL APPLICATIONS

When a transistor is used in a small-signal application, it is normally biased to a quiescent operating point in the active region, such as point 2 in Fig. 11-10(b). In digital applications, however, the transistor is driven between two well-defined operating points and used as a switch. The transistor can be biased to switch between saturation and cutoff [points 1 and 3 in Fig. 11-10(b)], or it might be switched between a quiescent operating point (2) and cutoff (3). In cutoff, the device behaves as an "off" switch and the output voltage V_{CE} is "high." In saturation, the device is "on" and the output voltage is "low." If, by some means, the collector saturation resistance can be reduced, the effect on the characteristic curves is shown by the dashed lines in Fig. 11-10(b). A reduction in collector saturation resistance will decrease the power dissipated in the device and lower the output voltage $V_{CE\ ON}$ when the device is in saturation.

The three basic digital logic families that are implemented with bipolar transistors are:

1. Transistor–transistor logic (T^2L).
2. Emitter-coupled logic (ECL).
3. Integrated-injection logic (I^2L).

These logic families are discussed in detail in Chapter 14. Since this chapter is entitled "IC Fabrication," however, it seems appropriate to discuss some

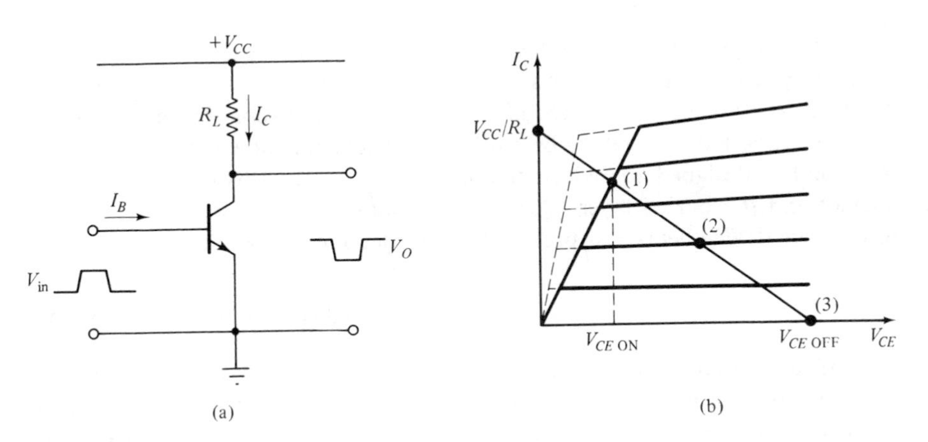

Fig. 11-10 (a) Basic transistor inverter. (b) Characteristic curves and load line.

variations in device fabrication that are used to improve transistor operation in T^2L and ECL circuits.

In standard T^2L logic circuits, the transistors are driven between saturation and cutoff. When a transistor is in saturation, both the emitter–base and base–collector junctions are forward-biased. Enough minority carriers are stored in the base region under these conditions to have a major effect on the propagation delay of standard T^2L circuits. When the base of the saturated transistor goes negative, the transistor can not turn off until the base charge is removed. A higher-speed version of T^2L circuits, called *Schottky T^2L,* was created by placing a *Schottky diode* (or Schottky barrier diode) between the base and collector of each saturating transistor. A Schottky diode turns on when the collector becomes 0.3 V or more negative with respect to the base. Since 0.5 to 0.7 V is required to forward-bias the base–collector diode, the saturation region is avoided. The typical propagation delay of a standard T^2L gate is about 9 ns, whereas that of a Schottky T^2L gate is about 3 ns.

In emitter-coupled logic circuits, constant-current sources are used in series with device emitters to keep transistors from saturating. Transistors are switched between a quiescent operating point [such as point 2 in Fig. 11-10(b)] and cutoff. Emitter-coupled logic provides the fastest switching times available with bipolar silicon devices. Propagation delays are on the order of 1 to 2 ns. Fairchild Camera and Instrument Corporation developed a fabrication technique called the Isoplanar[2] Process. With this process, silicon dioxide is used in place of the isolation diffusion to isolate silicon devices. As a result, devices with smaller geometries and smaller internal capacitances can be fabricated. When Isoplanar transistors are used in ECL circuits, gate propagation delays on the order of 650 ps can be obtained.

11-2-1 Deep n^+ Diffusions

Many digital-device manufacturers use deep n^+ collector diffusions to reduce the collector saturation resistance, as shown in Fig. 11-11. This deep n^+ diffusion is usually accomplished after the isolation diffusion, but may be done before it. It is done before the base diffusion. It extends down to the n^+ buried layer and is designed to reduce the vertical resistance between the collector contact and the buried layer. Deep n^+ diffusions are also used in I^2L circuits to enhance device betas and prevent lateral injection between devices.

11-2-2 Schottky-Clamped Transistors

If aluminum is brought into contact with the lightly doped n-type epilayer, a *Schottky barrier diode* (SBD) is formed. In *npn*-transistor processing, we have been using an n^+ diffusion under the collector contact to form a low

[2] Isoplanar is a Fairchild trademark.

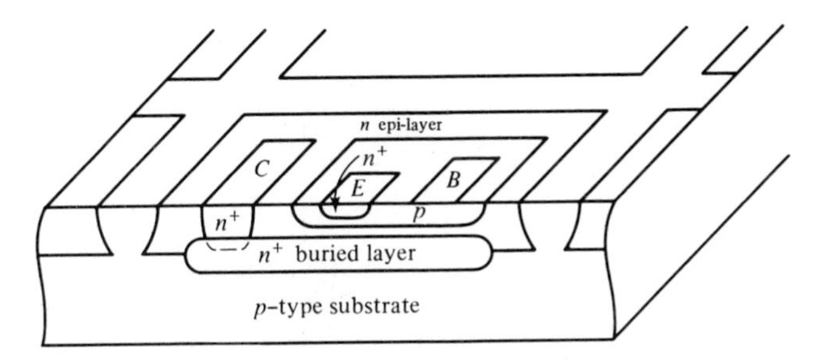

Fig. 11-11 Deep n^+ collector diffusion reduces collector saturation resistance.

or degenerate barrier Schottky diode which behaves as an ohmic contact. Schottky diodes can be used to advantage in T^2L circuits, however, to prevent transistors from saturating.

The structure of a Schottky barrier diode is shown in Fig. 11-12. Platinum silicide (Pt_5Si_2) is often used to restrict the diffusion of aluminum into the *n*-type epi-layer and thus improve yields. After the contact window is opened, platinum is evaporated onto the silicon surface and sintered at about 600°C. The platinum and silicon react to form platinum silicide. The excess platinum is then removed and aluminum is deposited to contact the platinum silicide. The Schottky barrier diode is formed at the Al–Pt_5Si_2–Si interface.

Schottky barrier diodes are metal semiconductor diodes in which current transport is caused by majority carriers. They have inherently fast response (storage times on the order of 0.1 ns). The knee of the forward bias curve occurs at 0.3 to 0.4 V, which is much lower than that of *p-n* junctions. They can be used between base and collector of standard *npn* transistors to "clamp" the voltage levels, as shown in Fig. 11-13. The Schottky barrier diode turns on before the base–collector junction can become forward-biased.

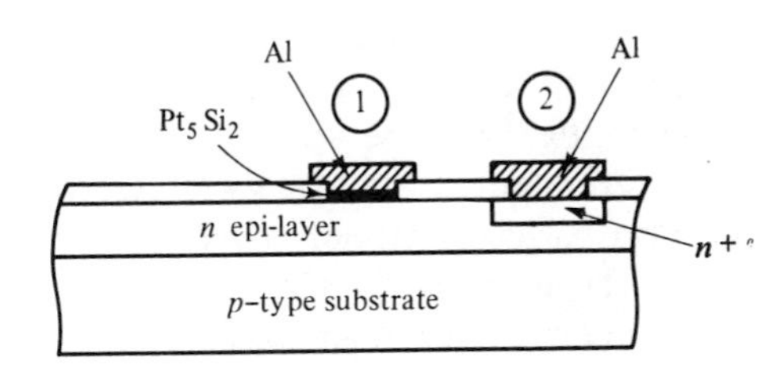

Fig. 11-12 Structure of a Schottky barrier diode.

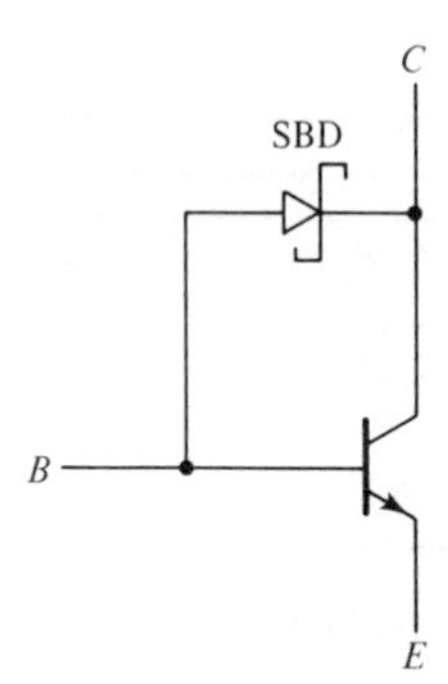

Fig. 11-13 Schottky-clamped transistor.

Figure 11-14 shows the construction of a Schottky clamped transistor. The base contact window is shifted over so that the base metalization contacts both the base region and the *n*-type epi-layer. The clamping diode then appears schematically as shown in Fig. 11-13.

11-2-3 The Isoplanar Process

In the Isoplanar process developed by Fairchild Camera and Instrument Corporation, silicon dioxide is used in place of the isolation diffusion to isolate silicon devices. This leads to a reduction in silicon area by more than a factor of 2 when compared to conventional *npn* transistors. The reduction occurs because the base diffusion can be butted up against the oxide isolation region. Figure 11-15 shows an *npn* transistor made by Fairchild's Isoplanar II process. Depending on the specific application, there are other versions of the Isoplanar II transistor. In the device structure shown in Fig. 11-15, the impurity profiles have been arranged to optimize high-frequency performance. This device structure is discussed in detail in reference [1].

Referring to Fig. 11-15, the transistor structure is completely surrounded by silicon dioxide. The n^+ collector diffusion makes contact with the n^+ buried layer. Both the base region and the base mask terminate at the oxide wall. Oversize base and base contact masks can be used with no adverse effects. Whereas the emitter must be separated from the ends of the base region, the emitter sides terminate at the oxide wall. Oversize emitter and emitter contact masks can also be used without adverse effects. Because the device structure terminates at the oxide walls and mask alignment is not critical, the device can be made much smaller than conventional planar transistors. The reduction in silicon area and the absence of sidewall capacitance cause a *reduction in collector–base capacitance.* This is the most significant feature of the Isoplanar structure because its major application is in high-speed switching circuits. The reduction in collector–substrate capacitance is also important in this application.

Isoplanar II devices with the structure shown in Fig. 11-15 were fabricated [1] with emitter sizes of 0.1 × 0.5 mil. The gain–bandwidth product of these

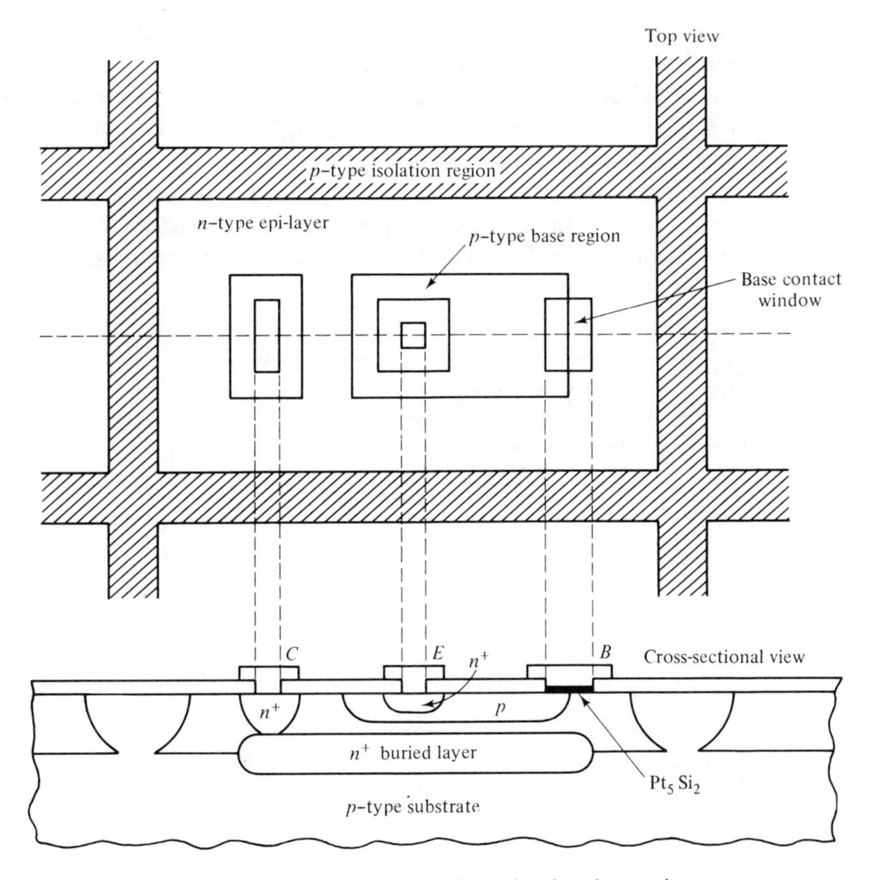

Fig. 11-14 Layout and structure of a Schottky-clamped transistor.

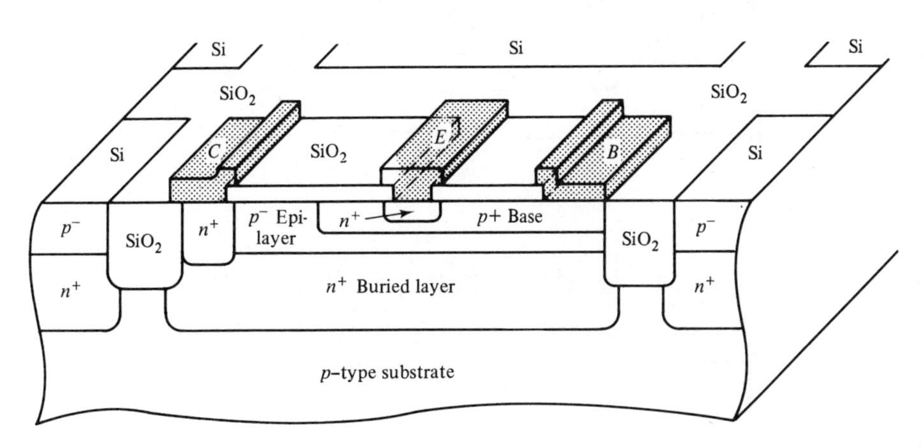

Fig. 11-15 *npn* Transistor made by Fairchild's Isoplanar II process.

devices was approximately 5 GHz. The collector–emitter breakdown voltage BV_{CEO} of these devices was approximately 5 V.

The major fabrication steps in the construction of an *npn* transistor using the Isoplanar II process are shown in Fig. 11-16. Beginning with a *p*-type substrate, buried n^+ layers are selectively diffused through oxide windows. The oxide layer is then removed and a *p*-type epitaxial layer (approximately 2 μm thick) is grown. A layer of thermal oxide is then grown, followed by the deposition of a silicon nitride layer (Si_3N_4) and an oxide layer. A wafer cross section

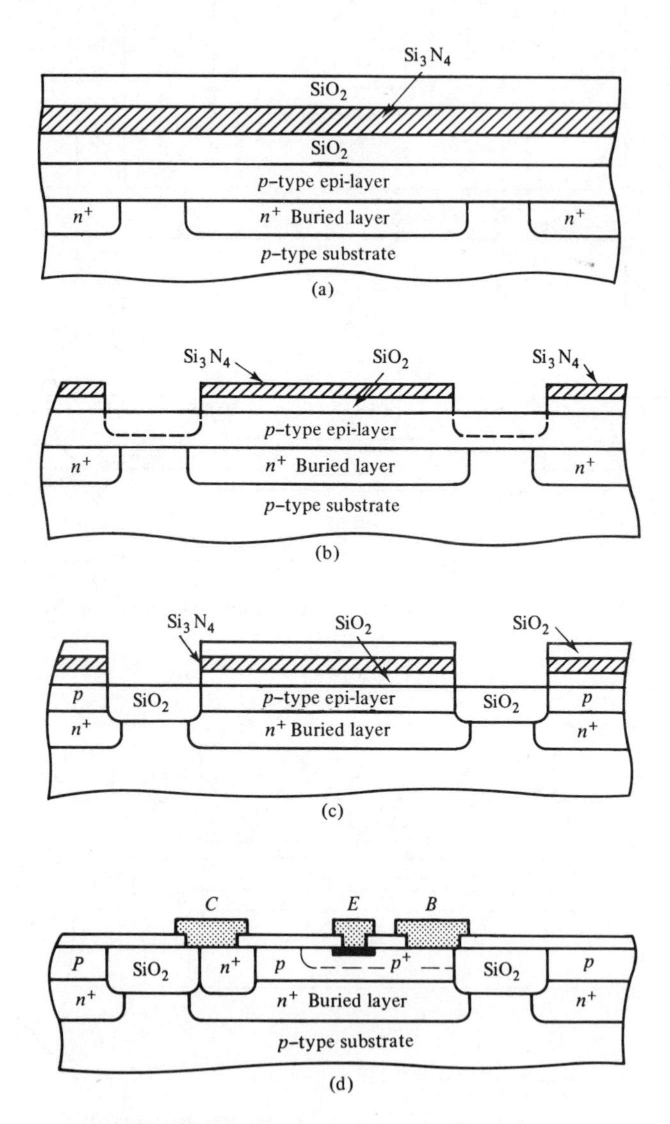

Fig. 11-16 Isoplanar II process sequence.

at this point is shown in Fig. 11-16(a). Photoprocessing is used to open windows in the oxide–nitride–oxide sandwich, so that silicon is exposed as shown in Fig. 11-16(b). Trenches are then etched in the *p*-type epi-layer as shown by the dashed lines in Fig. 11-16(b). This step is followed by an oxide growth, for which the silicon nitride acts as a mask. About 45% of the oxide growth is downward, through the remainder of the *p*-type epi-layer. About 55% of the oxide growth is upward, filling the trenches that had been etched. When this step is completed, the SiO_2 in the trenches is level with the original surface of the wafer. Downward, the oxide growth isolates the *p*-type epi-layers, as shown in Fig. 11-16(c). A thin oxide layer grows on the Si_3N_4 layer during this process step. With the wet-etching process, the top oxide layer is removed with buffered hydrofluoric acid. Boiling phosphoric acid is then used to remove the nitride layer. Conventional processing techniques are used for the deep n^+ collector diffusion, the p^+ base diffusion, and the n^+ emitter diffusion. The completed Isoplanar transistor is shown in Fig. 11-16(d).

11-3 MOS PROCESS TECHNOLOGIES

The purpose of this section is to summarize the dominant MOS technologies that are available and to describe some of the major processing techniques used to implement these technologies.

Although MOS devices are used in many nondigital applications, the major impact of the technology has been in digital systems. Large-scale and very-large-scale integrated (LSI and VLSI) systems have provided the major digital applications of MOS devices. We need to look at IC processing, not in terms of making price-competitive logic gates, flip-flops, registers, and so on, but in terms of fabricating building blocks for complex digital MOS/LSI structures, such as microprocessors and semiconductor memories. This difference in application goals is reflected in process complexity. Complex, expensive processing techniques, which attempt to maximize speed and packing density with minimum power dissipation, can often be justified for LSI and VLSI systems. In small- and medium-scale integrated circuits (SSI and MSI), however, the trade-offs may be very different.

The MOS transistor technologies can be classified as *p*-channel (PMOS), *n*-channel (NMOS), and complementary MOS (CMOS). CMOS employs series combinations of *p*- and *n*-channel MOS devices between power supply and ground.

PMOS, with aluminum gates and (111) silicon crystal orientation, was developed during the mid-1960s. At that time, mobile positive ion densities (sodium) in MOS oxide layers were causing problems with device threshold voltages. The negative gate voltages used with PMOS pulled the mobile ions up under the gate and away from the channel. Threshold voltage instability was more severe in NMOS devices because the mobile (+) ions were repelled to the SiO_2–Si interface. Eventually, mobile ion densities were decreased by using electron-beam deposition of aluminum rather than evaporation from tungsten filaments. PMOS devices were made with (100) silicon, which has a smaller surface charge density (Q_{SS}) at the $Si–SiO_2$ interface. Ion implantation systems became available and were used to adjust gate threshold voltages by implanting channel ions. The technique of using self-aligned polysilicon gates was developed to reduce device capacitances and channel lengths.

These techniques also made the fabrication of NMOS devices possible with reasonable yields. Electron mobilities are two to three times larger than hole mobilities. Since drift current density is proportional to mobility $(J_n = qn\mu_n E)$, the same current magnitudes could be achieved with a two- to threefold times reduction in device area; and size reductions allow faster switching speeds. As a result, NMOS has not only superseded PMOS, it has become the workhorse of the IC industry. Ion implantation, (100) silicon, and self-aligned silicon gates are in normal usage. Two levels of polysilicon are often used in MOS/LSI structures, with aluminum metalization runs above them. In comparison to NMOS, PMOS now provides a slower, larger, simpler, inexpensive, high-yield fabrication technology. It is used to advantage in applications where these are desirable attributes, such as in low-cost calculators.

CMOS employs series combinations of *p*- and *n*-channel devices between power supply and ground. In static operation, only one of the two series transistors is on at any given time. This leads to static power dissipation in the nanowatt range, which is one of the major advantages of CMOS. Power dissipation is proportional to frequency, however. At clock frequencies in the MHz range, CMOS power dissipation becomes comparable with other MOS technologies. A range of supply voltages between 3 and 15 V can be used. Noise immunity is typically in excess of 40% of the supply voltage. Also, CMOS can be used over a wide temperature range. In comparison to NMOS, it has traditionally been a more expensive technology with lower packing densities. A CMOS logic gate requires a minimum of $2N$ devices, where N is the number of logic inputs. NMOS requires a minimum of $(N + 1)$ devices. Additional devices cause an increase in chip area. When operating voltages of 5 V or more are used with CMOS, channel stoppers (or guard rings) are needed to isolate devices, with a corresponding increase in chip area. A wide variety of digital SSI and MSI CMOS circuits are available off the shelf. Because of its unique advantages in power dissipation, operating range, and noise immunity, CMOS provides a viable alternative to the T^2L logic families in this market area.

11-3-1 Metal Gate *n*-Channel Enhancement-Mode MOSFETs

A metal gate NMOS structure is shown in Fig. 11-17. The crystal orientation is (100) rather than (111) because it results in a lower magnitude of surface charge density and consequently a lower value of threshold voltage. $Q_{SS} = 1.4 \times 10^{-8}$ C/cm^2 for (100) silicon, whereas $Q_{SS} = 8 \times 10^{-8}$ C/cm^2 for (111)

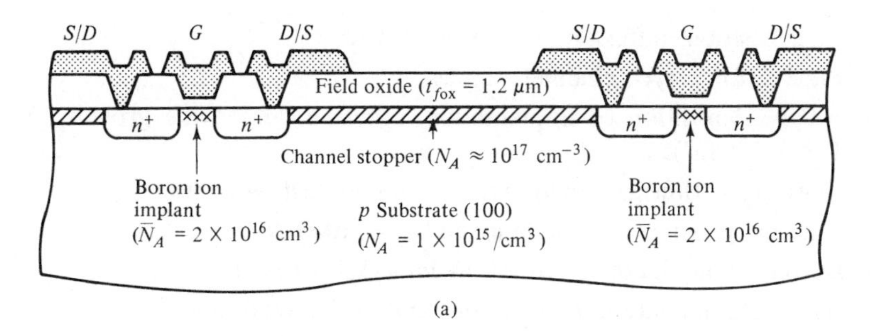

(a)

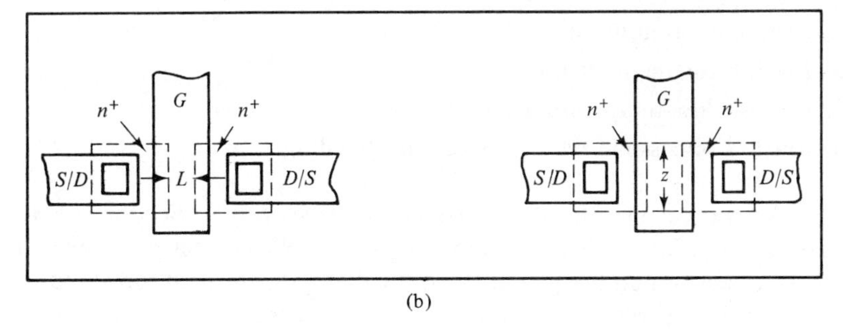

(b)

Fig. 11-17 Metal gate *n*-channel enhancement-mode MOSFETs: (a) cross section; (b) top view.

silicon. As shown in Fig. 11-17, a boron ion implantation is used to set the device threshold voltage V_T. Without this, an *n*-type inversion layer would exist in the channel (i.e., the device would be in the depletion mode). However, increasing the impurity concentration in the channel decreases electron mobility and increases the channel–substrate depletion layer capacitance. If these factors were critical, ion implantation could be avoided, and the threshold voltage could be set by reverse-biasing the substrate with respect to the source. This is called *substrate bias* and will be discussed later in this section. A *p*-type ion implantation creates a *guard ring* or *channel stopper,* which is used between devices for isolation. If this was not used, an *n*-type inversion layer would exist under the field oxide. Enough overlap must exist between the metal gate and the n^+ source and drain regions to compensate for alignment tolerances. The larger this overlap area, the larger the gate–drain and gate–source capacitances. These capacitances, in turn, limit the device switching speed.

A metal gate NMOS fabrication procedure that might be used for the device shown in Fig. 11-17 is as follows:

1. Start with a *p*-type (100) substrate ($N_A \approx 1 \times 10^{15}$/cm^3).
2. Grow a thick masking oxide.
3. Mask 1 for n^+ source and drain regions.

4. n^+ deposition, diffusion, and oxide growth.
5. Mask 2 for p-type channel stoppers.
6. p-Type deposition or implantation, diffusion, and oxide growth ($N_A \approx 1 \times 10^{17}/\text{cm}^3$).
7. Mask 3 for thin gate oxide regions and contact windows.
8. Thermally grow a thin gate oxide ($t_{ox} \approx 1000$ Å).
9. Boron ion implantation to set V_T for NMOS devices.
10. Anneal to activate implanted ions and heal crystal damage.
11. Mask 5 to open source and drain contact windows.
12. Evaporate aluminum.
13. Mask 6 for metal removal.
14. Deposit low-temperature pyrolytic glass.
15. Mask 7 opens pad windows for wire bonding.

Now we can use the device equations developed in Section 8-6 to find the operating parameters of the aluminum gate NMOS device shown in Fig. 11-17. We want to find the gate threshold voltage V_T, the field threshold voltage V_{TF},[3] the drain current $I_{D\text{sat}}$, and the transconductance $g_{m\,\text{sat}}$. The equations needed to find these parameters are summarized below.

$$V_T = V_{\text{Si}} + V_{\text{ox}} - V_{SS} + V_{WF} \tag{11-9}$$

$$V_{\text{Si}} = 2V_f = \frac{2kT}{q} \ln \frac{N_A}{n_i} \tag{11-10}$$

$$V_{\text{ox}} = \frac{Q_B}{C_{\text{ox}}} = \frac{(2\epsilon_0 \epsilon_{\text{Si}} q N_A |2V_f|)^{1/2}}{C_{\text{ox}}} \tag{11-11}$$

where

$$C_{\text{ox}} = \frac{\epsilon_0 \epsilon_{\text{ox}}}{t_{\text{ox}}} \tag{11-12}$$

$$V_{SS} = \frac{Q_{SS}}{C_{\text{ox}}} \tag{11-13}$$

where

$$\begin{aligned} Q_{SS} &= 1.4 \times 10^{-8}\ \text{C/cm}^2 \qquad \text{for (100) silicon} \\ V_{WF}(\text{Al-Si}) &= 3.2 - (3.25 + 0.55 \pm V_f)\ \text{volts} \end{aligned} \tag{11-14}$$

[3] V_{TF} is the voltage above the field oxide that will cause an inversion layer in the underlying silicon. It is assumed that the positive supply voltage will be applied to an aluminum runner somewhere above the field oxide.

or

$$V_{WF}(\text{Al-Si}) = -0.6 \mp V_f \tag{11-15}$$

where V_f is (−) for p-type silicon and (+) for n-type silicon.

$$I_{D\text{sat}} = \frac{Z}{L}\mu_{ns}C_{\text{ox}}\frac{(V_G - V_T)^2}{2} \qquad \text{(see Eq. 9-27)} \tag{11-16}$$

$$g_{m\ \text{sat}} = \frac{Z}{L}\mu_{ns}C_{\text{ox}}(V_G - V_T) \qquad \text{(see Eq. 9-32)} \tag{11-17}$$

The NMOS device of Fig. 11-17 has a substrate doping $N_A = 1 \times 10^{15}/\text{cm}^3$. For the channel ion implantation, $N_A = 2 \times 10^{16}/\text{cm}^3$, whereas for the channel stopper diffusion, $N_A \approx 1 \times 10^{17}/\text{cm}^3$. The gate oxide thickness t_{ox} is 0.1 μm, and the field oxide thickness $t_{f\text{ox}} = 1.2\ \mu$m. From Eqs. (11-9) through (11-15), we have that

$$\begin{aligned} V_{\text{Si}} &= 0.73\text{V} & V_{SS} &= 0.40 \text{ V} \\ Q_B &= 7.04 \times 10^{-8} \text{ C/cm}^2 & V_{WF} &= 0.97 \text{ V} \\ C_{\text{ox}} &= 3.54 \times 10^{-8} \text{ F/cm}^2 & & \\ V_{\text{ox}} &= 1.99 \text{ V} & & \end{aligned}$$

Thus, $V_T = 0.73 + 1.99 - 0.40 - 0.97 = 1.35$ V. Assuming that the substrate is grounded, the device turns on when a gate voltage of +1.35 V is applied. Equations (11-9) through (11-15) can be used again to find the field threshold voltage.

$$V_{\text{Si}} = 0.052 \ln \frac{1 \times 10^{17}}{1.5 \times 10^{10}} = 0.82 \text{ V}$$

$$\begin{aligned} Q_B &= [(2)(8.85 \times 10^{-14})(12)(1.6 \times 10^{-19})(1 \times 10^{17})(0.82)]^{1/2} \\ &= 1.67 \times 10^{-7} \text{ C/cm}^2 \end{aligned}$$

$$C_{\text{ox}} = \frac{(4)(8.85 \times 10^{-14})}{1.2 \times 10^{-4}} = 2.95 \times 10^{-9} \text{ F/cm}^2$$

$$V_{\text{ox}} = \frac{Q_B}{C_{\text{ox}}} = 56.61 \text{ V}$$

$$V_{SS} = \frac{Q_{SS}}{C_{\text{ox}}} = 4.75 \text{ V}$$

$$V_{WF}(\text{Si}) = 3.25 + 0.55 + 0.41 = 4.21 \text{ V}$$

$$V_{WF} = +3.2 - 4.21 = -1.01 \text{ V}$$

$$V_{TF} = 0.82 + 56.61 - 4.75 - 1.01 = 51.7 \text{ V}$$

The channel stopper region between devices will not become conductive until the aluminum above the field oxide reaches 51.7 V.

Given that $Z = 12\ \mu m$, $L = 6\ \mu m$, $\mu_{ns} = 450\ cm^2/V\text{-}s$, and $V_G = +5$ V, we can find I_{Dsat} and $g_{m\ sat}$ for the device of Fig. 11-17. Thus, from Eqs. (11-16) and (11-17),

$$I_{Dsat} = \left(\frac{12}{6}\right)(450)(3.54 \times 10^{-8})\frac{(5 - 1.35)^2}{2} = 212\ \mu A$$

$$g_{m\ sat} = (2)(450)(3.54 \times 10^{-8})(3.65) = 116.3\ \mu A/V$$

Given that one of these devices is driving a load resistor $R_L = 100\ k\Omega$, the voltage gain will be approximately equal to $-g_m R_L$, or -11.62.

11-3-2 Silicon Gate NMOS Processing

The switching speed of an MOS device is related to channel doping, channel length, and device capacitance. To maximize switching speed, we want to use a light impurity concentration in the channel, minimize channel length, and minimize device capacitances. In general, scaling down the dimensions of MOS devices will increase packing densities *and* switching speed.

With the aluminum gate process, the channel length is defined by a masking step. Channel length is restricted by the lithographic process and the gate must overlap the source and drain regions to compensate for mask alignment tolerances. This overlap causes gate–source and gate–drain capacitances which limit device switching speeds.

To make smaller and faster NMOS devices, a fabrication process called the *self-aligned silicon gate* process was developed. With this process, the alignment of the channel under the gate is not mask-dependent. Although threshold voltages are still adjusted with gate and field implants, the process description of Fig. 11-18 was simplified by not showing these steps. In Fig. 11-18(b) and (c), polycrystalline silicon is vapor-deposited on the oxide surface and then selectively etched, leaving only gate stripes. In Fig. 11-18(d) and (e), the oxide is removed in the source and drain regions. An n^+ diffusion is used to create the source and drain regions and to heavily dope the polysilicon gate. Since the n^+ impurity cannot penetrate the gate oxide, the channel is self-aligned under the gate. In Fig. 11-18(f), a thick phosphorus-doped oxide layer is vapor-deposited and subsequently reflowed to smooth the topography. If this oxide layer had been thermally grown, it would have used the polysilicon gate for oxide formation. In Fig. 11-18(g) and (h), contact windows are opened and aluminum metalization patterns are established. Figure 11-19 shows a top view of the silicon gate NMOS device.

Now let us look at how the silicon gate affects the calculation of device threshold voltage V_T. We will start by using the same gate–oxide thickness,

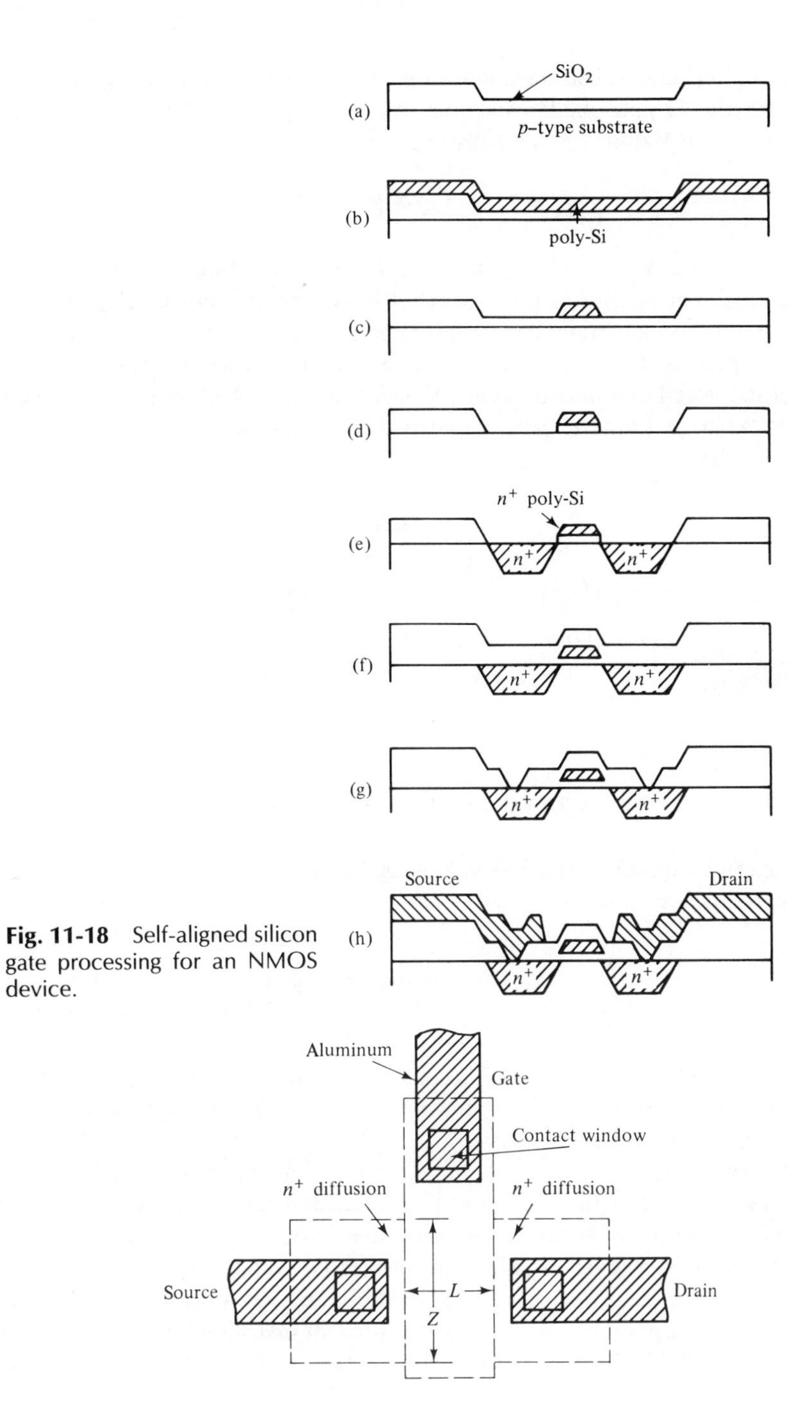

Fig. 11-18 Self-aligned silicon gate processing for an NMOS device.

Fig. 11-19 Top view of a silicon-gate *n*-channel MOS-FET.

channel doping, and substrate doping that we used for the metal gate: $t_{ox} =$ 0.1 μm for the Si gate–oxide thickness, $N_A = 2 \times 10^{16}/\text{cm}^3$ for the channel, and $N_A = 1 \times 10^{15}/\text{cm}^3$ for the substrate.

$$V_T = V_{\text{Si}} + V_{\text{ox}} - V_{\text{SS}} + V_{WF} \tag{11-9}$$

In Eq. (11-9), the values of V_{Si}, V_{ox} and V_{SS} do not change from the metal gate example. The work-function voltage V_{WF} is different, however, because of the silicon gate. As an engineering approximation, we will treat the polycrystalline silicon gate as if it has the energy band structure of single-crystal silicon.[4] This method specifies work-function voltages that agree well with experimental evidence. Assume that the gate impurity concentration is N_A (gate) $= 1 \times 10^{19}$ cm^{-3}. Then

$$V_f(\text{Si gate}) = \frac{kT}{q} \ln \frac{N_A(\text{gate})}{n_i} = 0.53 \text{ eV}$$

$$V_{WF}(\text{Si gate}) = 3.25 + 0.55 - 0.53 = 3.27 \text{ V}$$

$$V_{WF}(p\text{-channel}) = 3.25 + 0.55 + 0.37 = 4.17 \text{ V}$$

$$V_{WF} = 3.27 - 4.17 = -0.90 \text{ V}$$

Thus,

$$V_T = 0.73 + 1.99 - 0.40 - 0.90 = 1.42 \text{ V}$$

The device turns on when the gate voltage is +1.42 V, assuming that the source and substrate are grounded.

11-3-3 Processing Sequence for an Enhancement Driver/Depletion Load Inverter

In this section we discuss the construction of an NMOS enhancement driver/depletion load inverter, as shown in Fig. 11-20. The operation of this circuit is discussed in Section 9-1-4.

The processing sequence for this circuit is shown in Fig. 11-21(a)–(j). This sequence is useful to us because it includes ion implantation, self-aligned silicon gates, and local oxidation. Above the substrate, the starting wafer has a thin oxide layer, a layer of silicon nitride (Si_3N_4), and a top resist layer. A masking step is used to selectively etch Si_3N_4 and SiO_2 windows. A *p*-type channel stopper implantation is then used to prevent inversion layers from occurring under the field oxide (i.e., to isolate NMOS inverter stages). The wafer

[4] The existence of energy bands is intimately related to the periodicity of *single-crystal* lattice structures.

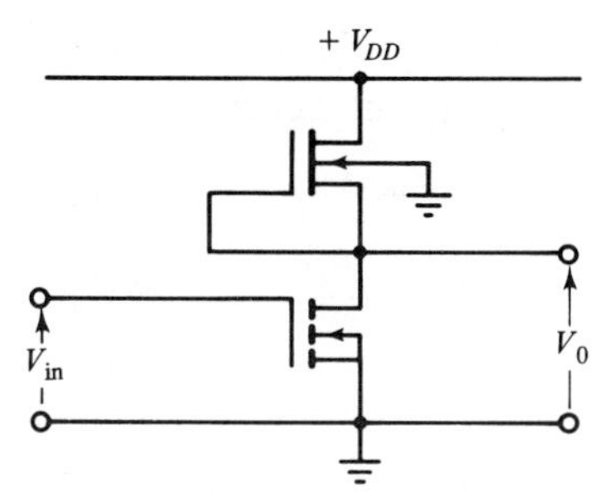

Fig. 11-20 NMOS inverter.

at this point is shown in Fig. 11-21(b). Next, a thick, phosphorus-doped oxide layer is thermally grown. As shown in Fig. 11-21(c), the oxide growth is restricted to regions in which the silicon nitride has been removed. Since oxide growth consumes silicon, the SiO_2 layer grows in both vertical directions. If the oxide layer were 10,000 Å thick, approximately 4400 Å would be below the original surface, with 5600 Å above the original surface. Ultimately, aluminum metalization runners must pass over steps in thick oxide regions. The glass layer is reflowed to control the slopes of these steps. This minimizes yield problems that could occur because of aluminum thinning or cracking at oxide step edges. Since the silicon nitride layer was only used for selective oxidation, it is now removed with hot phosphoric acid. The thin oxide layer shown in Fig. 11-21(c) is also removed and the final gate oxide layer is grown. Boron ions are then implanted through the gate oxide, as shown in Fig. 11-21(d). The purpose of this ion implantation is to adjust the threshold voltage (V_T) of the enhancement driver. With a masking step, photoresist is now used to protect the enhancement-driver regions of the wafer [as shown in Fig. 11-21(e)] while phosphorus ions are implanted to adjust threshold voltages of the depletion load devices.

At this point, a thin layer of polycrystalline silicon is deposited on the wafer. A mask-and-etch step is then used, and polysilicon is left only where silicon gates are needed, as shown in Fig. 11-21(f) and (g). The thin oxide layers adjacent to the polysilicon gates are removed. The gates protect the underlying oxide layers. It was necessary to implant channel ions prior to the polysilicon deposition; otherwise, the ions would not penetrate the polysilicon gates.

Next, a thick layer of phosphorous-doped glass is deposited over the surface of the wafer. When the wafer is brought to diffusion temperatures, the phosphorus acts as a diffusion source, doping the polysilicon gate n^+ and creating n^+ source and drain regions, as shown in Fig. 11-21(h). The channel regions are self-aligned under the gates. There are other ways to perform the n^+ deposition-diffusion step. Some manufacturers use arsenic rather than phosphorus, depositing phosphorus-doped glass after the diffusion. In any case, the channels are self-aligned under the gates, and the diffusion temperature acts to anneal the crystal damage caused by ion implantation. The phosphorous doping allows the glass to reflow, smoothing out any sharp edges to prevent thinning and

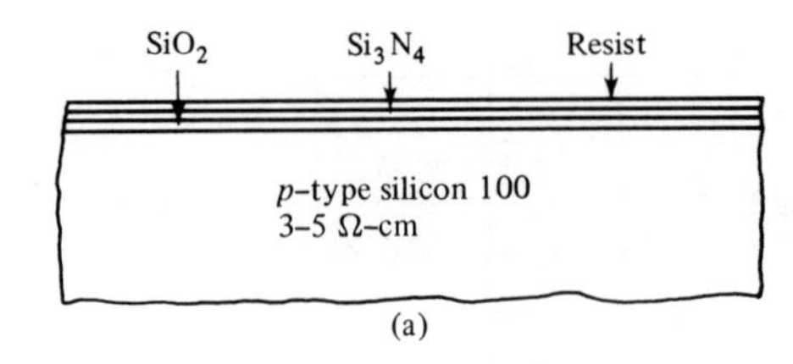
SiO2
Si3N4
Resist
p–type silicon 100
3–5 Ω–cm
(a)

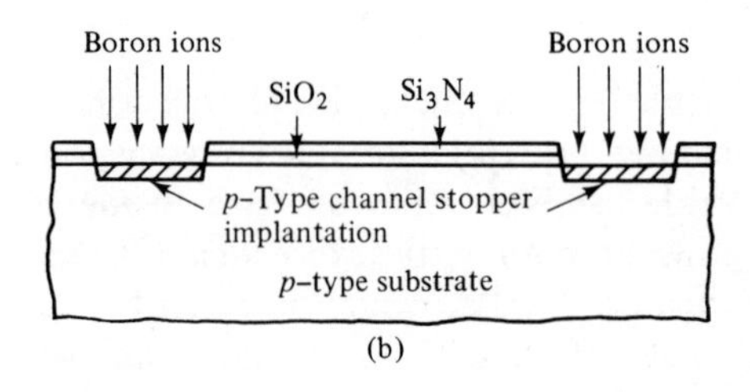
Boron ions
SiO2
Si3N4
Boron ions
p–Type channel stopper
implantation
p–type substrate
(b)

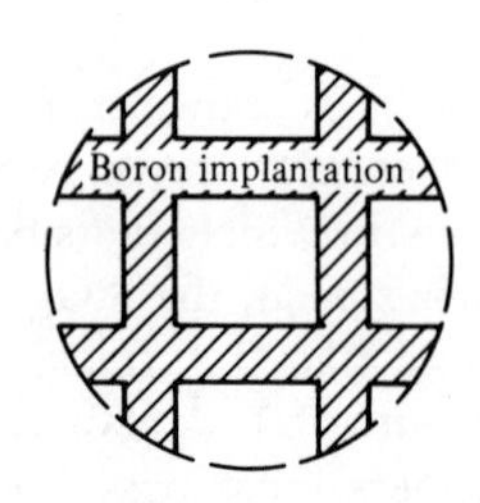
Boron implantation

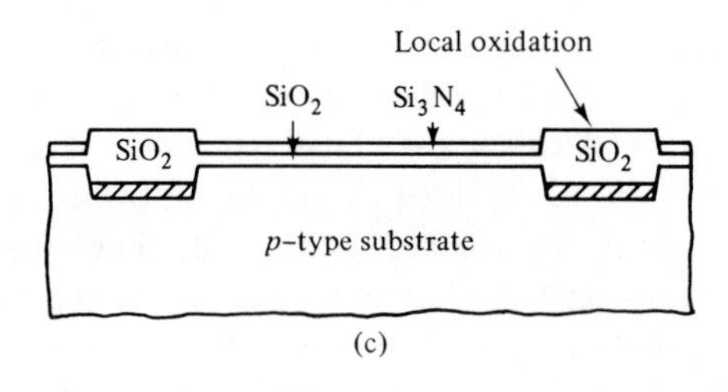
Local oxidation
SiO2
Si3N4
SiO2
SiO2
p–type substrate
(c)

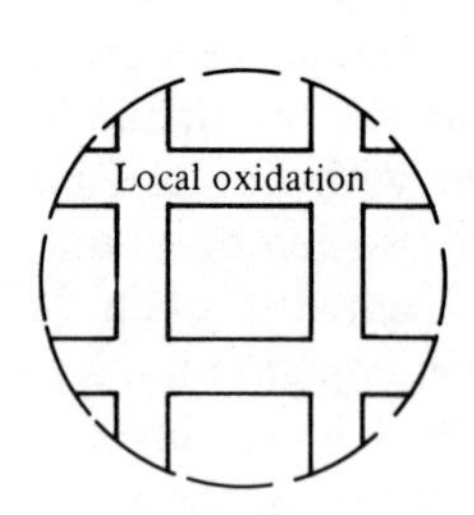
Local oxidation

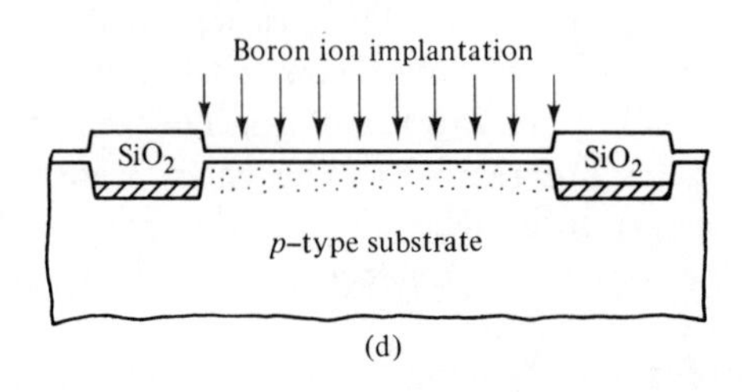
Boron ion implantation
SiO2
SiO2
p–type substrate
(d)

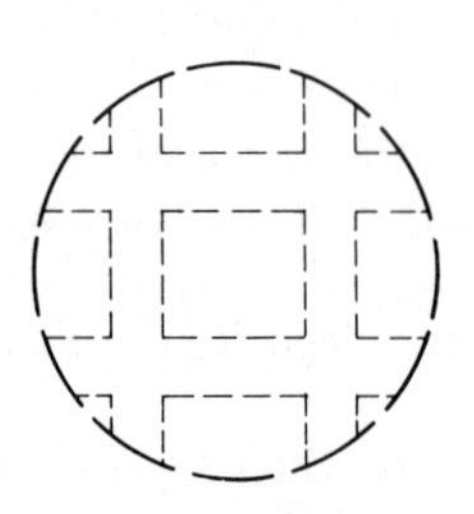

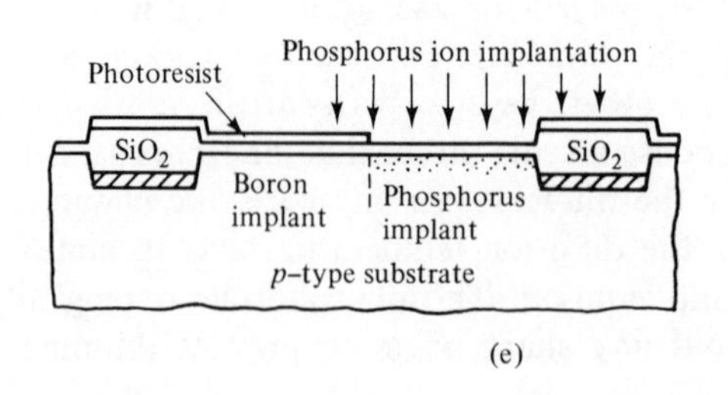
Photoresist
Phosphorus ion implantation
SiO2
SiO2
Boron
implant
Phosphorus
implant
p–type substrate
(e)

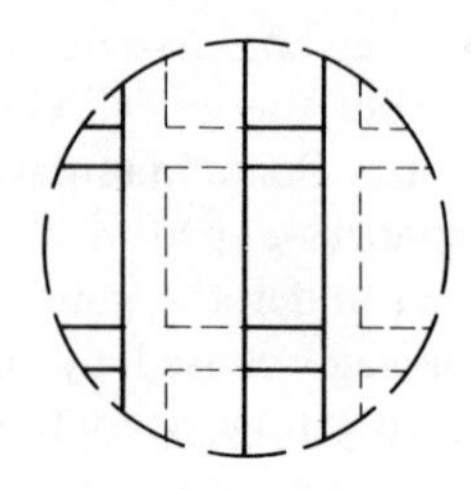

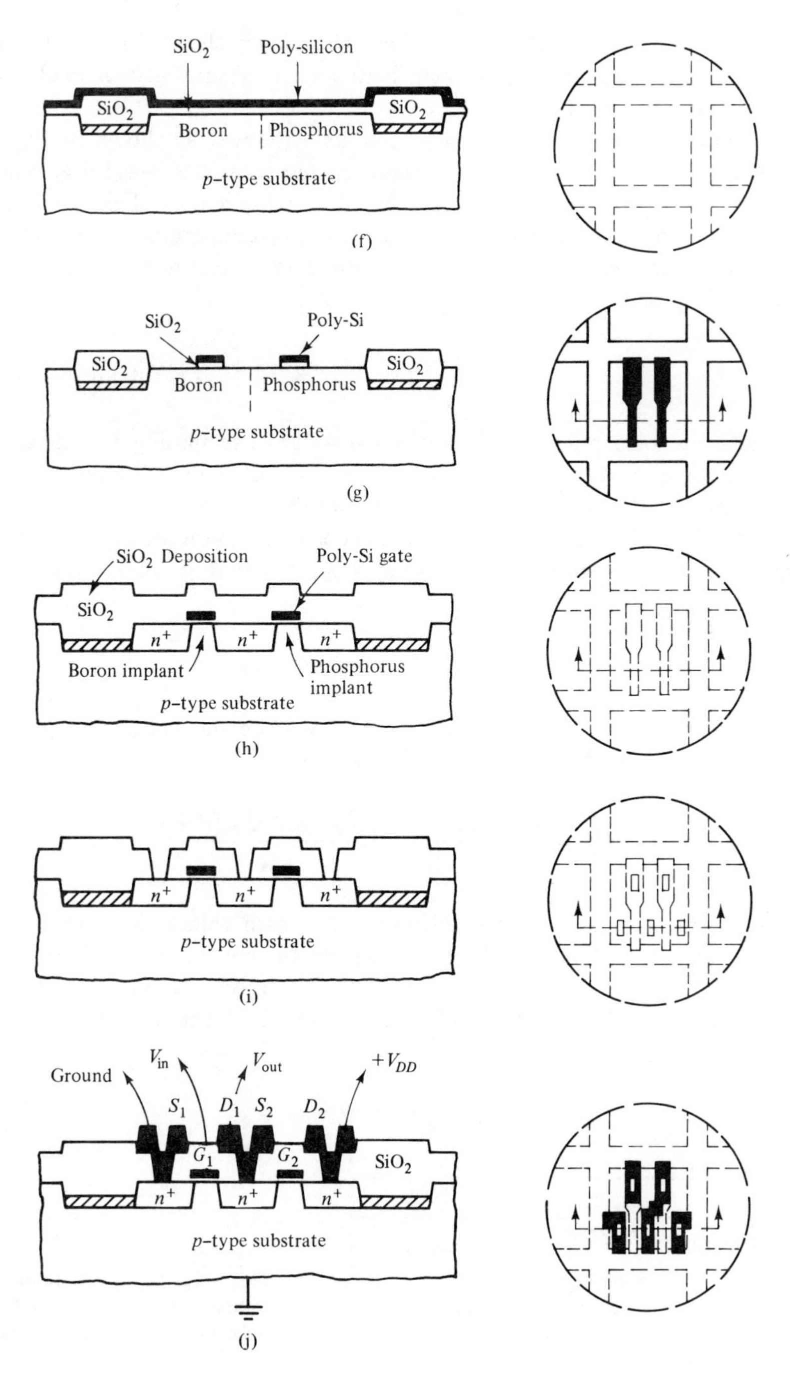

Fig. 11-21 Processing sequence for an n-channel enhancement driver/depletion load inverter.

cracking of aluminum runners. The thick glass layer shown in Fig. 11-21(h) must be deposited rather than grown, because a thermally grown oxide layer would consume the polysilicon gates.

A masking step is used to open contact windows, as shown in Fig. 11-21(i). Aluminum is deposited, and a masking step is used to selectively remove aluminum, as shown in Fig. 11-21(j). Finally, a microalloy diffusion is used to ensure good metal–silicon contacts, a layer of low-temperature pyrolytic glass is deposited on the wafer for scratch protection, and contact windows are opened in the bonding pad areas.

11-3-4 Substrate Bias

Substrate bias refers to the technique of reverse biasing the substrate–channel diode of a MOSFET, as shown in Fig. 11–22. This technique is widely used to improve the performance of MOS circuits.

Let us begin by discussing the effect of substrate bias on the device threshold voltage V_T. If we assume for now that $V_{\text{sub}} = 0$ in Fig. 11-22, then V_T is given by

$$V_T = V_{\text{Si}} + V_{\text{ox}} - V_{SS} + V_{WF} \tag{11-9}$$

The only component of V_T affected by V_{sub} will be the voltage drop across the gate–oxide layer. With $V_{\text{sub}} = 0$, this is given by

$$V_{\text{ox}} = \frac{Q_B}{C_{\text{ox}}} = \frac{qN_A W}{C_{\text{ox}}} = \frac{(2\epsilon_0\epsilon_{\text{Si}} qN_A|2V_f|)^{1/2}}{C_{\text{ox}}} \tag{11-11}$$

Starting with $V_G = 0$, as an increasing positive gate voltage is applied to the device in Fig. 11-22, holes will be repelled from the channel region, uncovering trapped (−) boron ions. The strong inversion model assumes that when $V_G = V_T$, the channel–substrate depletion layer width reaches a maximum value,

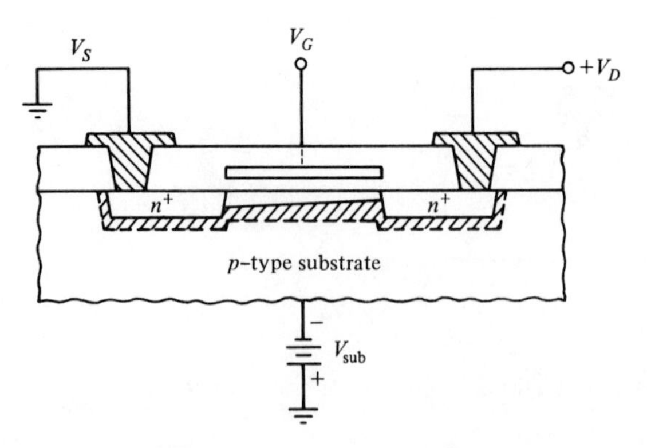

Fig. 11-22 *n*-Channel enhancement-mode MOSFET. Substrate bias reverse-biases the substrate with respect to the source, *n*-type channel, and drain.

and the density of trapped ions is given by Q_B in Eq. (11-11). For V_G greater than V_T, the depletion layer width remains constant, and mobile (−) electrons are pulled into the channel from the n^+ source.

Now, how does the substrate bias voltage change these conditions? If we reverse-bias the substrate with respect to the n-type source, channel, and drain regions, the width of the channel–substrate depletion layer will increase. The density of trapped ions in this depletion layer will also increase. The reverse substrate bias voltage shown in Fig. 11-22 will add to the channel–substrate junction potential. The density of bound charge Q_B in the depletion layer will then be given by

$$Q_B = [2\epsilon_0\epsilon_{\text{Si}}qN_A(|2V_f| + V_{\text{sub}})]^{1/2} \tag{11-18}$$

The voltage drop V_{ox} across the gate–channel oxide layer will increase correspondingly. The threshold voltage of our n-channel MOSFET will become more positive as the substrate bias voltage is increased.

For a perfectly self-aligned MOSFET, in which the effects of gate–drain, gate–source, and drain–substrate capacitances are negligible, the maximum frequency of device operation is limited by the transit time of charge carriers crossing the channel. This maximum frequency is given by Eq. (9-36), which is repeated here for convenience:

$$f_{\max} = \frac{g_{m\ \text{sat}}}{2\pi C_{GS}} \approx \frac{\mu_{ns}(V_G - V_T)}{2\pi L^2} \tag{9-36}$$

where

$$C_{GS} \approx C_{\text{ox}}ZL$$

and

$$g_{m\ \text{sat}} = \frac{Z}{L}\mu_{ns}C_{\text{ox}}(V_G - V_T)$$

To maximize the frequency response and transconductance of the device, we need to maximize channel mobility and minimize gate–channel capacitance and channel length L. Both $f_{\max}$ and g_m are directly proportional to the mobility of the carriers in the active channel region. Mobility is increased by decreasing impurity concentration; but if channel impurity concentrations below 10^{16} impurity atoms/cm^3 are used, the corresponding NMOS device will operate in the depletion mode (with $-V_T$), because the positive surface state voltage (V_{SS}) and the gate–substrate work-function voltage (V_{WF}) will tend to create an n-type channel. This is where substrate bias becomes useful. Substrate (and channel) doping levels on the order of 10^{14} to 10^{15} impurity atoms/cm^3 can be used, and then substrate bias can be applied to adjust the threshold voltage V_T to an acceptable positive value. Substrate bias will also reduce the capacitances between the *n*-diffused regions and the substrate.

Packaged ICs with substrate bias ordinarily require an additional pin and an external power supply. Some manufacturers are generating the substrate bias voltage on the chip to eliminate these requirements. One method of achieving on-chip substrate bias uses a self-starting oscillator that is capacitively connected to the substrate. The oscillator frequency and substrate voltage are designed to compensate for variations in threshold voltages caused by processing.

11-3-5 Double-Diffused MOS Transistors (DMOS)

MOSFET channel lengths on the order of 1 μm or less can be fabricated by performing a *p*-type diffusion through an oxide window and then using the same oxide window for the n^+ source diffusion. The effective channel length is determined by the distance between the two lateral diffusions. The channel length is thus determined by the *p* and n^+ diffusion profiles, and is automatically self-aligned to the source region. Precise channels can be obtained which are independent of masks, etching, photolithography, or mask alignment tolerances. The threshold voltage of the resulting device is very sensitive to the final surface concentration of the *p*-type impurity, however, so that this must be carefully controlled. Double-diffused MOS devices are commonly referred to as DMOS devices. The double-diffused technique is used primarily to increase the frequency response or the breakdown voltage of *n*-channel enhancement-mode MOSFETs.

One process technique that is used for relatively low voltage devices is shown in Fig. 11-23. The starting material is a *p* substrate ($N_A \approx 1 \times 10^{14}$ atoms/cm^3). A window is opened in the oxide layer and boron is implanted or deposited and then diffused, as shown in Fig. 11-23(a) ($N_A \approx 1 \times 10^{16}$ atoms/cm^3). Virtually no oxide is grown during the diffusion. An n^+ source diffusion is then performed through the same oxide window and an oxide layer is grown [Fig. 11-23(b)]. Conventional processing is used to define the n^+ drain region [Fig. 11-23(c)] and to complete the device. Figure 11-23(d) shows the final device. It is assumed here that conventional MOS devices would be fabricated on the same chips as the DMOS devices. The channel diffusion and source diffusion shown in Fig. 11-23(a) and (b) would only be performed for the DMOS devices.

Because of the surface charge $(+Q_{SS})$ and the low level of substrate doping, the area under the gate electrode next to the *p*-diffused channel will be heavily inverted. Thus, the actual channel length is just the difference in lateral diffusion between the diffused *p*-region and n^+ source. The reduced channel length leads to a significant increase in switching speed. Since the silicon under the field oxide is also heavily inverted, however, some means, such as channel stoppers

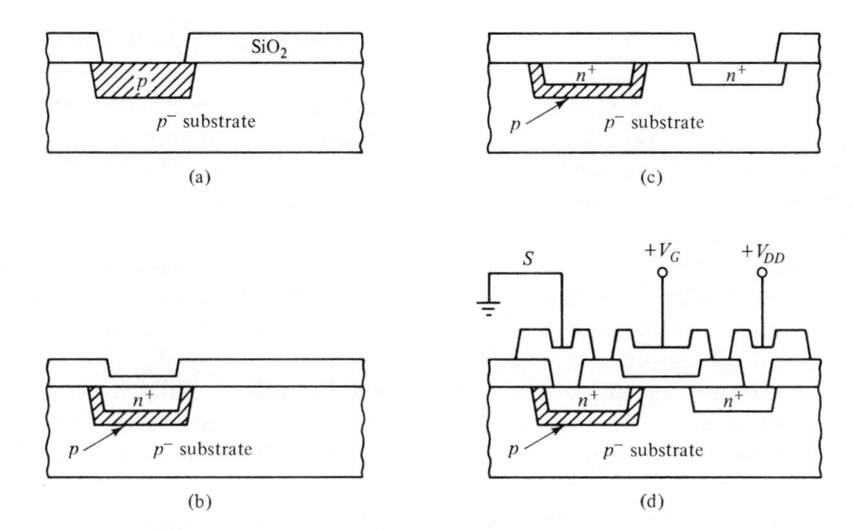

Fig. 11-23 Fabrication technique for DMOS transistors: (a) p-type channel diffusion; (b) n^+ source diffusion through the same window; (c) n^+ drain deposition and diffusion; (d) completed device.

or guard bands, must be used to isolate devices. This is not shown in Fig. 11-23.

DMOS devices are not symmetrical (i.e., identical I–V characteristics are not obtained when the source and drain connections are interchanged). A commonly used symbol for DMOS devices includes a darkened semicircle at the source terminal as shown in Fig. 11-24.

Because of the narrow channel, very high values of transconductance and channel conductance can be obtained. Low values of drain–gate capacitance can be achieved with a p-substrate because low-level substrate doping can be used.

n-Channel depletion loads are commonly used with DMOS drivers for the fabrication of high-speed digital inverters, logic gates, and so on. Switching times on the order of 1 ns can be achieved in this fashion with a +5 V power supply.

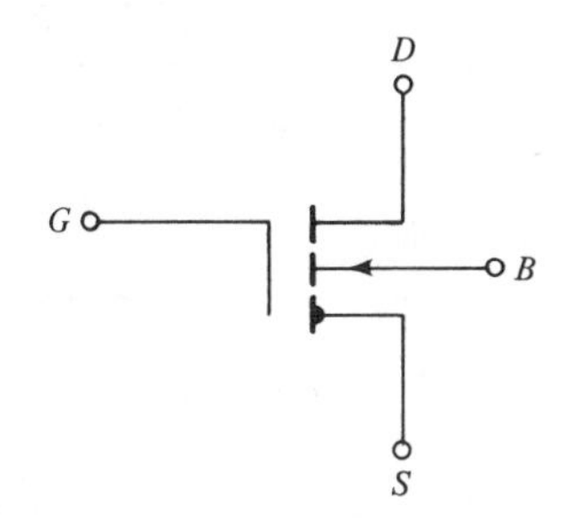

Fig. 11-24 Symbol for a DMOS transistor.

11-3-6 Complementary MOS (CMOS) Inverter

The CMOS inverter, as described in Section 9-1-4, is a basic building block for most CMOS circuits. Self-aligned silicon gates, ion implantation, and local oxidation are all used to manufacture CMOS circuits. Through the use of these process technologies, packing densities and yields are improved, and circuit-time delays and power dissipation are reduced.

In the interest of brevity, we will restrict our CMOS discussion to the metal gate CMOS inverter shown in Fig. 11-25. Referring to Fig. 11-25(b), the impurity concentration of the *n*-type substrate is on the order of 1×10^{15} cm^{-3}, and N_A for the *p*-type well is on the order of 3×10^{16} cm^{-3}. With these concentrations, the *p*- and *n*-type MOSFETs will normally work in the enhancement mode. As long as supply voltages on the order of 5 V or less are used, the surface regions under the field oxide will act to isolate adjacent devices. In other words, the *p*-type well is doped heavily enough that regions under the field oxide will remain *p*-type. The *n*-type substrate regions surrounding

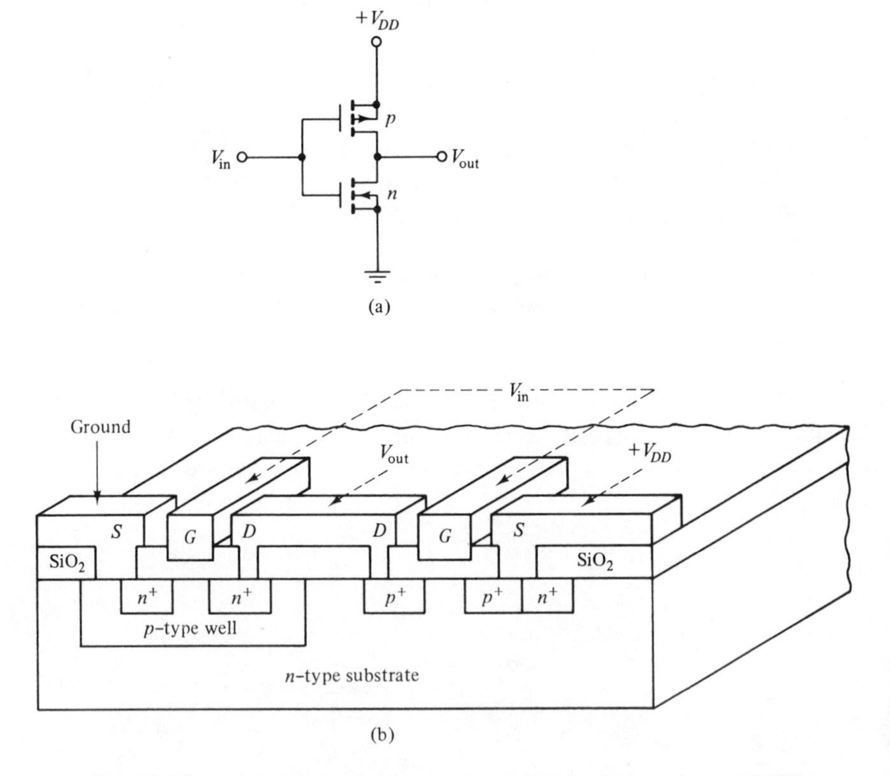

Fig. 11-25 Schematic and cross-sectional sketch of a metal gate CMOS inverter.

the PMOS device will tend to become n^+ due to surface states, the metal–semiconductor work function, and positive voltages sitting on top of the field oxide.

When supply voltages greater than ±5 V are to be used in CMOS circuits, *guard bands* or *channel stoppers* are used to prevent inversion layers under the field oxide. This prevents low-resistance paths from occurring between adjacent devices. Figure 11-26 shows a cross-sectional sketch of a CMOS inverter with n^+ guard bands surrounding p-channel devices and p^+ guard bands around n-channel devices. The n^+ guard band in Fig. 11-26(b) connects the n-type substrate to $+V_{DD}$ and the p^+ guard band connects the p-type well to ground. The device sources can be brought into contact with their adjacent guard bands because they are at the same potential. The drain regions must be isolated from the guard bands by lightly doped regions, however, to prevent avalanche

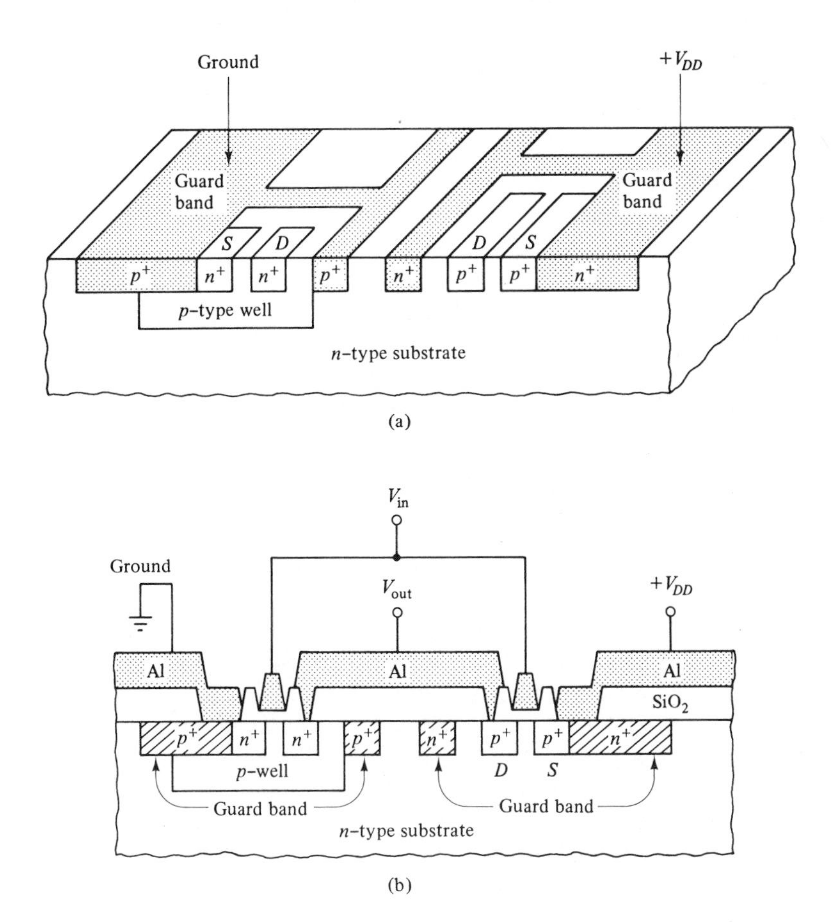

Fig. 11-26 (a) Source, drain, and guard-band structures for a CMOS inverter. (b) Interconnections for a CMOS inverter.

or zener breakdown. With the substrate tied to the positive power supply terminal, and the *p*-type well tied to ground, the well–substrate diode is reverse-biased.

The PMOS device in Fig. 11-26(a) is shown with approximately twice the area of the NMOS device. This is done in critical high-speed applications to compensate for the difference between hole and electron mobility. In noncritical applications, PMOS and NMOS device areas are often kept the same. Since mismatching is bound to occur between devices in complex logic circuits anyway, careful matching of device areas is often regarded as overdesign.

11-3-7 Silicon-on-Sapphire CMOS Circuits (SOS-CMOS)

One of the most fundamental characteristics of monolithic integrated circuits has also been a major disadvantage in high-speed applications. When devices are diffused and/or implanted into bulk single-crystal silicon, how do you isolate them? What do you do with the silicon between devices? The dominant answer to this question has been to use reverse-biased *pn* junctions for isolation. The development of the Isoplanar method led to a combination of reverse-biased junctions and SiO_2 layers to isolate devices. The remaining problems of junction capacitance, junction leakage currents, parasitic *pnpn* devices, and field inversion motivated a sustained search for isolation alternatives. The most promising alternative to bulk silicon devices was to deposit single-crystal epitaxial silicon on an insulating substrate, and then selectively remove the silicon where it was not needed. With separate silicon devices seated on a nearly perfect insulating substrate, parasitic capacitances and leakage currents could be greatly reduced, and field-inversion problems and parasitic *pnpn* devices could be eliminated.

It was found that single-crystal silicon films on the order of 1 μm thick (with 100 orientation) could be epitaxially grown on single-crystal sapphire substrates (Al_2O_3). Masking and etching techniques were used to define silicon islands. Conventional masking and doping techniques were then used to form *p*- and *n*-channel MOSFETs, as shown in Fig. 11-27. SOS-MOS construction begins with the growth of a *p*- or *n*-type epi-layer on a clean, polished sapphire substrate. Masking and etching techniques are used to define silicon islands. A preferential etch is used to taper the islands. This allows aluminum or polysilicon runners to enter and leave the islands with a minimum step height and therefore with higher yields. Using conventional masking techniques, impurities are deposited or ion-implanted and then diffused completely through the silicon epi-layers. Depending on the application, many different processing techniques are used to make SOS-MOS circuits. These include ion implantation, self-aligned silicon gates, and local oxidation.

Since sapphire is an extremely good insulator, leakage currents and capacitances between adjacent devices, and between device and substrate, are virtually eliminated. Metalization patterns can sit directly on the substrate. Metalization-

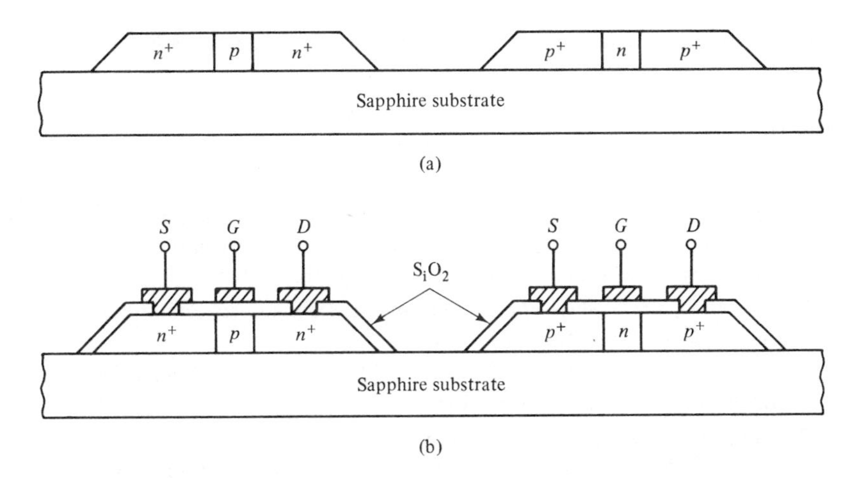

Fig. 11-27 *p* and *n*-Channel MOSFETs on a sapphire substrate.

substrate capacitances are eliminated. Thick oxides are not usually required and no field-inversion problems exist. Since no electrical connection to the substrate is possible, the device channel between source and drain floats, as shown in Fig. 11-27.

Because of surface states at the upper and lower surfaces of the thin silicon islands and the increased dislocation densities within the lattice structure, minority carrier lifetimes in SOS structures are very short (on the order of a few nanoseconds). This makes it impractical to construct *bipolar* SOS circuits. Minority lifetime is not a significant factor in MOS device operation, however, because MOS devices depend on the drift of majority carriers.

The major digital circuit technology in use with SOS structures is CMOS. CMOS circuits require more devices per logic gate than do the other MOS technologies. This leads to a corresponding increase in chip size. The use of SOS structures minimizes lateral diffusion and eliminates field-inversion problems, however, so that SOS-CMOS structures can be made roughly 30% smaller than their bulk CMOS counterparts. The reductions in size and device capacitance increase switching speed and reduce power dissipation.

As shown in Fig. 11-27, the SOS-CMOS technology provides a minimum *p-n* junction area. The influence of radiation on semiconductor devices depends on the exposed junction area. Reduced junction area minimizes the magnitudes of radiation-generated junction currents, making SOS-CMOS a very useful technology in high-radiation environments.

The major disadvantages of the SOS-CMOS technology are the cost of sapphire substrates and the stringent requirements on sapphire surface preparation. Whereas the long-range crystal order of sapphire permits the growth of epitaxial silicon, silicon and sapphire do not have the same crystal structure. Only an imperfect lattice match can be obtained at the interface. This factor increases silicon crystal dislocations and decreases carrier mobilities. Careful preparation of the sapphire surface is needed to optimize the silicon–sapphire interface.

BIBLIOGRAPHY

[1] DHAKA, V. A., J. E. MUSCHINSKE, AND W. K. OWENS, *Subnanosecond Emitter-Coupled Logic Gate Circuit Using Isoplanar II. IEEE Journal of Solid State Circuits,* Vol. SC-8, No. 5, October 1973, pp. 368–372.

[2] RICHMAN, P., *MOS Field-Effect Transistors and Integrated Circuits.* New York: John Wiley & Sons, Inc., 1973.

[3] CARR, W. N., and J. P. MIZE, *MOS/LSI Design and Application.* New York: McGraw-Hill Book Company, 1972

[4] GLASER, A. B., AND G. E. SUBAK-SHARPE, *Integrated Circuit Engineering—Design, Fabrication, and Applications.* Reading, Mass.: Addison-Wesley Publishing Company, Inc., 1977.

[5] HAMILTON, D. J., AND W. G. HOWARD, *Basic Integrated Circuit Engineering.* New York: McGraw-Hill Book Company, 1975.

[6] *Status 1978: A Report on the Integrated Circuit Industry.* Integrated Circuit Engineering Corporation, Scottsdale, Ariz.

[7] *Status 1979: A Report on the Integrated Circuit Industry.* Integrated Circuit Engineering Corporation, Scottsdale, Ariz.

[8] FAGGIN, F., AND T. KLEIN, *Silicon Gate Technology. Solid State Electronics,* Vol. 13, 1970, pp. 1125–1144.

PROBLEMS

11-1 A slice clean is used to remove contaminants from silicon wafer surfaces. Name three types of contaminants found on water surfaces, and provide an example of each type.

11-2 What is the purpose of placing n^+ buried layers under silicon devices? What types of impurities are commonly used for buried layers?

11-3 Consider using an n-type diffusion into silicon as an alternative to growing an epitaxial layer. Why is an epi-layer used?

11-4 Aluminum, boron, gallium, and indium are all candidates for p-type doping. Why is only boron used?

11-5 What are dummy wafers, and how are they used during wafer processing?

11-6 Why is a Gaussian diffusion used for transistor bases rather than an error function diffusion?

11-7 Describe the technique used to provide an ohmic contact where p-type aluminum touches the n-type collector region of an npn transistor.

11-8 Why is a pyrolytic oxide layer deposited on completed bipolar wafers?

11-9 Count and name the masks that are typically used to fabricate and interconnect bipolar transistors.

11-10 Why are deep n^+ collector diffusions often used for digital *npn* transistors?

11-11 Explain how Schottky barrier diodes increase the switching speeds of bipolar *npn* transistors.

11-12 List the advantages and disadvantages of using Isoplanar transistors versus conventional planar transistors in digital bipolar circuits.

11-13 Compare the PMOS, NMOS, and CMOS digital technologies in terms of application areas.

11-14 Why is (100) rather than (111) silicon crystal orientation normally used for MOS devices?

11-15 Explain the difference between the gate-threshold voltage V_T and the field threshold voltage V_{TF} of a MOS device.

11-16 Compare silicon gate NMOS devices to metal gate NMOS devices. Indicate the applications areas in which you would use each of these devices.

11-17 Silicon nitride is often used during NMOS processing and then removed. Why use it at all?

11-18 What are the advantages and disadvantages of using substrate bias in NMOS circuits?

11-19 What are the advantages and disadvantages of using DMOS devices in MOS circuits?

11-20 Why are guard bands needed in CMOS circuits when supply voltages are greater than 5 V?

11-21 Compare CMOS circuits and SOS-CMOS circuits, indicating the advantages and disadvantages of each.

hybrid-circuit fabrication

12

The intent of this chapter is to elaborate to some extent on the processes and materials involved in the fabrication of hybrid thick-film and thin-film circuits. It must be kept in mind that although the general approach to the fabrication of hybrid circuits is very similar from one facility to another, there will be substantial differences in equipment, design rules, and opinions as to what works best. Much of the information presented here must be considered as illustrative and, when specific rules or design criteria are stated, they are meant only as examples and must not be considered to be absolute.

Thick-film circuits consist of printed patterns of conductor, resistor, and sometimes capacitors on a ceramic substrate. The thickness of these patterns is 12 to 25 μm. Thin-film circuits consist of patterns of conductors, resistors, capacitors, and microwave delay lines deposited on ceramic or glass substrates usually by vacuum techniques. The thickness is generally around 0.01 to 0.10 μm. Hybrid circuits are either thick-film or thin-film circuits with active devices, transistors, ICs, and possibly some capacitors added.

Thick-film circuits are particularly useful when a large number of resistors are required and/or where a large number of discrete components or ICs must be interconnected in a fairly small package. Thin-film passive networks are especially well suited for microwave applications. Delay lines and distributed *RC* networks can be fabricated directly on the substrate because of the extremely fine line definition obtainable in thin-film networks.

12-1 HYBRID THICK-FILM CIRCUITS

Hybrid thick-film circuits are most beneficially used where a large number of resistors are needed, possibly requiring tight tolerances or precise ratios (see Fig. 12-1). Active devices, transistors, and ICs must be added. Such circuits have several advantages over similar circuits utilizing discrete components. They are smaller by at least a factor of 10, are generally more reliable, and in high-volume production, they are usually cheaper to produce than discrete circuits designed for the same function. Cost savings are primarily due to the advantage of functional trimming of resistors, which allows greater tolerances in remaining components. In this context, a discrete circuit is one in which individual resistors, capacitors, transistors, diodes, inductors, and possibly thermistors have been used. The comparison is not at all simple or clear cut in the case of dual-in-line (DIP) integrated circuits mounted on a printed circuit board with some discrete components added to complete the functional circuit. Decisions on how best to fabricate a given circuit must be based on knowledge of the capability of existing facilities and personnel, economics, volume of production, and the nature and eventual use of the circuit.

Fig. 12-1 Thick-film resistors with precise ratios.

12-2 THICK-FILM DESIGN AND PROCESS STEPS

Figure 12-2 is a flow diagram depicting the steps involved from circuit design to testing of the completed circuit for the thick-film fabrication process. Considerable time and effort may be expended before the actual circuit design stage is reached in determining whether the circuit can or should be made as a hybrid thick-film system.

Once the economic factors, as well as the availability of expertise and facilities have been weighed and the decision is made to go ahead with a thick-film design, the work can begin. The circuit designer can breadboard a thick-film circuit using discrete components but must be cognizant of design limitations imposed by the technology. The designer must consider total power dissipation

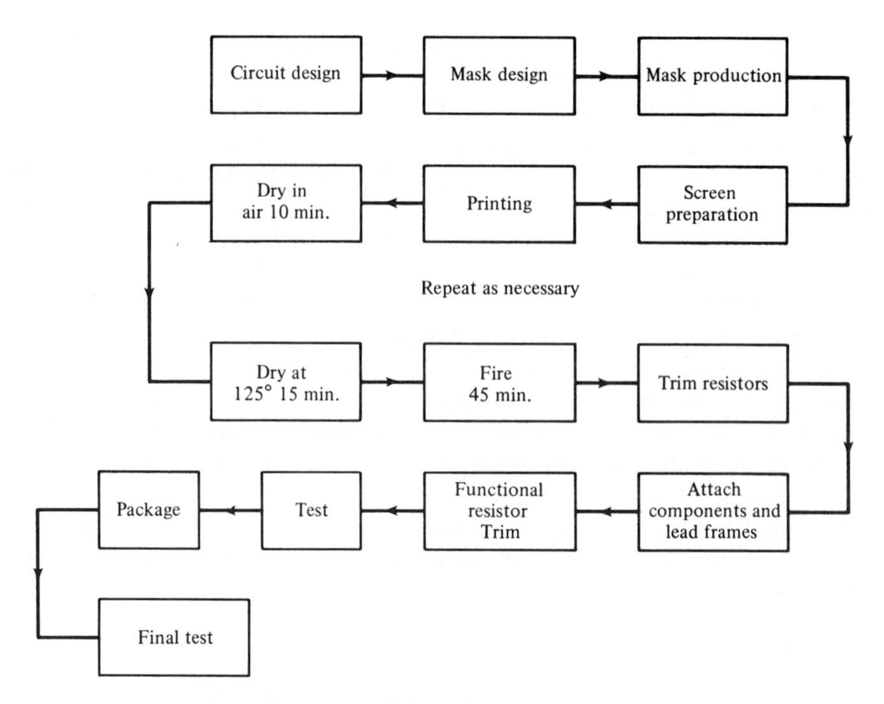

Fig. 12-2 Flow diagram for design and fabrication of hybrid thick-film circuit. Drying steps may not be required and times and temperatures will vary.

as well as power density for individual components. The designer must also consider such things as component packing density, the range of component values, dimensional limitations, and environmental conditions. Hence, the designer must be knowledgeable about the materials, components, and processes involved in thick-film fabrication. Resistors can be trimmed, but the fewer the better, so a good design employs resistor ratios where possible and/or functional trimming. Functional trimming refers to the process of adjusting one or more resistors of the completed circuit to bring the circuit functionally within tolerance.

Once the circuit has been designed and breadboard tested, it must be translated into a set of masks, one for each printing that is required. The masks are usually cut out of rubylith 10 times actual size, or greater. When great precision is required, the cutting is done on a coordinatograph (flat-bed plotter). Some coordinatographs are computer-controlled for this process.[1] The rubylith master is photographically reduced to produce a transparency that serves as a mask to expose an emulsion on a stainless steel wire mesh screen. (Other types of screens are discussed later in this chapter.) The screen is masked, exposed,

[1] Computer-aided mask generation is discussed at greater length in Chapter 10.

developed, and rinsed to produce openings in the emulsion. The openings are the pattern that is to be printed. Conductor patterns are printed and fired and then resistors, dielectrics, and thermistors are printed, dried, and fired simultaneously. Figure 12-2 gives typical drying and firing times. Conductors are fired at about 850°C and the remaining materials are fired at around 750°C. Resistors are air-abrasive or laser-trimmed, and the components and lead frame are then attached. Functional trimming is carried out (if required). Testing, packaging, and final functional test complete the process.

Various mechanical and thermal tests may be performed on part or all of the assembly at various points in the processing on a sample lot or a 100% basis. For example, tests are performed to determine the strength of wire and die bonds and the peel strength of resistors and conductors. Vibration, thermal cycling, and humidity tests may be performed to determine the mechanical endurance and the electrical stability of the assembly.

12-3 MATERIALS

The basic materials involved in the thick-film process are the substrates on which the circuits are printed, the inks, the screens used for the printing, and solder (solder ink and conventional solder). Other items, such as the printer squeegee material and design, are also important, but will be considered in the section on printing.

12-3-1 Substrates

The most commonly used substrate material is 96% alumina (Al_2O_3). Substrates of this type contain 4% MgO and SiO_2. Beryllia (BeO) substrates (99.5% BeO) are used to a lesser extent whenever exceptional thermal properties are desired. BeO dust is extremely toxic, so great care in handling this material is required. The thermal conductivity[2] of 99.5% BeO is 0.55 cal/s-cm-C° at 25°C compared with 0.084 cal/s-cm-C° for 96% Al_2O_3 at 25°C. However, BeO substrates have lower flexural strength and are considerably more expensive than 96% Al_2O_3 substrates. Both have good electrical characteristics (i.e., high resistivity, low dielectric constant, and high breakdown strength). BeO substrates may be used where high power dissipation is required.

For all substrates, certain surface and dimensional tolerances must be maintained to ensure good printing quality. Specifications will vary, but a typical set might include the following:

1. Surface finish variation (peak-to-valley height variation) not to exceed 0.6 μm rms.

[2] The thermal conductivity of a material is the rate of heat flow per unit area per unit thermal gradient.

2. Camber (variation in flatness per centimeter of edge length) not to exceed 0.004 cm/cm.
3. Random voids (pits in surface) in surface not to exceed 25 μm in diameter.
4. Length and width dimensions $\pm 1\%$.
5. Thickness dimension $\pm 10\%$.

Substrates may be cleaned before printing if not clean as received from the manufacturer. This may be done by washing in trichloroacetic acid, followed by an acetone rinse. Ultrasonic cleaning in deionized water is effective for laboratory work.

Figure 12-3(a) is an SEM micrograph of the edge of a resistor on a 96% Al_2O_3 substrate. The granular, fairly open structure of the surface of the Al_2O_3 should be noted, as this plays an important part in the bonding of the inks to the substrate.

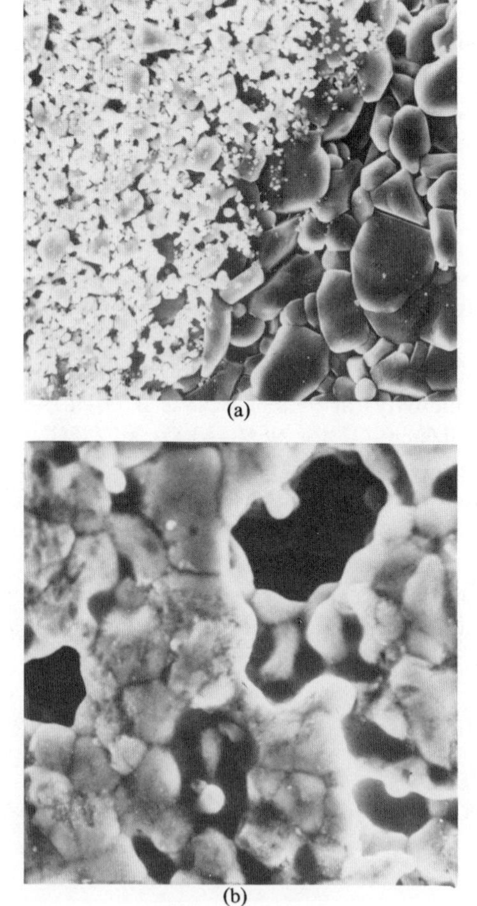

Fig. 12-3 (a) SEM micrograph of thick-film resistor (upper) and alumina substrate (lower). (b) SEM micrograph of thick-film conductor.

Thick-film inks are compositions of glass particles, metal and/or metal oxide particles, and organic solvents and viscosity control agents which are printed on alumina substrates by an adaptation of the silk-screen printing process. Thick-film inks are available for a wide variety of passive functions. The types of ink available are: conductor, resistor, dielectric for capacitors, dielectric for crossovers, glass for encapsulation, glass for solder dams, thermistor, and solder. Figures 12-3(a) and (b) show SEM micrographs of a thick-film resistor and conductor, respectively, after firing. The fired resistor consists mainly of glass with metal–metal oxide particles interspersed. The fired conductor is mostly metal–metal oxide particles sintered together with glass at the bottom to provide a bond with the substrate.

The conductor ink compositions, prior to firing, contain metal–metal oxide particles, a glass frit, an organic solvent, a wetting agent, a viscosity[3] control agent, and some resin. Two widely used compositions are Pt–Ag and Pd–Ag, providing a good compromise of cost, solderability, and printability. There are numerous other compositions available, some of which are listed in Table 12-1 together with a rating of fine-line printing capability, solderability, die-bonding capability, compatibility with resistor compositions, and wire bondability.

TABLE 12-1 Properties of Gold, Platinum–Gold, Paladium–Gold, and Palladium–Silver Conductor Compositions*

	Au	Pt/Au	Pd/Au	Pd/Ag
Line resolution (μm)	50–400	50–200	13–200	50–200
Solderability	Poor	Excellent	Excellent	Fair to excellent
Die bonding	Yes	Yes	Yes	No
Wire bonding	Yes	Yes	Yes	TC† to gold Ultrasonic to aluminum or gold
Resistor Compatibility	Fair to excellent	Excellent	Good to excellent	Poor to excellent

* Variations are due to different compositions and printing techniques.
† TC, thermocompression bonding.

Reference to Table 12-1 makes it clear that the choice of conductor material must be made carefully and that even within one composition type, such as Pd–Ag, considerable choice as to properties is available. This applies to cost as well as physical properties. Other criteria that may be considered in selecting a conductor ink are sheet resistance, which varies from 0.003 to 0.100 $\Omega/\square$

[3] Viscosity gives a measure of how easily a material flows. It is defined as the shear stress per unit shear rate in dyne-s/cm^2, which is called a poise. One one-hundredth of a poise, or centipoise, is a frequently used unit.

for different materials, acid resistance, compatibility with crossover and capacitor dielectrics, and adhesion to the substrate.

The rheology of a thick-film ink is a very important characteristic. It is especially significant for conductor inks, because these require the tightest tolerances on line width and line separation. Thick-film inks are designed to be pseudoplastic. Basically, this means they have a viscosity that is dependent on the rate of shear, and they have a definable yield point. The shear rate versus shear stress is depicted for a pseudoplastic material in Fig. 12-4. The change in shear stress per unit change in shear rate (viscosity or inverse slope of Fig. 12-4) decreases with increasing shear rate. Inks with too low a yield-point flow too soon. The decreasing viscosity with increasing shear rate (shear thinning) is essential for uniform screen metering and substrate wetting. Figure 12-5 is a graph of viscosity vs. time for a typical ink throughout the printing cycle.

Trease and Dietz generated this graph from viscosity measurements made at various shear rates. This shows a variation from 10^5 P during manual stirring (point *a* to *b*) to 2500 P when the ink is being forced through the openings in the screen (point *c* to bottom of curve). Trease and Dietz have defined a "screen viscosity index" (SVI) as the ratio of the high to low viscosity. A ratio of 100 or more is necessary to maintain the high-quality and fine-line printing of the ink.

Some of the printing defects that can occur when an ink composition is improper are:

1. Screen markings along the edges, probably due to too high a viscosity or too high a squeegee pressure.
2. Line widening, too low viscosity or yield point.
3. Screen pattern visible in the line, due to too high a yield point.
4. Bleedout on edges, improper wetting of metal, dielectric and/or glass frit, or particles too large.

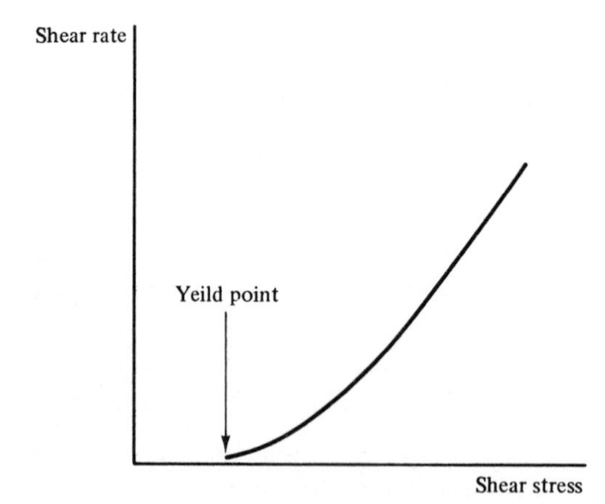

Fig. 12-4 Shear rate vs. shear stress for a pseudoplastic liquid.

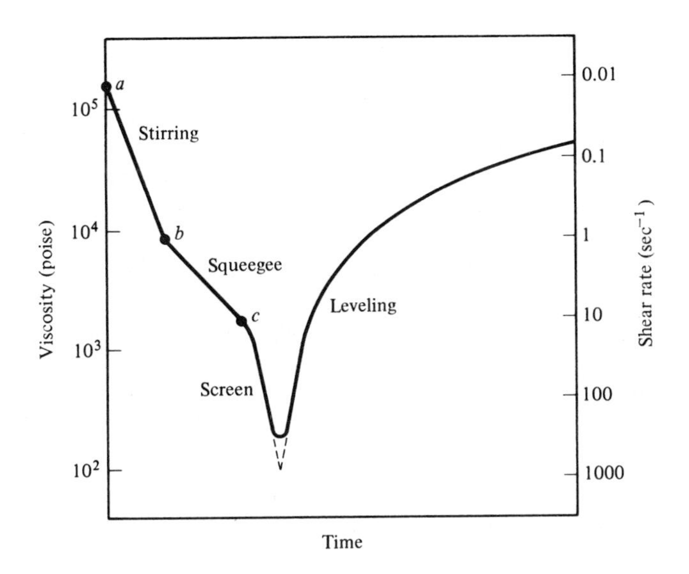

Fig. 12-5 Viscosity changes during screen printing. (Adapted from R. Trease and R. Dietz, "Rheology of Pastes in Thick-Film Printing," *Solid State Technology,* January 1972.)

5. Smearing, too much wetting agent, resulting in overwetting of substrate.

These defects and their causes by no means constitute an exhaustive list, but they do represent most of the common problems associated directly with the inks. Defects due to printing variables will be discussed later in this chapter.

Resistor ink compositions are similar to those of conductor inks but with conductive metal–oxide particles such as ruthenium dioxide, bismuth ruthenate, lead ruthenate, or iridium dioxide in low concentration. Cermet[4] inks employing refractory metals, such as molybdenum and manganese, are available for high-temperature applications.

As mentioned previously, conductor inks have sheet resistances from 3 to 100 mΩ/□. Resistor compositions vary from a few ohms per square to 10 MΩ/□. The range of any one composition is not this great, but 10 Ω/□ to 1 MΩ/□ is not an extreme range for a particular type of composition. The sheet resistance varies rapidly with variation in the metal concentration. A few percent change in Ag content in an Ag system causes a 100-kΩ/□ change in sheet resistance. The effect of varying the Pd concentration in a Pd system is much less. Consequently, a mixed Pd–Ag system has quite desirable characteristics with regard to sheet resistance variation with metal concentration.

Resistor inks of similar composition may be blended together to obtain intermediate sheet resistances. However, they do not blend linearly, so it is

[4] A cermet is a combination metal–ceramic.

desirable to generate blending curves (i.e., sheet resistance versus percentage of one ink with respect to the other). It is not recommended that inks differing by more than one decade of sheet resistance be blended because of the difficulty involved in achieving a uniform dispersion of the metal particles. A blending curve is easily generated for most inks. The sheet resistances of the two inks to be blended are plotted on the log axis of semilog paper and percentage by weight of one of the constituents on the other axis (illustrated in Fig. 12-6). This gives the 0% and 100% points. These points are connected by a straight line. The proper percentage mixture for any intermediate sheet resistance can be read directly from the graph. The actual sheet resistances of the two starting inks under in-house printing conditions must be known. As will be seen later, even in-house conditions are far from constant.

The parameters that should be considered when selecting a resistor material are:

1. Screenability (quality of the print and the degree of difficulty in obtaining good prints)
2. Passivation requirements (i.e., will the resistors need a protective coating?)
3. Firing requirements (e.g., reducing[5] or oxidizing atmosphere, temperature profile)
4. Reproducibility
5. Noise
6. Stability (long-term changes in TCR, VCR, noise, and resistance)
7. Temperature coefficient of resistance (TCR)
8. Voltage coefficient of resistance (VCR)
9. Power density capacity
10. Cost
11. Trimability (ruthenium oxide materials are not easily trimmed by lasers)

TCRs range from 100 to 200 ppm/C° and vary with temperature. Drift, measured under 90% of rated load, is usually less than 1% in 1000 h. Noise may range from −20 to +20 dB. A simple method for determining TCR is to measure the resistance while the circuit is dipped in melted paraffin. Knowing the resistance change and temperature difference from room temperature, the TCR can be calculated.

Dielectric inks are available for high and moderate dielectric constant capacitor applications and low loss/low dielectric constant crossovers. High-dielectric-constant inks contain $BaTiO_3$ and exhibit dielectric constants as high as 2000. When printed, they provide capacitances up to 40,000 pF/cm^2. When high-Q-value capacitors are required, crystallizing glass works quite well.

[5] Used only with nonprecious metal resistors and conductors.

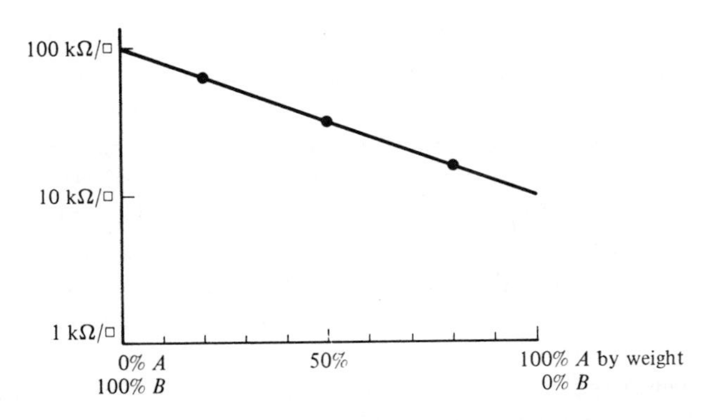

Fig. 12-6 Blending curve for obtaining sheet resistances intermediate to a 100-kΩ/□ ink and a 10-kΩ/□ ink.

When crossovers are necessary (one conductor run must pass over another), an insulating layer with low dielectric constant, high resistance, and high breakdown voltage is printed between the two conductor layers. These inks exhibit insulation resistance in excess of 10^{12} Ω and breakdown voltages over 800 V for a 25-μm thickness. Some of these inks are vitreous (noncrystalline) and others crystallize upon firing.

In addition to the properties above, the firing temperatures and compatibility with resistor and dielectric inks should be considered. Since a premium is paid for fine-line quality, cost is also a factor.

Glass encapsulation inks are available in various colors for ease of inspection. These are used to protect resistors from such hazards as chemical attack, dampness, and potting compounds. These inks are also used for circuit identification purposes and for solder dams (i.e., to prevent the spreading out of solder during reflow from an area where solder bonding is to take place).

Solder inks come in printable and nonprintable forms. They consist of solder particles in an organic vehicle and flux for removing any oxide layer that might exist on the conductor. Various alloys are available, including tin–lead, tin–silver, and tin–lead–silver compositions. A silver-bearing solder is often used to reduce solder leaching of silver from Pd–Ag or Pt–Ag conductors. Flow temperatures cover a wide range and can be selected to minimize heating of the circuit during processing. If more than one solder operation is required, each successive operation can be carried out at a lower temperature than the previous one to prevent undesired reflow of solder already on the circuit. These solder inks are so tacky prior to reflow that components can be mounted to stay in place if reasonable care is taken before the reflow operation is performed. This bonding technique permits some misalignment of the component with the bonding pads during the reflow of the liquid solder.

A brief discussion of solder and its use is presented here since solder processes are widely used in thick-film fabrication. Much of the component and the lead-frame attachment in thick-film circuits is done by conventional soldering processes. The solder is first added to the conductors of the circuit by solder dipping or a solder wave technique. In the latter process, the molten solder flows continuously over a small elevated dam, creating a wave of solder, and the substrate floats across the solder wave and in the process the conductor patterns are "tinned." Since solder only wets metals, the alumina substrate and the resistors are not wetted, but all exposed conductors are wetted. Although the resistors may have metal particles in them, the concentration is insufficient to provide adequate metallic surface area for solder wetting. The conductors, on the other hand, have a very high metallic content and the glass that is present is beneath the surface, where it performs the very important function of bonding the conductor to the substrate. Some of the parameters that should be considered when selecting solder for a given job will be presented briefly here.

There are many types of solders available. Au–Si and Au–Ge solders are used for chip attachment when preforms are used, the preforms generally being gold-plated or clad Kovar.[6] Au–Sn solder is similarly used and is also used for package sealing.

A major problem with any solder is leaching. Silver and gold are especially soluble in Pb–Sn solders. It is actually the solubility of gold and silver in the tin that creates the problem. Consequently, Pb–Sn solder may leach the Ag or Au out of inks based on these metals. Ways of getting around silver leaching is to add about 3% of silver to a Pb–Sn system. This essentially saturates the tin with silver and prevents serious leaching of silver from the conductor (migration of silver from the conductor into the tin). The addition of silver to solder also retards the leaching of gold from gold-bearing conductors. In the case of gold, even if all the gold is not leached out of the conductor, the solder is embrittled by the dissolution of gold and usually results in a very poor bond.

The most commonly used solder in thick-film applications is eutectic Sn–Pb solder, which is approximately 60% Sn–40% Pb. The modern Pd–Ag conductor compositions are very resistant to Ag leaching due to the alloying of silver and palladium. A eutectic Sn–Pb–Ag system is most often used when additional protection against silver leaching is required. Both of these solders melt at approximately 185°C. Reflow is carried out at 20 to 50C above the melting point temperature. When more than one soldering operation is required, the initial solder step is carried out with a moderately high temperature noneutectic solder composition. This temperature may exceed 300C. Subsequent solder steps are carried out at lower temperatures so as not to melt the solder used in previous steps.

Other parameters that may need to be considered when selecting a solder are its bond strength, plasticity, and thermal expansion.

[6] Kovar is an alloy of Ni, Co, and Fe designed to have thermal expansion characteristics similar to glass and ceramics.

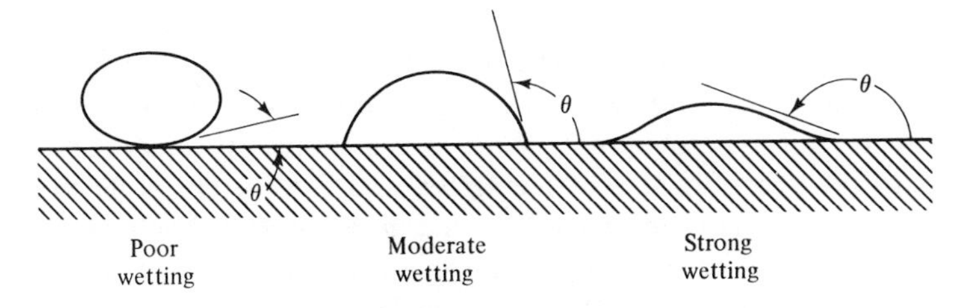

Fig. 12-7 Wetting of metal by solder.

The selection of a proper flux is as critical as the selection of the solder itself. The function of the flux is to remove oxides and tarnish[7] from the metallic surface. The reason the oxides and tarnish must be removed is so that the solder can wet the surface. The expression "wetting" is simply a convenient way of describing the interaction of the surface tension of the liquid solder and the adhesive forces of the surface material.

If the surface tension of the solder exceeds the adhesive forces of the ink and surface, poor wetting occurs. If they are about equal, good or moderate wetting occurs, and if the adhesive forces greatly exceed the surface tension of the solder, strong wetting occurs. Figure 12-7 depicts the various cases of surface wetting. When the wetting angle is nearly 180°, a solder dam is placed around the component, preventing the solder from dispersing into a thin layer and not forming a good bond.

Oxides or tarnish on a surface will diminish the forces of attraction between the solder and metal, and the solder's surface tension tends to pull the solder up into a sphere (a minimum-energy condition) somewhat flattened by gravity. The greater the wetting, the more intimate the contact between metal and solder, and the stronger the bond. Foreign materials other than oxides or tarnish must be removed by other cleaning techniques prior to flux and solder application. Thick-film solder flux can usually be removed with organic solvents, such as acetone or methyl ethyl ketone. Activated fluxes are occasionally used. These fluxes contain additives that undergo chemical change at elevated temperature and chemically attack the surface. The activator component is supposed to be harmless after cooling, but the chemical breakdown is not always complete, so careful cleaning is necessary. The additives can also adversely affect resistor properties.

12-3-4 Screens

Thick-film inks are printed using screens which have the desired pattern defined in them. These inks are forced through the pattern onto the substrate. Solder inks have been treated in Section 12-3-2. The various types of printing

[7] Tarnish refers to chemicals adsorbed on the surface of a metal, generally causing discoloration.

screens are direct emulsion stainless steel, indirect emulsion stainless steel, and direct and indirect metal masks. The most commonly used screens are direct emulsion stainless steel and for that reason most of the following discussion will pertain to this type of screen. With direct emulsion masks the emulsion is applied to the substrate side of the screen and the pattern is defined in the emulsion by photoprocessing. A method that gives improved pattern definition, but shorter screen life, is the indirect emulsion type. In this process, the pattern is defined in an emulsion attached to a Mylar backing and the emulsion is transferred to the screen by taping it on and peeling off the backing.

An indirect metal mask consists of a thin metal sheet with the desired pattern etched into it, bonded to the substrate side of a stainless steel screen. The direct metal mask consists of a metal sheet in which the desired pattern has been etched about halfway through from the substrate side and an array of holes has been etched through from the opposite side. Metal masks provide good line definition and long wear life, but they are expensive and time-consuming to make.

Stainless steel mesh screens are generally used in the off-contact printing mode, which means that 0.5 to 1.0 mm of space initially separates the screen from the substrate. The squeegee must deflect the screen to make contact with the substrate. Metal masks are used in the contact printing mode where 25 to 50 μm initially separate mask and substrate, and breakaway (separation of mask and substrate) is accomplished by movement of the entire screen-holder assembly. In the ensuing discussion, it will be assumed that we are dealing exclusively with stainless steel mesh off-contact screens.

The essential features of the screen that one must be concerned with are screen tension, mesh size, alignment of the pattern with the mesh (for fine-line printing), and emulsion thickness and uniformity. Screen tension and emulsion uniformity will vary with the life of the screen. When screen tension becomes too low to be compensated for by adjustment of the printer, the screen must be discarded. Some screen suppliers provide screens with adjustable tension, which can extend screen life somewhat. Screen tension that is too low results in irregular prints with regularly shaded gaps. An irregularly stretched screen will cause poor pattern definition.

Emulsion thickness is critical in controlling film thickness. Too thick an emulsion will result in poor line definition as well as lowered sheet resistance. Emulsion thickness for conductor, dielectric, and resistor inks should be about 25 μm. For solder and braze compositions, thicker emulsions of 0.1 to 0.25 mm are required. Alignment of the pattern with the mesh is critical in fine-line printing (50 to 75 μm), but not of serious consequence in applications where wide lines are used (0.25 to 0.4 mm).

Mesh size is very important because different types of inks must be printed with different mesh sizes. Conductor inks require mesh sizes ranging from 200 to 325, the fine mesh being for fine-line printing. Resistor inks require mesh sizes from 165 to 200, dielectric inks about 200 mesh, and solder and braze

TABLE 12-2 Mesh Number vs. Approximate Diameter and Mesh Opening Size

Mesh Number	Wire Diameter (mm)	Opening Size (mm)
80	0.094	0.118
165	0.051	0.107
200	0.053	0.074
325	0.028	0.051

compositions about 80 mesh. Table 12-2 is a list of mesh sizes versus approximate wire diameter and opening sizes for some common mesh numbers.

A number of manufacturers supply frames, frames with screens, frames with screens and emulsion, or if you send them the artwork (drawings of the desired patterns), they will supply screens ready for printing. They also supply the chemicals for sensitizing the emulsion. Hence, thick-film producers can do as little or as much of the screen preparation as is dictated by economics and facilities.

The details of mask design are discussed later in this chapter, so only a general description of how screens are exposed and developed will be presented here.

The appropriate positive mask is aligned on the substrate (emulsion) side of the screen by use of a microscope if necessary and taped in place. The emulsion is sensitized with a special chemical mixture poured on the squeegee side and dried and then the emulsion is exposed from the substrate side to intense ultraviolet light, such as from a xenon, mercury, or carbon arc lamp, for a few minutes. We have found that a 200-W incandescent bulb is satisfactory for our educational laboratory work. The mask is removed and the screen is soaked in hot water for a few minutes to harden the exposed emulsion and to soften the unexposed emulsion. The screen is then sprayed with warm water and air to clear the screen openings. The screen should be carefully checked over with an optical comparator to look for blockage and improper openings. Unwanted openings in the emulsion can be touched up with the sensitizing solution.

12-4 MASK DESIGN AND LAYOUT

To obtain a screen with the proper pattern to be printed defined in it, it is necessary to begin with a mask design and layout. The final layout is a 5× to 20× size drawing of the pattern to be defined in the screen.

In reading this section, it must be realized that the design criteria, such as line width and line separation, presented here are not sacred but are intended only for the purpose of illustrating what types of considerations are important. For the most part, the numbers used are conservative and apply to our own facility, which is an undergraduate teaching laboratory. In an industrial situation, tolerances would normally be tighter and mask design and generation is likely to be computerized.

After a circuit has been functionally designed, the configuration of the circuit diagram should be redrawn resembling as closely as possible the ultimate thick-film pattern layout. This means resistors should be drawn either parallel to the y direction (the printing direction) or perpendicular to it. Leads should be brought out to one side whenever possible. Crossovers should be avoided, if possible, because this requires a dielectric print and multiple conductor prints.

A composite drawing of the masks and components for an astable multivibrator thick-film circuit designed for student fabrication in our lab is presented in Fig. 12-8. For simplification of assembly, discrete plastic encapsulated transistors and capacitors are used. Crossovers are avoided through the use of these devices. A light-emitting diode (LED) is used to indicate the on condition. This would not be a satisfactory industrial design, but it is a suitable laboratory design for the purpose of illustrating printing, resistor trimming, and firing processes. The transistors, capacitors, LED, and leads can be soldered by hand with a small soldering iron, by solder dipping and reflow, or by solder printing

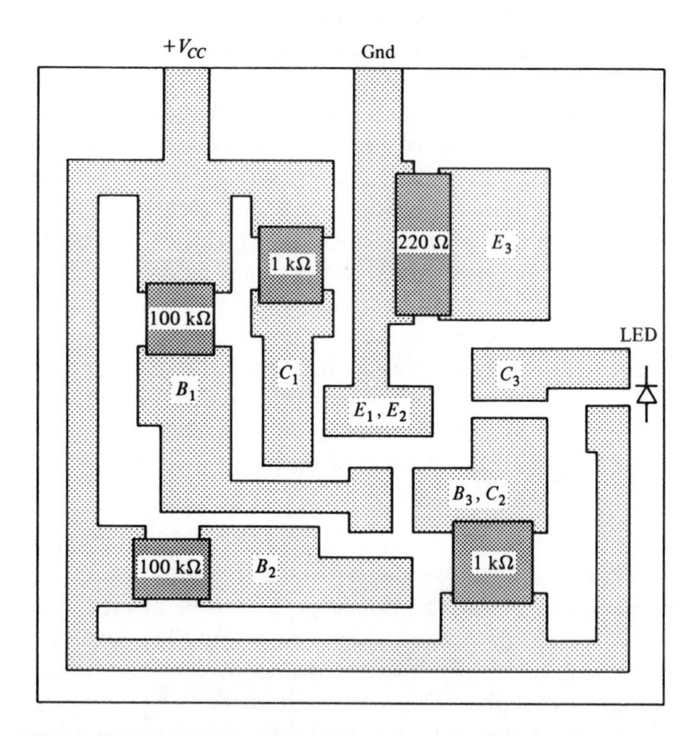

Fig. 12-8 Footprint diagram for an astable multivibrator. Light shading is conductor, dark shaded areas are resistors. Letters refer to emitter, collector, and bases of transistors.

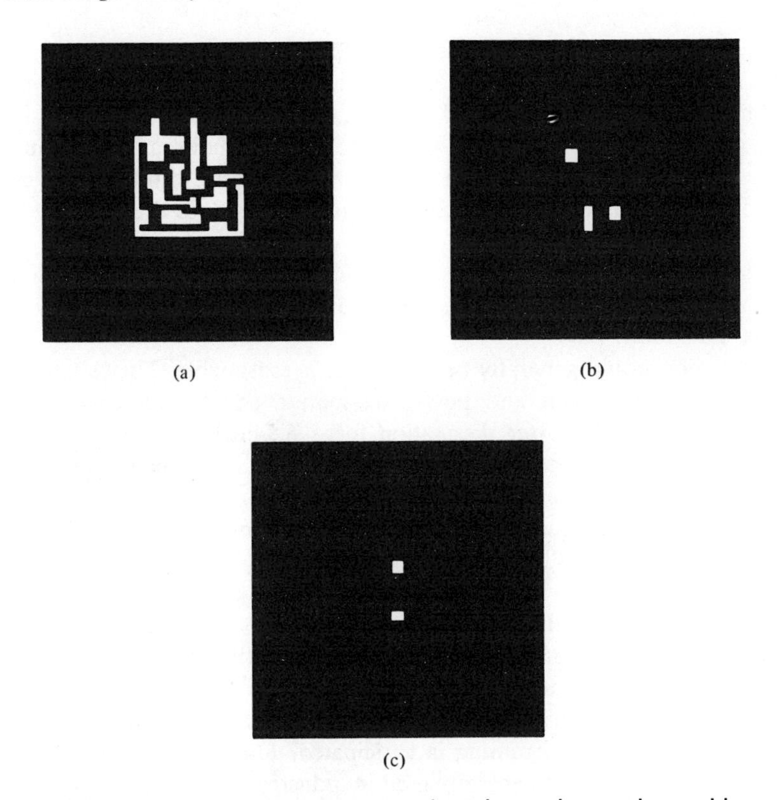

Fig. 12-9 Negative transparencies of masks used to make astable multivibrator: (a) Conductor mask; (b) 1-kΩ and 220-Ω resistor mask; (c) 100-kΩ resistor mask.

and reflow. Solder reflow can be accomplished by use of a hot plate or a hydrogen torch. This circuit is designed for a 2.54 × 2.54 cm substrate. The drawings of the masks were made five times actual size and sent out to a local photo-developing company for reduction to negative transparencies. The transparencies are shown in Fig. 12-9 approximately actual size. Such low-tolerance, low-volume fabrication is relatively simple. However, when millions of units are to be built, pennies per circuit become critical, and it is imperative that the circuit is made as compact as possible and the use of IC chips should be considered. The remainder of this section will be devoted to presenting a representative (albeit, somewhat conservative) set of design criteria.

12-4-1 Design Criteria

Masks are generally drawn or cut from rubylith, 10 to 20 times actual size. Special cases may require modification of this. Reduction is done by means of a high-quality reducing camera.

Resistors should be laid out parallel or perpendicular to the print direction. Zigzag resistors should be avoided. "Top hat" resistors (Fig. 12-10) may be used but are usually not required unless extremely close trimmed resistance tolerances must be held. It is also found that resistor noise and shifts in other resistor properties tend to increase with increased cross-cutting of the resistor. The "top hat" configuration reduces this problem, as does the L cut (Fig. 12-10).

Since closed-loop resistors cannot be measured, an open circuit must be maintained which can later be wire-bonded or, preferably, bridged by solder to close the circuit.

Resistors should generally be as large as is reasonable. They should never be less than 1 × 1 mm and power dissipation should not exceed 2 to 4 W/cm² or 2 W total power dissipation for a 6.5-cm² 96% Al_2O_3 substrate. Minimum resistor and conductor overlap is 0.4 mm. Whenever crossovers are required, the crossover insulation should overlap the conductors at least 0.4 mm. Ceramic chip capacitor terminations depend somewhat on the capacitor geometry. Values shown here are always minimum (see Fig. 12-11 for illustrations).

Conductor line widths should be a minimum of 0.5 mm. They should not be placed closer than 0.5 mm to the edge of the substrate except when they are used for pads for external connections.

Sufficient space must be left between resistors and other elements for resistor trimming. Where trimming is anticipated, 1 mm should be allowed for air abrasive trimming (not generally used in industry) to avoid overspray effects. When laser trimming is used, spacing will depend on the beam positioning accuracy. Pads for back-bonded chips should be at least 0.25 mm larger than the chip. When feasible, the pad may be made double-size to accommodate a

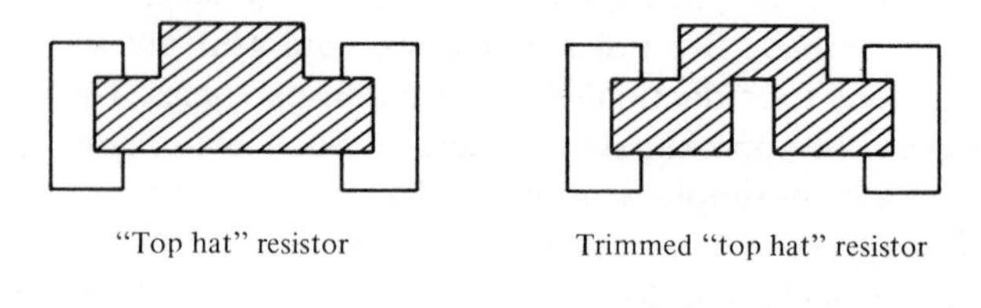

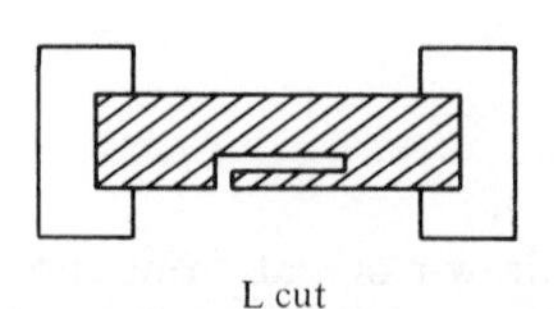

Fig. 12-10 "Top hat" and L-cut resistor trimming.

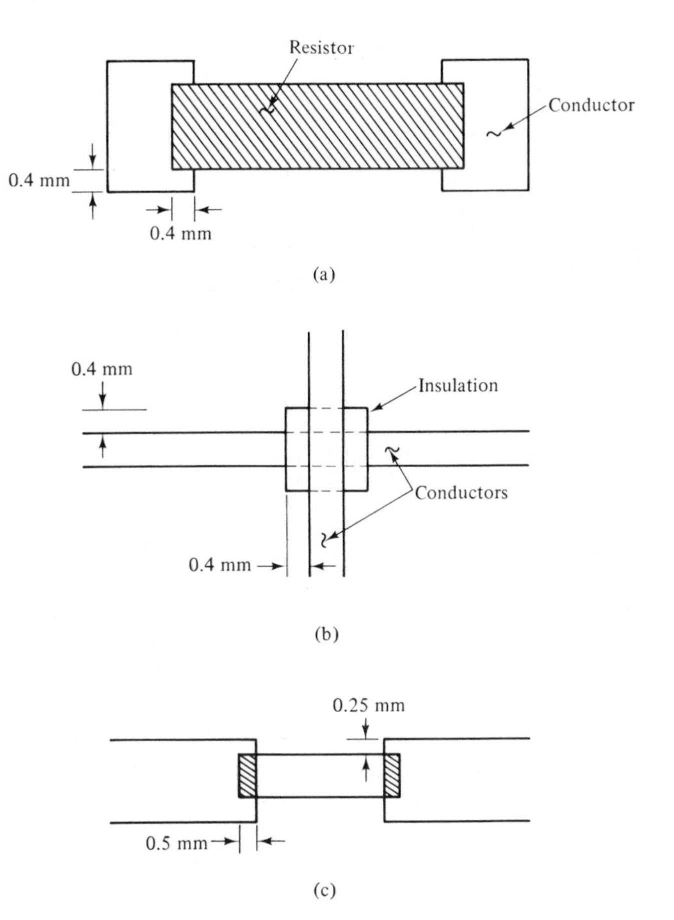

Fig. 12-11 (a) Resistor terminations. (b) Crossover. (c) Capacitor terminations.

second chip in case the first one fails. Flip chips[8] are used extensively and pad placement for the chip bonding bumps must be within 25 μm.

For thick-film capacitors, the dielectric completely overlaps the bottom electrode and the top electrode is smaller than the bottom one. See Fig. 12-12 for minimum dimensions. Wire bond pads should not be less than 0.5 × 0.5 mm. Wire length should not exceed 2.5 mm. Pads for external leads or lead frames generally should be 1.25 × 1.25 mm minimum.

A minimum of 0.25 mm of free space should be allowed around all elements. This will provide at least 0.5 mm between elements.

[8] Flip chips are ICs with bonding bumps placed on the circuit side of the chip, necessitating inversion of the chip to make contact to pads on the thick-film circuit. Flip chips are discussed further in Chapter 13.

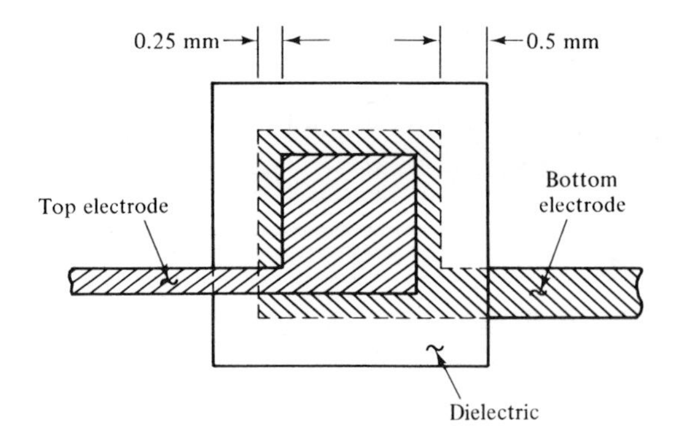

Fig. 12-12 Thick-film capacitor.

In designing masks for complex circuits, it may be desirable to start with a "footprint" diagram from which the individual masks may be traced. A "footprint" is simply the maximum dimensions of an element, including free space. An example of a resistor footprint is given in Fig. 12-13. If desired, footprints can be cut out of paper and arranged until the best design is achieved. Footprint patterns can also be computer-generated. Notice that allowance must still be made for conductor interconnections. The total footprint area should not exceed 50 to 60% of the substrate area.

If power transistors are required in the circuit, they should be mounted on a separate heat sink such as nickel-plated BeO. Several watts can be dissipated in this manner. A power Darlington may be seen on a heat sink in Fig. 1-7.

When masks are designed by use of a computer, a composite is first generated using lines and possibly rectangles of various lengths and widths stored in memory and called up on an interactive CRT terminal. In this way, new patterns can be generated or old ones can be corrected and stored in memory. A computer-controlled plotter is used to produce the drawings or Rubyliths for photographic reductions. Pictures of such a system can be seen in Chapter 10. The mask designs are stored digitally on magnetic tape for future use. Very often, design modifications require relatively simple changes in the masks. The design is called up on the interactive graphics terminal, corrected, and the computer-controlled plotter cuts a new mask. Turnaround time for the design and

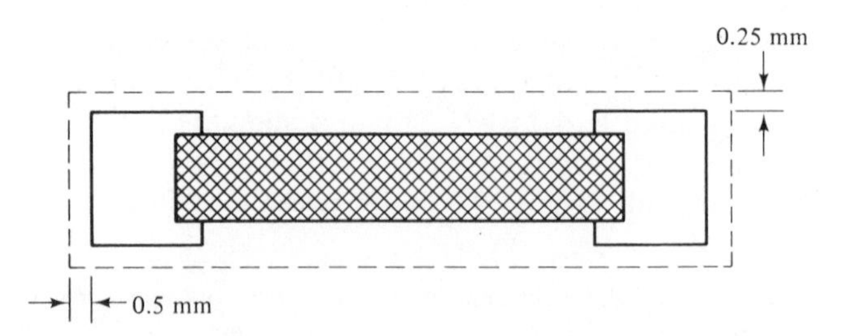

Fig. 12-13 Resistor "footprint."

production of masks has been reduced from days to hours by the use of computer graphics.

12-5 PRINTING PROCESS

The basic features of a thick-film printer are substrate holder, squeegee, screen, screen holder, and carriage. The carriage positions the substrate holder under the screen for printing. Means for varying squeegee pressure, screen height, breakaway distance, and squeegee travel speed are provided. The substrate holder must allow for fast, accurate placement of the substrate under the screen. The carriage to which the substrate holder is attached should allow for *x*-*y* and angular micrometer adjustment of the substrate position. Carriage operation may be manual or automatic. In some cases, there are several substrate holders arranged in a circular fashion for rapid printing and a cartridge-type substrate feed. Figure 12-14 is a picture of a fully automated industrial printer.

The squeegee, schematically depicted in Fig. 12-15, is a critical part of the printer.

The screen was discussed to some extent earlier, but must be considered an integral part of the printer in the printing process. The printer must possess

Fig. 12-14 Fully automated industrial printer. (Courtesy of Delco Electronics.)

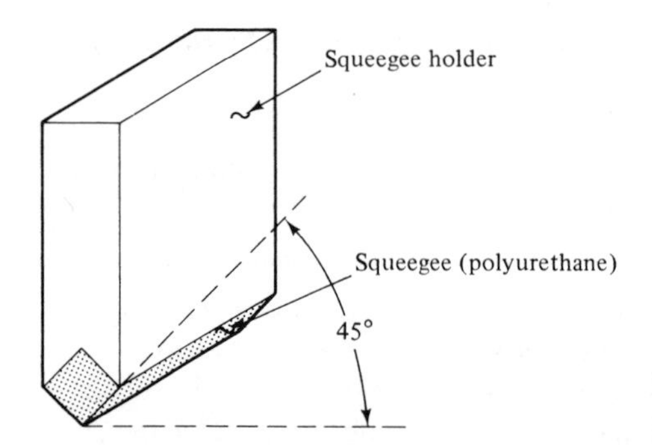

Fig. 12-15 Squeegee configuration.

the capability for rigidly holding the screen in place while allowing for adjustment of screen height above the substrate.

The printing process is basically simple. The substrate is placed on the substrate holder, which has been previously aligned with the screen pattern. The carriage positions the substrate under the screen, the squeegee moves down and across, wiping ink in front of it and through the openings in the screen onto the substrate. Figure 12-16 is an illustration of this process.

In the case of off-contact printing (see Fig. 12-17), the squeegee deflects the screen to make contact with the substrate and the screen breaks away from the substrate behind the moving squeegee. In contact printing, little or no screen

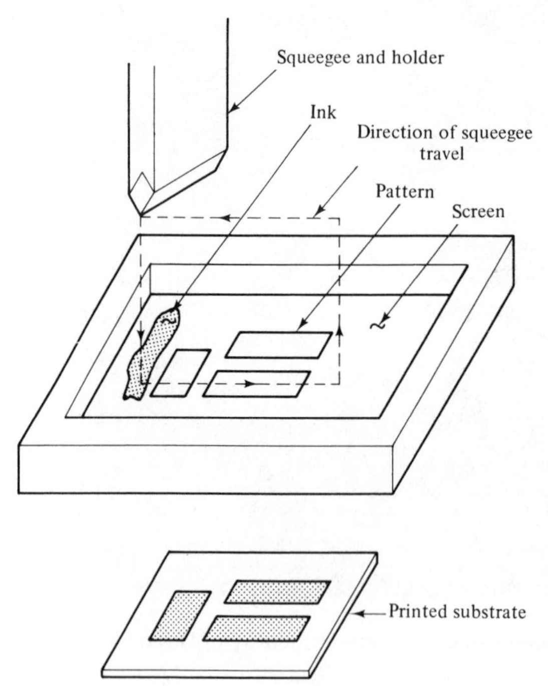

Fig. 12-16 Screen printing process.

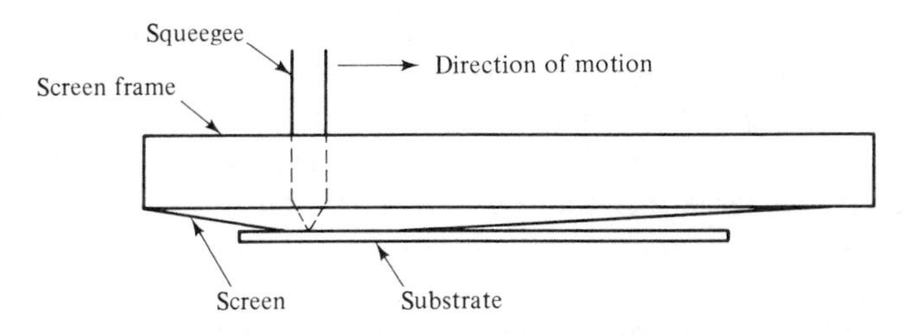

Fig. 12-17 Off-contact printing.

deflection occurs, so breakaway is achieved either by dropping the substrate holder or by raising the screen a little after each print.

Printers vary from the semiautomatic, hand-operated carriage types used in labs and low-production work to fully automatic and multisubstrate dial carriage types. Printers may be purchased with automatic substrate feed capability, such as the cartridge feed arrangement that can be seen in Fig. 12-14. It is often desirable, however, to design in-house substrate-handling equipment best suited to the particular application.

Printers can be obtained with on-contact and off-contact printing capability. Both unidirectional and bidirectional capabilities are available, often in the same printer. In unidirectional printing, a wiper is provided to drag the ink back after each print. In unidirectional printers, the squeegee lifts up at the end of each pass and comes back down on the side of the ink away from the pattern, so as to retain an adequate ink supply in front of the squeegee.

12-5-1 Process Variables

The process variables (and their effects on print quality and resistance) incurred during the printing phase will be discussed here.

Figure 12-18 contains a sketch of a squeegee in the process of printing

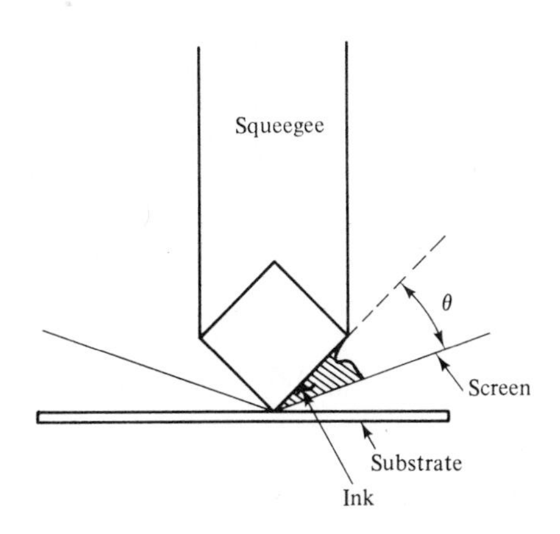

Fig. 12-18 Ink being printed.

ink through a screen onto a substrate. The ink is forced through the pattern openings because of the hydraulic pressure built up on the ink in front of the squeegee. The greater this pressure, the more ink is forced through the screen. This pressure can be increased either by decreasing the angle of attack, θ, or by increasing the force that the squeegee exerts on the screen and substrate. Once the screen is in contact with the substrate, increased squeegee pressure simply results in deformation of the squeegee tip, thereby decreasing θ. Any increase in ink deposited will, of course, reduce sheet resistance.

The hardness of the squeegee is important, because this will determine the amount of tip deflection. Large tip deflection decreases the effective angle of attack, thereby increasing the amount of ink deposited. This is illustrated in Fig. 12-19.[9] Most squeegees are polyurethane, which has moderate hardness, is fairly abrasion-resistant, and has fair resistance to hydrocarbons and fluorocarbons. However, oxygen containing solvents, alcohols, ketones, or acids should never come in contact with a polyurethane squeegee.

The faster the squeegee moves across the screen, the greater the pressure exerted on the ink; however, less ink is deposited on the substrate because of the reduced time that the screen and squeegee remain in contact with the substrate. This results in an increase in sheet resistance. Squeegee speeds vary between 5 and 25 cm/s.

The tension of the screen in its frame is an important consideration. The greater the tension, the less ink is printed. This is due to the fact that the angle of attack is less, due to reduced screen deflection (see Fig. 12-20). This has to be compensated for through lower breakaway distance. (Breakaway distance is the separation of the screen and substrate before printing.) The high-tension screen could be stretched as much as the low-tension screen, to give the same angle of attack, but this would require excessive squeegee pressure

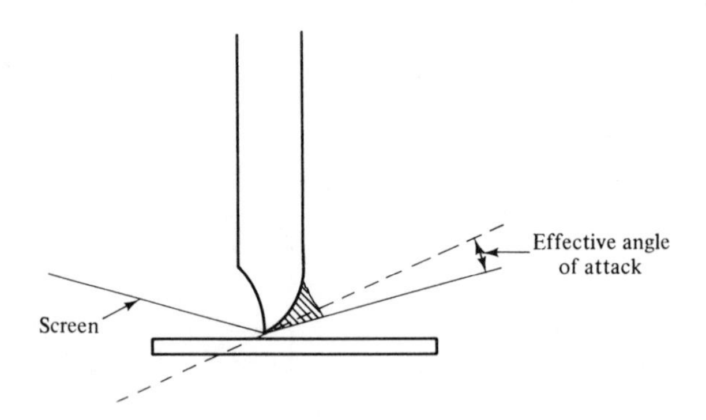

Fig. 12-19 Squeegee tip deflection.

[9] A worn squeegee will tend to increase the amount of ink deposited for the same reason and will result in uneven ink distribution on the screen.

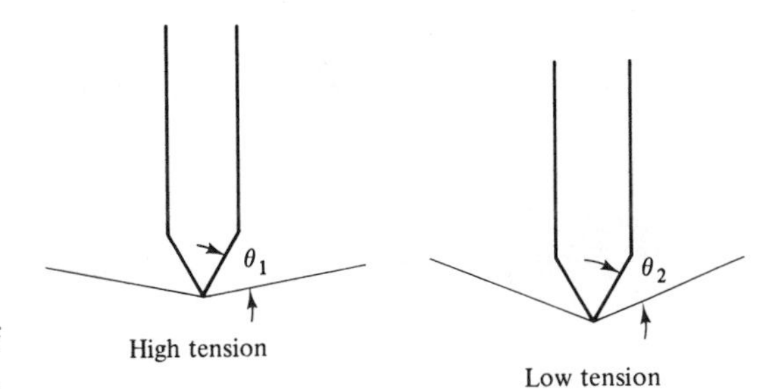

Fig. 12-20 Comparison of high- and low-tension screens.

and result in poor print definition. As a screen ages, it loses tension and eventually must be stretched or discarded because of loss of print definition. Too much stretching will degrade the print as well. It is possible to obtain 10,000 to 20,000 prints per screen, although this depends strongly on the degree of definition required in the application.

Poor screen parallelism with the substrate can cause uneven print thickness. If the breakaway distance is too great, lines can vary in width since screen deflection will not be uniform for different positions of the squeegee on the substrate. There is an optimum breakaway distance at which the resistor value spread will be minimum. This optimum must be determined for each set of printing conditions.

12-6 DRYING AND FIRING PROCESS

Printed patterns may be allowed to set in ambient conditions for a few minutes to ensure complete "healing" of screen marks. Resistor, dielectric, and thermistor patterns are dried at about 125°C in either an infrared oven or a forced air oven for 10 to 15 min. The oven drying step is intended to boil off solvents without damaging the print quality. If all the patterns are not at the same temperature for the same length of time, variations in sheet resistance and other parameters could be induced. The time from printing to drying should also be the same for all patterns. Since the process of removing all organic materials is not completed in the drying step, it is important to ensure that all circuits experience the same condition.

The firing process is accomplished in a conveyor furnace such as the one shown in Fig. 12-21. Actually, firing involves three parts: (1) burnout of organic materials left after the drying process; (2) the sintering period, in which the glass softens, bonds to the substrate, and forms a matrix for the metal or dielectric particles; and (3) the cool-down period, in which the circuits are returned to room temperature in such a way as to minimize oxidation and cracking of the glass.

Furnaces vary in size and detail, but generally consist of four or more separately controlled resistively heated zones, a quartz muffle, high-temperature

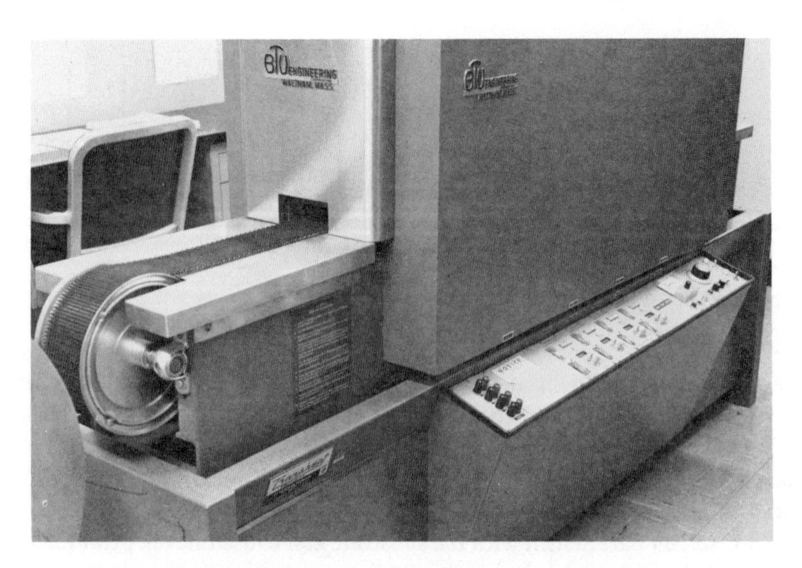

Fig. 12-21 Laboratory thick-film conveyor furnace.

controllable speed belt, separate atmospheric control for burnout and firing sections, control thermocouple readout, and overheat protection. Furnaces range in size from 7 ft in length for lab models to 18 ft and over for production models. The maximum operating temperature should be at least 900°C, since most ink compositions fire at a maximum of 850°C.

Figure 12-22 is a furnace profile (temperature versus time or distance along the furnace) for a thick-film resistor material obtained by attaching a thermocouple directly to the belt and allowing it to travel through the furnace in the same manner as a printed substrate.

There are three distinct regions notable in the profile: burnout, firing, and cool-down regions. The burnout section of the furnace is usually partially isolated from the firing section by a baffle. This section should be atmospherically isolated from the firing section so that gases released during burnout do not contaminate the final product. Atmospheric isolation can be achieved by piping clean air into the firing and cool-down sections separately from the burnout

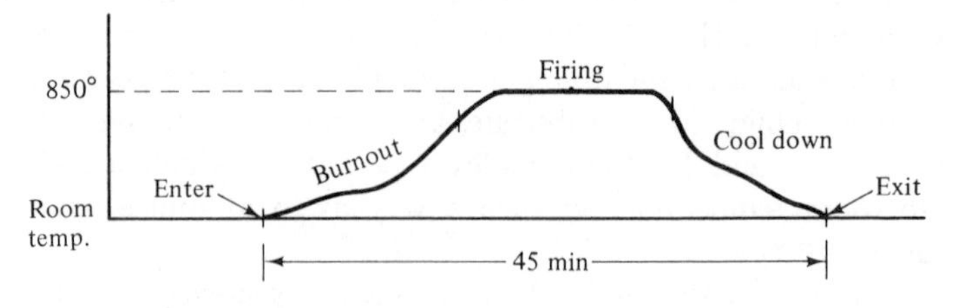

Fig. 12-22 Furnace profile for resistor ink.

section and exhausting the atmosphere through appropriately located stacks. Venturi tubes can be placed in the stacks to reduce pressure, thereby aiding in the exhaust of the furnace atmosphere. Some older furnaces were inclined 2° or so to make use of natural convection. In this manner, room air came in at the exit end and was exhausted through a stack between the firing and burnout sections, thus preventing gases from the burnout section from getting into the firing section. It is very important that whatever method is used, the air be clean so that the product and furnace muffle do not become contaminated, especially with oil.

With a good furnace, you should be able to change the profile or turn the furnace off and then on again at a later date and return to within 1 or 2°C of the original temperature at any point along the furnace. This accuracy is needed when tight tolerances are being held on resistor or capacitor values. In large furnaces, temperature variation across the belt may be a problem.

Suppliers of inks provide what they consider to be optimum firing profiles and allowed temperature ranges for their inks. Economics or special design considerations may, however, dictate the use of nonoptimum firing conditions. In any case, the final proof of the acceptability of a given firing profile is the quality of the product. The firing cycle affects the quality of the product in several ways. If burnout is too rapid, a rough surface will be created due to bubbles of entrapped gases bursting out rather than diffusing out slowly. Insufficient burnout will leave organic materials trapped in the print. This will add to resistor noise, degrade temperature coefficients of resistance (TCRs) and voltage coefficients, alter sheet resistances, and cause resistor drift. The peak firing temperature and the time spent at the peak firing temperature affect sheet resistance and TCR. TCR is also affected by the slope of the cool-down curve. Ink suppliers provide information on sheet resistance and TCR variation with peak temperature and time at peak temperature. The amount of dependence varies with nominal sheet resistance and may increase, remain fairly constant, or decrease with increasing temperature and firing time, such that no general statement can be made as to how sheet resistance and TCR or other parameters depend on these variables.

Too rapid a cool-down will cause the glass to crack, leading to poor terminations and poor-quality resistors and capacitors. On the other hand, too slow a cool-down could cause oxidation of palladium in Pd–Ag ink, which diminishes the solderability of the conductor. There are Pd–Ag conductor compositions which are resistant to oxidation.

The quality of resistor–conductor terminations and bonding of inks to the substrate may be seriously affected if too much deviation from the supplier's recommended profile is permitted. For economic reasons, it is very desirable to cofire (simultaneously fire) resistors and conductors. Unfortunately, this does not always result in satisfactory terminations. It requires that either the resistor or conductor ink, or both, be fired at nonoptimum temperatures, since most resistor and conductor inks have different optimum firing temperatures. Cofiring may lead to unsatisfactory bonding to the substrate as well as poor terminations and is not generally practiced in high-volume production.

12-7 THICK-FILM RESISTOR TRIMMING

The nature of thick-film printing and the inks that are employed makes the printing of resistors with very precise values a virtual impossibility. What can be expected is that the printed resistors will vary in value according to a normal distribution. Under good conditions, the 2σ[10] points may lie within ±15% of the average value. Unfortunately, the average value may differ from the desired value due to variation in print conditions, firing conditions, and inks from day to day. To get around this problem, resistors are designed and printed to have resistance values 20 to 40% below the desired value. The resistors are then trimmed, either by the air abrasive technique or by laser. A cut is made into the resistor, effectively reducing its cross-sectional area and increasing its resistance until the desired resistance is attained. In this fashion, a ±5% tolerance is easy to achieve, ±1% is moderately difficult, and under carefully controlled conditions ±0.1% can be achieved.

12-7-1 Air-Abrasive Trimming

In this technique, a stream of air and abrasive particles, such as alumina of about 50 μm particle size, is directed through a nozzle onto the resistor. The resistance of the resistor being trimmed is measured in a bridge network and when the desired value is reached the trimmer automatically shuts off. Substantial overshoot occurs, so the shut-off point is set below the desired value by an empirically determined amount. Figure 12-23 is a picture of an S. S. White air-abrasive trimmer. In this operation, the nozzle moves forward into the resistor, abading the resistor material away. A vacuum dust collection system serves to suck up the abrasive material and debris. Trimming rates of several inches per minute can be achieved. However, accuracy is sacrificed for speed. With a slow cutting rate and resistors that require minimal abrading, a tolerance of ±0.1% can be achieved. The trimmer shown in Fig. 12-23 is set up for manual laboratory operation, but it can be fully automated for parts handling and automatic trimming.

The air-abrasive technique is basically a "dirty" operation and great care must be taken to protect the operator from inhalation of the abrasive dust and debris. The width of the cut (kerf) is nearly 0.4 mm and the sides of the cut are ragged and the inside of the resistor is exposed. This can lead to long-term drift and noise problems due to the adsorption of gases or chemicals on

[10] Sigma (σ) is the standard deviation (root mean square); 95.45% of all values in a normal distribution lie within $\pm 2\sigma$ of the average.

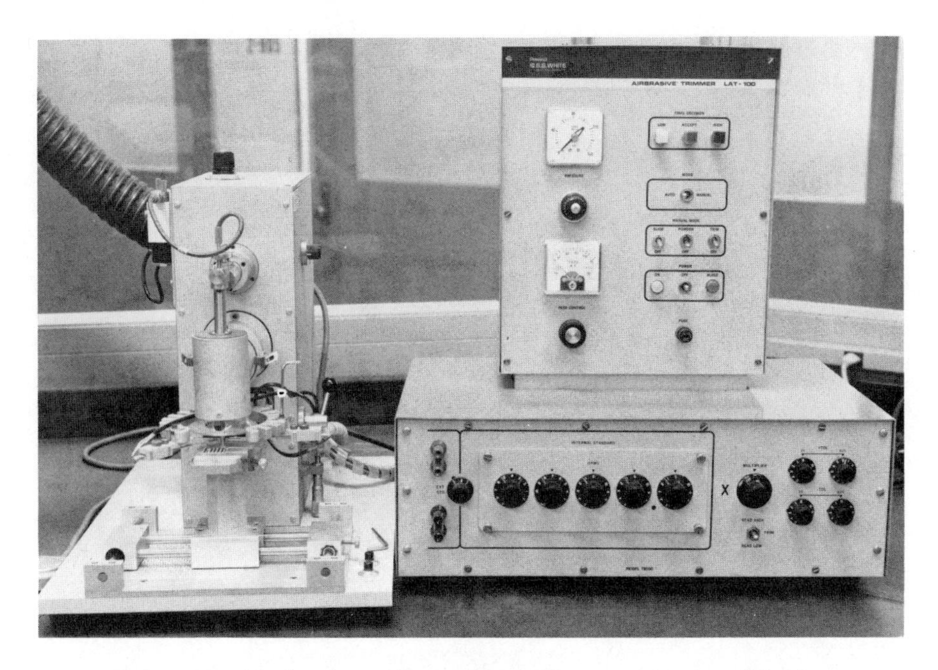

Fig. 12-23 Air-abrasive resistor trimmer. The control unit and bridge are on the right, trimming nozzle on the left.

the exposed surface area. These impurities can diffuse into the resistor, causing chemical reactions to take place and subsequent changes in resistor parameters. Consequently, it may be necessary to use an overglaze or encapsulant to protect air-abrasive-trimmed resistors.

12-7-2 Laser Trimming

A laser is a source of highly monochromatic light which can be focused to a small spot size. CO_2 and Nd^{3+} yttrium–aluminum–garnet (Nd–YAG) lasers are used for thick-film resistor trimming. Figure 12-24 shows an industrial Nd–YAG trimming system. The CO_2 laser emits radiation at 10.6 μm, the Nd–YAG at 1.06 μm. The size of the spot that a laser beam can be focused to is given approximately by

$$d = \frac{4\lambda f}{\pi D} \tag{12-1}$$

where d is the spot size, D the diameter of the laser beam, λ the wavelength, and f the focal length of the lens. Hence, it can be seen that the kerf or spot size for Nd–YAG laser trims can be as much as 10 times smaller than for

Fig. 12-24 Industrial Nd–YAG trimming system. (Courtesy of Delco Electronics Division, GMC.)

the CO_2 laser. The smaller spot size also means higher power densities (W/cm^2) for lower output power. CO_2 lasers used for resistor trimming and substrate scribing are around 50 W continuous power output (this is called continuous wave CW), whereas Nd–YAG lasers designed for the same purpose are about 2 to 3 W CW. In both cases, the lasers are Q-switched to achieve higher peak power. Q-switching is a process whereby the optical cavity of the laser is interrupted to prevent lasing by a device such as a mechanical chopper or an acousto-optical modulator. Interrupting the optical cavity allows the population inversion to build up (basically a technique of storing up energy). When the cavity is no longer blocked, lasing action takes place and the energy is emitted in a short pulse, usually about 100 μs, rather than continuously. Peak power of 300 times the CW power is common. Power densities in these short pulses may reach millions of watts per square centimeter. This easily vaporizes thick-film resistor or capacitor material and with larger dwell times punches holes in alumina substrates so that they can be snapped (broken apart) along a line of such holes. With repetition rates of 1000 pulses per second or more, trim rates of several inches per second are achieved.

CO_2 lasers are considerably more efficient than Nd–YAG lasers, 10 to 20% for CO_2 versus 0.1 to 1% for the Nd–YAG. However, the reflectivity of metals to the 10.6-μm radiation is much higher than for 1.06-μm radiation. Consequently, considerably more power is required to initiate the vaporization of a metal with the CO_2 laser. Once vaporization begins, the reflectivity drops rapidly. Because of the high reflectivity, CO_2 lasers are not suitable for thin-film resistor trimming. However, they do work well on thick-film resistors because of the high glass content. The major drawback of Nd–YAG lasers has been the life of the lamps used for excitation. This technological problem has been sufficiently overcome to warrant reconsideration of Nd–YAG lasers for thick-film applications because of their inherent advantages; namely, smaller spot size, use of conventional optics, and smaller physical dimensions. CO_2 lasers, because of the long wavelength, must use special water-cooled optics such as germanium or zinc–selenide lenses. Because of the high refractive index of these materials, nonreflective coatings are required. Copper or gold front-surfaced mirrors can be used for reflection.

12-7-3 Trim Geometry

There are various types of trim cuts that can be used. The notch or plunge cut is generally employed with air-abrasive trimming. The length of notch can be varied by the use of different-shaped nozzles. Both circular and rectangular orifices are used. Figure 12-25 is an illustration of several possible air-abrasive cuts.

The longer the notch L, the smaller the depth of cut d. For a notch length L much smaller than the resistor length, the resistor will have to be cut well over halfway through before any significant resistance change will occur. This can result in serious hot spots in trimmed resistors where current crowding

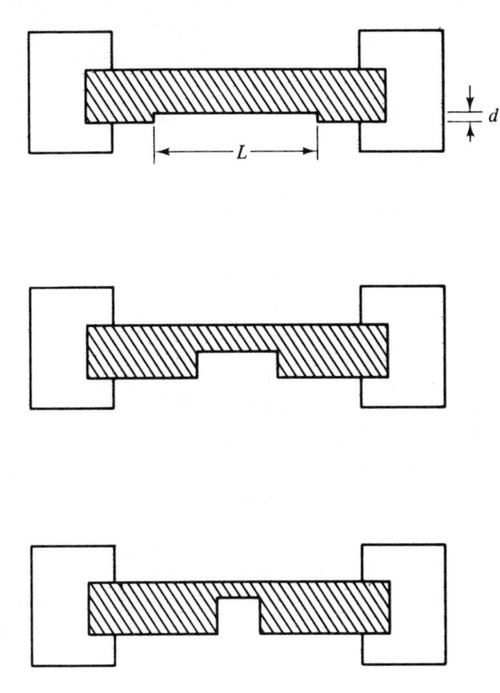

Fig. 12-25 Air-abrasive notch cuts.

occurs. As pointed out previously, "top hat" geometry may be used and this helps alleviate the problem mentioned here.

Since the laser trims are necessarily narrow, about 25 μm for the Nd–YAG laser, most resistors would have to be cut nearly clear through if the plunge cut were used. Other techniques have been developed, such as the multiple plunge cut and the *L* cut depicted in Fig. 12-26. For high-aspect-ratio (length/width-ratio) resistors, laser trimmers will not work unless the multiple cut, *L* cut, or "hat" technique is used. In both cases of Fig. 12-26, the trim consists of a rough cut and a fine tuning cut. In the multiple plunge cut, the resistance change with depth of cut is much slower for the second or later cuts; hence, very accurate control over the resistance change can be accomplished. The same statement applies to the parallel cut of the *L*-shaped trim. In any case, the power-handling capability of the resistor is diminished because the current must pass through a smaller area.

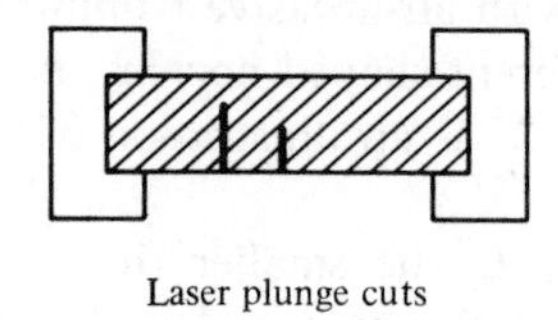

Laser plunge cuts

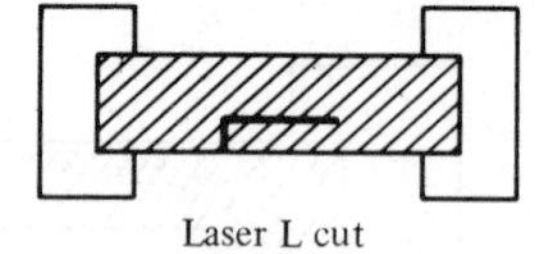

Laser L cut

Fig. 12-26 Laser trim cuts.

12-7-4 Resistor Trimming Considerations

When a resistor is abrasively trimmed, the TCR is affected because the positive TCR glossy material tends to be concentrated near the edges. Since more of the positive TCR material is removed from the bulk of the resistor, the TCR for the resistor tends to go negative. Trimmed resistors tend to be more noisy than untrimmed ones due to the rough edges left by the trim. Since laser trimming vaporizes the material, the glass near the kerf is literally melted and reflows to form a seal. In effect, the "wound" is cauterized. Hence, laser-trimmed resistors tend to be less noisy. The greater the distance the current travels parallel to a cut, the worse the noise problem; thus, *L* cuts and "hat" resistors can be a problem in this respect. Long-term effects on resistance, TCR, and noise can also occur because of the exposure of the inside of the resistor to contamination. This is less of a problem with laser trims as a result of the cauterizing effect. Laser trimming is rapid with six or more resistors on a substrate trimmed per second.

Although thick-film capacitors have not been discussed in detail, they, too, can be trimmed. The same problems that beset trimmed resistors also plague trimmed capacitors. Thus, a drift can be expected with changes in temperature coefficient of capacitance (TCC) and increased capacitance.

12-8 THIN-FILM TECHNOLOGY

Thin-film technology will not be covered as thoroughly as thick-film technology, for two compelling reasons. First, thin-film circuits have not captured a large share of the microelectronics market, and hence a great deal of detail would not be of great interest to most undergraduate students. This statement in no way implies a lack of importance of thin-film circuits. In the telephone and space industries, where small, highly reliable circuits are required, thin-film technology is extensively applied. In the microwave field, thin-film circuits have made a substantial impact because of their versatility, reliability, and the precision with which circuits can be made. Second, thin-film technology is essentially more complex than thick film because of the wide variety of deposition techniques employed and the multiplicity of materials available for thin-film work. The vacuum technology itself, with the various deposition techniques, vacuum pumps, vacuum gauges, and so on, that are used in thin-film work would make for an extensive treatise. Consequently, it is felt that detailed coverage, which would result in a considerably longer text, cannot be justified. It is hoped that sufficient information will be presented here to acquaint the reader with the capabilities and complexities of thin-film technology. For the reader with a deeper interest, we refer you to the more extensive works listed at the end of the chapter.

12-8-1 Vacuum Technology

The primary methods for depositing thin films on electronic circuits are electron-beam evaporation, vacuum evaporation, and sputtering. Therefore, a short discussion of vacuum technology seems in order. A typical vacuum system is pictured in Fig. 12-27. The vacuum chamber consists of a glass or metal bell jar, which can be raised, and a feed-through collar for electrical, mechanical, gas, and water-cooling entry to the chamber. The chamber is first evacuated by a mechanical forepump down to about 1 torr[11] and can then be pumped to the 10^{-7}-torr range under ideal conditions by a diffusion pump. Lower pressures can be obtained with the aid of cryogenic traps, gettering (chemisorption), and a variety of other techniques. The seal between the bell jar and the collar is made by a neoprene or Viton O-ring covered with vacuum grease that has a very low vapor pressure.

[11] A torr is a unit of pressure equal to 1 mm of Hg.

Fig. 12-27 Vacuum system.

For the reader who is unfamiliar with vacuum technology, some representative numbers may be in order. At a pressure of 1 torr and temperature of 300 K, the density of molecules is about 3.2×10^{16} cm^{-3}. At 10^{-7} torr, this reduces to 3.2×10^{9} cm^{-3}, still a pretty large number. This number is independent of the type of gas, and values for other temperatures and pressures can be found since density is inversely proportional to temperature and directly proportional to pressure. The mean free path,[12] λ, is dependent on the type of gas, but for air, $\lambda = 5 \times 10^{4}$ cm at $T = 300$ K and $P = 10^{-7}$ torr. λ is inversely proportional to pressure. The average rate of collision of air molecules per unit area with a surface is 3.8×10^{13} cm^{-2}/s at $p = 10^{-7}$ torr and $T = 300$ K. This rate, R, is directly proportional to pressure, so values at other pressures are easily calculated. It takes about 22 s to form a monolayer of air on a surface at $P = 10^{-7}$ torr and $T = 300$ K if each molecule striking the surface sticks. Actually, the sticking coefficient K, which is the ratio of molecules that stick to the total number incident, may range from 0.1 to 1 and is dependent on the type of surface, cleanliness, and temperature.

There are many types of vacuum pumps, all of which can be classified as either roughing pumps (fore-pumps) or high-vacuum pumps. Only the eccen-

[12] Mean free path is the average distance that molecules travel between collisions.

tric rotary oil pump and the diffusion pump will be considered here. In the rotary oil pump an eccentric rotor makes a seal between the inlet and outlet ports. Additional seals are provided for by spring-loaded vanes. This is schematically illustrated in Fig. 12-28. Molecules entering the inlet are trapped and pushed around by the vane and have to exit at the outlet. Seals are accomplished by an oil film. Putting pumps like this in series can extend their useful range. The primary thing that limits their range is the quality of the seals. Contaminants, such as water vapor in the oil, can have a disastrous effect because of its very high vapor pressure.

In a diffusion pump, a liquid, usually a low-vapor-pressure oil, is boiled and the vapor rises in a stack and exits from downward directed nozzles. The high-velocity oil molecules drive gas molecules from the vacuum system down and out a tube called the fore-arm to the fore-pump. The oil molecules condense on the water-cooled wall of the pump housing and drain back into a reservoir at the bottom. A schematic illustration of a diffusion pump is given in Fig. 12-29.

The ultimate limit of a diffusion pump is the vapor pressure of the liquid, usually below 10^{-8} torr at 300 K. Backstreaming, diffusion of oil molecules into the chamber, and migration of molecules condensed on the walls leading into the chamber make it difficult to get below 10^{-7} torr. Migration and backstreaming can be minimized by installing cooled baffles in the line of sight of the oil molecules entering into the chamber. Also, a nitrogen cold trap helps to condense backstreaming molecules. A cold trap can be placed between the mechanical pump and diffusion pump to prevent mechanical pump oil or contaminants from reaching the diffusion pump. Baffles slow down a system but decrease the ultimate pressure that can be attained in the system. Diffusion pumps operate inefficiently above 10^{-3} torr.

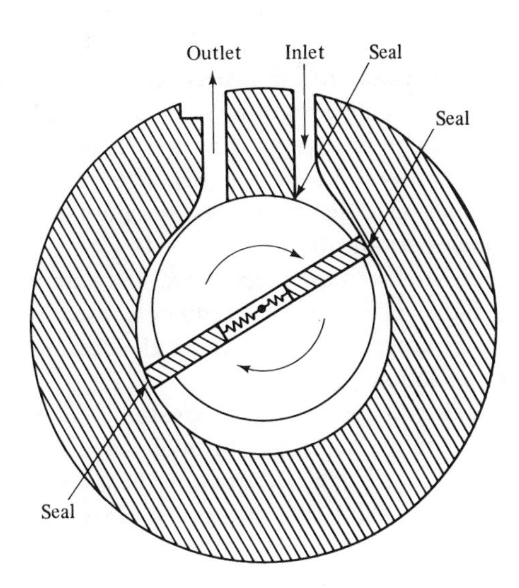

Fig. 12-28 Mechanical rotary pump.

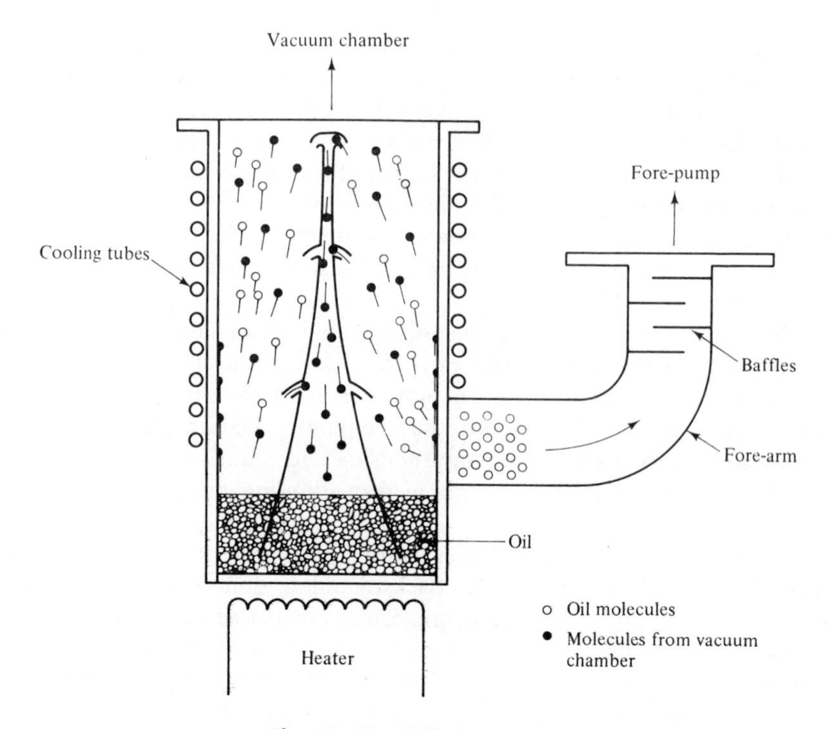

Fig. 12-29 Diffusion pump.

Other common methods of achieving high vacuum are cryogenic pumping, chemisorption, or gettering. In cryogenic pumping, molecules are condensed on walls held at very low temperature (e.g., liquid helium temperature of 4.2 K). Ultra-high vacuum can be achieved in this fashion. Chemisorption refers to the actual chemical reaction between a reactive metal surface (e.g., titanium) and the gas molecules, resulting in their removal from the system. This is referred to as gettering. Water vapor, O_2, and N_2 are commonly removed from vacuum systems this way. Titanium is deposited on removable plates by sublimation from an internal radiantly heated titanium source. The plates can be removed and cleaned periodically.

There are perhaps more types of vacuum gauges than there are vacuum pumps. Three of the more commonly used types will be discussed here for illustrative purposes. The three are the thermocouple gauge, the Pirani gauge, and the ionization gauge.

The thermocouple gauge consists of a thermocouple attached to a strip of metal, such as platinum, through which a constant current of around 50 mA is maintained. The Ohm's law heating (I^2R heating) of the metal strip constitutes a constant rate of heating. Heat removal from the strip by radiation is essentially constant. However, the heat loss due to the impinging gas molecules varies with the pressure of the gas. Hence, the *EMF* developed by the thermocouple can be calibrated in terms of absolute pressure. The calibration is different for different gasses. This type of gauge has a useful range from atmospheric pressure to about 10^{-3} torr.

The Pirani gauge consists of a wire heated by passage of a current through it, acting as one leg of a Wheatstone bridge. Variations in heat removal from the wire with the pressure of the gas causes resistance variations which are detected as an imbalance of the bridge. A wire, identical to the measuring wire, is placed in an enclosure at constant pressure to act as a reference. The range of utility of this type of gauge is about 1 to 10^{-3} torr and its calibration depends on the gas.

The ionization gauge employs thermionic emission of electrons from a filament, the cathode, which are accelerated through a high voltage toward the anode, usually a spiral winding. The electrons strike molecules on their way toward the anode, ionizing them. The ions are accelerated toward a wire running down through the center of the spiral anode. This wire is held at a negative potential relative to the anode. The resulting ionization current is proportional to the gas pressure. The electrons mostly miss the cathode and are absorbed by the tube surrounding the device. The electrons gain enough energy to cause ionization only after traveling far enough from the cathode that the ions will be accelerated to the collector and not the cathode. Figure 12-30 is a drawing of such an ionization gauge.

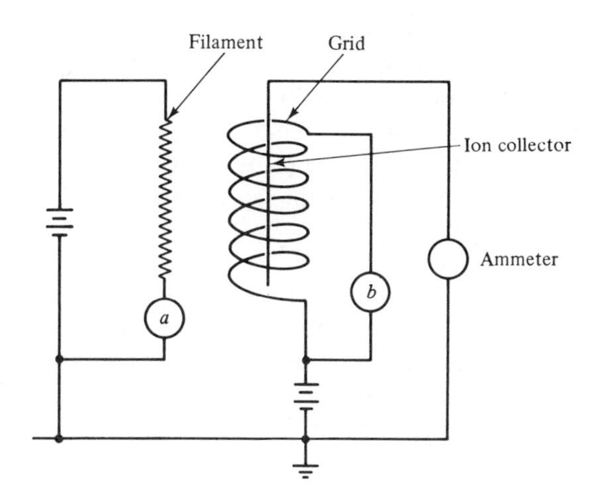

Fig. 12-30 Schematic of an ionization gauge. The power supply to the heat filament is located at a; the power supply for outgassing, at b.

12-8-2 Techniques of Film Deposition

There are many methods for depositing thin films, not all of which are vacuum techniques. Because of space limitations, only a few selected techniques will be discussed. These are the vacuum techniques: evaporation, sputtering, and electron-beam evaporation; and the nonvacuum techniques: electrochemical plating (including anodization), electroless plating, and chemical vapor plating. Chemical vapor plating has been covered to some extent in the discussion of epitaxial growth and will be treated very briefly here.

In any type of vacuum deposition, the importance of an initial pump down to high vacuum cannot be overestimated. System cleanliness is an extremely important factor. Below 10^{-6} torr, outgassing, that is, the removal of adsorbed gases, is the major source of "virtual" leaks in a tight system. The significance of system cleanliness and care in handling parts to be placed in the vacuum system is exemplified by the fact that the vapor pressure of a fingerprint is 10^{-8} torr. A thin layer of adsorbed gas on the surface on which a thin film is to be deposited can have serious adverse effects on the sticking coefficient and the properties of the film, especially its adherence to the surface. One common technique used to remove monolayers of adsorbed gases in a vacuum system is called ion scrubbing. In this process a glow discharge of an inert gas such as Ar is established throughout the entire vacuum enclosure. The argon ions impinging on the surface "scrub" them clean by knocking off adsorbed molecules, which can then be carried off by the high-vacuum pump.

Vapors are produced for thin-film deposition by two basic techniques. The first is simply to heat the material to be vaporized by passing a high current through a refractory metal filament, such as tungsten, for vaporizing wires, strips, or slugs of the material. The second is by impingement of a high-energy beam of electrons on the material (electron-beam evaporation).

Substrates are usually placed above the evaporant so that chunks or drops that come off will not fall on them. If the evaporant is essentially a point source of vapor molecules, the substrates are arranged on the surface of a sphere with the evaporant at the center for uniform deposition on all substrates. If the evaporant is more like a plane emitting surface, the evaporant should be placed on the inside surface of the same imaginary sphere as the substrates, for optimum deposition.

The energy that is required to release molecules from an evaporant is just the heat of vaporization plus their average kinetic energy, which is $\frac{3}{2}\,kT$. The average kinetic energy at 300 K is only a few hundredths of an electron volt and is negligible compared with the heat of vaporization. It is generally accepted that the vapor pressure of the evaporant, for good evaporation deposition, should be maintained at about 10^{-2} torr. The pressure difference between the vapor pressure of the evaporant and the system pressure (about 10^{-6} torr) controls the rate of evaporation and deposition. Since vapor pressure increases rapidly with increasing temperature, care must be taken to control the temperature of the evaporant or the film thickness must be monitored during deposition to get the desired thickness. Materials that sublime (change phase directly from a solid to a gas without a liquid phase) can be plated on filaments or vaporized in a boat. Outgassing of the evaporant is important to prevent contamination of the deposited film and to prevent splattering as trapped gases bubble out.

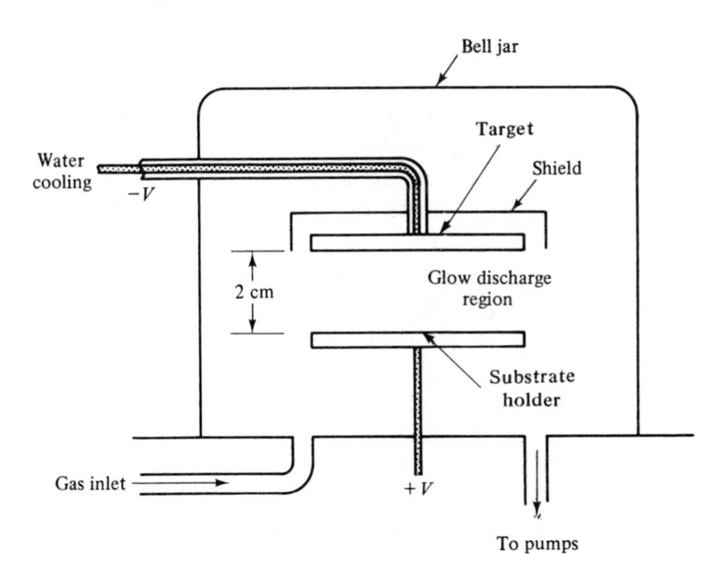

Fig. 12-31 Simplified diagram of a dc sputtering system.

Placing a shutter between the evaporant and substrates for a period of time to achieve outgassing is a good technique.

Evaporant molecules travel in all possible directions in the line of sight of the filament or boat. If they approach a substrate within a few atomic diameters, weak van der Waals forces[13] of attraction pull them in. The molecules may bounce off (reflect) or stick. Sticking may be by van der Waals forces or chemisorption for a reactive surface. The binding energies for these forces are about 10 kcal/mol and 100 kcal/mol, respectively. As the film builds up, the desorption energy (energy required to remove an adsorbed molecule) approaches the heat of vaporization of the bulk material, so that atoms are much more likely to stay on the surface. Of course, if the temperature of the substrate is too high, adsorption may be slowed or prevented.

In electron beam evaporation, the heat of vaporization is supplied by the impact of a beam of high-energy electrons. Electron-beam deposition is ideally suited for depositing refractory metals and insulators because of the high energy of the impinging electrons. The temperature rise is localized, hence minimizing outgassing, and no filaments that can add contaminants to the system are required. Electron-beam deposition has found wide acceptance because of its inherent cleanliness and versatility.

Probably the first vacuum deposition technique ever tried was sputtering. However, it is only in recent years that this technique has become acceptable for electronic thin-film deposition. Figure 12-31 is a simplified drawing of a

[13] Van der Waals forces are weak forces of attraction between neutral atoms or molecules due to the mutual inducement of dipole moments in their charge distributions.

dc sputtering system. A target, the cathode, and a substrate holder, the anode, are separated by about 2 cm. A gas, such as argon, is leaked into the system at about 10^{-3} torr, after the system has initially been pumped down to the 10^{-6} torr range. A glow discharge is established between the plates due to excitation of ionized argon. Ions of the sputtering gas are accelerated into the cathode. Apparently, as a result of direct momentum transfer, atoms of the target material are knocked out and many of them impinge on the grounded substrate holder, forming a film on anything supported by the holder. In dc sputtering, the target must be a conducting material. By the use of radio-frequency (RF) excitation, virtually any material can be sputtered, including ceramics.

Sputtering occurs for voltages in the approximate range 100 V to 10 kV. Above 30 kV, the ions embed themselves too deeply in the target to cause sputtering. Below 25 V, the energy of the ions is insufficient to cause sputtering. In RF sputtering, the frequency used is generally 13.56 MHz, which is one of the frequencies assigned by the Federal Communications Commission (FCC) for industrial applications. No license is required to operate RF equipment at that frequency. The equipment manufacturer is required to have FCC certification that the equipment is properly designed to operate at only the industrial frequencies. In RF systems, the total power consumption is around 600 W at several kilovolts. Coils are placed around the bell jar to provide a small magnetic field between and perpendicular to the plates. This causes the electrons to follow spiral paths and consequently increases their path length and the probability that they will cause ionization of the gas atoms.

The gases that produce the highest sputtering yield are the noble gases Ne, Ar, Kr, and Xe. Argon is the most commonly used. A reactive gas may be used or mixed with the noble gas to deposit a chemical compound of the sputtering target material and the reactive gas. As an example, SiO_2 films can be deposited by sputtering a Si target in an O_2 atmosphere.

Figure 12-32 is a picture of a laboratory RF sputtering setup.

In addition to the fact that almost anything can be sputtered, these films have better adherence properties and are less sensitive to contamination of the substrate. This is because the sputtered molecules or atoms strike the substrate with considerably more kinetic energy than in the case of evaporation deposition. Furthermore, alloys can be sputtered in essentially their bulk composition even if the constituents have different sputtering yields. For example, in Nichrome (80% Cr, 20% Ni), the chromium has a higher sputtering yield than Ni does. Initially, more Cr is removed from the target than Ni and the substrate film interface is Cr-rich. This is desirable because Cr has excellent adherence to almost all substrate materials. The target becomes enriched in Ni until an equilibrium is reached at which the exact ratio of the constituents in the original bulk target is being sputtered. Hence, the remainder of the film has an 80% Cr, 20% Ni composition.

Contamination can be a serious problem in sputtering since diffusion pumps are inefficient in the 10^{-3}-torr range. Considerable backstreaming can lead to

Fig. 12-32 RF Sputtering system removed from the vacuum system bell jar.

diffusion pump oil contamination. This can be minimized with cooled baffles and cold traps. One way of cutting backstreaming is to throttle the inlet (i.e., decrease its cross-sectional area). Since this slows the pump, the system must be free of real or virtual leaks. Gettering by clean metal surfaces such as titanium may be used to remove some of the common gases, such as H_2O, O_2, and N_2.

One of the remarkable features of reactively sputtering metal oxides is that by varying the concentration of the reactive gas in the noble gas, the deposited film can be varied from pure metal, to semimetal, to semiconductor, to insulator. It is also found that the solid solubility of the reactive gas in the metal may be significantly exceeded before compound formation takes place. In tantalum (one of the most imporatnt electronics thin-film materials), the solubility is exceeded by at least a factor of 2 before oxidation occurs.

In some sputtering applications, the substrate is negatively biased. This results in the acceleration of ionized target atoms or molecules into the target. This bombardment can cause significant differences in film properties. Tantalum normally has a body centered cubic (bcc) structure both in bulk and thin-film form. With negative biasing, Ta films form a tetragonal crystal structure referred to as β-Ta. The β-Ta has quite different electrical properties than those of bcc Ta.

Some electroplating is done in the production of thin films, particularly copper. An electroplating setup is schematically represented in Fig. 12-33. In this process, a cathode (connected to the negative battery terminal) and an anode (connected to the positive side of the battery) are immersed in an electro-

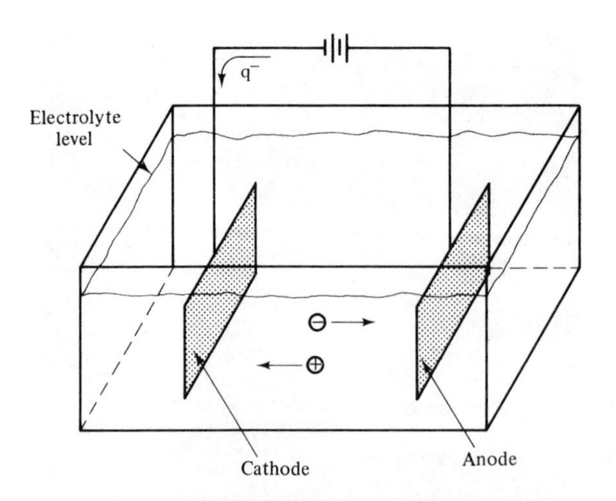

Fig. 12-33 Electroplating.

lytic solution. The anode is made of the material to be deposited and the cathode is the substrate upon which the film is to be deposited. The electrolyte is either an acidic or an alkaline solution in which molecular dissociation occurs, producing ions and thus rendering the solution conducting. Metal atoms go into solution at the anode and dissociate into positive ions and electrons. The metal ions are attracted to the cathode while the electrons return to the anode and travel around the conducting path through the battery to the cathode. At the cathode metal ions and electrons recombine and plate out. Hydrogen is evolved at the cathode as a result of recombination of hydrogen ions and electrons. Excessive hydrogen bubble-off can degrade the quality of the plated surface. Proper control of the pH[14] of the electrolyte will minimize this problem. The cathode must be extremely clean to ensure good film adherence and quality. Wetting agents may be added to the solution to improve the wettability of the cathode by the plating material. A rough cathode will lead to poor uniformity of the plated film because protrusions plate thicker than crevices or hollows.

Anodization, used for trimming Ta thin-film resistors and producing Ta_2O_5 for use as a capacitor dielectric, is an electrochemical process. In this case, an inert cathode is used and the metal of the anode reacts with hydroxide ions in the solution to form an oxide layer. In the case of Ta, the following reaction can be written:

$$2Ta + 10(OH^-) \rightarrow Ta_2O_5 + 5H_2O + 10e \tag{12-2}$$

In this reaction, as with niobium (Nb), Al, and tungsten (W) anodization, the reaction can take place at the metal–metal oxide interface or at the metal–

[14] The term "pH" refers to the negative of the base 10 logarithm of the hydrogen ion concentration in g-atom/liter. g-atom refers to gram atomic mass, the mass of an Avogodro's number of hydrogen ions in this case.

oxide surface because both ions can migrate through the oxide layer. The resistance of the oxide layer increases with oxide thickness so, if the current is to be held constant, the voltage must increase. In the production of Ta thin-film capacitors, care must be taken to prevent breakdown at pinholes in the Ta_2O_5 dielectric. One method is to put the Ta anode with its Ta_2O_5 layer in an electrolytic solution in which Ta is soluble. The Ta around the pinholes is dissolved and conducted away to the cathode. The anode is then anodized again for a short time to ensure a high-quality oxide film.

Electroless plating is a process whereby a material can be plated out of solution onto a surface that behaves as a catalyst. Nickel is electroless-plated in a hypophosphite (H_2PO_2) bath. Catalysts are Ni or any more reactive metal than Ni.

Chemical vapor deposition, as has been discussed previously, is a process whereby gases are passed over an inductively or radiantly heated substrate and as a result of chemical reactions in the gas stream and at the surface of the substrate a film is grown. This can be an epitaxial layer on a single-crystal substrate of Si, Si on a sapphire substrate, or a dielectric such as amorphous SiO_2 on a substrate such as Si.

$$SiCl_4 + 2H_2O \rightarrow SiO_2 + 4HCl \tag{12-3}$$

The reaction is an example of the growth of an amorphous dielectric film (SiO_2) by chemical vapor deposition. Another very important type of insulating film grown by this technique is silicon nitride (Si_3N_4), which has been discussed previously. Also, high-quality tungsten metalizations have been deposited by the chemical vapor process.

12-8-3 Thin-Film Design and Mask Layout

The discussion of thin-film circuit design and layout will be minimized since the process for defining resistors, conductors, and capacitors is basically a lithographic technique and this has been covered extensively for ICs and thick-film circuits previously. Thin-film transistors will not be considered.

As in thick-film circuit design, the designer must be cognizant of the limitations and capabilities of thin-film circuits. This includes sheet resistance, power-handling capability of resistors and substrates, TCR, capacitance per unit area capability of dielectrics, breakdown strength of dielectrics, substrate resistance, and for high-frequency applications, substrate dielectric constant. Since there are so many different thin-film materials and substrates, it is not feasible to cite simple rules of thumb as is sometimes possible for thick-film designing. Some of the basic mask design rules will be outlined here. Keep in mind, when numbers are given, they are only order of magnitude.

Although the absolute limit for photolithographic production of thin-film lines is about 1 to 2 μm, the actual line width of resistors is seldom less than

250 μm and conductor paths are usually about the same width. These dimensions are dictated by sheet resistance limitations which will be discussed later when thin-film materials are discussed. Line separation can be quite small, 25 μm or less, depending on the accuracy of the mask-generating procedure. Drawings of thin-film masks are usually made at least 20 times actual size. They can be drawn by standard drafting techniques where low accuracy is required. Another method is to use cut tape, as is done in the fabrication of printed circuits. The artwork can also be produced by one of the computer-generated artwork systems: that is, either by digitizing a master drawing and then generating the mask directly or by cutting a master out of Rubylith with a computer-controlled coordinatograph. By the use of interactive graphics, the circuit layout can be designed and produced directly with the aid of the computer.

The thin-film patterns are usually generated on the substrate by one of two methods. The first consists of depositing the film, which may be multilayer, over the entire substrate and then etching away the film where it is not wanted, protecting the desired parts with one of a number of possible photoresists. The second technique involves masking the substrate during the deposition process by a thin foil mask that has the desired pattern etched into it. The latter technique has the advantage that more than one masking can be done without breaking the vacuum by means of mechanical feed-throughs, and the photoresist development and rinse steps are eliminated except for the initial production of the mask.

A few of the etchants that can be used with thin-film materials are HCl for Nichrome, Cu, and manganese oxide (MnO); HF for tantalum oxide and Ti; aqua regia for Au and Pd; and HF + HNO_3 for Ta. Photoresists have been discussed previously. Metal masks are often made of thin foil molybdenum for high-temperature stability.

12-8-4 Thin-Film Materials

A few of the materials used in thin-film electronics will be elaborated on here, but this will be, by no means, an exhaustive list.

As pointed out in the introduction to this chapter, Nichrome and gold have been used successfully for a long time to produce resistors and conductors. Nichrome is sputtered or vacuum-evaporated to produce an initial layer, and gold is similarly deposited on top of the Nichrome. Gold is then etched away except where conductor paths are desired. Then the Nichrome is etched away everywhere except where covered by gold or where resistors are desired. Since Cr evaporates and sputters faster than Ni, the deposition of Nichrome results in a Cr-rich interface between the film and substrate. This is desirable because of the superior adherence of Cr on most substrates. Titanium also has good adherence properties, especially on glass, and is sometimes used for the initial layer. One serious difficulty with Nichrome evaporation is that it reacts with tungsten above 1500°C, so care must be taken not to permit hot spots when using tungsten filaments.

Aluminum can be deposited by any vacuum technique. Paladium can be deposited by either of the techniques and can be used as a protective layer because of its high resistance to oxidation. Refractory metals, such as W, Ta, and Mo, which melt at 3380, 2996, and 2610°C, respectively, must be deposited by either electron-beam evaporation or sputtering. As mentioned previously, Ta can have either its bulk fcc structure or a tetragonal (β-Ta) structure when the Ta bombards the substrate as a result of negative biasing. The β-Ta phase can occur in very thin films when bombardment is not used. Tantalum can be reactively sputtered with nitrogen or oxygen to produce oxide or nitride films of Ta. The TCR and resistivity vary with the partial pressure of N_2 or O_2 in the sputtering gas. For example, when starting at low N_2 vapor pressure, N is dissolved in the Ta film. At higher partial pressures Ta_2N is formed, and at even higher partial pressures, TaN is formed.

Complete Ta circuits can be constructed by using Ta for the conductors and nitrided, oxided, or anodized Ta for the resistors and capacitor dielectrics. To stabilize resistors, a preaging process is used, such as 5 h at 250°C or a higher temperature for a shorter time. This accelerates migration, chemical reaction, and oxidation effects that are self-limiting but continue for some time after a film is deposited, causing drift in resistor properties.

A common capacitor dielectric is silicon monoxide (SiO). This material is deposited by evaporation. Sometimes manganese oxide (MnO) is used between the capacitor dielectric and the metal electrodes. MnO is a semiconductor, but the high electric field at the edges of pinholes in the dielectric cause MnO to become insulating, thus preventing the pinholes in the dielectric from forming conducting paths between the electrodes. When MnO is not used, a capacitor discharge may be used to burn away the metalization around the edges of pinholes, thus preventing formation of a conducting path.

The substrates that are used in thin-film applications must have a better surface smoothness than thick-film substrates. Glass and ceramic substrates such as BeO, Al_2O_3, or sapphire (single-crystal Al_2O_3) are used. Alumina substrates are manufactured with sufficiently smooth surfaces (<1 μm rms surface variation) without glazing (i.e., they are satisfactory as fired). As pointed out earlier, cleanliness of the substrates is of the utmost importance for the proper adherence of thin films.

BIBLIOGRAPHY

[1] TOPFER, M. L., *Thick-Film Microelectronics.* New York: Van Nostrand Reinhold Company, 1971.

[2] AGNEW, J., *Thick-Film Technology.* Rochelle Park, N.J.: Hayden Book Company, Inc., 1973.

[3] HAMER, D. W., AND J. V. BIGGERS, *Thick-Film Hybrid Microcircuit Technology.* New York: Wiley–Interscience, 1972.

[4] RIKOWSKI, R. A., *Hybrid Microelectronic Circuits—The Thick-Film.* New York: John Wiley & Sons, Inc., 1973.

[5] HUGHES, D. C., JR., *Screen Printing of Microcircuits.* Somerville, N.J.: Dan Mor Publishing Company, 1967.

[6] MILLER, L. F., *Thick-Film Technology and Chip Joining.* New York: Gordon and Breach, Science Publishers, Inc., 1972.

[7] HARPER, C. A., ED., *Handbook of Thick-Film Hybrid Microelectronics.* New York: McGraw-Hill Book Company, 1974.

[8] HOLLAND, L., ED., *Thin-Film Microelectronics.* New York: John Wiley & Sons, Inc., 1965.

[9] BERRY, R. W., P. M. HALL, AND M. T. HARRIS, *Thin-Film Technology.* New York: Van Nostrand Reinhold Company, 1968.

PROBLEMS

12-1 Make a footprint diagram, composite layout, and a drawing of each required mask 5× scale for the circuit shown. Assume this circuit is to be placed on a 2.54 × 2.54 cm substrate and that a lead frame with 0.254 cm center-to-center spacing and wide contact pads will be used for input, output, ground, and V_{CC} connections. Available resistor inks are 2 kΩ/□ and 100 kΩ/□. Assume that the transistor to be used is a 1 × 1 mm *n-p-n* chip with the collector common with the back of the chip, the base contact in the middle, and emitter contact surrounding the base. The circuit is a simple one-stage amplifier.

$$V_{CC} = 20\text{ V}$$
$$R_C = 10\text{ k}\Omega$$
$$R_E = 1\text{ k}\Omega$$
$$R_1 = 184\text{ k}\Omega$$
$$R_2 = 16\text{ k}\Omega$$

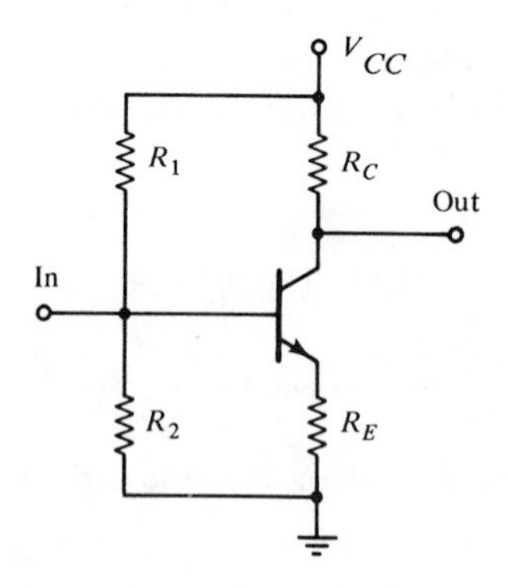

12-2 (a) Calculate the number of molecules in a vacuum system at pressures of 5×10^{-6} torr and 1×10^{-4} torr for a temperature of 300 K.

(b) Calculate the mean free path for air under the conditions given in part (a).

(c) For a sticking coefficient of 0.3, calculate the rate of buildup of molecules on a surface for the conditions given in part (a).

bonding, packaging, and testing

13

Integrated circuit chips are either packaged separately or are incorporated into a hybrid circuit. In either case, a general knowledge of techniques for bonding chips and making connections to other circuitry or the outside world is important. After bonding and interconnections are completed, some form of packaging is required to give the circuit mechanical and environmental stability. Before a chip is bonded and packaged, it must be thoroughly tested functionally prior to being separated from the rest of the wafer. After bonding and packaging, many additional tests—electrical, mechanical, and environmental are required.

A few common bonding and packaging techniques are described in this chapter, together with some general concepts of testing. Testing of microelectronics devices and circuits depends on the nature and application of the device or circuit, and as there are literally hundreds of tests that can be performed, only a few of the usually essential ones will be discussed. The chief emphasis is on silicon monolithic ICs, but where appropriate, discussions of hybrids are included.

13-1 BONDING

The term *bonding,* as used here, refers to the attachment of silicon IC chips (dice) to lead frames or other types of substrates and wire bonding of IC chips for external (to the chip) connections.

13-1-1 Die Bonding

Integrated-circuit chips are bonded by a wide variety of techniques. Silicon chips can be ultrasonically or eutectically bonded to a gold-plated or clad header; a gold preform can be utilized as a bonding pad as well. Often the backs of the chips are gold-plated to enhance the bondability. In this technique, energy is supplied by heating the die and the surface to which it is being bonded to the eutectic temperature of the Au–Si system, 370°C. Additional energy may be supplied by vibration of the die collet (see Fig. 13-1) as the chip is put in place. In thermocompression bonding, a slight mechanical scrubbing action is often used to achieve intimate contact between the bonding surfaces as force is applied normal to the bonding surfaces. The bonding takes place in a nitrogen atmosphere to prevent oxidation.

Silicon dice can be back-bonded by solder reflow. When this is done, the chip and/or the surface to which it is being bonded may be nickel-plated, since nickel has excellent adherence to most surfaces.

Epoxy die bonding is another technique for placing die on various surfaces. A spot of epoxy is put down first and then the die is placed on top of it. Numerous epoxy compounds are available, including one- and two-component systems, of both conducting and insulating types. The epoxy may be cured at an elevated temperature to speed up the curing process, but the temperatures are low enough to prevent damage to the device.

To avoid wire bonding, flip chips are often used. These are IC chips on which metal bumps have been placed to make electrical contact with appropriate pads on the substrate to which the chip is to be bonded. The device is therefore mounted inverted, hence the name "flip chip." Because these dice are mounted face down, special care must be taken to assure proper orientation.

The metallurgy of flip chip bumps is fairly complicated. Some general considerations for placing bumps on ICs will be presented here rather than a description of a specific type.

In producing flip-chip bumps, contact to Si can be made with Al as usual. A thin barrier layer such as chromium (Cr) is used. Vapor-deposited chromium adheres well to almost everything and consequently makes a good interface

Fig. 13-1 Ultrasonic or thermocompression bonding.

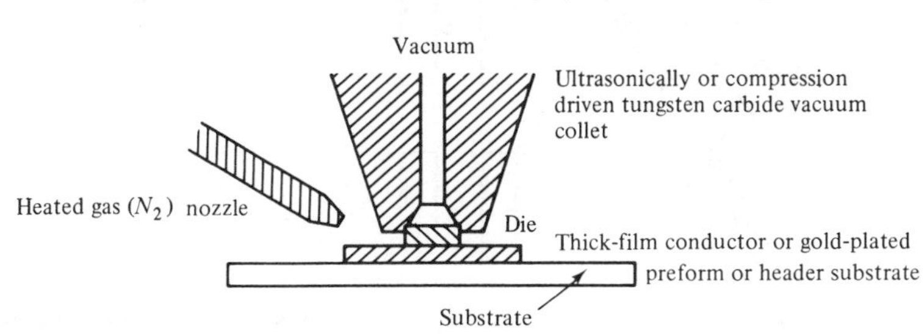

material. Next, nickel (Ni), which is mechanically strong and a good conductor, can be electrolessly plated to form the bump if the bump is to be a Sn–Pb solder type. Dipping the entire wafer in a low-temperature Sn–Pb solder coats the nickel with solder. A final electroless-plated thin coating of Au may be used to prevent oxidation of the solder. This gold is leached away during the reflow attachment process. In some cases a mixture of Cr and Cu is used in place of the Ni, and Sn–Pb solder is deposited and reflowed to form a spherical bump. For chips being attached to substrates that have been solder-dipped or solder-printed, electroplated silver may serve as the major component of the bump, since it solders well with Sn–Pb solders. For gold-plated or gold-bearing substrate bonding pads or lead frames, a Au–Sn solder can be used. The eutectic temperature of an 80% Au–20% Sn by weight mixture is 280° C. Such a mixture bonded to a gold pad absorbs gold and stays partially solid and partially liquid up to the melting point of Au at 1063° C. When chromium's high resistivity can be tolerated, a thicker layer (not absorbed by other layers) serves as an excellent diffusion barrier to prevent Au or Ni from diffusing to the Al and through to the underlying Si.

Beam-lead devices (see Fig. 13-2) are produced by depositing chromium or gold in the appropriate areas on the circuit side of the wafer and then selectively etching away the silicon from the backside to separate the chips and leave the beam leads.

Leadless inverted devices eliminate the need for wire bonding at the stage of production where the device is placed in the circuit. These are ceramic packages in which the die is initially bonded; this involves die bonding to a substrate and wire bonding to make contact to external lead pads. A picture of such a device is presented in Fig. 13-3. This is essentially a packaging technique, but LIDs can be used in place of wire-bonded dice, flip chips, and so on, in microelectronic circuits, where space and cost are not factors.

An important bonding approach for the high-production packaging of ICs is a beam-tape bonding. In this technique, metal leads are produced on a continuous metal or plastic tape by a combination of photolithographic etching and/or plating process. Flip chips are then bonded (thermocompression, Pb–Sn or Au–Sn solder) to the tape leads. Since the tape can be rolled up on a

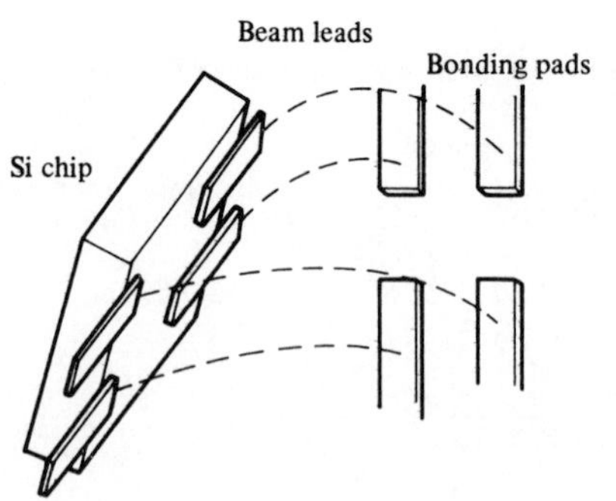

Fig. 13-2 Beam-lead device.

Fig. 13-3 Leadless inverted device (LID).

reel, this serves as an excellent means of facilitating high-speed parts handling. The tape is reeled out, cut, and the leads on the tape are bonded to leads on a printed circuit board, terminal frame, or thick/thin-film circuit substrate.

13-1-2 Wire Bonding

Microelectronic wire bonding is done by thermocompression, ultrasonic, or thermosonic bonding. The wire is either gold or aluminum with about a 25-μm diameter. In spite of the apparent advantages of direct interconnection bonding techniques such as flip chips, wire bonding is a widely used and highly successful process. Automatic wire bonders (see Fig. 13-4) have greatly improved

Fig. 13-4 Automatic wire bonder station. (Courtesy of Delco Electronics Division of GMC.)

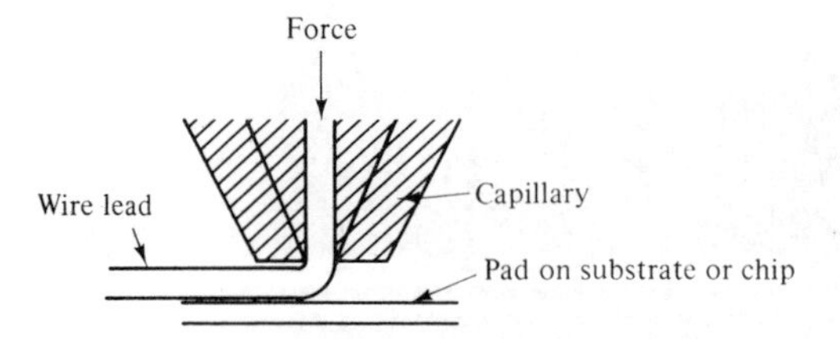

Fig. 13-5 Thermocompression and/or ultrasonic wirebonding.

the economical competitiveness of this process, and the inherent delicacy of the wires and the bonds is overcome through the use of conformal coatings or potting compounds combined with plastic or ceramic packages or injection molding of the device into a plastic dual-in-line package (DIP).

The process of attaching a wire from a pad on the silicon chip to an external bonding pad or post by the thermocompression or ultrasonic technique is called stitch bonding. The wire is fed through a hole that exits underneath a wedge tool. A tail of the wire extends out from under the wedge and essentially parallel to the bonding surface. The wedge is brought down on the wire forcing it onto the bonding pad. In thermocompression bonding, heat and pressure are applied to achieve the bond. In ultrasonic bonding sufficient energy is supplied by the ultrasonic vibration to achieve an excellent bond. A cover gas may be used to prevent oxidation. A schematic representation of thermocompression and/or ultrasonic bonding is depicted in Fig. 13-5. The setup in ultrasonic bonding is very similar to thermocompression bonding except for the absence of heating, and force is required normal to the bonding surfaces to achieve a

Fig. 13-6 SEM micrograph of 25-μm Au wire ultrasonically bonded to an Al pad.

bond, together with an input of ultrasonic energy. Au–Al and Al–Al bonds can be made by either of these processes. Au–Au bonds are made by thermosonic bonding (a combination of ultrasonic and heat). Figure 13-6 is a scanning electron microscope (SEM) picture of 25-μm Au wire ultrasonically bonded to an Al bonding pad.

Ball bonding is a technique that is effective only with Au wire. The wire is cut by an oxyhydrogen torch or electric spark leaving a ball on the end of the wire. The surface tension of Al is not high enough to form a good ball. The Au ball is thermocompression-bonded to a pad on the die. The wire is then fed out as the tool moves to the external (to the die) pad or post, whereupon a conventional thermocompression wedge bond is made. After this bond is made, the tool lifts and breaks the wire from the wedge bond. The end of the wire is melted by the torch or electric spark, leaving a ball for the next bond to die. Figure 13-7 is an SEM picture of a ball bond.

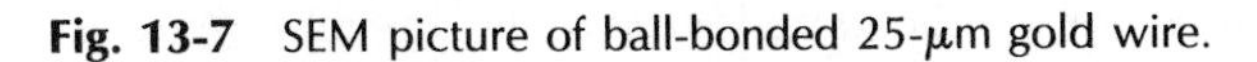

Fig. 13-7 SEM picture of ball-bonded 25-μm gold wire.

13-1-3 Hybrid Bonding

Bonding in hybrid circuits is essentially the same as described in Sections 13-1-1 and 13-1-2. Die bonders and wire bonders are adaptable for bonding in most types of packages, including thick-film and thin-film substrates. In thin-film hybrids the chips and the wire bonds are usually made to gold pads. In thick-film hybrids, care must be taken to ensure the bondability of both chips and wires to the thick-film conductor. Table 13-1 is a list of the bonding characteristics of some common thick-film conductor systems for thermocompression and ultrasonic bonding. When direct bonding of silicon devices to hybrid sub-

TABLE 13-1. Bondability of Thick-Film Conductor Materials

Thick-Film Conductor	Die bond		Wire bond	
	Au-Backed	Bare Si	Au	Al*
Au	—	Yes	Yes	Yes
Pt–Au	—	Yes	Yes	Yes
Pt–Pd–Au	—	Yes	Yes	Yes
Pd–Au	—	Yes	Yes	Yes
Pd–Ag	Yes	No	Yes	Yes
Ag	Yes	No	Yes	Yes

*Al wire cannot be ball-bonded and is not readily thermocompression-bonded because of surface oxides. It is usually ultrasonically bonded.

strates is not practical, LIDs, flat packs, or DIPs can be used. Gold-plated or clad Kovar preforms may be used to facilitate die bonding. Solder paste is extensively used for die bonding, particularly flip-chip types, as discussed in Chapter 12.

13-2 PACKAGING

To cover all the variety of microelectronics packaging techniques in detail would require a very large volume in itself. A brief description of some of the more common packages will be presented here. The LID has already been mentioned, TO-can types, dual-in-line package (DIP) types, and flat packs will be discussed in this section.

In TO-can-type packages, such as shown in Fig. 13-8, the silicon device is die-bonded to a gold-plated header and is wire-bonded to posts for external

Fig. 13-8 TO-5 package.

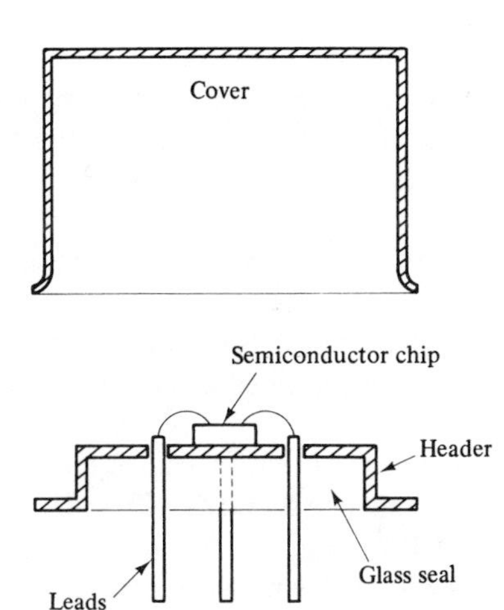

Fig. 13-9 TO-type package illustrating the materials and design.

connections. The cap is then affixed by resistance welding. If a hermetic seal is required, the cap is sealed in an inert atmosphere such as nitrogen. Figure 13-9 is a diagram of a TO-can-type package showing the details of its construction and the materials used in its fabrication.

The DIP is an extremely popular package because of its ease of handling and convenience for use in printed circuits. Figure 13-10 is a photograph of a ceramic and a plastic DIP with a socket. Ceramic DIPs have a gold-plated ceramic substrate for chip attachment and gold-plated Kovar (or a similar alloy)

Fig. 13-10 DIPs and plug-in socket.

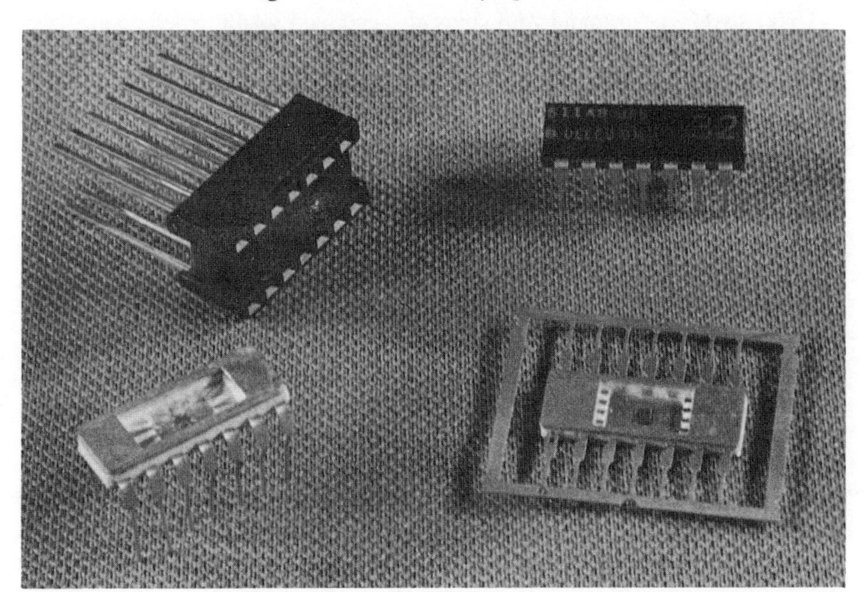

leads to which the chip is wire-bonded. The ceramic lid is attached by the use of sealing glass or epoxy. For laboratory use it is convenient to simply glue on a transparent lid in place of the ceramic lid. A flexible potting compound may be placed over the chip and wire bonds to ensure mechanical stability in either type of DIP.

Plastic or epoxy DIPs may be of the type shown in Fig. 13-10, or they may be molded with the bonded silicon chip as an integral part of the package. In the latter case, the chip is first bonded to a lead frame and wire-bonded and then epoxy is transfer-molded around the lead frame to form an essentially solid package.

Figure 13-11 is a picture of a lead frame with chips attached, and a lead frame with the molded epoxy DIPs. This figure illustrates the basic process for making solid DIPs.

A problem associated with nonhermetic packages is the tendency for Al to grow dendrites in the presence of moisture. This phenomenon, known as electromigration, can lead to shorts between metalizations or depletion of the metal to the point than an "open" can occur. This problem can be minimized by "glassivation" (pyrolytic deposition of glass over the metal) and by keeping the current density low in the Al.

Keep in mind that the pictures and illustrations included in this chapter are only samples of the many varieties available. They are shown here to serve as examples and to provide insight into the problems and complexities involved in microelectronics packaging.

Packaging of hybrid circuits is about as varied as the number of hybrid

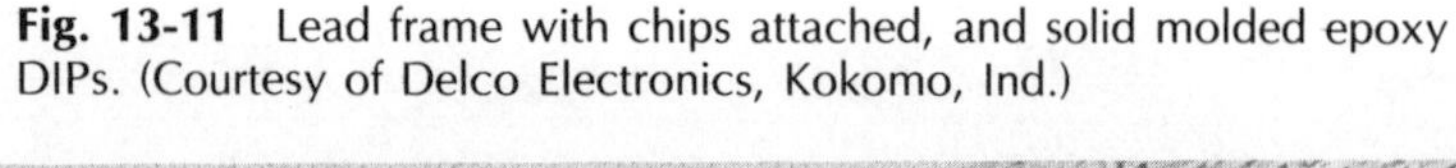

Fig. 13-11 Lead frame with chips attached, and solid molded epoxy DIPs. (Courtesy of Delco Electronics, Kokomo, Ind.)

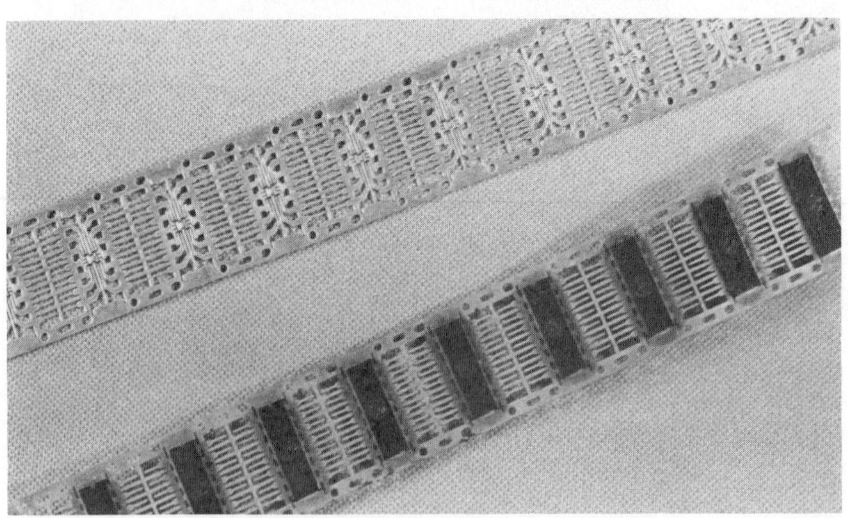

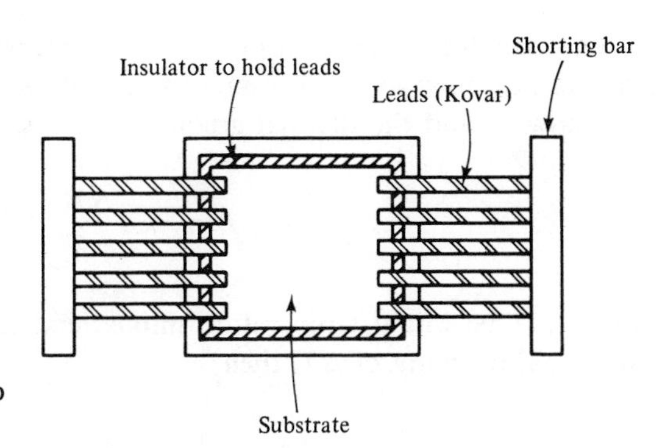

Fig. 13-12 Flat pack with top removed.

manufacturers. Most packaging is not hermetic.[1] When hermetic sealing is required, a flat-pack design is often used. A schematic of such a package is depicted in Fig. 13-12.

In some cases, the substrate is placed on a rigid metal platform and sealed with a vertical wall cover. Others utilize a bottom which has one or more sides bent at 45° with a lip for top sealing. Leads, generally Kovar or similar alloy, are passed through glass–metal seals. Flat packs may be either all Kovar, all Corning 7052 glass, a Kovar–Corning 7052 glass combination, a Corning 7052 glass–ceramic combination, a Kovar–ceramic combination, or all-ceramic. Many different types of lead seals are also employed. One common type is a glass frit which is sealed in a conveyor furnace in an inert atmosphere.

Where complete hermeticity is not required, hybrids may simply be covered with conformal coatings, potting compounds, or a rigid plastic coating. Plastic packages and combination metal and plastic or metal and ceramic may be used when a heat sink is required. The hybrid is usually covered with a flexible plastic coating to ensure mechanical integrity and to protect the circuit against vibration and mechanical shock.

13-2-1 Thermal Considerations

The upper temperature limit of operation for silicon ICs is 125 to 175°C. High temperatures can cause damage to hybrid components as well as short- and long-term variation in device parameters. Hence, the user and designer of microelectronics devices must be familiar with the effect of package design and ambient conditions on the temperature of IC or hybrid circuits. A rigorous analysis can become quite complicated in most cases. What will be presented here is a simplified approach which is intended to provide a basic understanding of the problem and also serves as a means of making a reasonable estimate of device temperature under given operating conditions.

[1] A hermetic package is one with a leak rate of less than 1×10^{-8} cm^3/s of helium under certain specified test conditions.

The one-dimensional steady-state heat-transfer equation, (13-1), relates the rate of heat flow to the thermal conductivity k, area normal to the direction of heat flow A, and the thermal gradient $\Delta T/\Delta X$:

$$P = -kA\frac{\Delta T}{\Delta X} \tag{13-1}$$

If we let ΔT be initial temperature minus final temperature, rather than the conventional meaning of ΔT, then

$$\Delta T = \theta P \tag{13-2}$$

where $\theta = \Delta X/kA$ is called thermal resistance. Equation (13-2) is analogous to Ohm's law, $V = RI$. Pursuing the analogy, it is clear that for parallel heat paths the equivalent thermal resistance will be given by

$$\frac{1}{R_{eq}} = \frac{1}{R_1} + \frac{1}{R_2} + \ldots \tag{13-3}$$

and for series heat paths,

$$R_{eq} = R_1 + R_2 + \ldots \tag{13-4}$$

Strictly speaking, the heat flow in microelectronic circuits is seldom one-dimensional. However, the rate of heat flow can always be linearly related to a temperature difference, so that the Ohm's law analogy is still valid. In actual practice, it may be quite difficult to determine a theoretical expression for the thermal resistance. Approximate thermal resistances for various paths are usually known or can be determined experimentally.

The thermal resistance of IC chips may vary from about 6°C/W for a 1-mm² chip to 0.7°C/W for a 9-mm² power chip. The thermal resistance of a TO-5 can is about 120°C/W. Hence, the total thermal resistance, IC chip to ambient, for a 1-mm² chip in a TO-5 can is around 126°C/W, since these resistances are in series.

The total, or equivalent, thermal resistance of a system can be broken down into numerous components. These include junction resistance, IC chip resistance exclusive of the junctions, chip-to-header bond resistance, substrate resistance (such as in hybrid circuits), and package resistance. Of course, the package resistance can be subdivided further into header and lead resistance. Usually, it is only when dealing with fairly high power devices [e.g., Darlington power transistors) that a complete analysis may be necessary. Most often, the operating temperature of the device (junction temperature) can be estimated with sufficient accuracy if the ambient temperature is known along with reasonable estimates of the package and chip thermal resistances. For example, if the thermal resistance of an IC chip is known to be 80°C/W and it dissipates 100 mW in a package that has a thermal resistance of 100°C/W at an ambient temperature of 30°C, what is the temperature of the junction?

$$T_J - 30°\text{C} = 0.100\ \text{W}\ (80°\text{C/W} + 100°\text{C/W})$$

or

$$T_J = 30°\text{C} + 18°\text{C} = 48°\text{C}$$

13-3 TESTING

Testing in microelectronics is such a vast subject that an entire textbook devoted to the area would probably just scratch the surface. Consequently, only some rather general observations will be made here. The intention is to provide the reader with some insight into the problems of quality control and reliability in microelectronics and to provide at least a starting point in determining what tests are necessary.

The term "testing," as used in this context, refers to a wide range of topics, from visual inspection of silicon wafers and circuits to complex electronic testing. This includes environmental tests such as thermal, humidity, vibration, mechanical shock, and corrosive atmospheres. It includes electrical tests of dc parameters, ac parameters, functional tests, operating life tests, and many others. For hermetic packages it includes various leak tests. Some types of tests are destructive and others are not. Diagnostic testing is important and includes such techniques as scanning electron microscopy (with and without voltage contrast and X-ray diagnostics), X-ray fluoroscopy, infrared (IR) microscopy, phase contrast and interference contrast microscopy, and IR thermography. Capacitance–voltage *(C–V)* testing of MOS configurations has been covered in Chapter 8 and will not be treated here.

13-3-1 Wafer Tests

With the advent of high-resolution photolithography (3 to 4-μm line widths), dimensional integrity of silicon wafers has become an extremely important consideration. The types of deformities that a wafer can have are chiefly warp, dish, bow, and taper. A warped wafer is one with various concave and convex regions. A dished wafer has one major concave or convex surface. A bowed wafer is one whose cross section in one direction would be concave or convex, and the cross section at right angles would be straight. Taper is the result of the two faces of the wafer not being parallel. In general, all these deformities are present in a wafer to some degree.

In lithography such deformities lead to loss of resolution. When the method used calls for contact between mask and wafer, areas not in contact will have poor line definition, owing to diffraction and divergence of the light source. In the off-contact process, line definition will also be reduced, because of poor focusing. The problems of warp, dish, and bow have been corrected to a large extent by better quality control of wafers and the use of wafer stages in mask aligners that automatically flatten the wafer.

Wafer deformities are measured either with a profilometer or interferometrically by placing an optical flat over the wafer and observing the curvature of the fringes. Noncontact capacitance measurements are used and have about a 0.25-μm resolution. Gauges using air pressure from a small orifice directed onto the wafer surface are also used but lack sufficient accuracy for high-resolution mask work.

Imperfections in the crystal itself may be a problem. Modern wafers are relatively defect-free but may contain dislocations and/or stacking faults. Stacking faults are particularly bothersome in epitaxial layers. The various types of crystal defects have been discussed in Chapter 2 and will not be stressed here. It is worth noting, however, that wafers can be easily damaged by mechanical handling. Large concentrations of dislocations are introduced in the regions where wafers touch the boat when placed vertically in a diffusion furnace and where handled by tweezers. Spikes, which are sharp protrusions from the epitaxial layer, are a problem, particularly in contact mask alignment. These are apparently caused by contamination on the wafer surface, which leads to excessively fast growth in the epitaxial reactor in a small region around the contaminant, thus causing a spike that protrudes well above the normal epitaxial surface. These defects are easily seen in an ordinary optical microscope.

Numerous electrical tests may be performed on wafers before processing begins. One of these is measurement of sheet resistance, usually by use of a four-point probe. Spreading resistance is also measured to determine the doping level at various distances into the wafer. This is done by beveling the wafer at a slight angle and using a two-point probe to measure the resistivity.

The quality of the oxide on a wafer may be tested by measuring breakdown voltage and leakage resistance. (C–V testing has been discussed in Chapter 8.) Pinholing is often a problem in oxide layers, especially the thin layers used in MOS devices. The distribution of pinholes over the surface may be determined by placing a gridwork of metal gates on the oxide and probing each gate separately. Capacitance, breakdown voltage, and isolation resistance can be measured at each site.

13-3-2 Electrical Testing of Processed Wafers

Test patterns may be probed at various stages of IC processing to determine if doping levels and diffusion processes are being properly maintained. For bipolar devices it is possible to tell by measuring the sheet resistance of a base-diffused resistor if the base diffusion is correct. The base–collector breakdown voltage indicates the quality of the base–collector junction. After emitter deposition and diffusion, β can be measured. If the diffusion has not gone far enough, the wafer(s) can be returned to the emitter-diffusion furnace tube. If the emitter diffusion has gone too far, thus wiping out the bases or resulting in a very high β with consequent low breakdown, the wafer(s) are scrapped before the metallization process is begun. The probing operation is carried out on an instrument of the sort shown in Fig. 13-13. This is a picture of a manually operated prober. A large number of probes can be lowered onto the die at one time. Once a wafer is completely processed, each die must be tested. This is generally carried out by an automated prober interfaced to a computer that is programmed to perform the appropriate tests and store or record data pertaining to each wafer. If the wafer contains transistors or transistor quads, then, as far as bipolar devices are concerned, the beta, leakage current, and breakdown voltage of each transistor is measured, and if a transistor is not within the specified tolerance placed on these values, it is automatically ink-marked so that it can be rejected when the wafer is scribed and broken into individual dice. If the dice contain complete circuits, a variety of electrical tests may be performed, depending on

Fig. 13-13 Manual IC probing station.

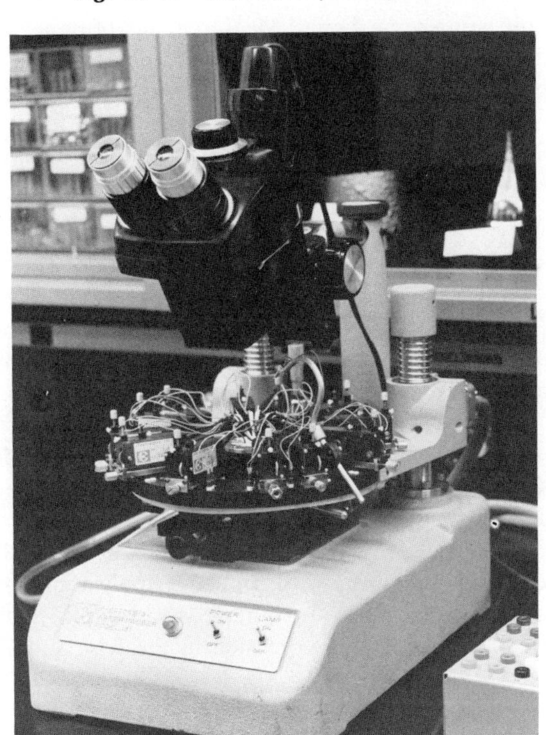

Fig. 13-14 Production IC probing station. (Courtesy of Delco Electronics Division of GMC.)

whether the circuit is linear or digital. Measurements of dc levels, leakage currents, input and output currents, gain, and frequency response are just a few of the tests made on linear circuits. For digital circuits, checks for proper response to inputs of 1's and 0's, and checks on logic levels, are a few tests that might be performed. Figure 13-14 is a picture of a production IC testing station.

13-3-3 Mechanical Testing

There are a variety of mechanical tests that are performed on ICs. The shear strength (force required to shear a bonded die from its bonding pad or header) may be tested by applying a known force on the side of the chip parallel to the bonding surface. This test may be destructive if the shear force is increased to the shearing point, or nondestructive if the maximum shear force is set at some upper level for a pass–fail type of measurement. Shear forces required to separate die from pad may range from several hundred to several thousand grams,[2] depending on the type of bond.

Wire bonds are tested by placing a small hook under the wire in the middle of the lead and pulling upward. In some applications, wire bonds are 100% tested on a pass–fail basis. In other cases, samples may be tested to determine the breaking strength. The tension in the wire and the normal and

[2] The gram, although not a legitimate unit of force, is conventionally used to indicate both shear force and wire pull force in integrated-circuit bonding.

shear forces on the bonding pads are a function of the length of the wire compared with the distance separating the bonds. If the bond separation is L and the wire length is L', then from Fig. 13-15,

$$\frac{L}{L'} = \cos \theta \tag{13-5}$$

From symmetry $F = 2P_N$ and

$$F' = F \sin \frac{\theta}{2} \tag{13-6}$$

which for $\theta = 45°$ means that

$$F' = 0.707\text{F}$$

The shear force P_s is given by

$$P_s = F' \cos \theta = F \cos \frac{\theta}{2} \tag{13-7}$$

Fig. 13-15 Diagram for wire pull test.

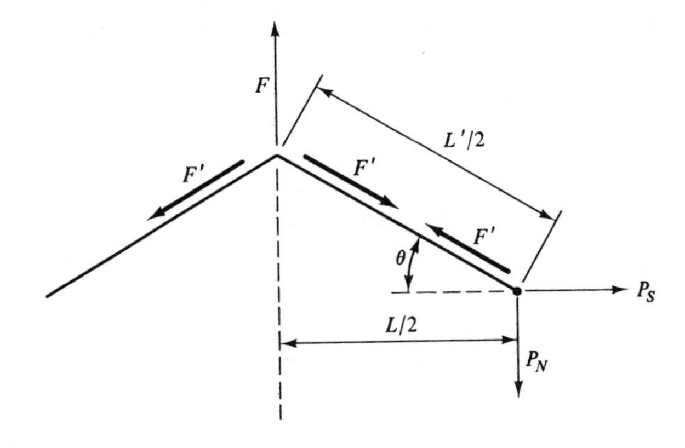

and the normal force on the bond is $F/2$. As can be seen from Eqs. (13-6) and (13-7), F' and P_s can become considerably greater than F when the wire length is not much greater than the distance L. For $\theta = 45°$, F may range from a few grams to tens of grams for 0.125-mm gold wire.

Failure of the wire pull test usually occurs at the wire bond due to insufficient bonding (underbonding), excessive bonding (overbonding), or some sort of plague in Au–Al bonds after excessive thermal exposure. Overbonding results in the wire or ball being squashed too much, causing the wire to neck down and become weakened. Underbonding results in poor metallurgical contact between pad and wire or ball. The plague problem, variously referred to as "purple plague" and "tan plague," is apparently due to the formation of vacancies on the Al-rich side of the Au–Al bond as a result of the more rapid diffusion of Al. These vacancies collect on microcracks formed at high temperature or during fairly high temperature thermal cycling to form voids. These voids, called Kirkendall voids, result in reduction of the strength of the bond and also can greatly increase the electrical resistance of the contact, even when the bond passes the mechanical test. For this reason, Au–Al bonding is done at relatively low temperatures and subsequent thermal cycling for other testing purposes (−55 to 175°C maximum) is kept low enough to minimize the problem. If high temperatures, 300°C or higher, are unavoidable, then Al–Al or Au–Au bonds should be used.

Numerous mechanical tests may be performed on completed packages. Some of them are listed here. Bending and tensile tests of package leads may be performed. Centrifuging up to 20,000 times the acceleration due to gravity to check for poor wire and die bonds, shorts, cracked substrates, and foreign matter in the package is often done. Variable frequency vibration with monitoring of output voltages, and noise output usually 10 to 2000 Hz with a maximum acceleration of 30 times gravity serves to detect faults similar to the centrifuge test under lower but variable stress conditions. Mechanical shock may be employed with up to 1500g applied in 0.5-ms intervals and repeated up to 25 times in any one direction. All these tests are carried out in each of three orthogonal directions, one normal to the package substrate and two in the plane of the substrate.

13-3-4 Thermal and Humidity Tests

Modern ICs are finding more and more applications in extremely harsh environments. These environments often involve extremes of temperature and humidity. Consequently, the proper electrical and mechanical behavior of ICs under such conditions must be assured.

Thermal shock cycling is used to accelerate the effects of thermal mismatch, cracks in the substrate, or high moisture content in the package. The cycling is carried out between −65 and 150 to 200°C, and electrical checks are made to determine if damage has occurred because of the cycling.

High-temperature storage tests may be performed. This involves nonoperating storage at 150°C for up to 1000 h. Operating-life tests are usually carried out at 125°C for 1000 h with a 50% duty cycle. For linear circuits this just means switching V_{CC} on and off at from 1 to 100 kHz, and for digital circuits 1's and 0's are applied to the inputs at 1 to 100 kHz.

If humidity is likely to be a problem, tests are conducted in a humdity-control chamber at 30 to 60°C while humidity is controlled at 90 to 98%

relative humidity. This is an operating-life test and usually runs 1000 h with a 50% duty cycle.

13-3-5 Electrical Tests

As with the types of testing that have been discussed thus far, the variety of electrical tests performed on ICs and hybrid circuits is huge. Only a small sampling of the possible tests will be covered, and undoubtedly many that are essential in some applications will not be included.

Most devices are put through a stabilization bake or "burn in" prior to final electrical testing. This is done to accelerate the stabilization of electrical parameters that undergo some drift due chiefly to the redistribution of charge in the oxide layer. This is done at something like 150°C for 50 h or 250°C for 10 h. The devices are not operating during this "burn-in" period.

DC parameters are generally checked. This might include dc voltage levels at various points in the circuit, input and output current and voltage levels, high and low voltage levels for digital circuits, and fan-in and fan-out for digital devices. AC parameters are checked, such as gain and bandwidth. Threshold levels may be determined as well as leakage currents. Power dissipation is often determined, as is input capacitance and device to package isolation resistance. The latter should be at least 15 MΩ at 0.5 V dc. It may be desirable to apply a high voltage momentarily to check for shorts in the oxide layers. Digital LSI circuits contain thousands of crossovers and a single oxide short can be disastrous. All of these tests, except the isolation resistance and oxide short tests, should be performed under the maximum operating stress levels of temperature and power supply voltages.

Functional tests include noise immunity and noise feedthrough for logic gates and common-mode rejection, balance and input offset levels for linear circuits. Dynamical tests such as rise and fall times, propagation delay times, and response to a step pulse are a few of the tests performed on digital circuits. Phase response and bandwidth are a couple of dynamical tests performed on linear circuits. Most electrical testing is handled automatically by digital data processors, which may be programmed to perform several hundred tests on a single IC.

Operating-life tests are often used to overcome the problem of infant mortality in microelectronic devices. These tests are carried out under maximum temperature and power conditions. The probability of failure is highest during the first few hours of operation because the serious problems of shorts, contamination, moisture, cracked substrate, poor bonds, and too thin metalization that may open due to heating, and many other possible defects, cause failure at an early operating age. The devices that survive the infant period then have a high probability of having a satisfactory operating life, usually considered to be at least 10,000 h.

13-3-6 Diagnostic Testing

Diagnostic tests range from the simple process of determining the quality of aluminum adhesion, to an IC or thin-film adhesion to a substrate by pressing plastic tape on it and peeling it off to see if the metal or thin film comes with it, to scanning electron microscope (SEM) voltage contrast inspection of LSI circuits for mask defects. The conventional optical microscope is adequate for most oxide and metalization inspection work. Since silicon is transparent to the infrared (IR), IR transmission microscopy can be used to detect many types of defects in silicon devices, especially concentrations of dislocations possibly introduced by handling. Also, IR thermography, a technique of thermally mapping the surface of an IC by means of the IR radiation emitted from it, can be helpful in locating hot spots and in analyzing heat flow in the circuit. However, the resolution of such techniques is not sufficient to resolve details in modern LSI circuits. The X-ray fluoroscope, basically the same thing used in the medical profession for taking X-rays, is extremely useful for internal inspection of completed packages. Such things as improper markings, poor bond between chip and package, foreign matter, and extra-long wires that could cause shorts can be detected. Both Si and Al, however, are transparent to X-rays that will penetrate the package. The image may be displayed on a TV monitor or a picture may be taken.

The state of the art of modern optical lithography applied to LSI digital circuits has resulted in such narrow line widths that inspection for oxide or metalization faults pushes the limits of capability of optical microscopy. The SEM provides a fast and convenient technique for oxide and metalization inspection. Oxide faults such as diffusion spikes (areas where a dopant has rapidly diffused through the oxide due to porous oxide), cracks, pinholes, and discontinuities over steps are easily spotted. Metalization faults such as undercutting, burnout, opens due to electrostatic discharge (these show up as dark spots in the metalization and are especially prevalent in gates of MOS devices), and cracks at oxide steps are conveniently observed. If shorts or open circuits are especially difficult to spot, voltage contrast can be employed. Voltage contrast involves applying a positive voltage to various parts of the circuit, which may involve normal dc operating conditions or other signals that would produce known conditions. In the usual mode of operation of the SEM, the electrons in the primary beam are accelerated from 2.5 kV to 30 kV, but the electrons that are detected are secondary electrons ejected from the sample by the primaries. The majority of the secondary electrons have only a few electron volts of energy. Consequently, a small positive voltage, even as low as 0.6 V, will result in an observable contrast difference due to the reduction of secondary electron emission in the positive voltage regions. If one knows what regions of the circuit should be light or dark under the applied voltage, shorts and open circuits can be spotted even when not apparent without voltage contrast.

Fig. 13-16 SEM micrograph of a bipolar transistor in voltage contrast.

Figure 13-16 is an SEM micrograph of a bipolar transistor operating in the "on" condition.

BIBLIOGRAPHY

[1] WARNER, R. M., JR., ED., *Integrated Circuits, Design Principles and Fabrication.* New York: McGraw-Hill Book Company, 1965.

[2] FOGIEL, M., *Modern Microelectronics.* New York: Research and Education Association, 1972.

[3] MADLAND, G. R., R. L. PRITCHARD, ET AL., *Integrated Circuit Engineering.* Cambridge, Mass: Boston Technical Publishers, Inc., 1966.

[4] PHILOFSKY, E., "Intermetallic Formation in Gold–Aluminum Systems," *Solid State Electronics,* Vol. 13, 1970, pp. 1391–1399.

PROBLEMS

13-1 An IC chip with a thermal resistance of 3°C/W is enclosed in a DIP with a thermal resistance of 80°C/W. Estimate the junction temperature if the IC dissipates 150 mW and the ambient temperature is 20°C.

13-2 Calculate the normal force and shear force on bonding pads of the same height for a wire pull test given the following conditions:

$$\text{bond separation} = 2\ \text{mm}$$
$$\text{wire length} = 3.5\ \text{mm}$$
$$\text{pull force} = 5\ \text{g}$$

Calculate the shear and normal stresses on the wire (force per unit area) if the wire diameter is 0.05 mm.

13-3 Briefly describe die bonding.

13-4 List and describe three techniques used with IC chips to make contact to the outside world.

13-5 Discuss two types of wire bonding and describe the different types of bonds that are made and which materials they apply to and why.

13-6 Describe three types of commonly used IC packaging techniques.

13-7 Discuss problems associated with silicon wafers which can lead to difficulty in mask alignment and mask life.

13-8 In a very general manner, list the types of tests that are performed on ICs.

13-9 List three diagnostic techniques used in fault analysis and suggest what types of failures can be detected by each.

digital logic families

14

14-1 BASIC LOGIC FUNCTIONS

The symbols shown in Fig. 14-1 describe hardware and relate the voltages at input terminals to the voltages at output terminals. The functions described by these symbols can be implemented with different types of hardware, which, in turn, use different voltage levels. Thus, rather than assigning specific voltage levels to hardware terminals, it is more general and more convenient to refer to them as "asserted high" or "asserted low." The voltage levels associated with these terms can then be provided in conjunction with the specific hardware implementation. Logic gate input terminals can be asserted high or low by applying appropriate voltages to them; then the output terminals will be asserted high or low, depending on the logic function to be performed. The inverter of Fig. 14-1(b) "inverts" the signal. If the input terminal is asserted low, the output will be asserted high, and vice versa. A "bubble" (or circle) shown at the input or output of a logic gate represents an inverter, and will change a high assertion to a low assertion, and vice versa. For the sake of simplicity, we will show two input terminals on our logic gates for the time being, but they could have many more. Only two voltage levels will be allowed at the terminals of digital logic gates: one that represents a "lo" voltage, and one that represents a "hi"

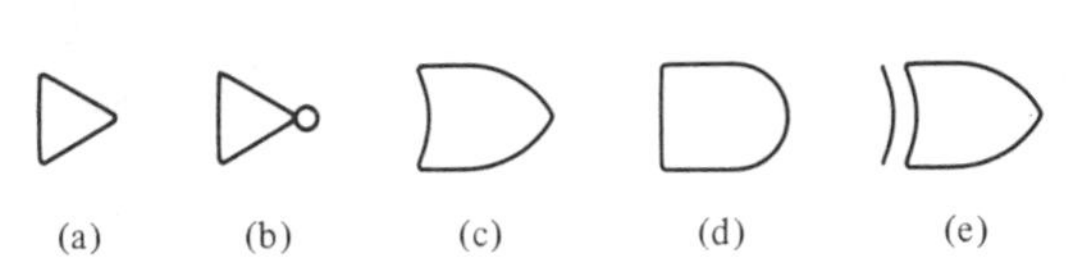

Fig. 14-1 Logic gate hardware symbols: (a) noninverting amplifier or buffer; (b) inverter; (c) (inclusive) OR gate; (d) AND gate; (e) exclusive OR gate.

voltage. Such things as a third voltage level or an open circuit remain undefined. The responses of these gates to signals other than "hi" or "lo" are also undefined, since they depend exclusively on the hardware used to implement the gates.

The operation of a logic gate is defined by a "truth table" which relates the voltage level at the output terminal to the voltage levels at the input terminals. A truth table for a two-input OR gate is shown in Fig. 14-2(a). It states that "if input terminal *A* OR *B* (or both) is asserted hi, output terminal *C* will be asserted hi." Figure 14-2(b) shows the hardware symbol that provides this input/output relationship. The same truth table can be represented by the statement: "If input terminals *A* AND *B* are asserted lo, output terminal *C* will be asserted lo." Then another hardware symbol, which provides the same input/output relationship and the same truth table, is shown in Fig. 14-2(c). In translating hardware symbols into words or truth tables, it is convenient to think "lo" when you see a bubble. OR gates detect the presence of hi inputs because if any of the inputs are hi, the output will be hi.

Fig. 14-2 (a) OR gate truth table. (b) Hardware symbol for an OR gate. (c) AND gate connected to provide the OR gate truth table.

A	*B*	*C*
lo	lo	lo
lo	hi	hi
hi	lo	hi
hi	hi	hi

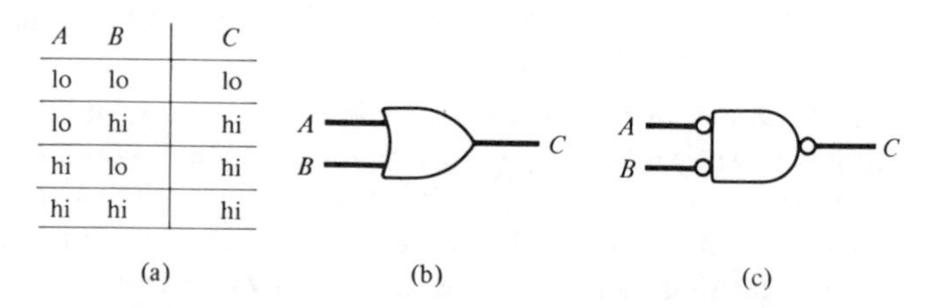

A two-input AND gate is represented by the truth table of Fig. 14-3(a). The hardware operation specified by the truth table can be implemented with the AND gate of Fig. 14-3(b) or with the OR gate of Fig. 14-3(c). Figure 14-3(b) says: "Output *C* will be hi if inputs *A* AND *B* are asserted hi." We can put Fig. 14-3(c) into words by reading the lo's off the truth table: "Output *C* will be asserted lo if input *A* OR *B* is asserted lo." We generally use the symbol with the least number of bubbles unless there is reason to do otherwise.

Fig. 14-3 (a) AND gate truth table. (b) Hardware symbol for an AND gate. (c) OR gate connected to provide the AND gate truth table.

A	*B*	*C*
lo	lo	lo
lo	hi	lo
hi	lo	lo
hi	hi	hi

A, B → C (b) A, B → C (c)

(a) (b) (c)

Figure 14-4 shows the truth table and hardware symbols for an *OR gate whose output is asserted lo.* This is commonly referred to as a NOR gate. Referring to Fig. 14-4(a) and (b), the output is asserted lo when input *A* OR *B* (or both) is asserted hi. As shown in Fig. 14-4(c), this is equivalent to saying that the output is hi when inputs *A* AND *B* are asserted lo.

Fig. 14-4 (a) NOR gate truth table. (b) Hardware symbol for a NOR gate. (c) AND gate connected to provide the NOR gate truth table.

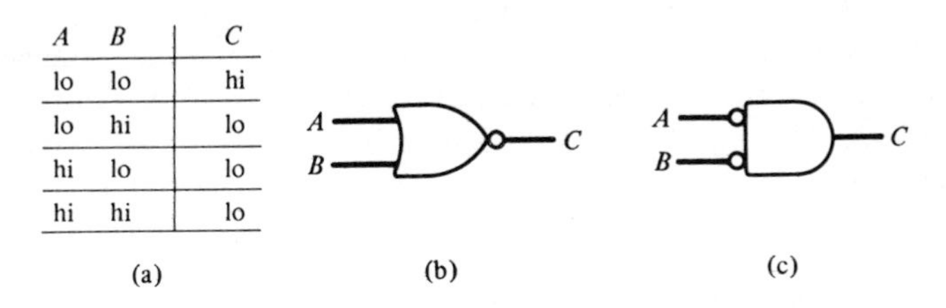

A	*B*	*C*
lo	lo	hi
lo	hi	lo
hi	lo	lo
hi	hi	lo

Figure 14-5 shows the truth table and hardware symbols for a two-input NAND gate (an AND gate whose output is asserted lo). If inputs *A* AND *B* are asserted hi, the output is asserted lo. This can also be stated: "If input *A* OR *B* is asserted lo, output *C* will be asserted hi."

Fig. 14-5 (a) NAND gate truth table. (b) Hardware symbol for a NAND gate. (c) OR gate connected to provide the same digital function.

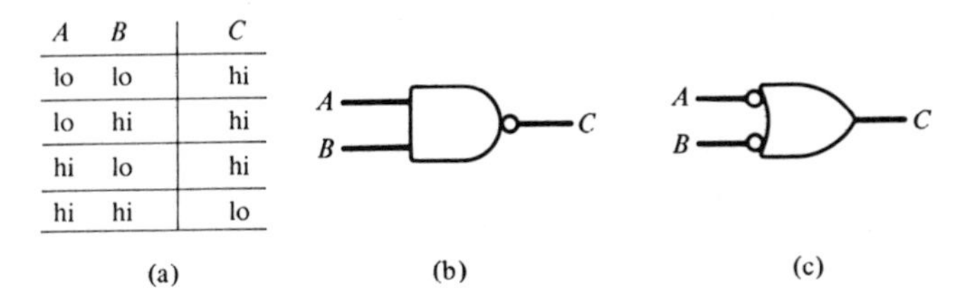

A	*B*	*C*
lo	lo	hi
lo	hi	hi
hi	lo	hi
hi	hi	lo

Generally, inverting logic gates such as NOR and NAND gates are easier to realize with integrated circuits. Because hardware realizations are simpler, they save on cost and power dissipation and are thus preferred over AND and OR gates. Although hobbyists purchase inverting logic gates and digital engineers design with them, it is more useful to think of them as AND and OR gates whose inputs and outputs are asserted hi or lo. For example, a logic gate that recognizes a specific address and has both inverting and noninverting outputs is shown in Fig. 14-6(a). The output *F* is asserted high if, and only if, *A* is hi, *B* is lo, *C* is hi, AND *D* is lo. Output *E* is asserted lo only if the foregoing input conditions apply. By switching bubbles and symbols, the same function is performed by the hardware represented in Fig. 14-6(b).

As shown in Figs. 14-2 through 14-6, a logic function can be implemented

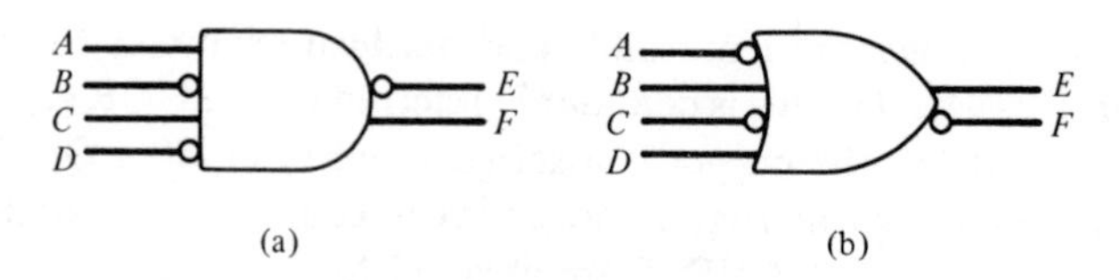

Fig. 14-6 Logic gate that recognizes a specific address: (a) AND gate implementation; (b) implementation with OR gates.

with OR or AND gates combined with inverters (bubbles). Given a logic *circuit,* this fact can be used to make the logic *function* more easily understood. For example, NAND gate 3 in Fig. 14-7(a) can be redrawn as shown in Fig. 14-7(b). Since two series bubbles cancel, the circuit can also be drawn as shown in Fig. 14-7(c). The digital function implemented by Fig. 14-7 can then be stated: "Output E is asserted hi if *A* AND *B* are hi, OR if *C* AND *D* are hi."

Fig. 14-7 Redrawing logic circuits to make logic functions more easily understood: (a) circuit with NAND gates; (b) gate 3 is redrawn; (c) series bubbles are removed.

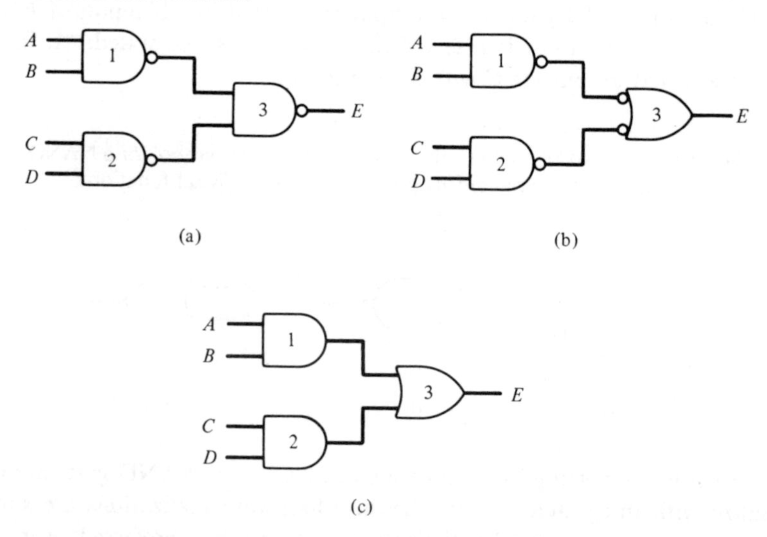

The truth table and logic symbol for a two-input EXCLUSIVE OR gate are shown in Fig. 14-8. The output *C* is asserted hi if *A* is hi OR *B* is hi (but not both). Since EXCLUSIVE OR gates are available as digital components, the NAND gate implementation shown in Fig. 14-8(c) is provided simply as a tutorial aid.

Figure 14-9 shows the truth table and logic symbol for an EXCLUSIVE NOR gate. As shown, output *C* is asserted hi if the inputs are the same.

By using two logic states (hi and lo), it becomes possible to design *accurate computers with inaccurate components.* One range of voltage levels is used to

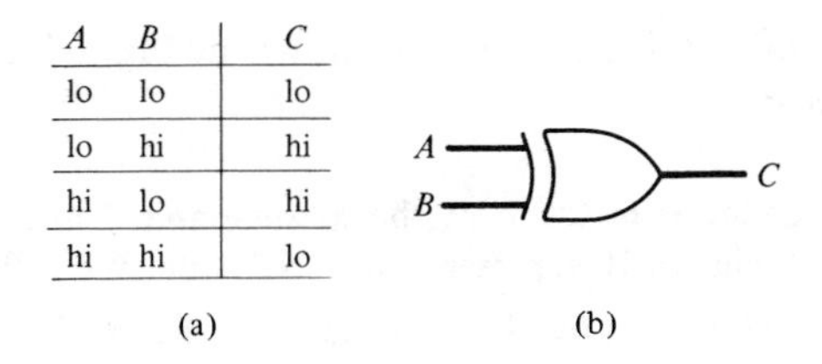

A	B	C
lo	lo	lo
lo	hi	hi
hi	lo	hi
hi	hi	lo

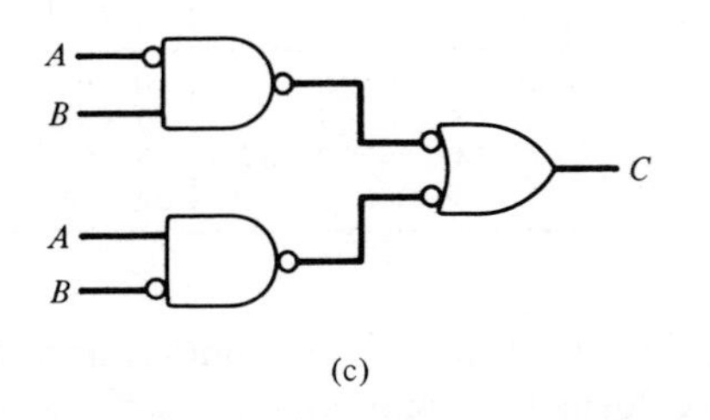

Fig. 14-8 Two-input EXCLUSIVE OR gate: (a) truth table; (b) logic symbol; (c) implementation with NAND gates.

A	B	C
lo	lo	hi
lo	hi	lo
hi	lo	lo
hi	hi	hi

(a)

(b)

(c)

Fig. 14-9 Two-input EXCLUSIVE NOR gate: (a) truth table; (b) logic symbol; (c) implementation with NAND gates.

define the lo state and another voltage range defines the hi state. As long as the component and system tolerances are such that the specified voltage ranges can be held at the inputs and outputs of digital circuits, no computer errors are made. For a given logic family, we then have to define the range of voltage levels corresponding to hi and lo logic states. Suppose, for example, that we are operating a hypothetical logic family between +5 V and ground. Suppose further that the gate input and output voltages for this logic family are either between 0 and 0.5 V or between 3 and 5 V. The truth table for a specific member of this logic family is shown in Fig. 14-10. We now have a choice of definitions:

> If we define *0 to 0.5 V* as the *hi state* and *3 to 5 V* as the *lo state,* the truth table of Fig. 14-10 represents an AND gate. We call this *negative logic.*
>
> If we define *0 to 0.5 V* as the *lo state* and *3 to 5 V* as the *hi state,* the truth table represents an OR gate. This definition of hi and lo is called *positive logic.*

A (V)	*B* (V)	*C* (V)
0 − 0.5	0 − 0.5	0 − 0.5
0 − 0.5	3 − 5.0	3 − 5.0
3 − 5.0	0 − 0.5	3 − 5.0
3 − 5.0	3 − 5.0	3 − 5.0

Fig. 14-10 Truth table for a hypothetical logic gate.

When designing a system, it is sometimes convenient to alternate between positive and negative logic to economize on parts or to simplify the overall circuit. The names and functions of the logic families are normally based on *positive* logic, however. Only positive logic is discussed in the remainder of this chapter. With positive logic, it is conventional to let the hi state equal binary "1" and the lo state equal binary "0." A truth table that uses 1's and 0's to summarize the basic logic function outputs is shown in Fig. 14-11.

14-2 OPERATING PARAMETERS FOR LOGIC GATES

Logic gates have to work over a range of temperatures, power supply voltages, and loading conditions. These operating parameters are specified for each logic family by the manufacturer. Since logic gates are mass produced, some variations in transistor and resistor parameters are to be expected.

Given a logic family, one of the first things a designer has to know is the worst-case range of input voltages that a gate will recognize as a lo assertion (logic 0) and a hi assertion (logic 1). Given these values, the worst-case output voltages must be known. We must know the worst-case output voltages to expect when a gate output terminal is asserted hi (logic 1) or asserted lo (logic 0). Typical and worst-case input and output voltage levels can be observed by connecting logic gates as inverters and plotting the voltage transfer characteristics. NOR and NAND gates can be connected as inverters, as shown in Fig. 14-12. The connections shown in Fig. 14-12(a) and (c) are generally preferred because V_{in} sees only one input load. The input current to or from the previous stage will be smaller with these connections.

Typical and worst-case voltage transfer characteristics for logic gates connected as inverters are shown in Fig. 14-13. When the input voltage is asserted hi, the output voltage is asserted lo, and vice versa. The worst-case voltage parameters V_{IL}, V_{IH}, V_{OL}, and V_{OH} are as shown. These parameters are specified

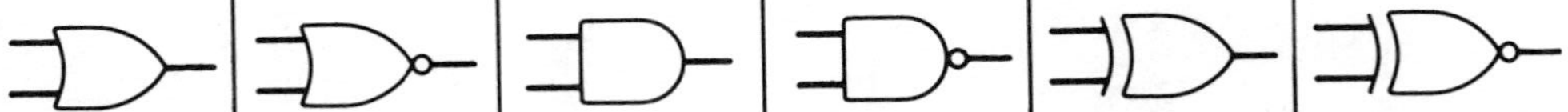

A	B	OR	NOR	AND	NAND	EX-OR	EX-NOR
0	0	0	1	0	1	0	1
0	1	1	0	0	1	1	0
1	0	1	0	0	1	1	0
1	1	1	0	1	0	0	1

Fig. 14-11 Truth table showing the outputs of the basic logic functions using positive logic.

to account for manufacturing tolerances and temperature, power supply, and loading ranges. They are defined as follows:

V_{OL} is the *maximum* voltage at the gate output terminal when the output is asserted lo (logic 0).

V_{OH} is the *minimum* voltage at the gate output terminal when the output is asserted hi (logic 1).

V_{IL} is the *maximum* gate input voltage that will be recognized as a lo assertion (a logic 0).

V_{IH} is the *minimum* gate input voltage that will be recognized as a hi assertion (a logic 1).

Fig. 14-12 Inverter connections for NAND and NOR gates. Connections (a) and (c) are generally preferred because V_{in} sees only one input load.

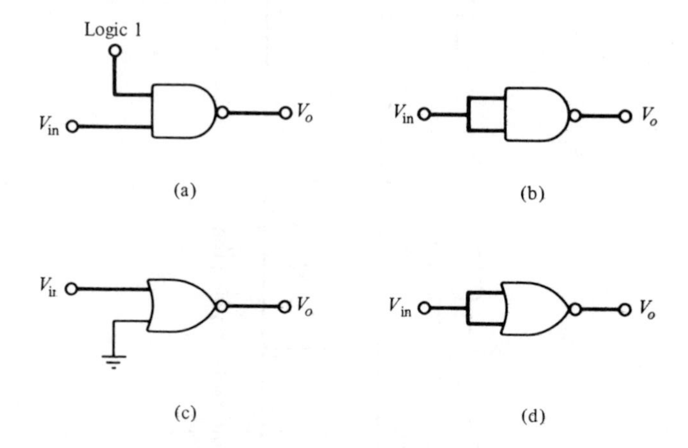

Fig. 14-13 Voltage-transfer characteristic for a digital gate or inverter.

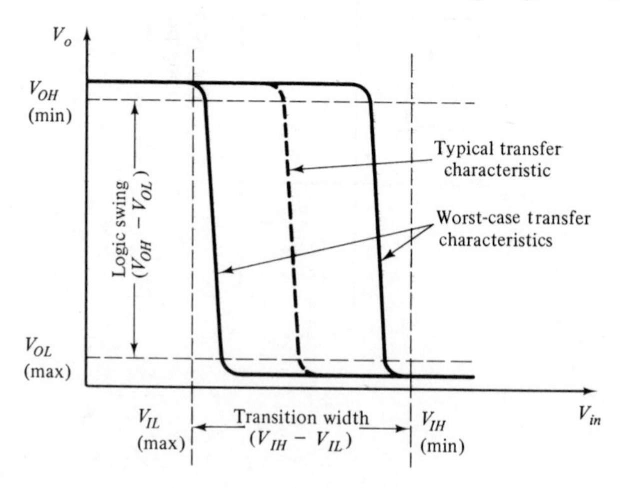

As an example, the logic voltage specifications for standard T²L logic gates are:

$$V_{OL} = 0.4 \text{ V (maximum)} \qquad V_{IL} = 0.8 \text{ V (maximum)}$$
$$V_{OH} = 2.4 \text{ V (minimum)} \qquad V_{IH} = 2.0 \text{ V (minimum)}$$

Referring to Fig. 14-13, the logic swing and transition width for standard T²L logic are then given by:

$$\text{minimum logic swing} = V_{OH} - V_{OL} = 2.4 - 0.4 = 2 \text{ V} \tag{14-1}$$
$$\text{minimum transition width} = V_{IH} - V_{IL} = 2.0 - 0.8 = 1.2 \text{ V} \tag{14-2}$$

Keep in mind that for an actual T²L gate, V_{OH} may be 5.0 V and V_{OL} may be 0.02 V, giving an actual logic swing of 4.98 V. Worst-case values are specified by the manufacturer, and they are generally conservative.

14-2-1 Noise Margins

When we connect the output of one inverting logic gate to the input of another, it is desirable that $V_{OH1} > V_{IH2}$ so that V_{OH1} always turns the second inverter on. In the case of a lo output from the first inverter, we want $V_{OL1} < V_{IL2}$, so that the second inverter stays off. The logic swing $(V_{OH} - V_{OL})$ should bracket the transition width and be larger. Figure 14-14 shows two series NAND gates (connected as inverters) with a noise generator connected between them. We might ask: How large a noise signal can we tolerate with a given logic family before inverter 2 enters the transition region incorrectly?

For a T²L NAND gate, $V_{OH} = 2.4$ V, $V_{OL} = 0.4$ V, $V_{IH} = 2.0$ V, and $V_{IL} = 0.8$ V. If V_{OL1} in Fig. 14-14 is 0.4 V and V_{IL2} has to be 0.8 V for NAND gate 2 to reach the transition region, we can stand a 0.4-V zero-to-peak noise signal before an error begins. The *low-level noise margin* (NM_L) is 0.4 V.

$$NM_L = V_{IL} - V_{OL} = 0.8 - 0.4 = 0.4 \text{ V} \tag{14-3}$$

Fig. 14-14 Series-connected NAND gates with a noise generator between them.

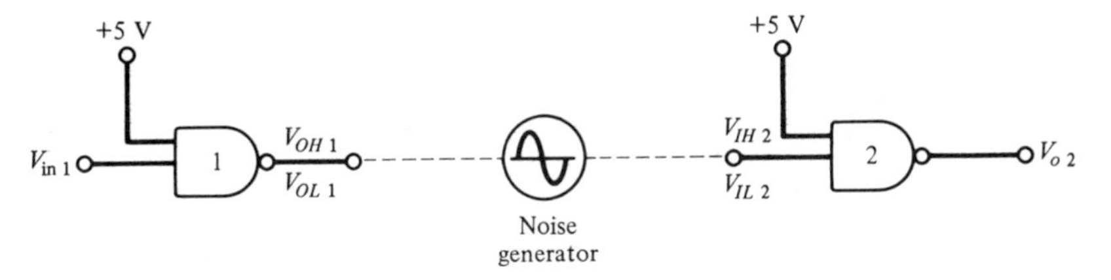

If V_{OH1} is 2.4 V and V_{IH2} has to be 2.0 V or more to keep NAND gate 2 on, we can stand a -0.4 V zero-to-peak noise signal before the second inverter reaches the transition region. The *high-level noise margin* (NM_H) is 0.4 V.

$$NM_H = V_{OH} - V_{IH} = 2.4 - 2.0 = 0.4 \text{ V} \tag{14-4}$$

14-2-2 Fan-Out

The word "fan-out" is generally used in two ways. The number of gates driven by a single gate is usually referred to as the fan-out of the single gate. In manufacturers' specifications, however, it has a more specific meaning. It refers to the *maximum number* of gates that can be driven by a single gate. We will use fan-out to refer to the maximum number of driven gates. *Fan-in* generally refers to the number of input terminals on a gate. A four-input gate has a fan-in of 4.

It is useful to describe the operating parameters for logic gates without getting deeply into the circuit operation of any specific logic family. We can do this by using a resistor-switch model as shown in Fig. 14-15. The switching device contains a low value of series resistance R_S, where $R_S << R_L$. While an input circuit is not actually shown, we will assume that a gate input voltage V_{in} gives rise to an input current I_{in}. When V_{in} is hi, the switch is closed and V_o is lo. When V_{in} is lo, the switch is open and V_o is hi. Circuits of this type function accurately over a wide range of R_L values. Using this model, Fig. 14-16(a) shows a two-input NAND gate and Fig. 14-16(b) shows a two-input NOR gate. The NAND gate can be connected as an inverter by holding input B hi. The NOR gate is connected as an inverter by holding B lo. Thus, we can use an inverter to look at the fan-out of logic gates or inverters.

In Fig. 14-17(a), a single gate is shown driving N other gates. Since the

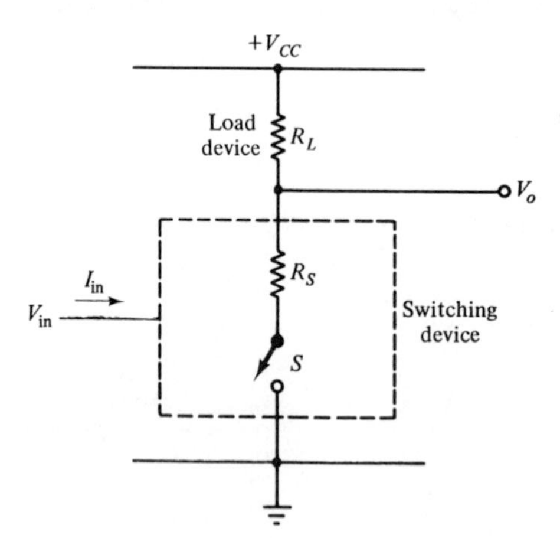

Fig. 14-15 Digital inverter modeled as a load device in series with a switching device.

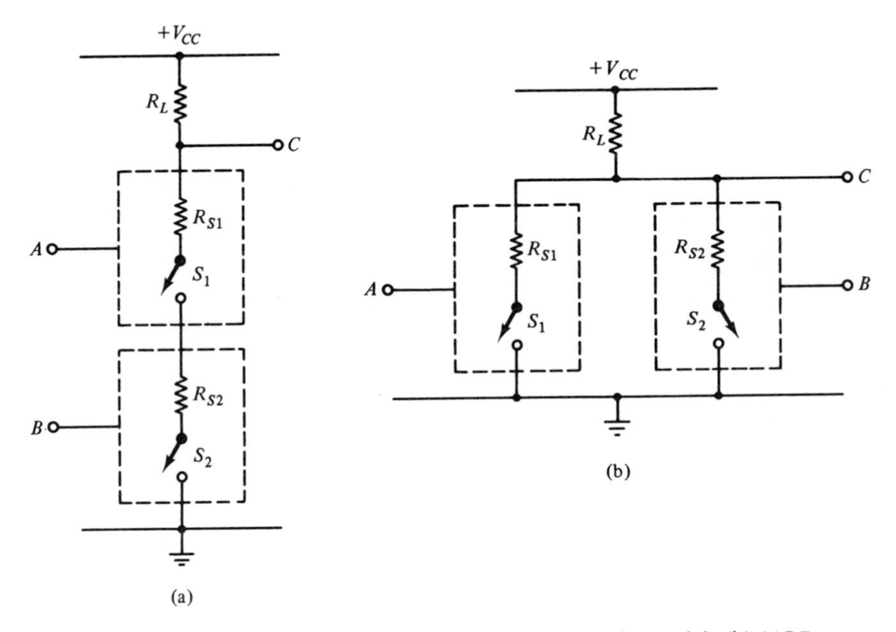

Fig. 14-16 (a) NAND gate with the resistor-switch model. (b) NOR gate with the resistor-switch model.

input to the single gate is lo, switch S_0 is open and the gate output is at V_{OH}.[1] The inputs to the N driven gates are all hi, so that the gate input currents $I_{in}(1)$ through $I_{in}(N)$ go through resistor R_L. Then, since the input currents are identical,

$$V_{CC} = NI_{in} R_L + V_{OH} \tag{14-5}$$

or

$$N = \frac{V_{CC} - V_{OH}}{I_{in} R_L} \tag{14-6}$$

Given V_{CC} and R_L, fan-out N depends on V_{OH} and I_{in}. As V_{OH} approaches V_{CC}, N decreases. As I_{in} increases, N decreases. The V_{OH} specification may involve a trade-off between the high-level noise margin $(V_{OH} - V_{IH})$ and fan-out. If V_{OH} was increased to improve the noise margin, then according to Eq. (14-6), fan-out would decrease.

In Fig. 14-17(b), the input to the driving gate is hi and switch S_0 is closed. The output voltage of the driving gate is V_{OL}, and the N driven gates are off. Let us assume that in this situation the input currents are negative [i.e., they are directed back through the closed switch S_0 to ground (this happens

[1] We are using the actual value of V_{OH} here rather than the worst-case value.

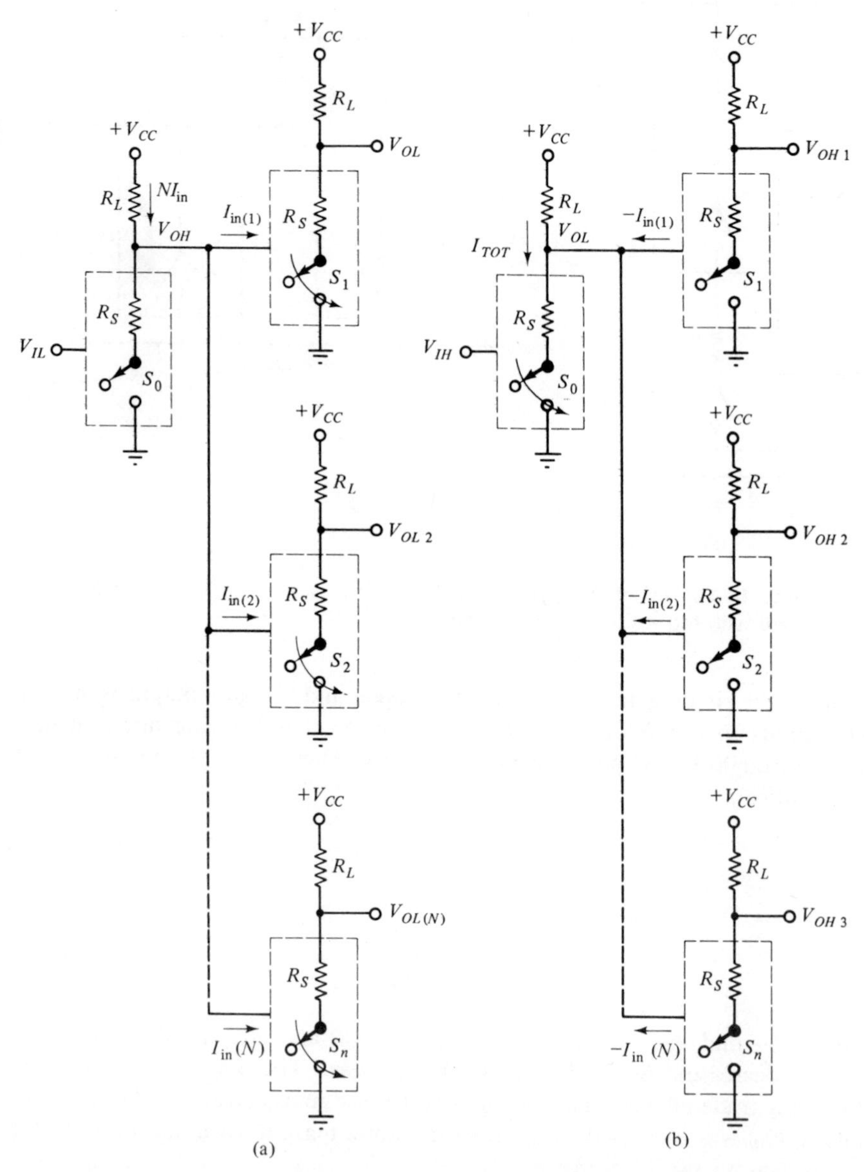

Fig. 14-17 Logic gate fan-out: (a) driving gate output is hi; (b) driving gate output is lo.

with T²L logic)]. Then the total current to ground through switch S_0 is given by

$$I_{\text{TOT}} = \frac{V_{CC}}{R_L + R_S} + NI_{\text{in}} \tag{14-7}$$

V_{OL} is given by

$$V_{OL} = \left(\frac{R_S}{R_L + R_S}\right) V_{CC} + NI_{\text{in}} R_S \tag{14-8}$$

Assuming that $R_S << R_L$, Eq. (14-8) can be written

$$N \approx \frac{V_{OL}}{I_{\text{in}} R_S} - \frac{V_{CC}}{I_{\text{in}} R_L} \tag{14-9}$$

Looking at Eq. (14-9), we want to keep I_{in} small and R_L much larger than R_S for a large fan-out *(N)*. The V_{OL} specification may involve a trade-off between fan-out and the low-level noise margin $N_{ML} = V_{IL} - V_{OL}$. If the V_{OL} specification was increased, fan-out would increase but N_{ML} would decrease. For standard T²L logic, $V_{CC} = 5$ V, $V_{OH} = 2.4$ V minimum, and $V_{OL} = 0.4$ V maximum. Maximum fan-out, as specified by standard T²L manufacturers, is 10. In effect, the manufacturers guarantee that V_{OL} in Eq. (14-8) and Fig. 14-17(b) will not rise above 0.4 V when $N = 10$. Generally, manufacturers' specifications on logic gates are very conservative.

14-2-3 Power Dissipation and Propagation Delay

In order to cover general parameters such as power dissipation and propagation delay, we first have to discuss gate complexity. A simple gate may contain 1 to 10 transistors. Complex digital gates may have 20 to 40 switching devices associated with one load device. Logic structures may contain input protection circuits to protect against damped oscillations and/or voltage surges. They may contain fail–safe circuitry, and they may be able to compensate for temperature, processing, and power supply variations. They may also contain circuits to generate their own stable bias voltages. We must make a distinction between *simple gates* and *complex gates.* When you see literature referring, for example, to 10,000 gates per chip, simple gates are implied. Our discussion of power dissipation and propagation delay will be restricted to simple gates.

In battery-operated equipment such as digital watches, and in VLSI circuits, where packing density has been maximized, power dissipation is of major importance. For example, you want your digital watch to run on a small battery for over a year. In very large computers, speed is critical. With literally millions of interconnected logic gates to contend with, we need to minimize the time it takes for a signal to propagate through any given logic gate. We need to minimize the *propagation delay.*

Within certain boundary conditions, the power dissipation and propagation delay of a given logic family are inversely related. By changing component values, it is sometimes possible to decrease propagation delay by increasing power dissipation, and vice versa. If larger currents are used in a given digital family, circuit capacitances can be charged more rapidly.

In most logic families, a variety of digital ICs are supplied off-the-shelf in dual-in-line packages. Manufacturers' data sheets specify operating conditions, including the average power dissipation of the gate. This average is generally obtained by assuming that the inputs of the gates are hi 50% of the time and lo 50% of the time. This is referred to as a 50% duty cycle. *Average power dissipation* (PD_{AV}) is then given by

$$PD_{AV} = \frac{PD(\text{inputs hi}) + PD(\text{inputs lo})}{2} \tag{14-10}$$

In logic families specifically designed for low-power, low-speed applications, the average power dissipation may be on the order of 10 nW per gate. In logic families specifically designed for high-speed applications, the average power dissipation may be on the order of 50 mW per gate. This represents a power dissipation range greater than six orders of magnitude.

Propagation delay time (t_{pd}) provides a measure of how long it takes a digital signal to propagate through a logic gate. This delay occurs because of charge carrier transit times within devices, and device and circuit capacitances. Figure 14-18 can be used to define gate propagation delay. The time t_{pd}(HL) is the propagation delay associated with an output transition from a hi to lo

Fig. 14-18 Propagation delay time in a logic gate.

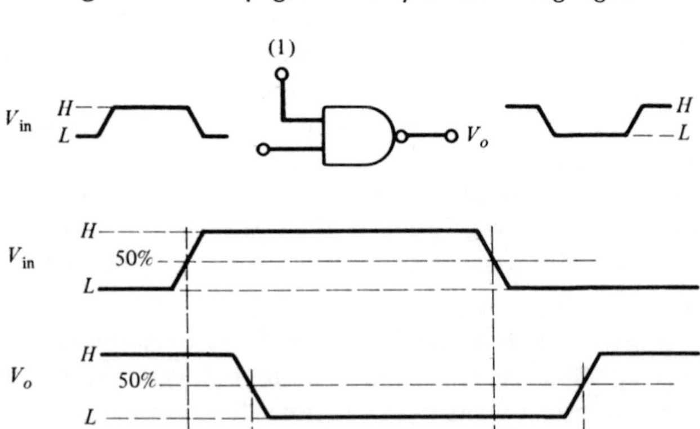

voltage state. The time t_{pd}(LH) is the delay associated with an output voltage transition from lo to hi. These two delay times are not usually equal. Propagation delay time is defined as the average of $t_{pd}\,(HL)$ and $t_{pd}\,(LH)$:

$$t_{pd} = \frac{t_{pd}\,(HL) + t_{pd}\,(LH)}{2} \tag{14-11}$$

The reference points used for the measurement of propagation delay are not standard. Although they vary depending on both the manufacturer and the logic family involved, they always relate to the input transition region $(V_{IH} - V_{IL})$. For our purposes, we will use the 50% point between a hi and a lo signal to define propagation delay, as shown in Fig. 14-18.

Gate inputs have capacitance associated with them. If a single gate drives 10 parallel gates, each with an input capacitance C_{in}, then the load capacitance on the single gate is 10 C_{in} plus wiring capacitance. This load capacitance must be charged and discharged by the output circuit of the single gate. Thus, propagation delay is related to fan-out. In the case of very fast logic gates, the propagation delay specification sets the upper limit on fan-out.

During output transitions from hi to lo and lo to hi voltage levels, the *rise time* (t_r) and *fall time* (t_f) should also be considered. These parameters represent the time interval between 10 and 90% points on the voltage waveform, as shown in Fig. 14-19. When logic gates are used at speeds where the propagation delay time begins to affect circuit performance, the magnitudes of t_r and t_f also become significant. An input voltage making a hi–lo transition has to go from a hi level to V_{IH}(min) before the circuit begins to react. Similarly, an input voltage making a lo–hi transition has to go from lo to V_{IL}(max) before the circuit reacts. Thus, rise and fall times affect propagation delay times. Since rise and fall times are sensitive to load and wiring capacitances, they can, in the upper limit, affect gate switching speeds. Typical propagation delay times for standard T^2L logic are in the vicinity of 10 ns. Typical values of t_r and t_f are about 8 and 5 ns, respectively.

Propagation delays range from approximately 0.5 ns per gate for the fastest logic families to about 25 ns per gate for the slowest.

Gate propagation delay times can be measured by connecting an odd number of inverters into a ring oscillator as shown in Fig. 14-20. Let us begin with switch S connected to ground as shown. Then A is lo, B is high, and C

Fig. 14-19 Rise-time and fall-time measurements.

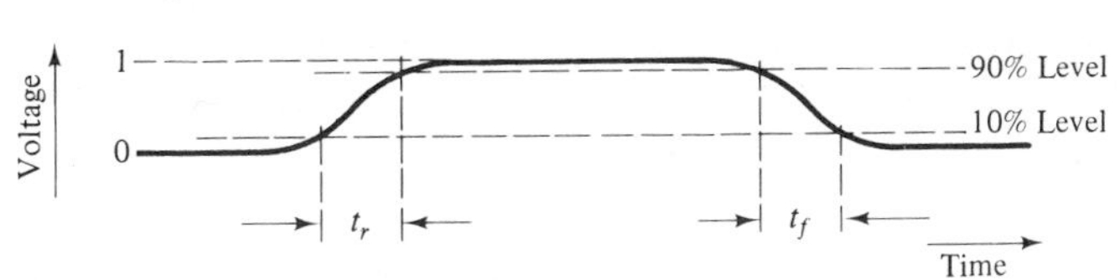

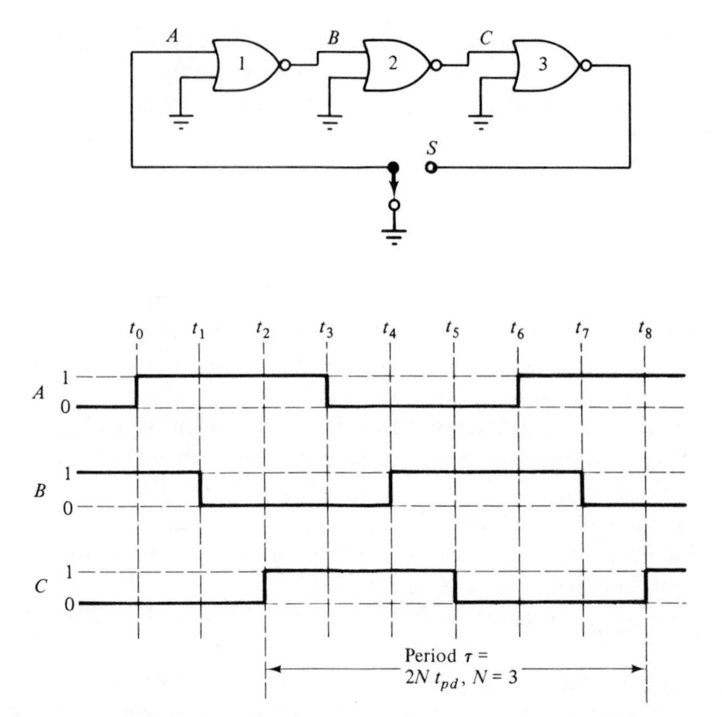

Fig. 14-20 Use of a ring counter to measure gate propagation delay time.

is lo. At time t_0 we will close the switch, connecting the output of gate 3 to the input of gate 1. Since the output of gate 3 was hi, A is forced hi. After the gate propagation delay $t_{pd} = t_1 - t_0$, point B goes lo at time t_1. After another propagation delay, C goes hi, and so on. For three logic gates, the period of the waveform $\tau = 6t_{pd}$. For a ring oscillator consisting of N logic gates (*N is an odd integer),* $\tau = 2Nt_{pd}$, or $t_{pd} = \tau/2N$.

The product of average power dissipation (PD_{AV}) and gate propagation delay (t_{pd}) is a useful figure of merit for a logic family. It is called the *power delay product* (or loosely the "speed–power" product):

$$\text{power–delay product} = (PD_{AV})(t_{pd}) \tag{14-12}$$

It permits a comparison of logic families and provides an aid in finding the best fit between digital technologies and application requirements. Typically, average power dissipation falls in the milliwatt range, and propagation delay time falls in the nanosecond range. Thus, power–delay products for most logic families fall in the picojoule range, between about 0.1 and 100 pJ.

Figure 14-21 shows a graph of propagation delay versus power dissipation for the digital logic families. Constant energy values appear as diagonal lines.

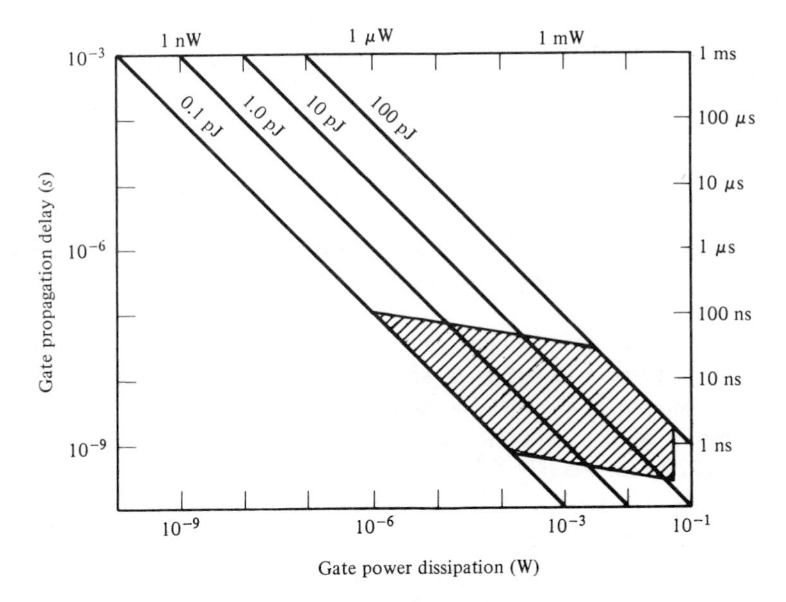

Fig. 14-21 Power–delay products for digital logic gates.

The shaded area in Fig. 14-21 shows the range of power dissipation, propagation delay, and power–delay product values for most of the logic families. After the digital logic families have been discussed, this graph will be shown again with each logic family in its appropriate niche within the shaded area.

Logic functions can be realized with bipolar or MOS transistors. If bipolar transistor circuits are used, the transistor circuits can be categorized as saturating or nonsaturating. If MOS transistors are used, we can use *n*-channel, *p*-channel or both. Given a set of technologies and an applications area, we try to find the best fit between them. Applications can be categorized according to the requirements on:

propagation delay	logic swing
power dissipation	noise margins
packing density	fan-out
cost	

The major hardware technologies that are used to realize the logic functions are:

- Transistor–transistor logic (T^2L)
- Emitter-coupled logic (ECL)
- Integrated-injection logic (I^2L)
- *n*-Channel MOSFET logic (NMOS)
- Complementary MOS logic (CMOS)

Now that we have defined the major operating parameters by which logic familes are categorized, we can discuss circuit operation for the logic families listed above.

14-3 TRANSISTOR–TRANSISTOR LOGIC (T^2L)

In order to simplify our discussion of digital transistor circuit operation, we will use first-order device models. We will assume that:

1. Conventional silicon diodes, and the base–emitter and base–collector diodes of transistors, will act as *ideal silicon diodes* when connected as diodes. For example, if $V_{BE} > 0.7$ V, an ideal silicon diode acts as a closed switch in series with a 0.7-Volt battery. For $V_{BE} < 0.7$ V, it acts as an open switch. Thus, $V_{BE}(\text{ON}) = V_D(\text{ON}) = 0.7$ V.
2. When a transistor is in saturation and not sinking external currents, the collector–emitter voltage $V_{CE}(\text{sat}) = 0.2$ V.
3. β_F, the transistor beta in the forward-active operating region, is 50 and remains constant throughout the active region.

Figure 14-22(a) shows a basic diode–transistor logic (DTL) NAND gate and Fig. 14-22(b) shows a basic T^2L NAND gate. Since the T^2L input transistor (Q_1) evolved from diodes D_1, D_2, and D_3 in Fig. 14-22(a), it is convenient to discuss the diode circuit before discussing the T^2L input circuit. To discuss logic gate operation, we will ordinarily assume that our inputs come from other gates and that our outputs go to other gates. If inputs V_1 and V_2 to the DTL gate are lo, it is because they are driven by the output transistors of previous gates that are in saturation. Thus, if $V_1 = V_2 = 0.2$ V, diodes D_1 and D_2 are on and the voltage $V_A = 0.9$ V. Because R_3 pulls the n side of diode D_3 down

Fig. 14-22 (a) Basic diode–transistor logic (DTL) NAND gate. (b) Basic transistor–transistor logic (T^2L) NAND gate.

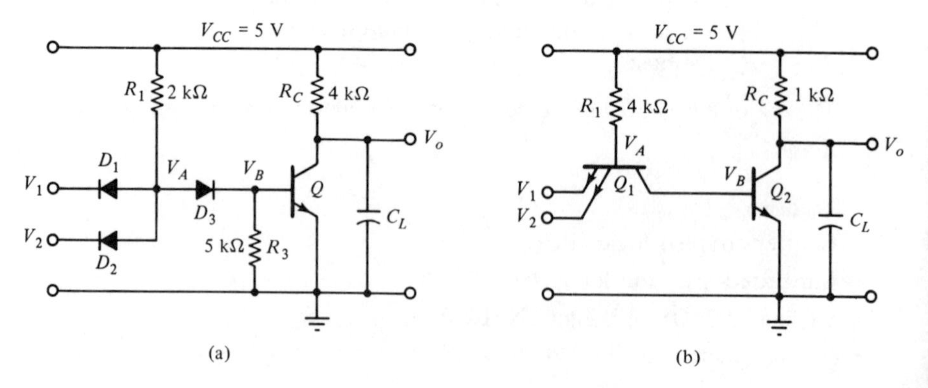

to ground (R_3 is a *pull-down* resistor), diode D_3 is on and $V_B = 0.2$ V. Thus, transistor Q is off, and under no-load conditions, V_0 is at 5.0 V. If one of the inputs goes hi and the other stays lo, one input diode is still forward-biased, so that the current and voltage levels within the circuit do not change. If inputs V_1 and V_2 both go hi, diodes D_1 and D_2 turn off and V_A is two diode drops above ground (i.e., $V_A = V_{D3} + V_{BE} = 1.4$ V). Transistor Q is driven into saturation and $V_0 = V_{CE}(\text{sat}) = 0.2$ V. Since the transistor has been driven into saturation, it contains stored base charge. If one of the inputs goes lo, the transistor cannot turn off until the stored base charge is removed. Resistor R_3 is in the circuit to provide a path to ground for stored base charge. In some cases, R_3 has been connected to a negative supply voltage ($V_{BB} = -2.0$ V) rather than ground to remove stored base charge more quickly and improve the propagation delay time. The load capacitance C_L is shown in Fig. 14-22 to model interconnection capacitance and the capacitance of driven gates. When the output of the DTL gate makes the transition from hi to lo, transistor Q turns on and discharges C_L. During the lo-to-hi output transition, however, the transistor turns off and C_L must be charged through the *pull-up* resistor R_C. The turn-off propagation delay t_{pd} *(LH)* is typically two to three times larger than the turn-on delay t_{pd} *(HL)* due to the $R_C C_L$ time constant.

With the T^2L gate shown in Fig. 14-22(b), diodes D_1, D_2, and D_3 are replaced by a transistor with two emitters. The base–emitter diodes of transistor Q_1 replace diodes D_1 and D_2 and the base–collector diode replaces diode D_3. This change enables a reduction in silicon area and a decrease in propagation delay time. The operation of the circuit is very similar to that of the DTL gate. If either or both inputs are held lo, the current through R_1 is directed toward the input. V_A is held at $0.2 + 0.7 = 0.9$ V, and transistor Q_2 is off. If both inputs are hi, V_A is at 1.4 V and Q_2 is in saturation. With both inputs hi, the base–emitter diodes of Q_1 are reverse-biased and the base–collector diode is forward-biased. The input transistor is operating in its *inverse active region.* Under these circumstances, an input current I_{in} *enters* the emitters of Q_1, loading the previous stage. Transistor Q_1 is deliberately designed to keep the reverse beta (β_R) small, however ($\beta_R \approx 0.02$). Then $I_{B1} = (V_{CC} - V_A)/R_1 = (5 - 1.4)\text{V}/4\ \text{k}\Omega = 900\ \mu\text{A}$, and $I_{\text{in}} = \beta_R I_{B1} = 18\ \mu\text{A}$. This minimizes loading on the previous stage when both inputs are hi.

Notice that there is no need for a base pull-down resistor (R_3) in the T^2L gate of Fig. 14-22(b). To explain this, let us assume that transistor Q_2 has been in saturation and that the input voltages now make a transition from hi to lo so that Q_2 is forced to turn off. When the base–emitter diodes of Q_1 become forward-biased, the stored base charge in Q_2 becomes the collector current of Q_1. This charge is swept out through Q_1, from collector to emitter. This charge removal mechanism gives T^2L gates a pronounced speed advantage over DTL gates.

The T^2L gate of Fig. 14-22(b) still has an unacceptably long propagation delay, t_{pd} *(LH)*, because of the $R_C C_L$ time constant. When transistor Q_2 turns

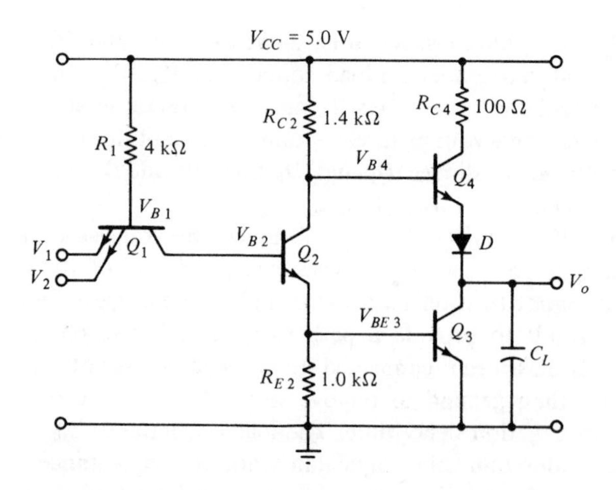

Fig. 14-23 Standard two-input T²L NAND gate.

off, the load capacitance must charge through R_C. What is needed is an *active pull-up circuit* to replace R_C. This active pull-up circuit might switch a small series resistance between C_L and the supply voltage while C_L is charging, and then switch off again. A new output stage should also have a low output impedance so that a large fan-out can be obtained.

Figure 14-23 shows a standard two-input NAND gate with an active pull-up circuit. The output circuit, consisting of R_{C4}–Q_4–D–Q_3 is imaginatively called a *totem-pole* output stage. Transistor Q_2 is used as a phase-splitter. Its collector and emitter voltages are out of phase. During voltage transitions, it tends to turn Q_3 on and Q_4 off, or vice versa. When Q_2 is in the forward-active region, it has a voltage gain of $\Delta V_{B4}/\Delta V_{B2} = -R_{C2}/R_{E2}$. When transistor Q_4 is in the forward-active region, it has the low output impedance of an emitter follower, which aids in charging C_L quickly.

When V_1 and V_2 are lo ($\approx$0.2 V) in Fig. 14-23, the base-emitter diodes of transistor Q_1 are on and $V_{B1} = 0.9$ V. From the base of Q_1, there are three series *p-n* diodes to ground: the base–collector diode of Q_1, the base–emitter diode of Q_2, and the base–emitter diode of Q_3. Since $V_{B1} = 0.9$ V, all of these diodes are off. Transistor Q_2 cannot turn on until $V_{B2} \geq 0.7$ V. Transistor Q_3 cannot turn on until $V_{B3} \geq 0.7$ V. With transistors Q_2 and Q_3 off, the output voltage is hi. We can assume for now that V_0 became hi because transistor Q_4 turned on until the load capacitance was charged. The hi output voltage is given by

$$V_0 = V_{CC} - I_{B4}R_{C2} - V_{BE4} - V_D \tag{14-13}$$

When the T²L gate output is hi, we are looking into the reverse-biased diodes of the fan-out gates and I_{B4} is negligibly small. Then

$$V_0 = V_{CC} - V_{BE4} - V_D = 5 - 0.7 - 0.7 = 3.6 \text{ V} \tag{14-14}$$

If V_0 drops significantly below 3.6 V, then $V_{BE4} + V_D > 1.4$ V, transistor Q_4 enters the forward-active region, and current is supplied to C_L. As long as one of the inputs, V_1 or V_2, is held lo, the existing current and voltage levels in the circuit will be maintained, and V_0 will be hi.

When V_1 and V_2 are both hi (greater than 1.4 V), V_{B1} will be at 2.1 V. Transistors Q_2 and Q_3 will be in saturation and the output voltage will be lo.

The transfer characteristic for a typical T²L gate is shown in Fig. 14-24. To develop this curve, let us assume that input V_2 in Fig. 14-23 is tied to a hi voltage level and V_1 is slowly increased from 0.2 V. As long as V_1 is smaller than 0.7 V, transistors Q_2 and Q_3 will be off, and $V_0 = 3.6$ V, as shown. When V_1 reaches 0.7 V, $V_{B1} = 1.4$ V, and $V_{B2} = 0.7$ V. This allows transistor Q_2 to enter the forward-active region such that $\Delta V_{B4}/\Delta V_{B2} = -R_{C2}/R_{E2} = -1.4$. Since $V_{BE3} < 0.7$ V, Q_3 remains in cutoff. If we assume that Q_4 is sourcing a small current to an external load, it will act as an emitter follower with a voltage gain of unity, and $\Delta V_0/\Delta V_{B2} = \Delta V_0/\Delta V_1 = -1.4$. Referring to Fig. 14-24, transistor Q_2 enters the forward-active region at point *A*, and the transfer characteristic has a slope of -1.4 until we reach point *B*, where $V_1 = 1.4$ V. At this point, transistor Q_3 enters its forward-active region. At point *C* on the transfer characteristic, transistor Q_3 enters saturation and $V_0 \approx 0.2$ V.

Now let us go the other way. With V_1 hi, Q_2 and Q_3 in saturation and V_0 lo, we suddenly pull the input voltage V_1 lo and watch what happens. The input base–emitter diode turns on, and Q_1 pulls the stored base charge

Fig. 14-24 Transfer characteristic for a typical T²L gate.

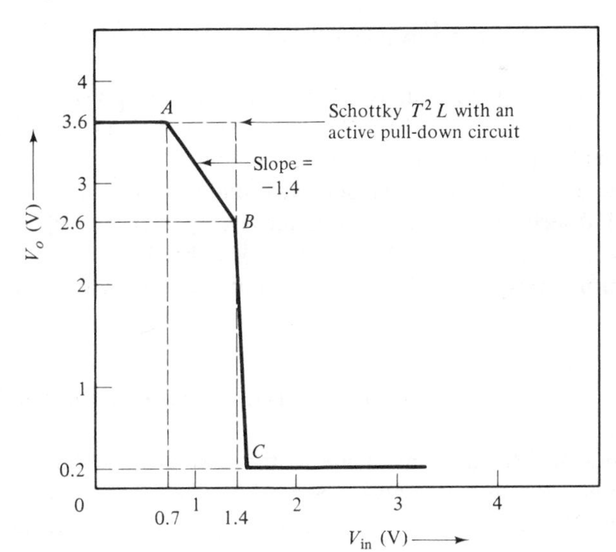

out of Q_2. As Q_2 turns off, its collector goes hi, turning Q_4 on. Q_3 is still on because it takes several additional nanoseconds to get rid of its stored base charge. The transient current to ground through Q_4 and Q_3 is then given by

$$I_{tr} = \frac{V_{CC} - V_{CE4}(\text{sat}) - V_D - V_{CE3}(\text{sat})}{R_{C4}} = \frac{(5 - 1.1)\text{ V}}{100\ \Omega} = 39\text{ mA} \quad (14\text{-}15)$$

Q_3 turns off as its stored base charge is removed, and the load capacitance charges through R_{C4} and Q_4. The 100-Ω resistor, R_{C4}, is in the circuit to limit the transient current of Eq. (14-15). With the output hi, Q_4 will turn on as needed to source current to the fan-out gates. In other words, whenever the output voltage falls below 3.6 V, V_{BE4} becomes forward-biased and Q_4 enters the active region.

In summary, the worst-case logic levels for the standard T^2L logic family are:

$$V_{OL}(\text{max}) = 0.4\text{ V}$$
$$V_{OH}(\text{min}) = 2.4\text{ V}$$
$$V_{IL}(\text{max}) = 0.8\text{ V}$$
$$V_{IH}(\text{min}) = 2.0\text{ V}$$

The noise margins are given by

$$NM_L = V_{IL}(\text{max}) - V_{OL}(\text{max}) = 0.4\text{ V}$$

and

$$NM_H = V_{OH}(\text{min}) - V_{IH}(\text{min}) = 0.4\text{V}$$

The specified maximum fan-out is 10, and the average power dissipation per gate (PD_{AV}) is approximately 10 mW. The average propagation delay (t_{pd}) is about 10 ns, and the power–delay product is about 100 pJ.

Manufacturers of T^2L gates normally make a 7400 series and a 5400 series. The 7400 series gates are intended for a temperature range of 0 to 70°C, and a supply voltage range of 4.75 to 5.25 V. The 5400 series are intended for a temperature range of −55 to 125°C and a supply voltage range of 4.5 to 5.5 V.

14-3-1 Schottky-Clamped T^2L Logic

The types of T^2L gates available to the user are standard, high power, low power, Schottky, and low-power Schottky. They differ principally in the trade-offs made between power dissipation and propagation delay. We have discussed standard T^2L gates. In the interest of brevity, we will now discuss Schottky T^2L gates and then move on to emitter-coupled logic.

Figure 14-25 shows a Schottky-clamped two-input NAND gate. This circuit dissipates about 20 mW (PD_{AV}), and has a propagation delay time of about 3 ns. The design changes that make this possible will now be discussed, one at a time.

1. With the exception of Q_4, all the *transistors are Schottky-clamped.* Metal–semiconductor diodes are built in between the transistor bases and collectors as discussed in Section 11-2-2. The Schottky diodes turn on whenever a base voltage becomes 0.3 V more positive than the collector voltage. Thus, the base–collector diodes cannot be forward-biased, the transistors cannot saturate, and the stored-base-charge problem is minimized. The current pulse that occurred in standard T^2L when Q_3 and Q_4 were both on is reduced by a factor of 5.

2. *Schottky diodes* D_1 and D_2 are tied between the inputs and ground. At high frequencies, connecting wires act as transmission lines. Sudden voltage transitions cause damped oscillations. When the damped oscillations at inputs V_1 and V_2 reach minus 0.3 V with respect to ground, diodes D_1 and D_2 turn on to discharge the lines and prevent further oscillation.

3. Comparing the Schottky T^2L gate of Fig. 14-25 to the standard T^2L gate of Fig. 14-23 shows that all the *resistor values* have been lowered. In effect, we are sacrificing power to gain speed. Small resistors and large current pulses permit device capacitances to charge and discharge quickly.

4. Transistors Q_5 and Q_4 are connected in a Darlington configuration.

Fig. 14-25 Schottky-clamped two-input T^2L NAND gate.

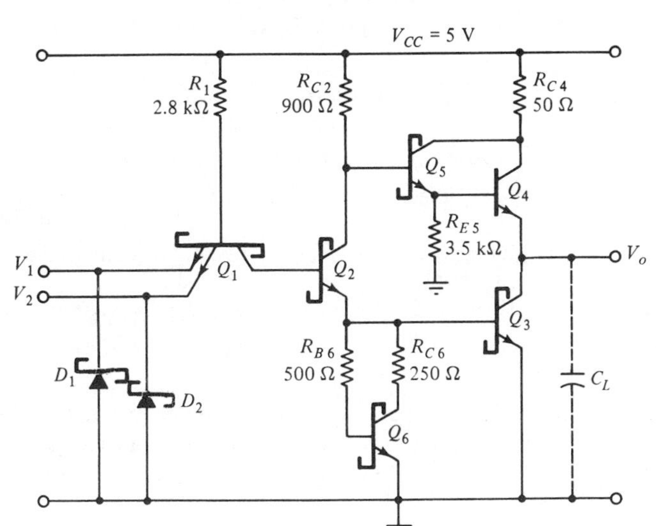

Referring back to Fig. 14-23, two diode voltage drops are needed between the collector of transistor Q_2 and the circuit output to prevent Q_4 from turning on when the output is lo. In the Schottky T²L gate, these two diode voltage drops consist of V_{BE5} and V_{BE4}. When inputs V_1 and V_2 go lo, transistor Q_2 turns off and its collector voltage goes hi. Transistor Q_5 enters the forward-active region and turns on Q_4. The output resistance of the Darlington pair is approximated by $R_{C2}/\beta_5\beta_4 + r_{e4}$, the emitter resistance of Q_4. This provides a low-resistance charge path for C_L. The base–collector diode of Q_5 clamps at 0.3 V. Then V_{CE5} is held at $V_{CB5} + V_{BE5} = -0.3 + 0.7 = 0.4$ V. Since $V_{CE5} = V_{CB4}$, the collector–base diode of transistor Q_4 is held positive at 0.4 V. Thus, Q_4 cannot saturate and a Schottky clamp is not needed.

5. Transistor Q_6, in conjunction with resistors R_{B6} and R_{C6}, constitute an *active pull-down* circuit. In effect, this circuit replaces the pull-down resistor R_{E2} in Fig. 14-23 with a nonlinear resistance that is designed to improve the switching speed of transistor Q_3. The active pull-down circuit also changes the T²L transfer characteristic, as shown by the dashed line in Fig. 14-24. If transistor Q_2 has a passive emitter resistor to ground, it enters the active region when V_{B2} is at 0.7 V. With Q_6 between the emitter of Q_2 and ground, as shown in the Schottky T²L gate, Q_2 cannot turn on until its base voltage reaches 1.4 V. This change improves the transfer characteristic and the noise margin of the T²L gate.

As specified by the manufacturers, the worst-case logic levels for Schottky T²L gates are given as

$$V_{OL}(\text{max}) = 0.5 \text{ V}$$
$$V_{OH}(\text{min}) = 2.7 \text{ V}$$
$$V_{IL}(\text{max}) = 0.8 \text{ V}$$
$$V_{IH}(\text{min}) = 2.0 \text{ V}$$

The worst-case noise margins are then given by

$$NM_L = V_{IL}(\text{max}) - V_{OL}(\text{max}) = 0.8 - 0.5 = 0.3 \text{ V}$$
$$NM_H = V_{OH}(\text{min}) - V_{IH}(\text{min}) = 2.7 - 2.0 = 0.7 \text{ V}$$

With regard to fan-out, a Schottky T²L gate can typically drive 12 standard T²L gates or 10 Schottky T²L gates.

Average power dissipation is 20 mW, and the propagation delay time is 3 ns. Thus, the power–delay product is 60 pJ. Rise time t_r and fall time t_f are both on the order of 3 ns.

The use of Schottky-clamped transistors reduces propagation delay time and improves the power–delay product by keeping the transistors out of saturation. Improvements in device fabrication are also used, however, to achieve propagation delays on the order of 3 ns. Schottky T²L circuits use shallower diffusions and smaller device geometries than standard T^2L gates. The fabrication and layout of Schottky-clamped transistors is discussed in Section 11-2-2.

14-4 EMITTER-COUPLED LOGIC (ECL)

Emitter coupled logic, also called *current-mode logic* (CML), is the fastest silicon logic family available. It is the most widely used digital family for applications in which propagation delays on the order of 1 ns are required. Naturally, an average power dissipation on the order of 50 mW per gate is needed to achieve this speed.

The design of ECL gates is based on a current switching arrangement as shown in Fig. 14-26. A fixed reference voltage is applied to the base of transistor Q_2 and an input voltage to the base of Q_1. The voltage levels are arranged so that V_{in} hi is larger than V_R, and V_{in} lo is equally smaller than V_R. With V_{in} hi $> V_R$, the base–emitter diode of Q_1 (V_{BE1}) is forward-biased. Since the emitters are connected and tied to a current source, V_{BE2} is less than 0.7 V. Transistor Q_1 is on and Q_2 is off. The current I_o goes through Q_1, and V_{o1} is lo. When V_{in} goes lo, V_{BE2} becomes greater than V_{BE1}, transistor Q_2 turns on, and Q_1 turns off. Current I_o goes through Q_2 and V_{o2} goes lo. As V_{in} is switched hi and lo, the current I_o toggles between transistors Q_1 and Q_2. Component values are arranged such that the transistors are driven between cutoff and a quiescent operating point in the forward-active region. Since they never approach saturation, the stored-base-charge problem does not exist. With very careful circuit design, subnanosecond propagation delays can be achieved.

An example of a complete ECL gate is shown in Fig. 14-27. This gate was used as a bipolar mask layout example in Section 10-1. It was made by Motorola and is designated as a member of the MECL II family of ECL circuits.

The current source I_o in Fig. 14-26 has been replaced by a large emitter resistor, R_E in Fig. 14-27. Input signals V_1 and V_2 are connected to the bases of transistors Q_1 and Q_2, which share the current through R_E when V_1 and

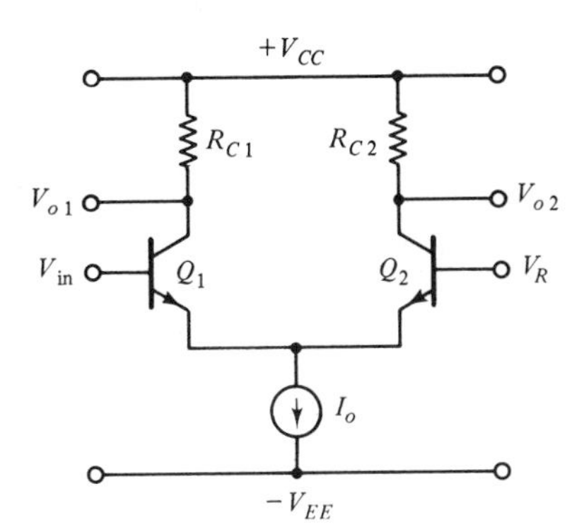

Fig. 14-26 Current switching is the basis for emitter-coupled logic.

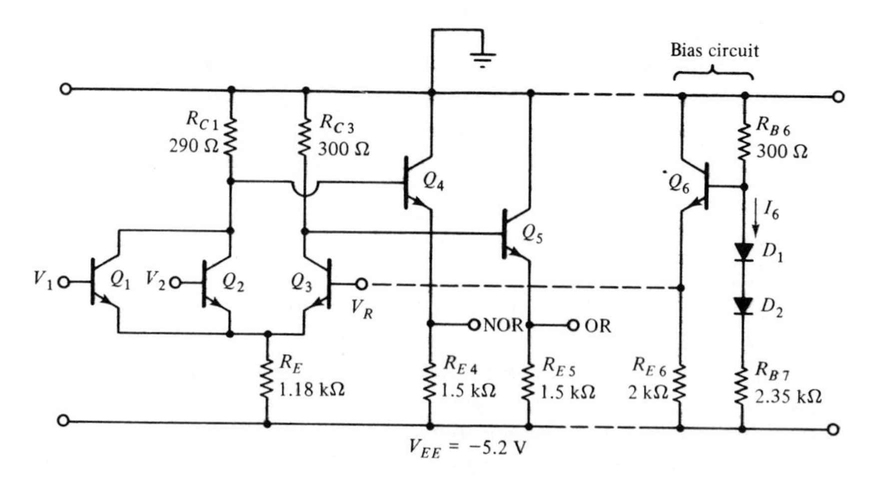

Fig. 14-27 ECL OR–NOR gate.

V_2 are both hi. The outputs are taken from emitter followers Q_4 and Q_5, which operate in the forward-active region. They provide a low output impedance to maximize fan-out, and also provide a V_{BE} drop, so that the input and output signals are of the same magnitude. When V_1 and V_2 are lo, $V_{C1,2}$ is hi and the NOR output is hi. Since Q_3 is on, V_{C3} is lo and the OR ouput is lo. If V_1, V_2, or both go hi, Q_3 turns off, the OR output is hi, and the NOR output is lo. Thus, the OR and NOR functions are realized at the outputs.

Kirchhoff's voltage law can be used on the bias circuit of Fig. 14-27 to find the reference voltage V_R. If we assume that base current I_{B6} is negligible with respect to I_6, then

$$I_6 = \frac{-V_{EE} - 2V_D}{R_{B6} + R_{B7}} \tag{14-16}$$

Then, $V_{B6} = -I_6 R_{B6}$ and $V_R = V_{B6} - V_{BE6}$. Combining terms, we have that

$$V_R = -\frac{R_{B6}}{R_{B6} + R_{B7}}(-V_{EE} - 2V_D) - V_{BE6} \tag{14-17}$$

Notice that V_R is dependent only on the supply voltage, diode voltage drops, and a resistor ratio. Because of the reduced transistor areas and the magnitudes of the currents used, it is practical to assume that all forward-biased diode voltage drops are 0.75 V for ECL gates. Applying numerical values to Eq. (14-17), we have

$$V_R = \frac{0.3\ \text{k}\Omega}{2.65\ \text{k}\Omega}(5.2 - 1.5)\ \text{V} - 0.75\ \text{V} = -1.17\ \text{V} \tag{14-18}$$

One bias circuit is normally used for a number of ECL gates within the same IC chip.

Now we can analyze the ECL gate proper to determine the output voltage levels. If we hold inputs V_1 and V_2 sufficiently negative with respect to V_R, transistors Q_1 and Q_2 will be off and Q_3 will be on. This allows us to find V_{OL}(OR) and V_{OH}(NOR). Since transistor Q_3 is on, its emitter voltage is $V_E = V_R - V_{BE3}$. Given that transistor betas are on the order of 100, we can ignore base currents in the ECL gate and assume that $I_E \approx I_C$. Then

$$I_{C3} \approx I_E = \frac{V_E - V_{EE}}{R_E} = \frac{V_R - V_{BE3} - V_{EE}}{R_E} \tag{14-19}$$

The collector voltage of transistor Q_3 is given by

$$V_{C3} = 0 - I_{C3}R_{C3} = -\frac{R_{C3}}{R_E}(V_R - V_{BE3} - V_{EE}) \tag{14-20}$$

The output voltage V_{OL} (OR) is then given by

$$V_{OL}(\text{OR}) = -\frac{R_{C3}}{R_E}(V_R - V_{BE3} - V_{EE}) - V_{BE5} \tag{14-21}$$

Notice that V_{OL}(OR) is dependent on a resistor ratio, diode voltage drops, and the supply voltage. Equation (14-21) has the same form as Eq. (14-17). Since the output of one gate is the input of the next, Eqs. (14-17) and (14-21) represent the signal input and the reference input to an ECL differential input pair. As temperature changes, V_R compensates for the change in V_{in}.

Applying numerical values to Eq. (14-21), we have

$$\begin{aligned} V_{OL}(\text{OR}) &= \frac{0.3\ \text{k}\Omega}{1.18\ \text{k}\Omega}(-1.17 - 0.75 + 5.2)\ V - 0.75\ V \\ &= -1.58\ \text{V} \end{aligned} \tag{14-22}$$

Since transistors Q_1 and Q_2 are off, $V_{C2} \approx 0$ V and

$$V_{OH}(\text{NOR}) = V_{C2} - V_{BE4} = 0 - 0.75 = -0.75\ V \tag{14-23}$$

Let us ignore worst-case values, noise margins, and fan-out for the time being and suppose that the V_{OL}(OR) and V_{OH}(NOR) logic levels that we just obtained can also be used as our hi and lo *input* logic levels. Then we can find the output voltages when an input voltage is hi. From Eq. (14-23) the hi input level is $V_{\text{in}}(\text{hi}) = -0.75$V. Input V_2 can be held lo (−1.58 V) to keep it out of the way. The emitter voltage of transistor Q_1 is then $V_E = V_{\text{in}}(\text{hi}) - V_{BE1}$, and the current I_{C1} is given by

$$I_{C1} = I_E = \frac{V_E - V_{EE}}{R_E} = \frac{V_{\text{in}}(\text{hi}) - V_{BE1} - V_{EE}}{R_E} \tag{14-24}$$

The collector voltage of transistor Q_1 is

$$V_{C1} = 0 - I_{C1}R_{C1} = -\frac{R_{C1}}{R_E}[V_{\text{in}}(\text{hi}) - V_{BE1} - V_{EE}] \tag{14-25}$$

The NOR output voltage at the emitter of Q_4 is then given by

$$\begin{aligned} V_{OL}(\text{NOR}) &= V_{C1} - V_{BE4} \\ &= -\frac{R_{C1}}{R_E}[V_{\text{in}}(\text{hi}) - V_{BE1} - V_{EE}] - V_{BE4} \end{aligned} \tag{14-26}$$

We would like $V_{OL}(\text{NOR})$ to have about the same numerical value as $V_{OL}(\text{OR})$, but $V_R = -1.17$ V and $V_{\text{in}}(\text{hi}) = -0.75$ V. This is compensated for by making $R_{C1} < R_{C3}$. Thus, $R_{C1} = 290\ \Omega$ and $R_{C3} = 300\ \Omega$. Plugging numbers into Eq. (14-26), we have that

$$\begin{aligned} V_{OL}(\text{NOR}) &= -\frac{0.29\ \text{k}\Omega}{1.18\ \text{k}\Omega}(-0.75 - 0.75 + 5.2)\ V - 0.75\ V \\ &= -1.66\ \text{V} \end{aligned} \tag{14-27}$$

Since transistor Q_3 is off, $V_{C3} \approx 0$ V. Then

$$V_{OH}(\text{OR}) = V_{C3} - V_{BE5} = 0 - 0.75 = -0.75\ \text{V} \tag{14-28}$$

In summary, with $V_{IH} = -0.75$ V and $V_{IL} = -1.58$ V, typical values for the hi and lo output voltages are given by

$$\begin{aligned} V_{OL}(\text{OR}) &= -1.58\ \text{V} \qquad & V_{OL}(\text{NOR}) &= -1.66\ \text{V} \\ V_{OH}(\text{OR}) &= -0.75\ \text{V} \qquad & V_{OH}(\text{NOR}) &= -0.75\ \text{V} \end{aligned}$$

To draw the transfer characteristic for the ECL gate of Fig. 14-27, we have to depart from the ideal ECL silicon diode model ($V_{BE} = 0.75$ V). The transition width of the transfer characteristic depends on the current–voltage characteristics of the base–emitter diodes. From the Ebers–Moll model of a transistor biased in the forward-action region [Eq. (7-54)], we have that

$$I_{E1} \approx -I_{ES} \exp\left(\frac{qV_{E1}}{kT}\right) \tag{14-29}$$

where V_{E1} represents the emitter–base diode voltage, and $kT/q = 0.026$ V at room temperature. Let us increase the base–emitter voltage to V_{E2}, such that

$$I_{E2} = 10 I_{E1} = -I_{ES} \exp\left(\frac{qV_{E2}}{kT}\right) \tag{14-30}$$

Then

$$\frac{I_{E2}}{I_{E1}} = 10 = \exp\left(\frac{q\Delta V_E}{kT}\right) \tag{14-31}$$

where $\Delta V_E = V_{E2} - V_{E1}$ and

$$\Delta V_E = \frac{kT}{q} \ln 10 = 60 \text{ mV} \tag{14-32}$$

A 60-mV change in V_{BE} causes the emitter current to change by a factor of 10. When $V_{\text{in}} = V_R$ in Fig. 14-27, the current through R_E splits evenly between transistors Q_1 and Q_3 (assuming that V_2 is held lo). When $V_{\text{in}} = V_R -$ 60 mV, practically all of the emitter current is through Q_3. When $V_{\text{in}} = V_R$ + 60 mV, practically all of the current is through Q_1. The actual transition width is then approximated by $V_{IH} - V_{IL} = 120$ mV.

Figure 14-28 shows the transfer characteristic for the ECL OR/NOR gate of Fig. 14-27. These curves were calculated from Eqs. (14-17) through (14-28). Notice that the output voltages are symmetrical about the reference voltage to keep the noise margins approximately equal. Notice also that V_{OL}(NOR) varies as a function of V_{in}(hi) while the other output voltages remain constant as V_{in} varies (outside the transition region).

Worst-case voltage specifications on ECL (MECL II) gates are

$$\begin{aligned}
V_{OH}(\text{max}) &= -0.7 \text{ V} \\
V_{OH}(\text{min}) &= -0.85 \text{ V} \\
V_{OL}(\text{max}) &= -1.325 \text{ V} \\
V_{OL}(\text{min}) &= -1.5 \text{ V} \\
V_{IH}(\text{min}) &= -1.025 \text{ V} \\
V_{IL}(\text{max}) &= -1.325 \text{ V}
\end{aligned}$$

The worst-case noise margins are then given by

$$\begin{aligned}
NM_H &= V_{OH}(\text{min}) - V_{IH}(\text{min}) = -0.85 - (-1.025) \\
&= 0.175 \text{ V} \\
NM_L &= V_{IL}(\text{max}) - V_{OL}(\text{max}) = -1.325 - (-1.5) \\
&= 0.175 \text{ V}
\end{aligned}$$

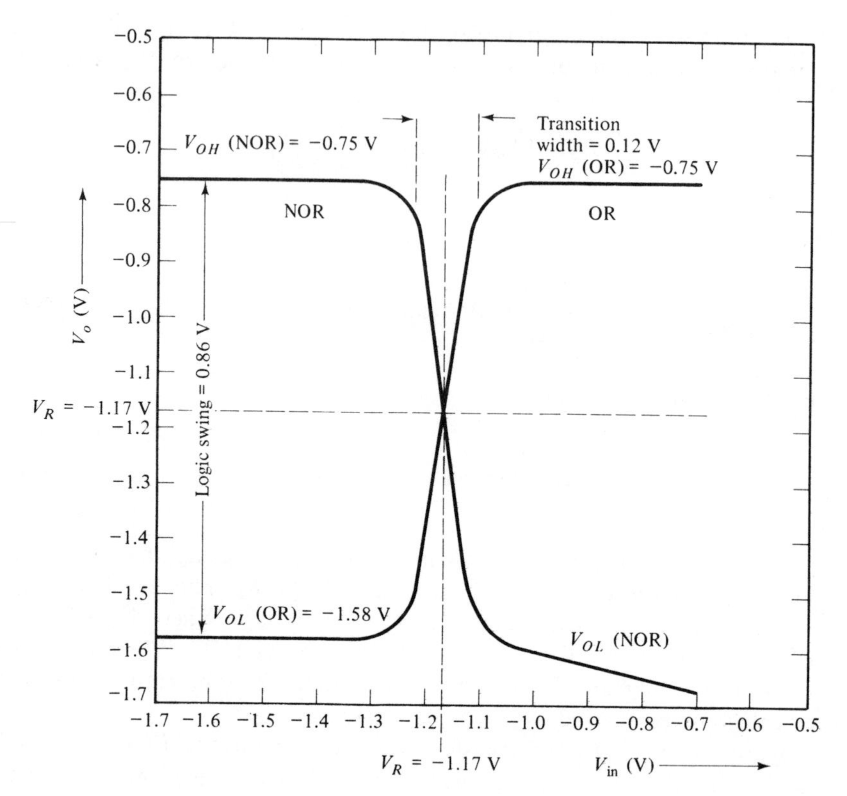

Fig. 14-28 Transfer Characteristics for the ECL Gate of Fig. 14-27.

Fan-out is typically specified at $N = 25$ or more for ECL gates, since the emitter followers provide low output impedance. Since fan-out is related to the noise margins and propagation delay, it is generally reduced in cases where the noise margins are critical or the gates are operated near maximum switching speeds. Input capacitance is on the order of 3 pF per gate. Allowing for wiring capacitance, each added fan-out terminal contributes about 5 pF of capacitance. Because the transistors do not saturate and the voltage swing is less than 1 V, propagation delays are not strongly dependent on the rise and fall times of input signals. The outputs follow the collector voltages of the current switches with very little delay. ECL propagation delays on the order of 1 ns are typical. Isoplanar transistor construction is used when subnanosecond delays are required, as discussed in Section 11-2-3.

Most commercially available ECL gates do not include the emitter follower pull-down resistors. The user provides these terminations, quite often in the form of 50-Ω transmission lines. This reduces line reflections and improves noise immunity.

14-5 INTEGRATED-INJECTION LOGIC (I^2L)

Integrated injection logic is a form of bipolar logic capable of much higher packing densities and lower power–delay products than other bipolar logic families. The basic unit of I^2L logic is a lateral *pnp* transistor which serves as a current source and a load, and a vertical *npn* transistor with multiple collectors, which serves as an inverter. High packing densities can be obtained because the *pnp* and *npn* transistors are merged together and no internal resistors are needed for either source or load functions.

The structure of the *pnp-npn* merged transistors is shown in Fig. 14-29.

Fig. 14-29 Basic I^2L gate structure: (a) top view; (b) cross section.

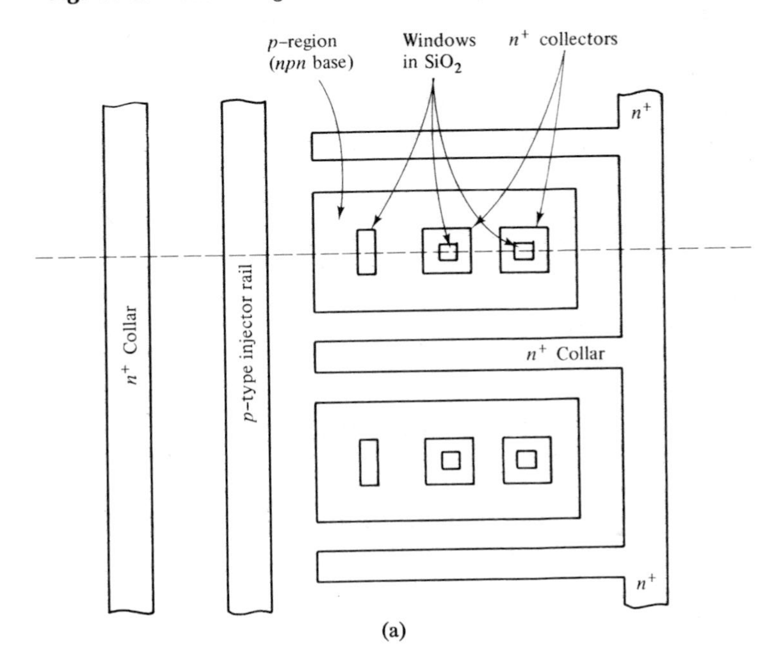

(a)

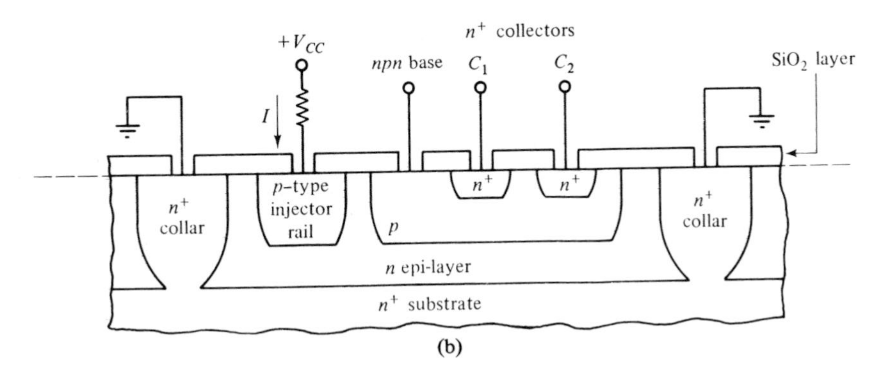

(b)

As shown in the top view, a grounded n^+ collar is used to isolate transistor structures. A single *p*-type "injector rail," which serves as the *pnp* transistor emitter, can be used to inject current into many *p*-type islands which serve as *pnp* collectors and *npn* bases. The *n*-type epi-layer serves as the grounded base for the lateral *pnp* transistor and the grounded emitter of the *npn* transistor. Compared to a conventional *npn* transistor, a vertical *npn* transistor of Fig. 14-29 operates in the inverse-active region. The n^+ collectors sit in *p*-type base regions, and the *n*-type epi-layer serves as the emitter. Because of the structure and the doping levels, beta values are on the order of 10. The vertical n^+ diffusions (the n^+ collars) provide isolation between devices. Because of the n^+ region which is used as a collector, the collector-to-base breakdown voltage is low and the collector-to-base capacitance is relatively large.

Since the *pnp* emitter–base diode is forward-biased, a diffusion current will always exist through this diode to ground. This injected hole current has lateral components, however, which diffuse to the *pnp* collector regions. Now the *pnp* collector regions are also the base regions of the vertical *npn* transistors. Thus, what has been described so far is a lateral *pnp* transistor acting to continuously inject constant current into the base regions of the vertical *npn* transistors. If the base contact of a given *npn* transistor is held near ground potential, the device will be in the cutoff region. The *npn* collectors will then take on voltage levels determined by their external connections. If the base contact is at a high impedance with respect to ground, however, the constant base current will drive the *npn* transistor into saturation. In this case, assuming that paths exist for collector current, the *npn* collector–emitter voltages will be at approximately 0.1 V.

Schematically, the basic I²L gate structure can be drawn as shown in Fig. 14-30. Figure 14-30(a) shows the merged transistor schematic and Fig. 14-30(b) shows the *pnp* transistor modeled as a constant-current source between base and ground. If the base is at a high impedance with respect to ground, the *npn* transistor will be in saturation, and the collector voltages will be lo ($V_{CE}\text{sat} \approx 0.1$ V). If the base voltage is lo, the current source in Fig. 14-30(b) will be routed to the left and the *npn* transistor will be off. It is convenient to

Fig. 14-30 Schematics for an I²L structure: (a) with lateral *pnp* transistor; (b) *pnp* transistor is replaced by a constant current source.

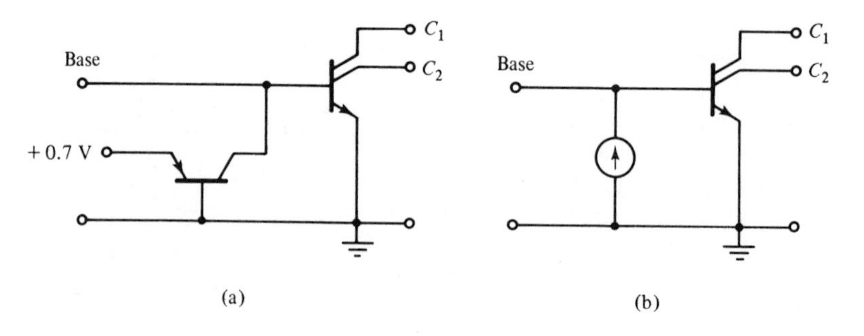

think of the I²L structure as an *npn* transistor with a source of base current (and more than one collector). Logical operations then become a matter of current steering (i.e., of finding or not finding alternative paths to ground for the current sources). If current paths other than the associated *npn* transistors are available, the transistors will be off. If not, the transistors will be on and their collector voltages will be lo. Logic functions are implemented by connecting the collector of one gate to the base of the next, and by interconnecting collectors. A single collector is not usually tied to more than one base. Any number of collectors can be connected to a single base, however. This is shown in Fig. 14-31.

Figure 14-32 shows a two-input I²L gate. Depending on where the inputs and output are chosen, it can be thought of as a NOR gate or a NAND gate.

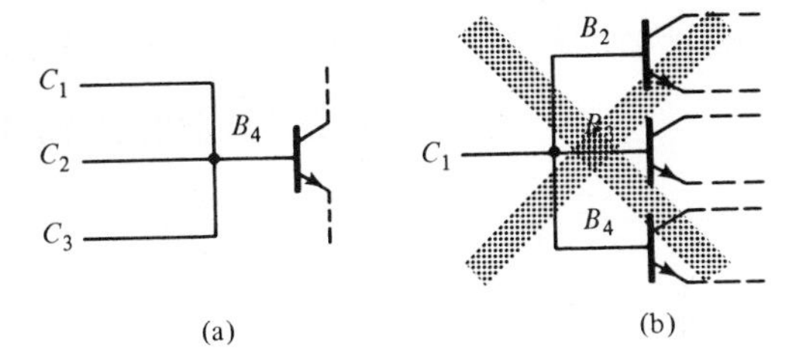

Fig. 14-31 (a) Permitted: multiple collectors can be used as inputs to an *npn* base. (b) Not permitted: single collector cannot be used as input to more than one base.

Fig. 14-32 I²L structures connected to form a logic gate: (a) circuit diagram; (b) truth tables and logic symbols.

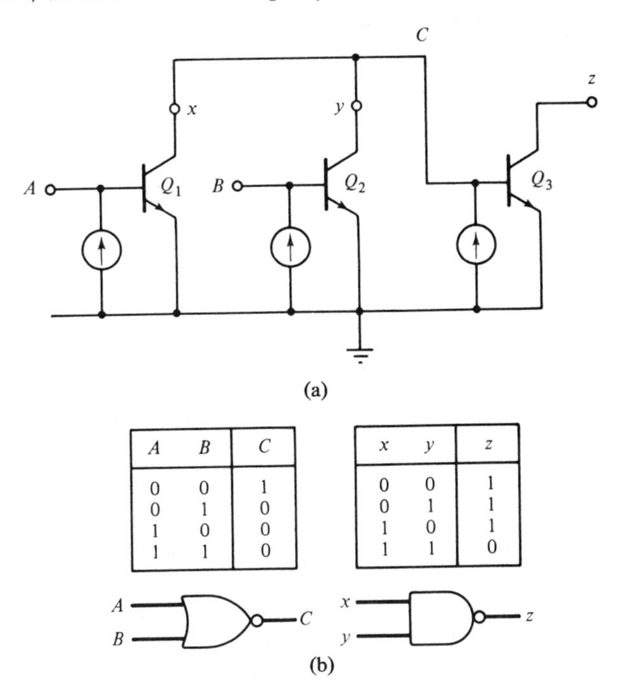

A	B	C
0	0	1
0	1	0
1	0	0
1	1	0

x	y	z
0	0	1
0	1	1
1	0	1
1	1	0

When inputs A and B are lo, transistors Q_1 and Q_2 are off and C is hi. If A and/or B goes hi, Q_1 and/or Q_2 is on, and C is lo. As shown, this represents the NOR function. In many applications, it is useful to consider x and y as the inputs and z as the output. Because x and y are wired together, if one of

Fig. 14-33 I²L example: from layout to logic diagram: (a) simplified mask layout; (b) circuit diagram; (c) NAND gate representation; (d) replacement of AND with OR; (e) simplified logic diagram.

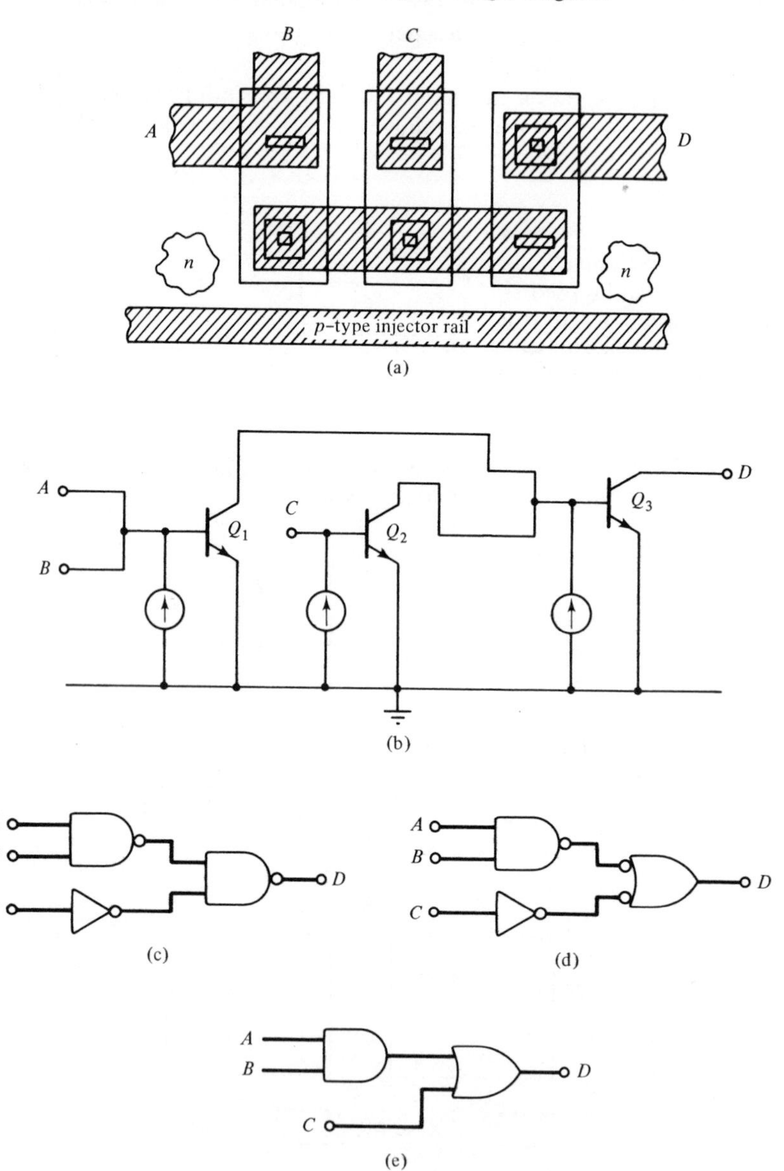

a depletion-load device can be modeled as a resistor and enhancement drivers and/or y is lo, transistor Q_3 is off and z is hi. If x and y are hi, Q_3 is on and z is lo. The x, y, z relation provides the NAND function. This is called a wired NAND because inputs x and y are wired together. It is possible to wire inputs together in I^2L because we are routing currents rather than working with discrete voltage levels.

As an example of I^2L logic, consider the I^2L layout shown in Fig. 14-33(a), where A, B, and C represent digital inputs and D is the output. The corresponding circuit diagram can be drawn as shown in Fig. 14-33(b), where transistors Q_1 and Q_3 act as NAND gates and Q_2 acts as an inverter. The logic diagram can then be drawn and simplified as shown in Fig. 14-33(c)–(e). Referring to Fig. 14-33(e), the output D is hi if A AND B are hi OR if C is hi.

The lateral hole current injected into the base region of an *npn* transistor must diffuse to a given collector region in order to effect a transition between cutoff and active operation. Thus, *npn* collectors can be turned on or off with propagation delays which depend on their distance from the injector rail. For example, suppose that the *npn* transistor base in Fig. 14-34(a) is connected to ground. Then transistor Q_1 will be off and collectors C_1 and C_2 will be high. If the ground connection to the base is now opened, collector C_1 will turn on before C_2 because the injected base current diffuses to C_1 first. In the layout of Fig. 14-34(b), both collectors will respond at the same time because they are equidistant from the injector rail. Maximum speed can be obtained from an I^2L gate by making the injector rail surround the *npn* base region as shown in Fig. 14-34(c).

Isolation diffusions are not required since I^2L structures are self-isolating. The n^+ buried layer provides a common, distributed ground and the injector rail provides a distributed positive voltage connection. Also, the lateral *pnp* transistors act as constant-current sources, which minimizes the need for resistors. All these factors tend to increase packing densities for I^2L structures, but there are drawbacks. If we look at NMOS logic structures for a moment,

Fig. 14-34 Layout techniques to control switching speeds: (a) collector C_1 turns on before C_2; (b) collectors C_1 and C_2 turn on at the same time; (c) a maximum speed configuration.

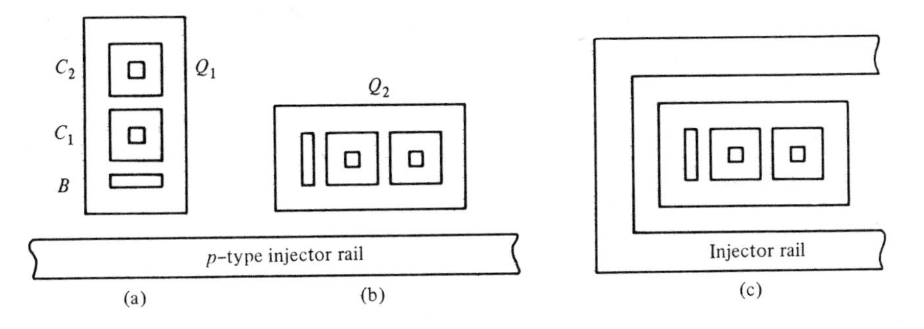

them is pulled lo by a transistor in saturation, they are both lo. Thus, if x can be modeled as switches. As shown in Fig. 14-35, switching devices can be stacked vertically (in series) and/or in parallel. Thus, with NMOS, complex logic functions can be realized directly, by using series–parallel arrangements of enhancement drivers in conjunction with one load device. Because of the common distributed ground used in I^2L logic, active components cannot be stacked vertically. The only way to realize complex logic functions is by interconnecting simple gates. Thus, many interconnections are required. Even with double-layer metalizations, additional chip area is required. Although the high packing densities inherent to the I^2L structure save chip area, some of this area must be given back in terms of interconnection space.

Another factor should be considered in comparing I^2L to other logic families. I^2L gates normally operate at low current and voltage levels (logic swing $\approx$ 700 mV), and have very limited fan-out capabilities. To make I^2L inputs and outputs compatible with peripheral electronic systems, buffering or interface circuits are normally provided on the chip. To take full advantage of I^2L packing densities, and to minimize interconnection and interfacing problems, I^2L is normally used in LSI applications (1000 to 10,000 gates/chip). On-chip interfacing is provided by processing conventional bipolar devices along with I^2L circuits, as shown in Fig. 14-36.

Fig. 14-35 Direct realization of complex logic structures by using series–parallel arrangements of switching devices. This is possible with NMOS, but not with I^2L structures.

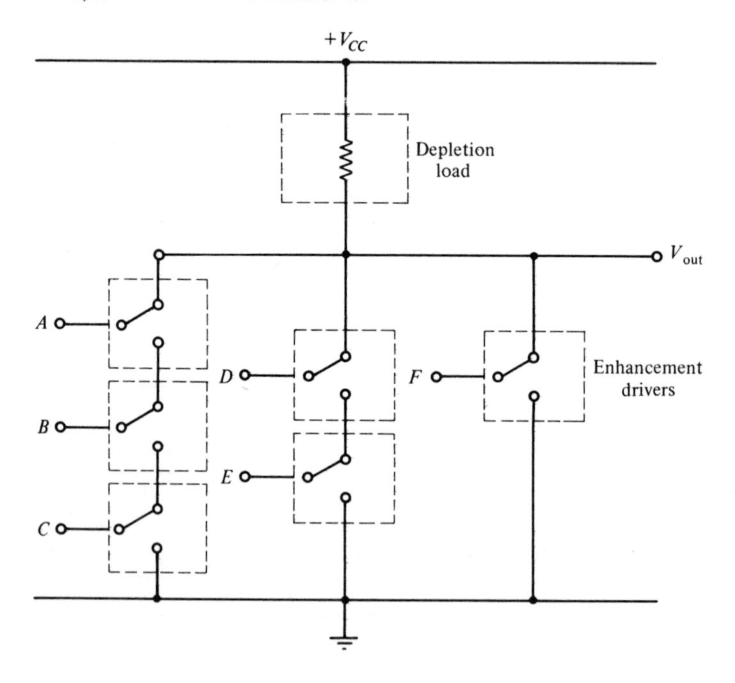

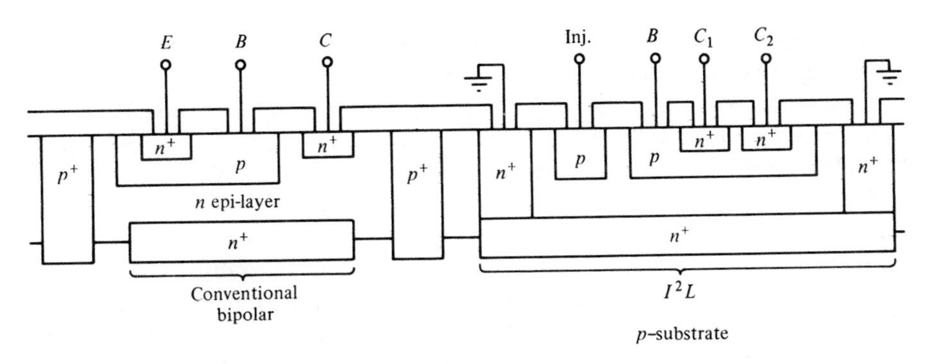

Fig. 14-36 Conventional NPN and I^2L circuits on the same chip.

Because I^2L processing is compatible with linear bipolar processing, efficient combinations of linear and digital circuits can be obtained on the same chip. Although I^2L cannot compete successfully with other technologies as a free-standing digital technology, it has been very successful in systems that require digital functions on linear chips, such as analog-to-digital converters and other digital–linear interface equipment.

Since I^2L is normally used in LSI circuits with bipolar interface circuits on the chip, manufacturers' specifications on noise margins, average power dissipation, and so on, are not directly available. Propagation delays (t_{pd}) for internal gates of I^2L products are in the range 10 to 50 ns per gate. Power–delay products for production I^2L circuits are in the range $\frac{1}{2}$ to 1 pJ per gate. Propagation delay and the power–delay product can be reduced by using oxide isolation between I^2L devices. This technique, also called the Isoplanar method, allows reduced device geometries and reduced capacitances and leakage currents. Isoplanar I^2L, or I^3L circuits, have propagation delays on the order of 4 ns per gate.

14-6 NMOS LOGIC

NMOS has become the "workhorse" of the silicon technologies. Because of the simplicity of the gate structures and the packing densities that can be achieved, it is widely used for large-scale and very-large-scale IC chips (LSI and VLSI).

In the 1960s masks were made by photoreduction. Contact printing and wet etching with negative photoresists were used to process ICs. Minimum line widths in the early 1960s were on the order of 10 to 20 μm. By the 1970s, projection printing, positive resists, and plasma etching were used to achieve line widths on the order of 5 to 6 μm. Innovations such as SEM lithography, ion implantation, local oxidation, and self-aligned gate structures have led to minimum line widths on the order of 2 μm in the early 1980s. By the 1990s it is anticipated that minimum line widths (and NMOS channel lengths) will be on the order of 0.2 μm.

Within the scientific disciplines, working with strict sets of boundary conditions, it is the business of engineers to predict the future. "If we do *A*, *B* and *C*, and those other items remain constant, *D* should happen." Within the NMOS technologies, a scaling factor *(S)* is generally used to study the impact of scaling down device geometries and interconnection lines on device and circuit operation. In modern industrial nomenclature, the term *NMOS* generally refers to 5 to 6-μm device geometries. ICs that are scaled down by a certain percentage from the 5- to 6-μm geometries are referred to as *scaled NMOS.* As lateral dimensions are reduced, diffusion depths can also be reduced. More accurate control of shallow source and drain diffusions can be achieved with arsenic than with phosphorus. NMOS ICs which combine scaling down with arsenic source and drain diffusions are referred to as HMOS technologies, where the term HMOS refers to *high-performance* NMOS. Further scalings are being categorized with the terms HMOS I, HMOS II and HMOS III.

As devices are scaled down, substantial improvements are achieved in gate propagation delays and power–delay products. Packing densities are increased and more complex, sophisticated circuits can be built into silicon chips. On the other hand, voltage breakdown, doping fluctuations, contact resistance, fringe fields and other second-order effects present severe challenges. To limit chip temperatures, power dissipation per gate must be continuously reduced as the number of gates per chip is increased.

Increasing memory capacity is an immediate benefit of VLSI. Beyond memory applications, however, the number of person-years required to design a VLSI chip is becoming a very severe boundary condition.

14-6-1 NMOS Inverter

Figure 14-37 shows the circuit diagram for an enhancement driver/depletion load inverter. Starting with the NMOS device equations, we want to develop equations for the inverter transfer characteristics (V_o vs. V_{in}). Then, using typical values for the device parameters, we can plot the transfer characteristics. They provide the basis for logic gate operation.

The *I–V* equations for NMOS devices are as shown below.

$$I_D = \beta[(V_{GS} - V_T)V_{DS} - \tfrac{1}{2}V_{DS}^2] \quad \text{(linear region)} \tag{14-33}$$

$$I_D = \tfrac{1}{2}\beta(V_{GS} - V_T)^2 \quad \text{(saturation region)} \tag{14-34}$$

$$\beta = \frac{Z}{L}\mu_{ns}C_{\text{ox}} \tag{14-35}$$

Although the foregoing equations are equally valid for the driver and load devices shown in Fig. 14-37, the width/length ratios *(Z/L)* will cause

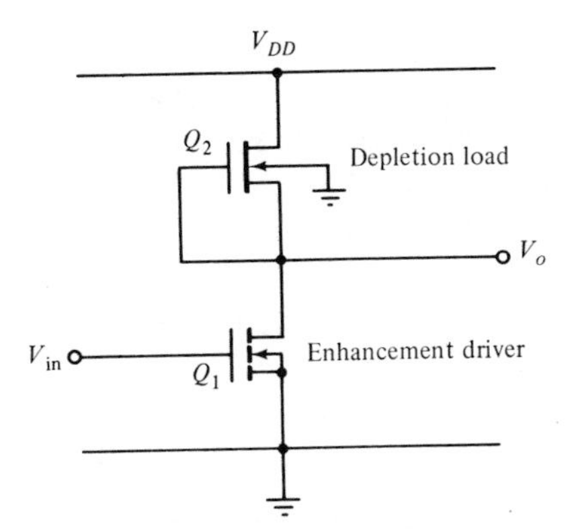

Fig. 14-37 Enhancement-driver depletion load inverter.

Q_1 and Q_2 to have different β values. Also, the load device is designed to be normally on, so that a negative value of threshold voltage (V_T) is required to turn it off. The driver device is normally off, and a positive V_T is used to turn it on.

From Fig. 14-37, the gate–source and drain–source voltages are given by

$$V_{GS}(Q_1) = V_{\text{in}} \tag{14-36}$$

$$V_{GS}(Q_2) = 0 \tag{14-37}$$

$$V_{DS}(Q_1) = V_o \tag{14-38}$$

$$V_{DS}(Q_2) = V_{DD} - V_o \tag{14-39}$$

Substituting Eqs. (14-36) through (14-39) in Eqs. (14-33) and (14-34), we have that

$$I_D(\text{load}) = \beta_L[(-V_{TL})(V_{DD} - V_o) - \tfrac{1}{2}(V_{DD} - V_o)^2] \quad \text{(linear region)} \tag{14-40}$$

$$I_D(\text{load}) = \tfrac{1}{2}\beta_L V_{TL}^2 \quad \text{(saturation region)} \tag{14-41}$$

$$I_D(\text{driver}) = \beta_D[(V_{\text{in}} - V_{TD})V_o - \tfrac{1}{2}V_o^2] \quad \text{(linear region)} \tag{14-42}$$

$$I_D(\text{driver}) = \tfrac{1}{2}\beta_D(V_{\text{in}} - V_{TD})^2 \quad \text{(saturation region)} \tag{14-43}$$

The L and D subscripts refer to the load and driver device, respectively.

If we were to set the supply voltage $V_{DD} = 5$ V in Fig. 14-37 and allow the input voltage to sweep from 0 to +5 V, both devices would make transitions between the linear and saturation regions of operation. Since the devices are in series, and load currents can be considered negligible, the drain currents will be equal [i.e., I_D(load) $= I_D$(driver)]. If we plot I_D vs. V_{DS} (Q_1), and then plot the characteristic curve for the depletion-load device on the same graph, the result will be as shown in Fig. 14-38. In this figure, the solid lines represent the family of enhancement driver characteristics and the single dashed line represents the depletion load. For any given value of input voltage V_{in}, the actual operating point will occur at the intersection between the dashed curve and one of the solid curves.

Fig. 14-38 *I–V* characteristics for enhancement driver and depletion load.

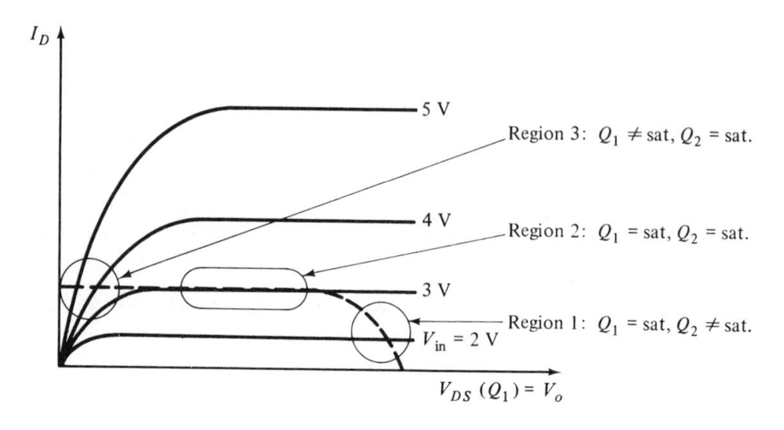

Our present goal is to use Figs. 14-37 and 14-38 in conjunction with Eqs. (14-40) through (14-43) to develop the transfer characteristic for the NMOS inverter. Before we do this, however, a few words about second-order effects in the modeling process are necessary. Referring back to the NMOS inverter schematic of Fig. 14-37, the substrate of the depletion-load device is connected to ground while its source is connected to V_o. In other words, the substrate of the depletion-load device is reverse-biased with respect to the source by the amount of the output voltage. This causes the threshold voltage of the load device (V_{TL}) to vary as a function of the output voltage:

$$V_{TL} = V_{TLO} + K_1(V_o + K_2)^{1/2} - K_1(K_2)^{1/2} \tag{14-44}$$

V_{TLO} represents V_{TL} with no substrate bias ($V_o = 0$). Now, given that V_{TLO} is negative for the depletion-load device, what happens to V_{TL} in Eq. (14-44) as V_o is decreased? A smaller positive number is subtracted from V_{TLO}, and the magnitude of V_{TL} increases. Equations (14-40) and (14-41) show that as the magnitude of V_{TL} increases, I_D(load) increases. Thus, the effect of substrate bias is to increase I_D(load) as V_o decreases. This effect is shown in Fig. 14-39. It causes the depletion-load curve to bend upward as V_o decreases.

Now consider the depletion-load device operating in the saturation region. As V_o decreases, the drain-to-source voltage increases [Eq. (14-39)]. This causes

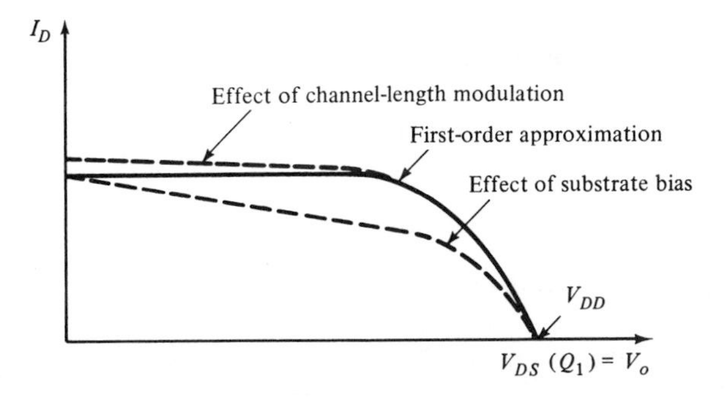

Fig. 14-39 Effects of substrate bias and channel-length modulation on the depletion-load *I–V* characteristics.

the drain–channel depletion region to become slightly wider and the channel becomes shorter. As a result, channel resistance decreases and I_D increases with decreasing V_o. This effect, called *channel length modulation,* is also shown in Fig. 14-39.

We can develop a useful, first-order engineering approximation for NMOS inverters and logic gates by treating both substrate bias and channel modulation as second-order effects and subsequently ignoring them. We therefore assign a fixed threshold voltage for the load device (V_{TL}), and ignore the fact that it varies somewhat with V_o.

Now, getting back to the I–V characteristics of Fig. 14-38, we can look at the NMOS inverter as operating in region 1, 2, or 3.

In *region 1,* Q_1 is operating in the saturation region and Q_2 is in the linear region. The equations for the load and driver devices are then given by

$$I_D(\text{load}) = \beta_L[-V_{TL}(V_{DD} - V_o) - \tfrac{1}{2}(V_{DD} - V_o)^2] \tag{14-45}$$

and

$$I_D(\text{driver}) = \tfrac{1}{2}\beta_D(V_{\text{in}} - V_{TD})^2 \tag{14-46}$$

Setting $I_D(\text{load}) = I_D(\text{driver})$, and solving for V_o, we have that for region 1,

$$V_{01} = V_{DD} + V_{TL} + \sqrt{V_{TL}^2 - \frac{\beta_D}{\beta_L}(V_{\text{in}} - V_{TD})^2} \tag{14-47}$$

This equation is valid for

$$V_{TL}^2 \geq \frac{\beta_D}{\beta_L}(V_{\text{in}} - V_{TD})^2, \qquad V_{\text{in}} \geq V_{TD} \tag{14-48}$$

In *region 2,* the load and driver devices are both operating in the saturation region. The device equations are given by

$$I_D(\text{load}) = \tfrac{1}{2}\beta_L V_{TL}^2 \tag{14-49}$$

and

$$I_D(\text{driver}) = \tfrac{1}{2}\beta_D (V_{\text{in}} - V_{TD})^2 \tag{14-50}$$

Setting the drain currents equal and solving for V_{in}, we have that

$$V_{\text{in}} = V_{TD} + \left(\frac{\beta_D}{\beta_L}\right)^{-1/2} |V_{TL}| \tag{14-51}$$

The fact that V_o does not appear as a factor in Eq. (14-51) implies that the slope of the transfer characteristic is infinite in this region. This is the consequence of using a first-order model. Taking substrate bias into account would lead to a finite slope and thus to an inverter with a finite amplification factor in the transition region. The input voltage at which the transition between V_o (hi) and V_o (lo) occurs can be predicted from Eq. (14-51). It is dependent on the device threshold voltages and the beta ratio (β_D/β_L). The beta ratio, in turn, is dependent on the relative dimensions of the two device channels.

Figure 14-38 shows that in *region 3* the load device is operating in the saturation region and the driver is in the linear region. The device equations are given by

$$I_D(\text{load}) = \tfrac{1}{2}\beta_L V_{TL}^2 \tag{14-52}$$

and

$$I_D(\text{driver}) = \beta_D[(V_{\text{in}} - V_{TD})V_o - \tfrac{1}{2}V_o^2] \tag{14-53}$$

Setting the drain currents equal and solving for V_{03}, we have that

$$V_{03} = (V_{\text{in}} - V_{TD}) - \sqrt{(V_{\text{in}} - V_{TD})^2 - \frac{\beta_L}{\beta_D} V_{TL}^2} \tag{14-54}$$

This equation is valid for

$$(V_{\text{in}} - V_{TD})^2 \geq \frac{\beta_L}{\beta_D} V_{TL}^2$$

In summary, for region 1,

$$V_{01} = V_{DD} + V_{TL} + \sqrt{V_{TL}^2 - \frac{\beta_D}{\beta_L}(V_{\text{in}} - V_{TD})^2} \tag{14-47}$$

for region 2,

$$V_{\text{in}} = V_{TD} + \left(\frac{\beta_D}{\beta_L}\right)^{-1/2} |V_{TL}| \tag{14-51}$$

and for region 3,

$$V_{03} = (V_{\text{in}} - V_{TD}) - \sqrt{(V_{\text{in}} - V_{TD})^2 - \frac{\beta_L}{\beta_D} V_{TL}^2} \tag{14-54}$$

The beta ratio $\beta_R = \beta_D/\beta_L$ plays an important role in obtaining a good, sharp transfer characteristic.

$$\beta_L = \frac{Z_L}{L_L} \mu_{ns} C_{\text{ox}} \tag{14-55}$$

and

$$\beta_D = \frac{Z_D}{L_D} \mu_{ns} C_{\text{ox}} \tag{14-56}$$

Z_L/L_L is the width/length ratio of the channel of the load device and Z_D/L_D is the width/length ratio of the driver device channel. The gate oxide capacitance per unit area (C_{ox}) is the same for both devices. We will make the approximation that the mobility values (μ_{ns}) are the same for both devices. Then the beta ratio,

$$\beta_R = \frac{\beta_D}{\beta_L} \approx \frac{Z_D/L_D}{Z_L/L_L} \tag{14-57}$$

can be controlled by adjusting the channel dimensions of the driver and load devices. Typically, a beta ratio between 10 and 20 is needed to obtain a sharp transfer characteristic for an NMOS inverter. A geometry that could be used to obtain $\beta_R = 20$ is shown in Fig. 14-40.

Given a reasonable set of device parameters and $\beta_R = 20$, the transfer characteristics obtained when Eqs. (14-47), (14-51), and (14-54) are plotted appear as shown in Fig. 14-41.

It will be instructive at this point to assign a set of device parameters and then develop the transfer characteristics for an NMOS inverter. Given

$$V_{DD} = 5\text{ V}$$
$$V_{TD} = +0.6\text{ V}$$
$$V_{TL} = -2.0\text{ V}$$
$$t_{ox} = 500\text{ Å} = 5 \times 10^{-6}\text{ cm}$$
$$\mu_{ns} = 700\text{ cm}^2/\text{V-s}$$
$$\frac{Z_L}{L_L} = \frac{1}{5}$$
$$\frac{Z_D}{L_D} = 4$$

then

$$\beta_R = \frac{\beta_D}{\beta_L} = 20$$
$$C_{ox} = \frac{\epsilon_0 \epsilon_{ox}}{t_{ox}} = 7.08 \times 10^{-8}\text{ F/cm}^2$$
$$\beta_L = \frac{Z_L}{L_L}\mu_{ns} C_{ox} = 9.91\ \mu\text{A/V}^2$$

and

$$\beta_D = \frac{Z_D}{L_D}\mu_{ns} C_{ox} = 198\ \mu\text{A/V}^2$$

The maximum drain current will occur when the load device is in saturation. Then

$$I_D(\text{load}) = \tfrac{1}{2}\beta_L V_{TL}^2 = 19.82\ \mu\text{A}$$

The equation for region 1 is given by

$$V_{01} = V_{DD} + V_{TL} + \sqrt{V_{TL}^2 - \beta_R (V_{\text{in}} - V_{TD})^2}$$
$$V_{01} = 5 - 2 + \sqrt{4 - 20(V_{\text{in}} - 0.6)^2}$$

The region 1–2 interface occurs when

$$V_{\text{in}} = V_{TD} + |V_{TL}|/\sqrt{\beta_R} = 0.6 + \frac{2}{\sqrt{20}} = 1.05\text{ V}$$

The V_{01} equation is valid over the range $0.6\text{ V} \leq V_{\text{in}} \leq 1.05\text{ V}$, and region 1 of the transfer characteristic can be plotted.

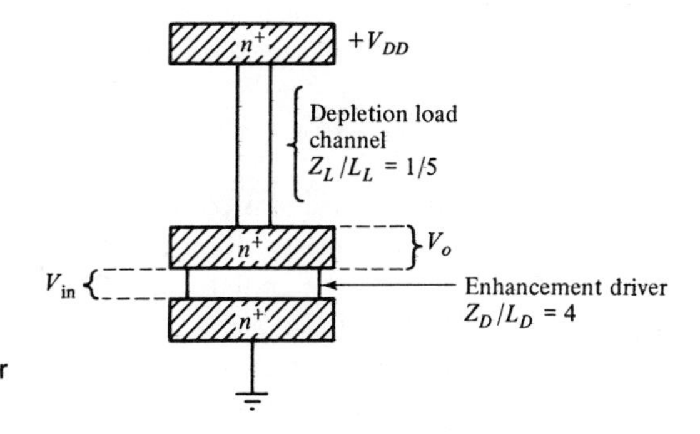

Fig. 14-40 NMOS Inverter geometry for $\beta_R = 20$.

Fig. 14-41 Typical transfer characteristic for an NMOS inverter.

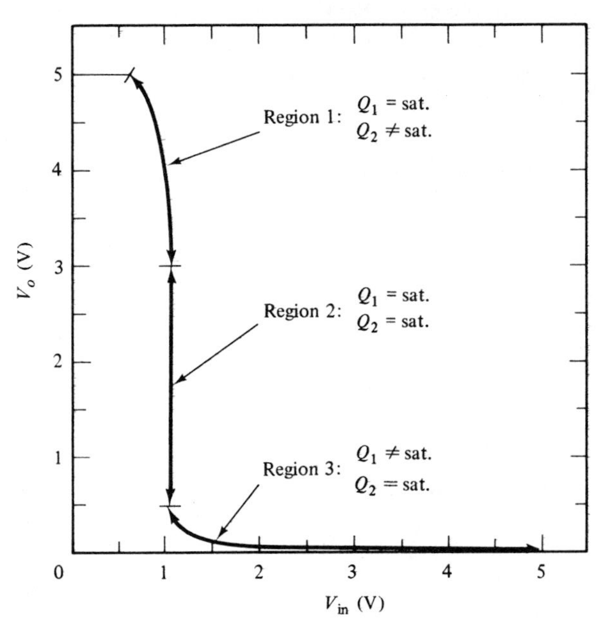

Region 2 occurs at $V_{in} = 1.05$ V. For this value, the radical in the V_{01} and V_{03} equations goes to zero. The region 1–2 interface then occurs at

$$V_o = V_{DD} + V_{TL} = 5 - 2 = 3 \text{ V}$$

The region 2–3 interface occurs at

$$V_o = V_{in} - V_{TD} = 1.05 - 0.6 = 0.45 \text{ V}$$

Thus, for region 2, $V_{in} = 1.05$ V and $0.45 \text{ V} \leq V_o \leq 3$ V.

The equation for region 3 is given by

$$V_{03} = (V_{in} - V_{TD}) - \sqrt{(V_{in} - V_{TD})^2 - \frac{V_{TL}^2}{\beta_R}}$$

$$V_{03} = (V_{in} - 0.6) - \sqrt{(V_{in} - 0.6)^2 - 0.2}$$

This equation is valid for $V_{in} \geq 1.05$ V.

The transfer characteristic for the preceding example is shown in Fig. 14-42. Without specifying the actual width/length ratios of the load and driver devices, transfer characteristics were also plotted for $\beta_R = 5$ and $\beta_R = 1$. To optimize the operation of NMOS systems, it is very desirable to have a sharp transfer characteristic, such as the $\beta_R = 20$ curve shown in Fig. 14-42.

When looking at transfer characteristics for NMOS inverters, device tolerances should be kept in mind. For example, a typical threshold voltage variation for an enhancement driver might be $V_{TD} = 0.60 \pm 0.25$ V. A typical depletion-load threshold variation might be $V_{TL} = -2.0 \pm 0.3$ V. Given that $V_{in} = \beta_R^{-1/2}|V_{TL}|$, this results in an input transition voltage variation from 0.73 to 1.36 V for the $\beta_R = 20$ inverter discussed above. As device sizes are scaled down, the gate oxide must become thinner. As a result, ion implants through the gate oxide are more predictable, and threshold voltages can be more tightly controlled. However, with scaling, larger variations take place in other device parameters.

Fig. 14-42 Transfer characteristics for an NMOS inverter as a function of β_R. ($V_{DD} = 5$ V, $V_{TD} = 0.6$ V, $V_{TL} = -2.0$ V, $t_{ox} = 500$ Å, $\mu_{ns} = 700$ cm²/V − s).

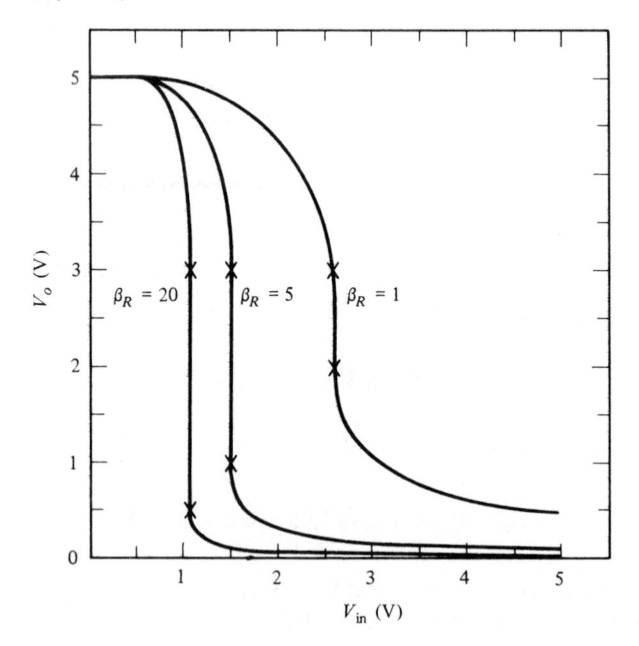

Figure 14-43 shows the basic two-input NAND and NOR gates and uses simpler device symbols. Gates with more inputs can be constructed by adding more series enhancement drivers to the NAND circuit of Fig. 14-43(b), or by adding more parallel enhancement drivers to the NOR gate of Fig. 14-43(c). In the case of the NAND gate, as enhancement drivers are connected in series,

Fig. 14-43 NMOS NAND and NOR gates using simpler device symbols: (a) simplified device symbols; (b) two-input NMOS NAND gate; (c) two-input NMOS NOR gate.

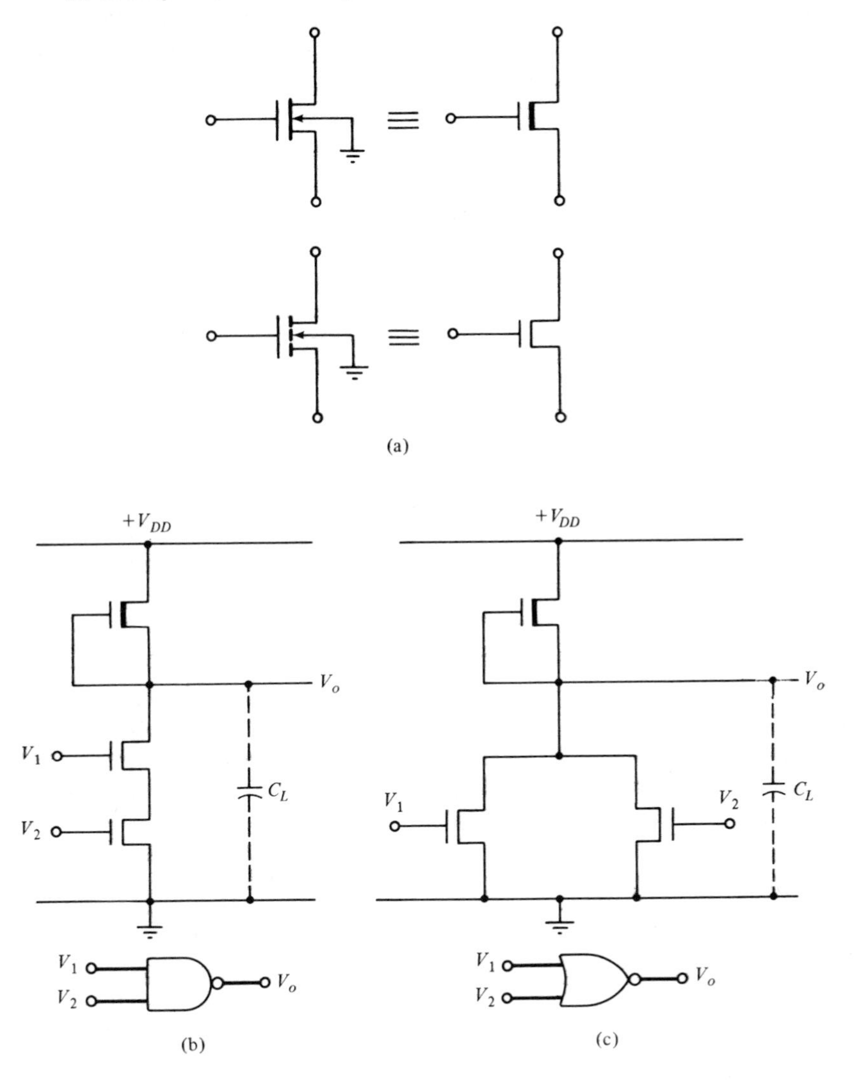

their Z/L ratios are increased to keep the total resistance from V_o to ground approximately the same as that of a simple inverter. It is important to note that if an NMOS logic gate has N inputs, $N + 1$ devices are needed. A gate needs a depletion load plus one device for every input. In contrast, CMOS logic gates require $2N$ devices for N inputs.

Referring to the NMOS gates of Fig. 14-43, as pulses are applied to the inputs and the enhancement drivers turn on and off, the load capacitance C_L must be charged and discharged. When driver devices turn on, they present low-impedance paths to discharge C_L. Thus, the fall time t_f of the output pulse, as V_o makes the transition from hi to lo, is relatively short. When the enhancement drivers have been on and are then turned off, however, C_L must be charged through the high-resistance depletion load device. Thus, the rise time t_r of the output pulse tends to be rather long. In cases where the fan-out and the corresponding load capacitance are large, inverting or noninverting buffers can be used so that rise times do not become prohibitive.

Fig. 14-44 (a) Inverting NMOS buffer. (b) Noninverting NMOS buffer.

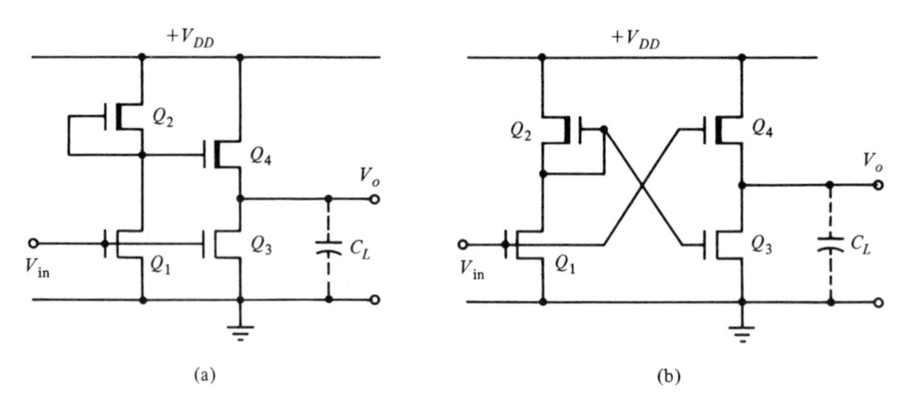

Figure 14-44 shows an inverting and a noninverting NMOS buffer. Referring to Fig. 14-44(a), let us start by assuming that V_{in} is hi, Q_1 and Q_3 are on, and V_o is lo. The gate of Q_4 is then held lo, and the Q_3–Q_4 circuit looks like a conventional NMOS inverter. When V_{in} goes lo, Q_1 and Q_3 turn off. Additional current is now supplied by Q_4 because its gate is hi. The initial load current used to charge C_L is given by

$$I_D(\text{load}) = \beta_L[(V_G - V_{TL})V_{DD} - \tfrac{1}{2}V_{DD}^2] \tag{14-58}$$

Since V_{TL} takes on a negative value, the gate voltage of Q_4 *adds* to the threshold voltage, thus providing a larger current to charge C_L. The operation of the noninverting buffer of Fig. 14-44(b) is essentially the same, except that the devices are cross-coupled so that the buffer is noninverting.

14-6-3 NMOS Transmission Gate

Consider the MOS circuit shown in Fig. 14-45. It consists of an enhancement-mode transistor in series with an inverter. Transistor Q_1 acts as the transmission gate. It is used to isolate and temporarily store charge on adjacent circuit nodes. The gate of transistor Q_1 is connected to a voltage ϕ which is either hi or lo. When ϕ is hi, Q_1 is conducting; then C_2, the normal gate capacitance of transistor Q_2, charges approximately to the input voltage V_{in}. When ϕ goes lo, Q_1 turns off and V_{in} is isolated from Q_2. The inverter will *temporarily store* the sampled input voltage V_{in} on the gate capacitance for a time interval on the order of 1 ms. As long as this type of circuit has cycle times considerably less than 1 ms, it can be used as a dynamic flip-flop, or storage element.

To explain the operation of the transmission gate, suppose that $V_{in} = 0$ V, $\phi = 0$ V, and the gate capacitor voltage $V_{C2} = 0$ V. If we now let $V_{in} = 5$ V, the capacitor remains uncharged and Q_2 remains off because Q_1 is off. Now let $\phi = 5$ V. Transistor Q_1 turns on and the direction of conventional current is from left to right, from V_{in} to capacitor C_2. The left side of Q_1 (the highest potential side) becomes the drain and the right side becomes the source. Q_1 conducts until the gate–source voltage ($\phi - V_{C2}$) is equal to the threshold voltage V_T. If V_T for the transmission gate is 0.6 V, the gate capacitance charges to $V_{C2} = \phi - V_T = 5 - 0.6 = 4.4$ V. If we now let ϕ go to 0 V, then Q_1 turns off and V_{in} is isolated from Q_2. C_2 temporarily stores its charge, Q_2 stays on, and V_o is lo. Now let us set $V_{in} = 0$ V, and let ϕ go to 5V. The right side of Q_1 (the highest potential side) becomes the drain, and the left side becomes the source. Conventional current through Q_1 is from right to

Fig. 14-45 NMOS transmission gate in series with an inverter.

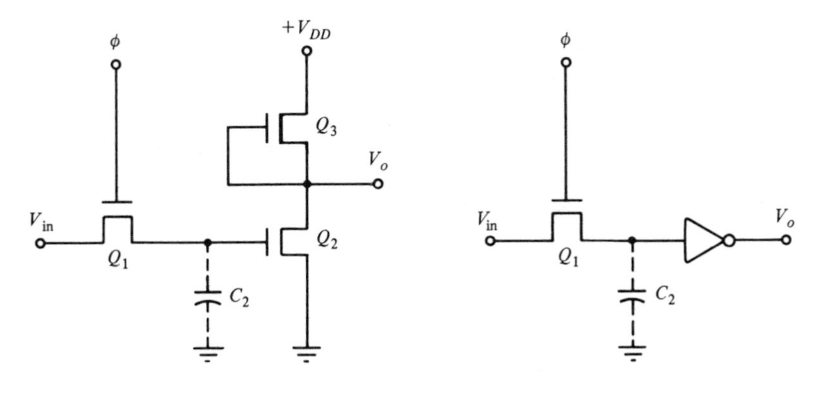

left, discharging C_2. Q_1 stays on as long as $(\phi - V_{in})$ is greater than 0.6 V. Since ϕ is at 5 V and $V_{in} = 0$ V, Q_1 remains on as long as $\phi = 5$ V, and the gate capacitance discharges to $V_{C2} = 0$ V. One limitation on an NMOS transmission gate is that gate capacitance can only charge to a voltage $(V_{in} - V_T)$. By using transmission gates in series with inverters, however, voltage levels can be restored. Even though V_{C2} may be 4.4 V in Fig. 14-45, transistor Q_2 is on and V_o is close to 0 V. When $V_{C2} = 0$ V, Q_2 is off and $V_o = 5$ V.

You can tell whether a digital system is operating in a *static* or a *dynamic* mode by observing the operating frequency. If system operation can be extended to very low frequencies, it is static. If not, the mode of operation is referred to as dynamic. Prior to the transmission gate, all the gates in this chapter operated in a static mode. With the circuit of Fig. 14-45, however, we are talking about temporary storage of charge on the gate capacitance. There are economic and performance benefits to be obtained from dynamic operation. For example, in solid-state memories, memory cells can be made from fewer components, but then a recirculating mode of operation must be used such that capacitors are periodically recharged to appropriate logic levels.

If the bulk of NMOS LSI and VLSI circuits were intended to work in a static mode, it would have made sense to set device threshold voltages and beta ratios such that inverters made transitions from hi to lo voltage levels at $V_{in} = \frac{1}{2} V_{DD}$. The noise margins would then be optimized. Notice, however, that V_T was set at 0.60 ± 0.25 V for enhancement-mode devices and that in Fig. 14-42, the inverter output transition took place at $V_{in} \approx 1.0$ V. The reason for this is that most NMOS systems are dynamic. Transmission gates and temporary charge storage are widely used. It is to our advantage that transmission gates turn on at relatively low threshold voltages in dynamic systems.

14-6-4 NMOS Logic Gate Specifications

The NMOS-scaled NMOS-HMOS technologies are in a rapid state of evolution and are used primarily in LSI and VLSI applications. As fabrication techniques improve, minimum channel lengths are shrinking from 25 μm in the late 1960s to below 1 μm in the 1980s. In planning MOS chips, it is often desirable to design specifically for high-speed paths internally and then design devices at the chip output interface to operate at higher currents. Under these circumstances, it is difficult to specify fixed numbers for gate parameters such as noise margins, fan-out, power dissipation per gate, and so on. If the reader is willing to interpret these figures somewhat liberally, however, some general ranges for gate parameters are provided below.

- Average power dissipation per gate is in the range 0.1 to 1.0 mW.
- Average propagation delay per gate is in the range 1.0 to 10.0 ns.
- Power–delay products are in the range 0.1 to 10 pJ.

CMOS consists of digital circuits in which complementary MOS devices (PMOS and NMOS) are always used in series between the power supply and ground. The source and substrate terminals of the PMOS devices are held positive with respect to the drain terminals. Source and substrate terminals of the NMOS devices are held negative with respect to the drain terminals. Thus, lo gate input voltages turn PMOS devices on, and hi gate input voltages turn NMOS devices on. For static conditions, the PMOS and NMOS units are not on at the same time, series currents from power supply to ground are in the nanoampere range, and power dissipation is minimized. Thus, CMOS is a very useful technology for applications in which power dissipation is a critical factor.

14-7-1 CMOS Inverter

Figure 14-46(a) shows the circuit diagram for a CMOS inverter, and Fig. 14-46(b) shows the drain characteristics for the CMOS devices.

To develop the equations for the transfer characteristic of the CMOS inverter, we can begin with a lo input voltage ($V_{\text{in}} = 0$), slowly increase it, and observe circuit operation. We will assume for this discussion that $V_{DD} = 5$ V, and the threshold voltage of Q_p (V_{TP}) is -1 V. As long as the gate of Q_p is 1 V or more negative with respect to the source, Q_p will be on. When $V_{\text{in}} > 4$ V, V_{GS} $(P) = (V_{\text{in}} - V_{DD})$ is more positive than -1 V, and Q_p is off. We will assume that the threshold voltage of Q_n, $V_{TN} = +1$ V. Thus, Q_n is on whenever V_{in} is greater than 1 V. As V_{in} increases, the operating modes of the inverter can be separated into five regions as follows:

Region 1 ($V_{\text{in}} \leq 1$ V). Transistor *Q_n is off and Q_p is on.* The output voltage $V_o = V_{DD}$.

Fig. 14-46 (a) CMOS inverter. (b) Drain characteristics for the CMOS devices.

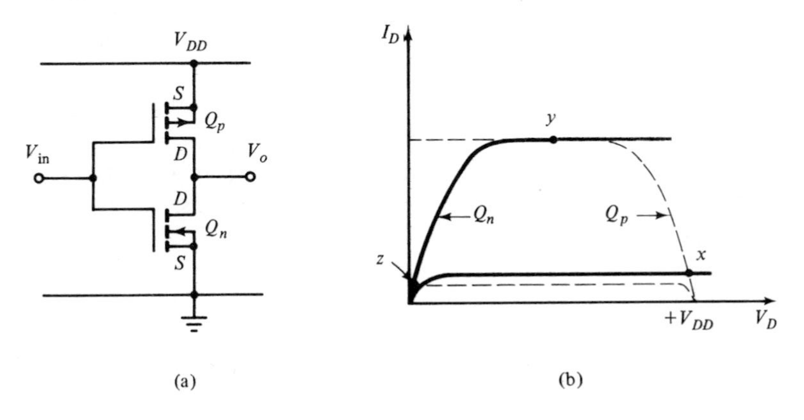

Region 2 ($1\text{ V} \le V_{\text{in}} \le 2.5\text{ V}$). Both transistors are on. A small series current exists between $+V_{DD}$ and ground. As shown in Fig. 14-46(b), the devices are operating in the vicinity of point X. *Q_n is in saturation and Q_p is operating in the linear region.*

Region 3 ($V_{\text{in}} \approx 2.5\text{ V}$). *Both devices are in saturation* as shown by point Y in Fig. 14-46(b). The transient series current between $+V_{DD}$ and ground is at a maximum.

Region 4 ($2.5\text{ V} \le V_{\text{in}} \le 4\text{ V}$). Both devices are on and operating in the vicinity of point Z in Fig. 14-46(b). *Q_p is in saturation and Q_n is operating in the linear region.*

Region 5 ($V_{\text{in}} \ge 4\text{ V}$). $V_{GS}\,(P) = (V_{\text{in}} - V_{DD})$ is more positive than -1 V (V_{TP}). Q_p is off and Q_n is on. Thus, $V_o = 0$ V.

Equations (14-33) and (14-34) for MOS devices are repeated below for convenience.

$$I_D\,(\text{linear}) = \beta[(V_G - V_T)V_D - \tfrac{1}{2}\,V_D^2] \tag{14-33}$$

$$I_D\,(\text{sat}) = \tfrac{1}{2}\,\beta(V_G - V_T)^2 \tag{14-34}$$

We can now apply these equations to the CMOS inverter of Fig. 14-46 to find expressions for V_o as a function of V_{in}.

Region 2 (Q_p is in the linear region, Q_n is in saturation)

$$I_{DP} = \beta_P[(V_{\text{in}} - V_{DD} - V_{TP})(V_o - V_{DD}) - \tfrac{1}{2}(V_o - V_{DD})^2] \tag{14-59}$$

$$I_{DN} = \tfrac{1}{2}\beta_N(V_{\text{in}} - V_{TN})^2 \tag{14-60}$$

Setting $I_{DP} = I_{DN}$ and solving for V_o, we have

$$V_o = (V_{\text{in}} - V_{TP}) + \sqrt{(V_{\text{in}} - V_{TP})^2 - 2(V_{\text{in}} - \tfrac{1}{2}\,V_{DD} - V_{TP})V_{DD} - \beta_R(V_{\text{in}} - V_{TN})^2} \tag{14-61}$$

where $\beta_R = \beta_N/\beta_P$. If $\beta_R = 1$ and $V_{TN} = -V_{TP}$, the radical in Eq. (14-61) goes to zero when $V_{\text{in}} = \tfrac{1}{2}V_{DD}$. In other words, the inverter is in the transition region when V_{in} is $\tfrac{1}{2}$ of the supply voltage regardless of the supply voltage magnitude. This is very desirable in terms of noise margins.

If we let $\beta_R = 1$, $V_{DD} = +5$ V, $V_{TN} = 1$ V, and $V_{TP} = -1$ V, then Eq. (14-61) reduces to

$$V_o = (V_{\text{in}} + 1) + \sqrt{15 - 6V_{\text{in}}} \tag{14-62}$$

This equation is valid for 1 V $\leq V_{\text{in}} \leq$ 2.5 V. When $V_{\text{in}} \leq$ 1 V, transistor Q_n is off. When V_{in} = 2.5 V, Q_p enters the saturation region.

Region 3 (Q_p and Q_n are in saturation)

$$I_{DP}(\text{sat}) = \tfrac{1}{2}\beta_P(V_{\text{in}} - V_{DD} - V_{TP})^2 \tag{14-63}$$

$$I_{DN}(\text{sat}) = \tfrac{1}{2}\beta_N(V_{\text{in}} - V_{TN})^2 \tag{14-64}$$

Equating the drain currents and letting $\beta_R = \beta_N/\beta_P$, we have

$$\sqrt{\beta_R}\,(V_{\text{in}} - V_{TN}) = (V_{DD} + V_{TP} - V_{\text{in}}) \tag{14-65}$$

If $\beta_R = 1$ and $V_{TN} = -V_{TP}$, then $V_{\text{in}} = \frac{1}{2}V_{DD}$ in the transition region between V_o = 5 V and V_o = 0 V. This maximizes the noise margins independently of the supply voltage. If the devices behaved exactly as described by Eq. (14-65), then $V_o \neq f(V_{\text{in}})$ and the gain in the transition region would be infinite. Since I_D does increase somewhat with V_D in the saturation region, the actual inverter gain is large but finite.

Region 4 (Q_p is in saturation, Q_n is in the linear region)

$$I_{DP} = \tfrac{1}{2}\beta_P(V_{\text{in}} - V_{DD} - V_{TP})^2 \tag{14-66}$$

$$I_{DN} = \beta_N[(V_{\text{in}} - V_{TN})V_o - \tfrac{1}{2}V_o^2] \tag{14-67}$$

Setting the drain currents equal and solving for V_o, we have

$$V_o = (V_{\text{in}} - V_{TN}) - \sqrt{(V_{\text{in}} - V_{TN})^2 - \frac{1}{\beta_R}(V_{\text{in}} - V_{DD} - V_{TP})^2} \tag{14-68}$$

If $\beta_R = 1$ and $V_{TN} = -V_{TP}$, the radical in Eq. (14-68) reduces to zero when $V_{\text{in}} = \frac{1}{2}V_{DD}$. Then $V_o = V_{\text{in}} - V_{TN}$.

If we again let $\beta_R = 1$, V_{DD} = 5 V, V_{TN} = 1 V, and V_{TP} = −1 V, then Eq. (14-68) reduces to

$$V_o = (V_{\text{in}} - 1) - \sqrt{6V_{\text{in}} - 15} \tag{14-69}$$

This equation is valid for 2.5 V $\leq V_{\text{in}} \leq$ 4.0 V. When V_{in} = 2.5 V, transistor Q_n is on the edge of saturation. When $V_{\text{in}} \geq$ 4 V, transistor Q_p is off.

Figure 14-47 shows the CMOS inverter transfer characteristics for V_{DD} = 5 V, $V_{TN} = -V_{TP}$ = 1 V, and $\beta_R = 1$. Essentially, this is a plot of Eqs. (14-61), (14-65), and (14-68) for a specific beta ratio and specific device voltages. The foregoing equations show that V_o is still dependent on β_R, the ratio of device channel dimensions. To illustrate this dependence, the CMOS transfer

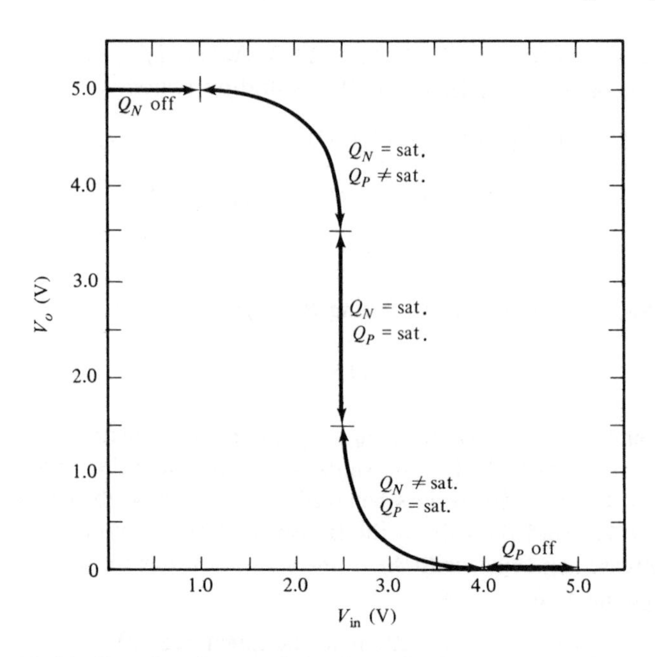

Fig. 14-47 Transfer characteristics for a CMOS inverter with $V_{DD} =$ 5 V, $V_{TN} = -V_{TP} = 1$ V, and $\beta_R = 1$.

characteristics have been plotted for $\beta_R = 0.1$, 1.0, and 10.0, as shown in Fig. 14-48. It is interesting to compare the CMOS curves of Fig. 14-48 to the NMOS curves of Fig. 14-42. For the NMOS inverter, a high value of β_R is needed. As β_R is decreased, the transition region becomes poorly defined. For the NMOS digital inverter, circuit performance is degraded. For the CMOS inverter, the transition region shifts to the right or left as β_R changes, but the output voltage transition remains sharp and well defined, so that the performance of the CMOS inverter is not necessarily degraded. If we were to load the CMOS inverter with an output capacitance, a beta ratio of about 1 would be desirable so that the times required to charge and discharge the load capacitance would be about equal.

The equations for β_P and β_N for the p- and n-type devices in CMOS structures are

$$\beta_P = \frac{Z_P}{L_P} \mu_{ps} C_{ox} \tag{14-70}$$

$$\beta_N = \frac{Z_N}{L_N} \mu_{ns} C_{ox} \tag{14-71}$$

The surface mobility of electrons in n-type devices (μ_{ns}) is two to three times larger than that of holes in p-type devices (μ_{ps}). Suppose that we want to keep

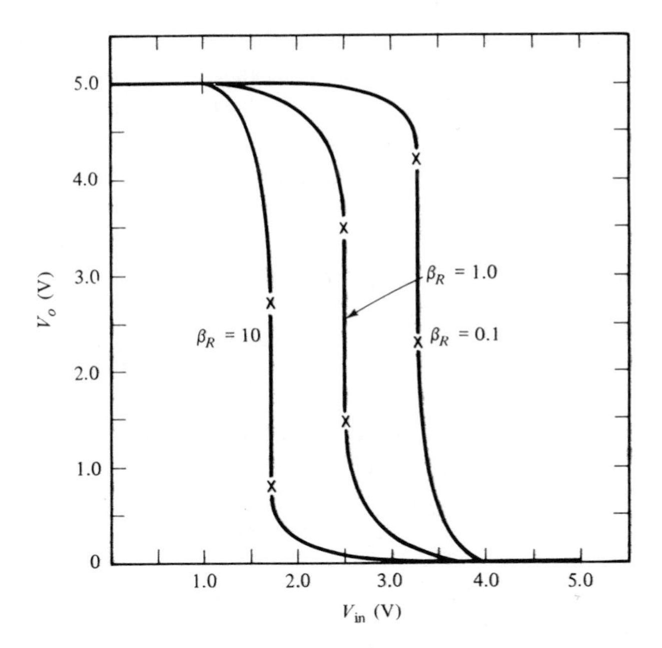

Fig. 14-48 Transfer characteristics for a CMOS inverter as a function of the beta ratio, $\beta_R = \beta_N/\beta_P$.

$L_P = L_N$, and set $\beta_P = \beta_N$ so that $\beta_R = 1$. Then we have to make Z_P two to three times larger than Z_N. The degree to which p and n units are matched ($\beta_P = \beta_N$) in CMOS structures depends on the application. In applications where switching times or noise margins are critical, p units are generally made with about twice the area of n units. If the application parameters are not critical, p and n device areas may be made the same size. We could argue that mismatching between devices is bound to occur in complex logic structures anyway, so why use the extra silicon area if it is not needed?

Referring to Fig. 14-49, power dissipation in a CMOS inverter can be described in terms of three components.

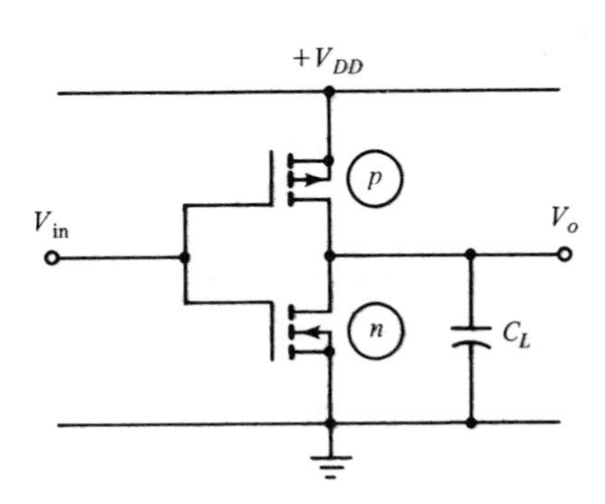

Fig. 14-49 CMOS Inverter with fan-out modeled as load capacitance.

1. Quiescent power dissipation is the product of the power supply voltage and the series leakage current. Since leakage currents are typically in the nanoampere range, this component is extremely small.
2. A dynamic component of power dissipation is caused by the through current from power supply to ground during the switching transient, when both n and p units are on. The magnitude of this through current depends on the channel resistances and the rise and fall times of the input pulses.
3. Another dynamic component of power dissipation is caused by the fact that the load capacitance must be charged and discharged during each cycle of the input waveform. This is generally the largest source of power dissipation in CMOS circuits.

To discuss item 3 further, let us apply a relatively large sine wave to the inverter input of Fig. 14-49, negative half-cycle first, and see what happens. As the input sine wave goes negative, the n unit is off and the p unit turns on. Assuming that the period of the input waveform is much longer than the $R_P C_L$ time constant, C_L charges from 0 V to $+V_{DD}$. The energy dissipated in the p unit during this half-cycle is $\frac{1}{2}C_L V_{DD}^2$. As the input sine wave swings positive, the p unit turns off and the n unit switches on. The capacitor discharges through the n channel from $+V_{DD}$ to 0 V. The energy dissipated in the n unit during the positive half-cycle is also $\frac{1}{2}C_L V_{DD}^2$. The total energy dissipated by the inverter during one full cycle of the input waveform is then given by $C_L V_{DD}^2$. The average power dissipated in the inverter during one cycle is given by the energy dissipated divided by the period, or since $T = 1/f$,

$$P = C_L \frac{V_{DD}^2}{T} = C_L V_{DD}^2 f \tag{14-72}$$

Thus, dynamic power dissipation increases linearly with load capacitance and frequency. It also increases as the square of power supply voltage. If, for example, $C_L = 50$ pF and $V_{DD} = 5$ V, a CMOS inverter will dissipate power in the nanowatt range at low frequencies. At $f = 8$ MHz, however, it will dissipate 10 mW of power. Thus, CMOS can be regarded as a low-power technology only within set frequency bounds.

14-7-2 CMOS NAND and NOR Gates

Like the T^{2}L, Schottky T^{2}L, and ECL logic families, digital CMOS building blocks are available off the shelf in SSI and MSI form. Like the NMOS logic family, CMOS is available in LSI and VLSI form (primarily as memory chips and microprocessor chips).

The properties of CMOS that make it uniquely valuable in the marketplace are:

1. Extremely low static power dissipation.
2. High noise immunity.
3. Variable supply voltage (typically +3 to +12 V).
4. High fan-out capability.

In LSI and VLSI applications, where CMOS is directly competitive with NMOS, conventional CMOS circuits have the following disadvantages:

1. CMOS has inherently lower packing densities. NMOS logic gates require $(N + 1)$ devices for N digital inputs, whereas CMOS gates require $2N$ devices for N digital inputs. Also, in CMOS circuits the *n*-channel units must be placed in *p*-type wells (see Figs. 11-26 and 11-27). This requires additional chip area.
2. Propagation delays are longer in CMOS circuits, partially because of the larger chip area needed per function.

To discuss CMOS logic gates, it is convenient to use an abbreviated device notation, as shown in Fig. 14-50. When using this shorthand notation, it is useful to remember that the substrate regions of all *p*-channel units are connected to $+V_{DD}$ and the substrate regions of all *n*-channel units are connected to ground.

Static NAND and NOR gates are developed from inverters by adding *p*-channel and *n*-channel devices in series–parallel pairs as shown in Fig. 14-51. Gate input voltages, such as V_1, are applied to a pair of transistors, one *p*-channel and one *n*-channel. *p*-Channel devices (or *p* units) turn on for lo gate voltages. *n* Units turn on for hi gate voltages. Referring to the NAND gate of Fig. 14-51(a), the path from power supply to ground is open regardless of the conditions of the two input signals. Whether the V_1–V_2 condition is hi–hi, hi–lo, lo–hi, or lo–lo, the series path from $+V_{DD}$ to ground is opened by an off device. Static power dissipation is thus determined by reverse-biased *p-n* junction leakage currents. CMOS devices do not have to be designed to limit dc currents.

The fact that the circuit of Fig. 14-51(a) performs the NAND function can be verified by verbally going through the truth table. If V_1 or V_2 is lo, a *p* unit is on and an *n* unit is off. A low-resistance path exists from V_{DD} to V_o.

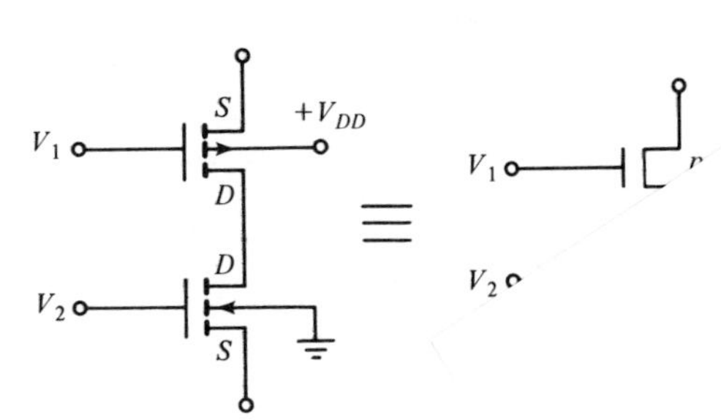

Fig. 14-50 Abbreviated device notation for CMOS logic gates.

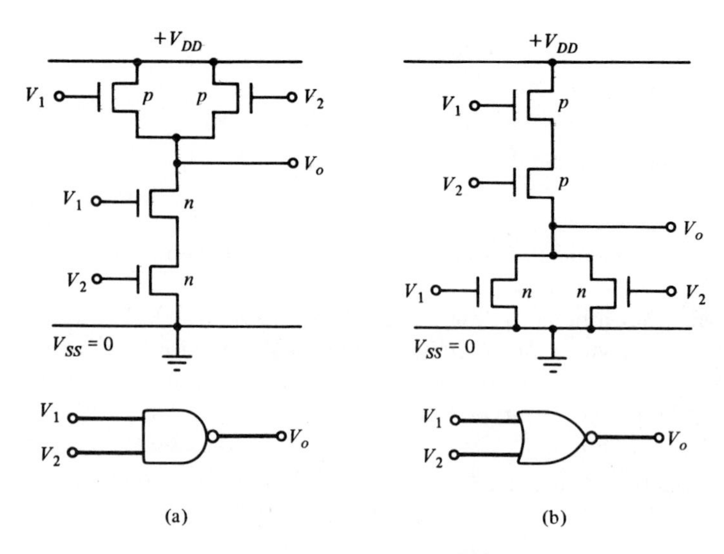

Fig. 14-51 CMOS (a) NAND and (b) NOR gates.

The off n unit results in a high-resistance path from V_o to ground. V_o is hi. When V_1 and V_2 are both hi, both n units are on and both p units are off. A low-resistance path exists from V_o to ground, and V_o is lo. The fact that the circuit of Fig. 14-51(b) performs the NOR function can be verified in a similar manner. To draw a three-input NAND gate we would use three parallel p units and three series n units.

14-7-3 CMOS Transmission Gate

A CMOS transmission gate, or *bilateral switch,* consists of an n-channel and a p-channel transistor connected in parallel, as shown in Fig. 14-52. A digital control pulse (ϕ) is applied to the gate of the n-channel device. The

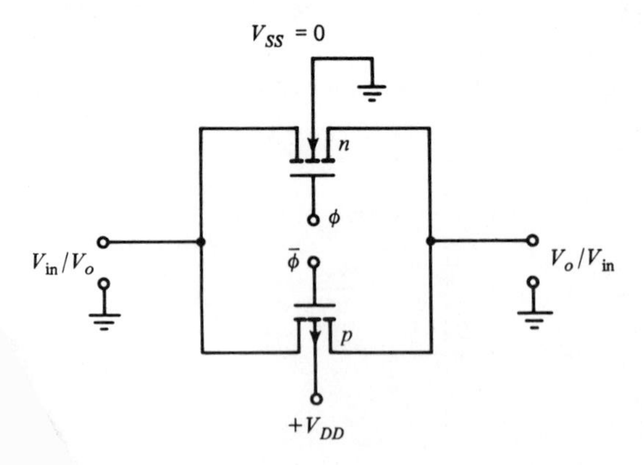

Fig. 14-52 CMOS transmission gate or bilateral switch.

inverse of this control pulse ($\overline{\phi}$) is applied to the gate of the *p*-channel device. Both devices are either on (acting as closed switches), or off (acting as open switches), depending on whether the control pulses are hi or lo. When ϕ is hi and $\overline{\phi}$ is low, both devices are on and signal information can be transmitted in either direction (left to right or right to left in Fig. 14-52). When ϕ is lo and $\overline{\phi}$ is hi, both devices are off.

The magnitudes of the control pulses are set at the power supply voltages, V_{DD} and V_{SS}. Transmission gates can be used in digital or analog signal-switching applications. In digital applications, V_{DD} may be set at +5, +10, or +15 V, and V_{SS} is normally set at 0 V. In analog applications, where V_{in} and V_o may swing (+) or (−), $+V_{DD}$ and $-V_{SS}$ are normally set at ±7.5 V or ±5 V, depending on the magnitude of the analog signal information.

To describe the operation of a CMOS transmission gate in a digital application, let us choose specific supply, signal, and device voltages as follows:

$$V_{DD} = +5\text{ V} \qquad V_{TP} = -0.6\text{ V}$$
$$V_{SS} = 0\text{ V} \qquad V_{\text{signal}} = +5\text{ V or }0\text{ V}$$
$$V_{TN} = +0.6\text{ V}$$

Figure 14-53(a) shows a transmission gate with $V_{in} = +5$ V and $V_o = 0$ V. C_L represents load capacitance due to circuits connected to the gate output. We will begin with both devices off ($\phi = 0$ V) and then suddenly let $\phi = 5$ V and $\overline{\phi} = 0$ V, so that both devices switch on. Since V_{in} is at a higher potential than V_o, the direction of conventional current through both devices is from left to right. Device polarities establish the drain of the *n* unit and the source of the *p* unit at the left, as shown in Fig. 14-53(a). When V_o reaches 4.4 V, $V_{GS}(N) = +0.6$ V and the *n* unit is at the point of cutoff. At this point, however, $V_{GS}(P) = -5$ V because $\overline{\phi} = 0$ V. Thus, the *p* unit stays on and C_L charges to the full 5 V.

Figure 14-53(b) shows a transmission gate with $V_{in} = 0$ V and $V_o = +5$ V. Starting with both devices off, we will again switch ϕ to +5 V and $\overline{\phi}$ to 0 V, so that both devices switch on. The direction of conventional current is now from right to left. The *n*-unit drain and the *p*-unit source are established at the right in Fig. 14-53(b). Capacitor C_L discharges through both devices ($V_{in} = 0$ V) until $V_o = +0.6$ V. At this point, $V_{GS}(P) = -0.6$ V. The *p* unit is at the cutoff point. Since $\phi = +5$ V, however, $V_{GS}(N) = +5$ V. Thus, the *n* unit stays on and C_L discharges to 0 V.

In summary, with ϕ hi and $\overline{\phi}$ low, there is always an on transistor and a low-resistance path between the input and output terminals. Typical values of "on" resistance for transmission gates range from 200 to 800 Ω, depending on supply voltages and load conditions. When the devices are off, leakage currents are in the 100-pA range. Propagation delay times (t_{pd}) are in the range 10 to 50 ns.

When individual transmission gates are provided on IC chips, inverters

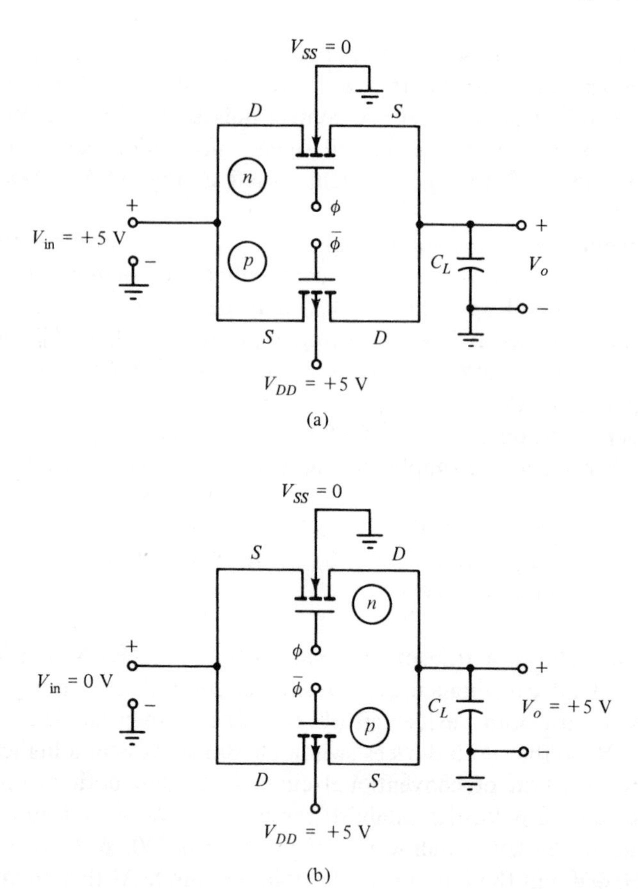

Fig. 14-53 (a) Charging C_L through a transmission gate. (b) Discharging C_L through a transmission gate.

are usually included, as shown in Fig. 14-54(a). An input terminal is made available for the control pulse ϕ and an inverter is used to generate $\overline{\phi}$ on the chip. Figure 14-54(b) shows the functional diagram for a CD4016 quad bilateral switch. Here, four transmission gates are provided in a 14-pin package. Pins 1 and 2 are the input and output terminals for transmission gate *A* (SW A). Pin 13 is the input terminal for ϕ (CONT A). Control pulse $\overline{\phi}$ is generated by an inverter within SW A.

14-7-4 CMOS Logic Gate Specifications

Because basic digital CMOS circuits are available as "off-the-shelf" items, manufacturers' specifications are provided. These specifications are similar to those provided for the T²L, Schottky T²L, and ECL logic families.

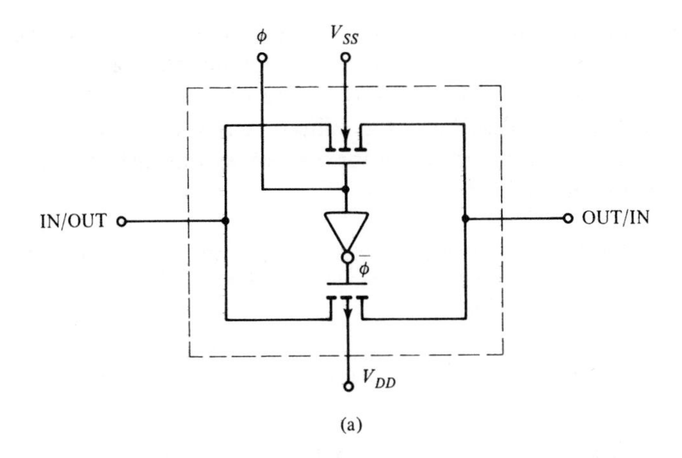

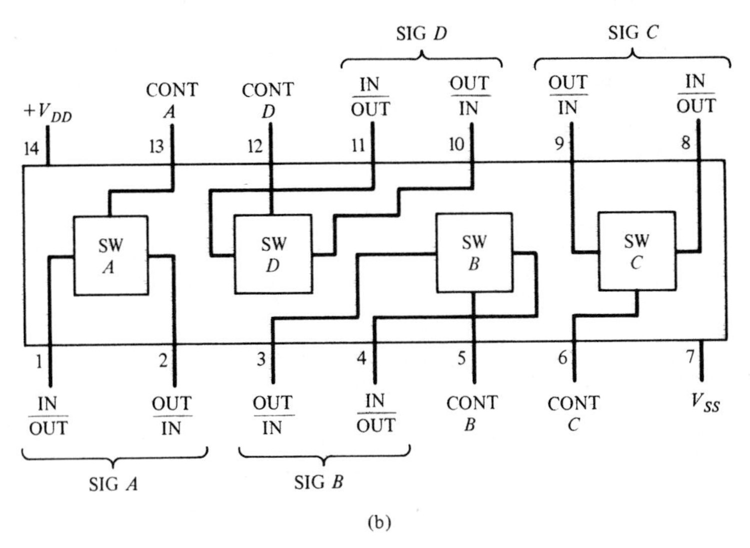

Fig. 14-54 (a) Single transmission gate with an inverter on the chip. (b) Four CMOS transmission gates in a package. The functional diagram for a CD4016 quad bilateral switch.

CMOS circuits can be fabricated within silicon substrates (bulk CMOS), or on insulating substrates such as sapphire (silicon on sapphire, or SOS-CMOS). Although SOS-CMOS is a very promising future technology, it is not presently price-competitive for most commercial applications. We will therefore restrict our present discussion to bulk CMOS structures.

The bulk CMOS logic families are normally given a 4000 series numerical designation by manufacturers. The 4000 number is normally followed by a letter indicating whether the circuit meets commercial or military voltage and temperature specifications. Another letter may be used to indicate the package type (i.e., a ceramic or plastic package). Letter designations vary from one manufacturer to another. For the convenience of being specific, we will call the commercial series 40XXC and the military series 40XXM. Typical supply voltage and temperature specifications for the CMOS 40XXM and 40XXC series logic families are shown in Table 14-1.

TABLE 14-1 Typical Supply Voltage and Temperature Specifications for the Bulk CMOS Logic Families

Characteristic	Military Series (40XXM)	Commercial Series (40XXC)*
Supply voltage range (V)	3–18	3–{12, 16, 18}
Operating temperature range (°C)	−55 to +125	−40 to +85

* The maximum voltage specification on the 40XXC series varies from one manufacturer to another.

Series-connected NOR gates can be used as a convenience in specifying operating voltages and currents for the CMOS logic families as shown in Fig. 14-55.

Fig. 14-55 Current and voltage levels for CMOS logic gates.

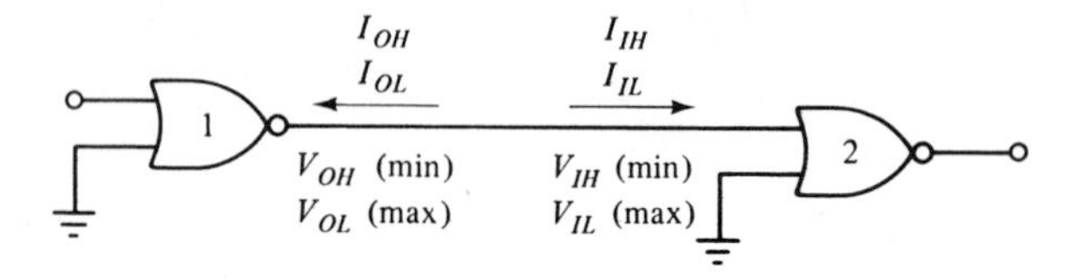

Typical input and output current levels are:

$$I_{OH} \approx -2 \text{ mA} \qquad I_{IH} \approx 10 \text{ pA}$$
$$I_{OL} \approx 1 \text{ mA} \qquad I_{IL} \approx -10 \text{ pA}$$

Typical minimum and maximum voltage specifications for CMOS logic gates are:

$$V_{OL} = V_{SS} + 50 \text{ mV (maximum)} \qquad V_{IL} = V_{SS} + 0.3\,V_{DD} \text{ (maximum)}$$
$$V_{OH} = V_{DD} - 50 \text{ mV (minimum)} \qquad V_{IH} = 0.7\,V_{DD} \text{ (minimum)}$$

If, for example, $V_{DD} = +5.0$ V and $V_{SS} = 0$ V, then the hi and lo voltage levels are given as:

$$V_{OL} = 0.05 \text{ V (maximum)} \qquad V_{IL} = 1.50 \text{ V (maximum)}$$
$$V_{OH} = 4.95 \text{ V (minimum)} \qquad V_{IH} = 3.50 \text{ V (minimum)}$$

The noise margins are then given by:

$$NM_L = V_{IL}(\text{max}) - V_{OL}(\text{max}) = 1.50 \text{ V} - 0.05 \text{ V} = 1.45 \text{ V}$$
$$NM_H = V_{OH}(\text{min}) - V_{IH}(\text{min}) = 4.95 \text{ V} - 3.50 \text{ V} = 1.45 \text{ V}$$

Referring to Fig. 14-55, if V_{OH1} is 4.95 V and V_{IH2} has to be 3.50 V or more to recognize V_{OH1} as a hi logic level, we can stand a −1.45 V zero-to-peak noise signal between gates 1 and 2 before the second inverter input is out of specification.

Most CMOS manufacturers specify worst-case noise margins (NM_L and NM_H) at 30% of V_{DD}, and specify typical noise margins at 45% of V_{DD}.

A typical value of static power dissipation for a CMOS logic gate is about 50 nW. As frequency is increased, power dissipation is given by Eq. (14-72) (i.e., $P = C_L V_{DD}^2 f$). Power increases as the square of supply voltage and increases linearly with load capacitance and frequency. Since a typical value of input capacitance for a CMOS gate is 5 pF, power dissipation per gate is linearly related to fan-out.

An interesting relationship exists between time delays (t_{pd}, t_r, and t_f) and the supply voltage in the CMOS logic families. Time delays *decrease* as the supply voltage is *increased*. To see why this is so, refer to Fig. 14-56(a). We

Fig. 14-56 (a) CMOS Inverter with capacitive load C_L. (b) Output circuit. The *p* unit is modeled as a current source in series with channel resistance R_P.

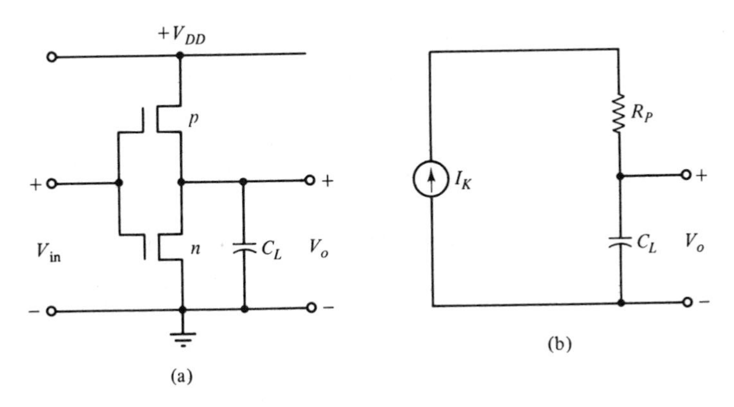

assume initially that V_{in} is hi. Then the p unit is off, the n unit is on, and V_o is low. Now we will let V_{in} go lo. Then the p unit turns on, the n unit turns off, and C_L begins to charge toward $+V_{DD}$ through the p unit. We can establish an approximate relationship between $+V_{DD}$ and the C_L charge time by assuming that the p unit stays in the saturation region during the entire charge time. Then

$$I_{DP}(\text{sat}) = \tfrac{1}{2}\beta_P(V_{GS} - V_{TP})^2 \tag{14-73}$$

where $V_{GS} = -V_{DD}$. If we further assume that the threshold voltage V_{TP} is much smaller than the supply voltage V_{DD}, we can state that

$$I_{DP}(\text{sat}) \approx \tfrac{1}{2}\beta_P V_{DD}^2 \tag{14-74}$$

Referring now to Fig. 14-56(b), the current source I_K can be written as

$$I_K = I_{DP}(\text{sat}) \approx \tfrac{1}{2}\beta_P V_{DD}^2 \tag{14-75}$$

Since the current charging C_L is assumed to be constant, V_o is then given by

$$V_o \approx \frac{I_K t}{C_L} \tag{14-76}$$

Solving for time t, we have that

$$t \approx \frac{C_L V_o}{I_K} = \frac{2C_L V_o}{\beta_P V_{DD}^2} \tag{14-77}$$

A circuit-time constant can be roughly approximated by letting $t = \tau$ when $V_o = V_{DD}$. Then

$$\tau = \frac{2C_L}{\beta_P V_{DD}} \tag{14-78}$$

This leads us to the conclusion that, to a first approximation, the rise time of a CMOS gate is inversely proportional to the supply voltage V_{DD}. As the supply voltage is increased, the rise time decreases.

The time required to charge and discharge the load capacitance is a dominant factor in determining fall time t_f and gate propagation delay t_{pd} as well as rise time t_r. Propagation delay, rise time, and fall time all have an inverse relationship with supply voltage V_{DD}. Typical values of t_{pd}, t_r, and t_f for CMOS logic gates are shown in Table 14-2. The delay times are specified for a load capacitance of 15 pF. These delay times increase linearly as fan-out is increased.

Although there is no direct limitation on CMOS fan-out, it causes a linear increase in power dissipation and circuit time delays. For these reasons, a fan-out of 50 is sometimes mentioned in the literature.

TABLE 14-2 Typical Delay-Time Values for a CMOS Logic Gate with $T_A = 25°C$, $C_L = 15$ pF, and Input $t_r = t_f = 20$ ns

Parameter	Conditions	Typical Value	Units
Average propagation delay time (t_{pd})	$V_{DD} = 5$ V	35	ns
	$V_{DD} = 10$ V	25	ns
Rise time (t_r)	$V_{DD} = 5$ V	65	ns
	$V_{DD} = 10$ V	35	ns
Fall time (t_f)	$V_{DD} = 5$ V	65	ns
	$V_{DD} = 10$ V	35	ns

14-8 COMPARISON OF THE DIGITAL LOGIC FAMILIES

In summary, the dominant digital logic families are listed below.

Bipolar Technologies

- Transistor–transistor Logic (T^2L)
- Schottky T^2L
- Emitter-coupled logic (ECL)
- Integrated-injection logic (I^2L)

MOS Technologies

- *n*-Channel MOS logic (NMOS)
- Complementary MOS logic (CMOS)

Using a technology or a set of technologies to fulfill application requirements is often a matter of finding the best fit between boundary conditions on technology parameters and boundary conditions on application requirements. Some factors that should be considered in fitting technologies to applications are listed below.

- Cost
- Availability of digital functions (SSI, MSI, LSI, VLSI)
- Speed (propagation delay)
- Power dissipation

- Power–delay product
- Packing density
- Noise margin
- Fan-out
- Operating temperature range
- Supply voltage requirements

A summary of approximate operating parameters for digital logic gates is shown in Table 14-3. The parameters in this table are, for the most part, intended as "ballpark" figures. Parameters such as power dissipation per gate, propagation delay per gate, and power–delay product, for example, vary with the manufacturer, the processes employed, and with the circuit function performed. Parameters such as temperature range, supply voltage, noise margin, and fan-out in Table 14-3 are representative of simple logic gates in the cases of T^2L, S-T^2L, ECL, and CMOS. In these technologies, simple logic gates and their specifications are available. It is more difficult to clearly specify operating parameters for I^2L and NMOS logic because they are primarily LSI and VLSI technologies. Simple logic gates and their specifications are not generally available. The data presented on I^2L and NMOS are the result of literature surveys and discussions with manufacturers.

It is appropriate at this point to discuss the major factors that have contributed to the continued success in the marketplace of the technologies listed above.

Transistor–Transistor Logic. T^2L is an old, proven, low-cost technology. A wide variety of general-purpose digital functions are available from a large number of manufacturers. Since T^2L operates at current levels on the order of 10 mA, it is useful in interfacing applications, where low-current, low-power LSI chips must be connected to the outside world. It is commercially available in SSI and MSI circuits.

Schottky T^2L. In comparison to conventional T^2L circuits, Schottky T^2L provides a significant performance improvement. As shown in Table 14-3, the propagation delay is reduced from about 10 ns to about 3 ns. The power–delay product is reduced from about 100 pJ to about 60 pJ. It is commercially available in SSI and MSI circuits.

Emitter-Coupled Logic. The continued success of ECL in the marketplace is due to propagation delay times on the order of 0.5 to 2 ns. It is the highest-speed digital technology presently available. ECL is commercially available in SSI and MSI circuits.

Integrated Injection Logic. I^2L has a high value of packing density (~100 gates/cm^2) and a low power–delay product (0.5 pJ). The I^2L process technologies

TABLE 14-3. Summary of Approximate Operating Parameters for Digital Logic Gates (1980)

Operating Parameter	Bipolar				MOS	
	T^2L	$S\text{-}T^2L$	ECL	I^2L	NMOS	CMOS
Operating temperature range (commercial) (°C)	0–70	0–70	0–75	0–70	0–70	−40 to +85
Operating supply voltage (commercial) (V)	5	5	−5.2	1.5 (minimum internal)	5	3–15
Worst-case noise margin (V)	0.4	0.3	0.17	~0.1 (internal)	Process-dependent	$0.3\,V_{DD}$
Fan-out	10	10	25	1	25	50
Power dissipation per gate (mW)	10	20	25–50	~50 μW	0.1–1.0	50 nW static (frequency-dependent)
Propagation delay per gate (ns)	10	3	0.5–2	10	1–10	10–50
Power delay product (pJ)	100	60	25	0.5	0.1–10	Frequency dependent
Packing density (gates/mm²) (1 mm ≈ 0.04 in.)	~15	~15	~15	~100	~150	~70
Number of components per two-input gate	9–12	~14	10–12	3–4	3	4

are compatible with the linear bipolar processes. As a result, I^2L logic can be put on the same chip with linear bipolar circuits. This makes I^2L very useful in linear to digital and digital to linear interface applications. I^2L is not commercially available as a "stand-alone" digital technology. It is widely used in MSI and LSI circuits, however, in conjunction with other process technologies.

NMOS (Enhancement Driver, Depletion Load). NMOS is often referred to as the "workhorse" of the technologies. As shown in Table 14-3, it compares very favorably with the other logic families listed. Rapid improvements in lithography, processing, and circuit design are spearheading a planned transition to smaller NMOS dimensions. NMOS is shrinking to *scaled* NMOS, which, in turn, is being scaled down to form HMOS I and HMOS II. LSI and VLSI NMOS circuits are widely used to produce microcomputer chips and memory chips, which improve every year. NMOS is not generally available in SSI and MSI circuits.

Complementary MOS Logic. Bulk CMOS has high noise margins and very low values of static power dissipation. It has captured a unique market in battery-operated electronic equipment. CMOS is in competition with T^2L and Schottky T^2L at the MSI level. A wide variety of general-purpose digital functions are available from a large number of manufacturers. Bulk CMOS packing densities are considerably lower than those for NMOS (about 70 gates/mm^2 for CMOS as compared to about 150 gates/mm^2 for NMOS). Thus, CMOS does not appear to be competitive with NMOS at the VLSI level at this time. CMOS is commercially available in SSI, MSI, and LSI circuits.

Some of the major limitations of silicon integrated circuits are caused by their very nature. *p* and *n* regions exist side by side within single-crystal silicon. Circuit elements are isolated from each other with reverse-biased *p-n* junctions. These isolating junctions act as capacitors, have leakage currents, have finite avalanche breakdown voltages, and create parasitic *npn* and *pnp* devices. If integrated circuits can be mass produced inexpensively by depositing single-crystal silicon islands on insulating substrates, then the circuit limitations caused by junction isolation can be removed. In this context, sapphire has been widely used as an insulating substrate and silicon-on-sapphire (SOS) CMOS circuits are seeing limited use in some specialized applications. However, the cost of sapphire is a limiting factor at present. The silicon island approach can provide significant increases in frequency response and packing density while reducing power dissipation. Whether sapphire remains the dominant insulating substrate or not, it is very likely that, in the long run, improvements will be made in the silicon island approach.

It is important to keep in mind that the digital technologies are in a rapid state of change. The number of years over which any discussion of digital operating parameters remains valid is open to question. This chapter was written in the summer of 1980. We would expect considerable change in the value

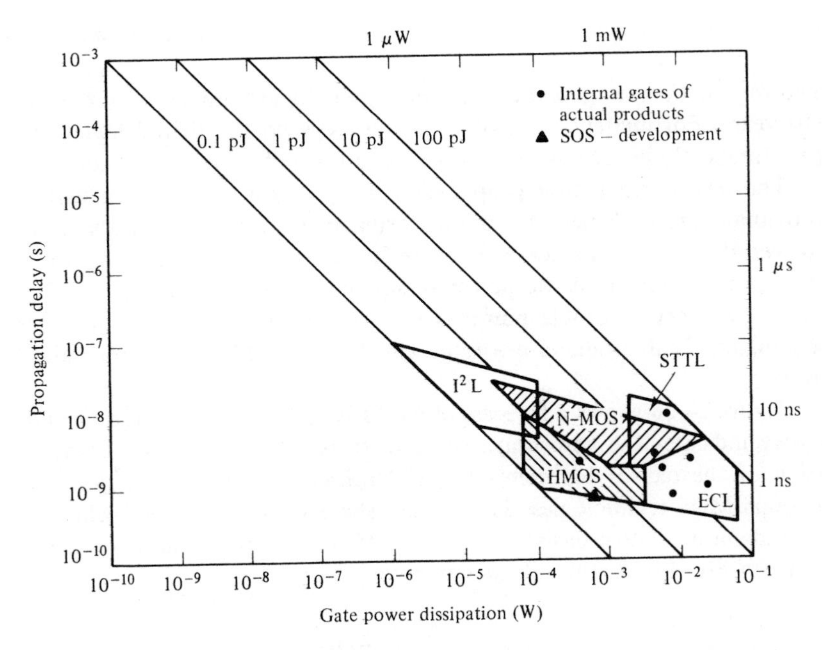

Fig. 14-57 Power–delay products for digital logic gates. (Reprinted from "Status '79; A Report on the Integrated Circuit Industry." Copyright 1979 by Integrated Circuit Engineering Corporation, by permission [2].)

Fig. 14-58 World IC consumption by technology. (Reprinted from "Status '79; A Report on the Integrated Circuit Industry." Copyright 1979 by Integrated Circuit Engineering Corporation, by permission [2].)

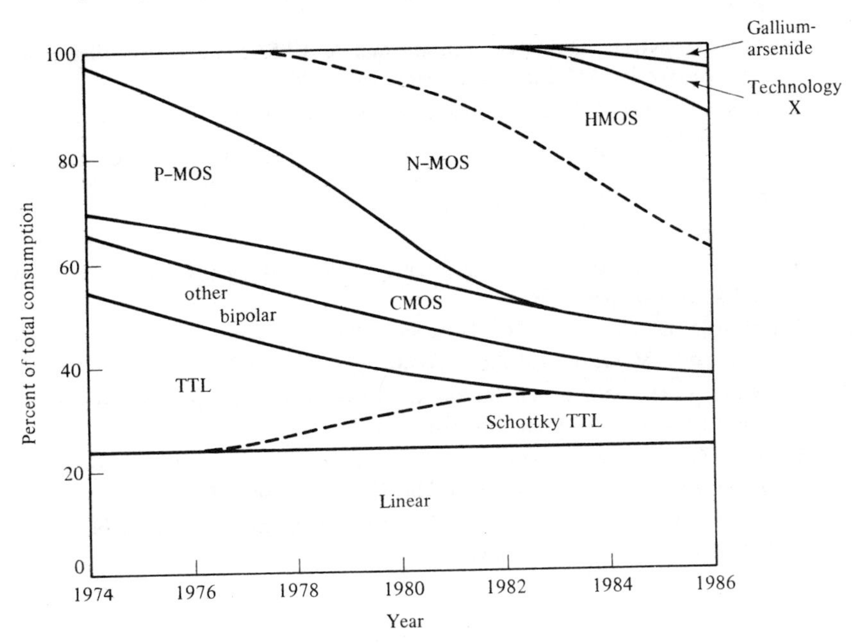

structure and operating parameters for digital logic families over the next 5 to 10 years. Fortunately, state-of-the-art surveys on the digital logic families appear frequently in periodic electronics literature and at electronics conferences.

The power dissipation–propagation delay product provides a useful but approximate figure of merit for the performance of digital logic gates. A range of power–delay products for a given technology shows up as a region on a plot of propagation delay vs. power dissipation, as shown in Fig. 14-57. This figure shows very favorable power–delay regions for the NMOS and HMOS technologies. Real products are represented in Fig. 14-57 by data points, as shown.

Figure 14-58 shows a forecast of world IC consumption. Each technology category indicates the percentage share of sales dollars. In general, the lower portion of the curves represents bipolar technologies and the upper portion represents MOS technologies. The market share held by a given technology is displayed on a relative basis. Since world IC consumption increases yearly, a flat line in Fig. 14-58 still indicates relative growth.

BIBLIOGRAPHY

[1] EKLUND, M. H., AND W. I. STRAUSS, "Status 1980; A Report on the Integrated Circuit Industry," Integrated Circuit Engineering Corporation, Scottsdale, Ariz., 1980.

[2] Integrated Circuit Engineering Corporation, "Status 1979; A Report on the Integrated Circuit Industry," Pitcher Technical Publications, Scottsdale, Ariz., 1979.

[3] MEAD, C., AND L. CONWAY, *Introduction to VLSI Systems.* Reading, Mass.: Addison-Wesley Publishing Company, Inc., 1980.

[4] EDITORIAL STAFF, "Future Technology," *Electronics Magazine,* April 17, 1980, pp. 531–583.

[5] GLASER, A. B., AND G. E. SUBAK-SHARPE, *Integrated Circuit Engineering: Design, Fabrication, and Applications.* Reading, Mass.: Addison-Wesley Publishing Company, Inc., 1977.

[6] TAUB, H., AND D. SCHILLING, *Digital Integrated Electronics.* New York: McGraw-Hill Book Company, 1977.

[7] HAMILTON, D. J., AND W. G. HOWARD, *Basic Integrated Circuit Engineering.* New York: McGraw-Hill Book Company, 1975.

[8] GRINICH, V. H., AND H. G. JACKSON, *Introduction to Integrated Circuits.* New York: McGraw-Hill Book Company, 1975.

[9] LUECKE, G., J. P. MIZE, AND W. N. CARR, *Semiconductor Memory Design and Application.* New York: McGraw-Hill Book Company, 1973.

[10] CARR, W. N., AND J. P. MIZE, *MOS/LSI Design and Application.* New York: McGraw-Hill Book Company, 1972.

PROBLEMS

14-1 Draw a logic diagram for an EXCLUSIVE NOR gate using only a minimum number of NAND gates. Assume that V_1 and V_2 are available as inputs (inverted inputs are not available).

14-2 Draw a logic diagram for an EXCLUSIVE OR gate using only a minimum number of NOR gates. Assume that V_1 and V_2 are available as inputs (inverted inputs are not available).

14-3 Assume that $V_{BE}(\text{ON}) = V_D(\text{ON}) = 0.7$ V and $V_{CE}(\text{sat}) = 0.2$ V in the T²L NAND gate of Fig. 14-23. Also assume no-load conditions ($R_L = \infty$).
(a) Find the power dissipated when $V_1 = V_2 = 3.6$ V.
(b) Find the power dissipated when $V_1 = V_2 = 0.2$ V.
(c) Find PD_{AV} for this gate.

14-4 Find the magnitude of the current through transistors Q_3 and Q_4 in Fig. 14-23 during the time interval when both devices are in saturation.

14-5 Find the average power (PD_{AV}) dissipated by the Schottky-clamped T²L gate shown in Fig. 14-25. Assume that:
(a) $V_{BE}(\text{ON}) = 0.7$ V.
(b) The forward diode voltage drop across a Schottky diode is 0.3 V.
(c) The gate is operating under no-load conditions.
(d) Base currents can be neglected.
(e) Transient currents can be neglected.

14-6 Determine the logic function performed by the T²L gate shown.

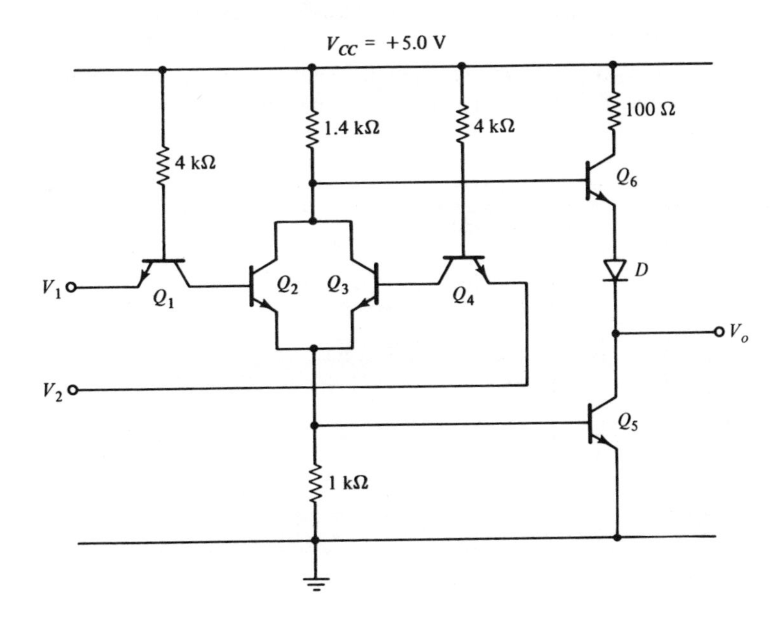

14-7 Determine the logic function performed by the T²L gate shown.

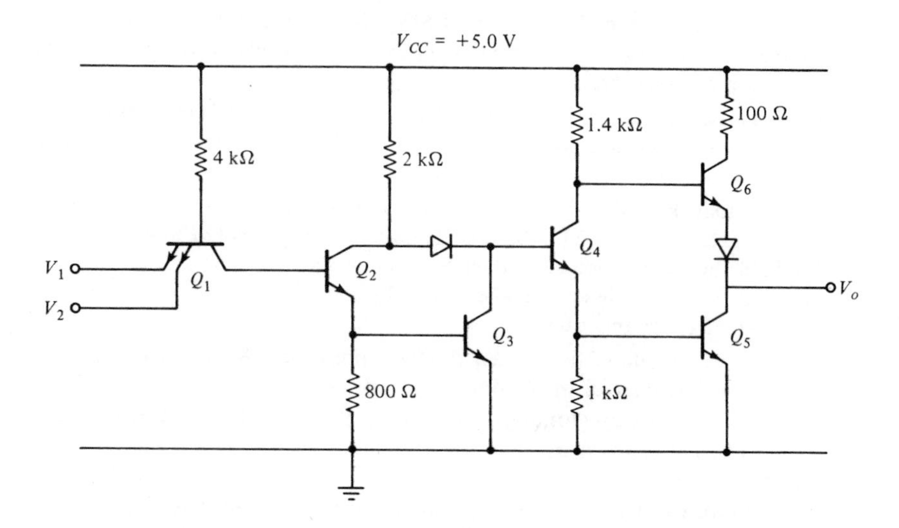

14-8 In the circuit of Fig. 14-27, let $V_{IL} = -1.325$ V, $V_{IH} = -1.025$ V, and $V_R = -1.170$ V. Assume that $V_{BE}(\text{ON}) = 0.75$ V, base currents are negligibly small ($I_E \approx I_C$), and neglect the power dissipation in the bias circuit.
(a) Find the average power dissipated when the inputs are lo.
(b) Find the average power dissipated when the inputs are hi.
(c) Find PD_{AV} for this gate.

14-9 Given that V_1 and V_2 are the input terminals and V_o is the output terminal, determine the digital logic functions performed by I²L gates (a) and (b) shown.

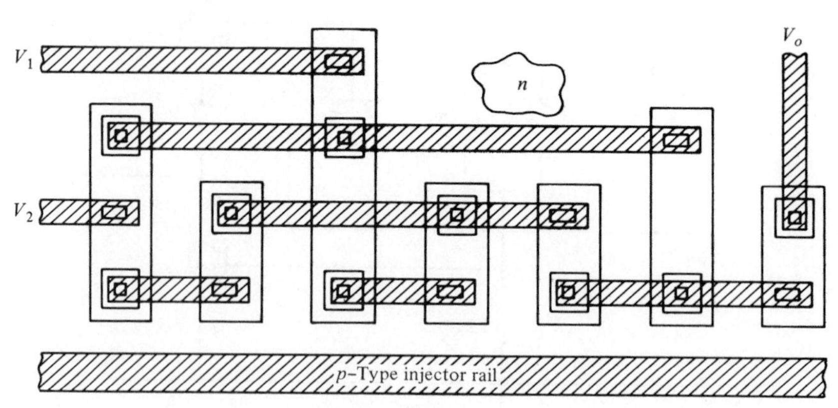

(a)

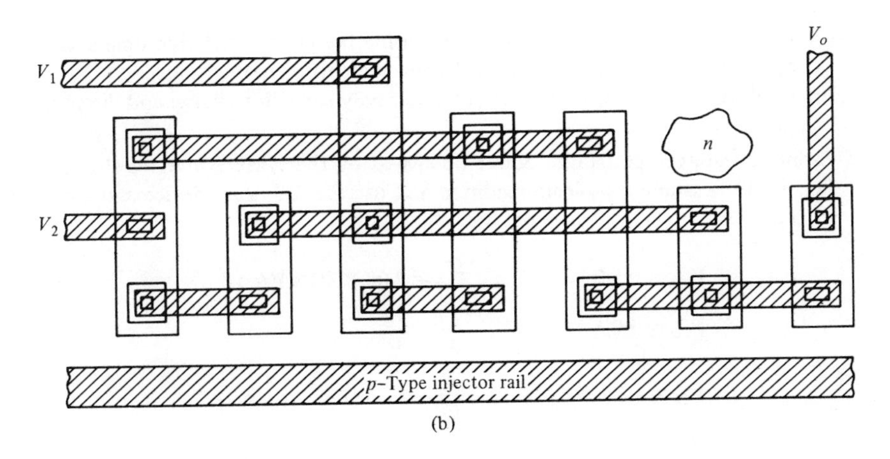

(b)

14-10 Using the specifications listed, plot the transfer characteristic for an NMOS inverter of the type shown in Fig. 14-37.

$$V_{DD} = +5.0 \text{ V} \qquad C_{ox} = 5.9 \times 10^{-8} \text{ F/cm}^2$$

$$V_{TD} = +0.7 \text{ V} \qquad \mu_{ns} = 700 \text{ cm}^2\text{/V-s}$$

$$V_{TL} = -1.5 \text{ V} \qquad \frac{Z_L}{L_L} = \frac{1}{4}$$

$$t_{ox} = 6 \times 10^{-6} \text{ cm} \qquad \frac{Z_D}{L_D} = 3$$

14-11 Estimate the time required to fully charge and discharge the load capacitance ($C_L = 0.05$ pF) when the input voltage is switched from 0 V to 5.0 V and back to 0 V in the NMOS inverter shown. Use the device and circuit specifications listed for Problem 14-10. Make the following simplifying assumptions:
(a) $V_o(\text{lo}) = 0$ V and $V_o(\text{hi}) = 5.0$ V.

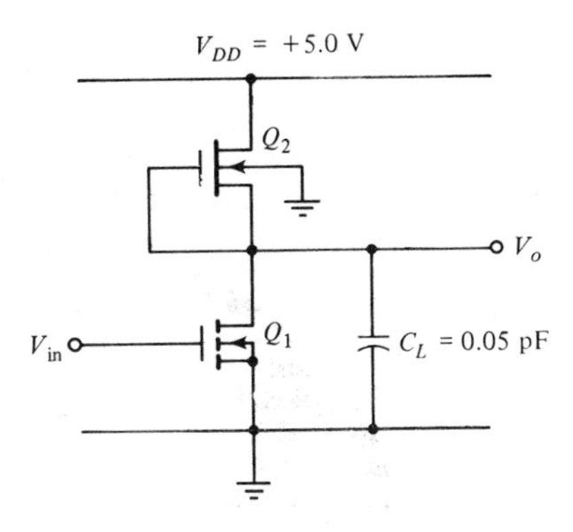

(b) Q_1 operates in the saturation region during the entire discharge time and Q_2 operates in the saturation region during the entire charge time.

(c) Neglect device transit times and consider only capacitor charge and discharge times.

14-12 Estimate the average (static) power dissipated by the NMOS inverter shown in Fig. 14-37. Assume a no-load condition and use the following device and circuit specifications:

$$V_{DD} = +5.0 \text{ V} \qquad C_{ox} = 7.08 \times 10^{-8} \text{ F/cm}^2$$

$$V_{TD} = +0.6 \text{ V} \qquad \frac{Z_L}{L_L} = \frac{1}{3}$$

$$V_{TL} = -2.0 \text{ V} \qquad \frac{Z_D}{L_D} = 4$$

$$\mu_{ns} = 700 \text{ cm}^2/\text{V-s}$$

14-13 Determine the logic function performed by the NMOS logic gate shown.

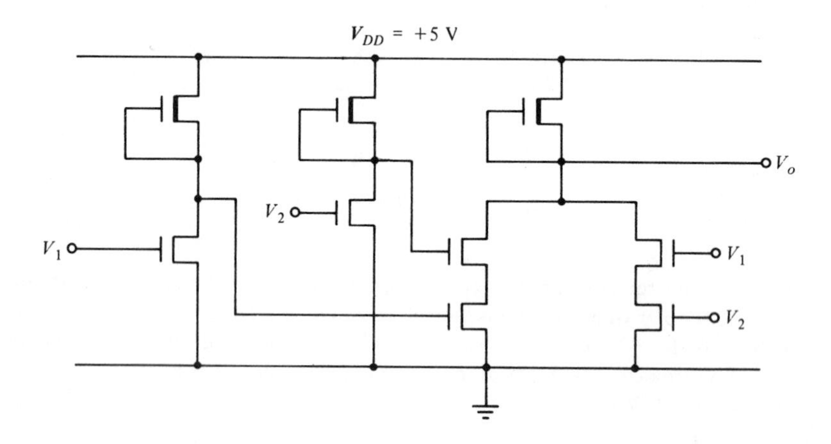

14-14 For the CMOS inverter shown in Fig. 14-46(a), $V_{DD} = +10.0$ V, $V_{TN} = +1.5$ V, $V_{TP} = -1.5$ V, and $\beta_R = \beta_N/\beta_P = 3$. Plot the inverter transfer characteristic for $0 \text{ V} \leq V_{in} \leq 10$ V.

14-15 Determine the logic function realized by the CMOS circuit shown. Draw a CMOS circuit using six devices that will realize the same logic function.

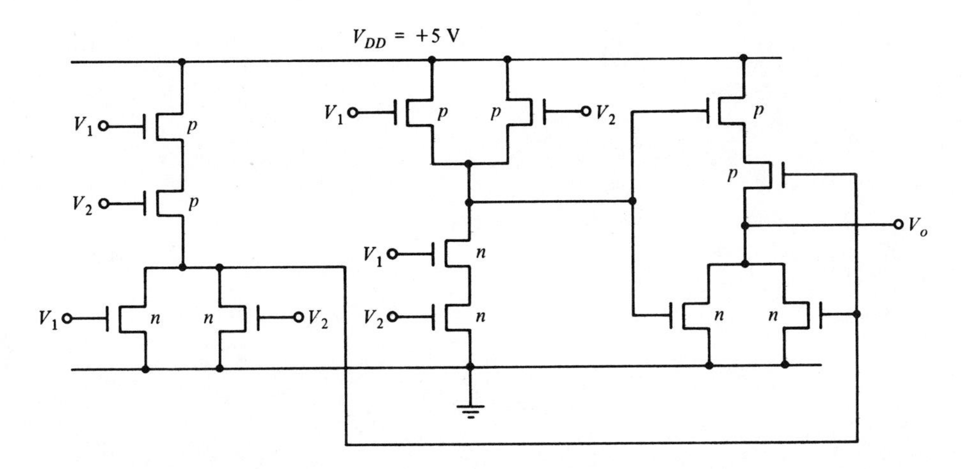

14-16 The digital technologies discussed in this chapter include T²L, ECL, I²L, NMOS, and CMOS. Indicate which technology provides:

(a) Minimum power–delay product.
(b) Minimum static power dissipation.
(c) Lowest cost in VLSI applications.
(d) Minimum propagation delay time (t_{pdAV}).
(e) Maximum packing density.
(f) The highest noise margins.

index

index

A

B

C

D

E

F

K

L

M

N